Fundamentals of Drug and Non-Drug Interactions

James O. Fajemiroye
Editor

# Fundamentals of Drug and Non-Drug Interactions

## Physiopathological Perspectives and Clinical Approaches

Springer

*Editor*
James O. Fajemiroye
Institute of Biological Sciences
Federal University of Goias
Goiania, Goias, Brazil

ISBN 978-3-031-80106-8 ISBN 978-3-031-80107-5 (eBook)
https://doi.org/10.1007/978-3-031-80107-5



This Springer imprint is published by the registered company Springer Nature Switzerland AG
The registered company address is: Gewerbestrasse 11, 6330 Cham, Switzerland

*To my wife (Ana Paula) and our lovely girls (Alice and Clara).*

# Preface

Understanding the impact of drug and nondrug use in humans demands herculean efforts. The principles of interactions in pharmacology are an interesting pillar of the medical curriculum. Safety or effective prescriptions often define the outcomes of interactions. This book delves into different topics—from physiopathological standpoints (first chapter) to the remedies against plausible adverse interactions (last chapter) to bolster the consciousness and understanding of readers on the interaction spectrum. This book presents a comprehensive interplay of drug and nondrug interactions involving components of diseased states, supplements, nutrients, and active principles from phytotherapy/herbal products beyond the drug-drug dogma of interactions. This book updates therapeutic advances with suitable illustrations of mechanisms and principles of interactions while drawing attention to cases of mild to life-threatening interactions from clinical to hypothetical points of view. The growing culture of drug and nondrug associations in the treatment of a spectrum of medical conditions demands awareness of possible interactions. The topics in this book intend to expose and prepare trainees, professionals, and undergraduate medical, nursing, nutrition, biomedical, and pharmacy students for better practical therapeutics, safety, and efficacy.

Goiania, Brazil James O. Fajemiroye

# Contents

# About the Editor

**James O. Fajemiroye, PhD,** Head of the Department of Pharmacology (2022–2024), Coordinator of Research for the Institute of Biological Sciences and Research Group on the Pharmacology of Natural and Synthetic Products at The Federal University of Goiás. With his notable presentation at the International Conference on Opioids held at Harvard Medical School as well as the world's leading Annual Meeting of the American College of Neuropsychopharmacology on the brain and psychotropic drugs as a recipient of the Prestigious ACNP 2020 Travel Award, he has contributed to different topics on pharmacology with a network of collaborators across the globe. He specializes in pharmacological screenings with an interest in drug and nondrug interactions.

# Chapter 1
# System Overview from Physiological and Pathological Standpoints

**Gustavo Rodrigues Pedrino, Luiza Toubas Chaul, Virgínia Farias Alves, Caroline Vitória de Lima Moreira, Anderson Luiz-Ferreira, and James O. Fajemiroye**

**Abstract** The inbuilt biological machinery from molecules to systems is responsible for a wide variety of complex body functions and controls that are often mimicked or disrupted by drugs among other xenobiotics. This chapter focuses on the function and dysfunction of biological systems especially cardiovascular and gastrointestinal systems (from healthy to diseased state) to provide a mechanistic background for drug and nondrug entities interactions (DNIs) with the overall impact on drug absorption, distribution, elimination, safety profiles, therapeutic success, or failure. The presence of food can interfere with the availability or excretion of drugs in the body. Interestingly, drugs can also interfere with nutritional status, increasing or decreasing appetite and thus affecting the amount of food consumed. As of now, providing a clinically helpful definition of DNIs remains challenging. In this chapter and subsequent ones, alterations in biological processes with or without nutrients, minerals, functional foods, herbs, isolated compounds, metabolites, vitamins, beverages, supplements, etc., are considered important components of nondrug entity interactions. Our grasp of DNIs is expected to change the conceptual framework and notions of drug interactions while keeping pace with the new and emerging findings on biological systems.

**Keywords** Biological systems · Cardiovascular system · Gastrointestinal tract · Interactions · Xenobiotics

---

G. R. Pedrino · C. V. de Lima Moreira · J. O. Fajemiroye (✉)
Institute of Biological Sciences, Federal University of Goias, Goiania, Goias, Brazil
e-mail: pedrino@ufg.br; jamesfajemiroye@ufg.br

L. T. Chaul · V. F. Alves
Faculty of Pharmacy, Federal University of Goias, Goiania, Goias, Brazil
e-mail: luizachaul@discente.ufg.br; virginia_alves@ufg.br

A. Luiz-Ferreira
Institute of Biotechnology, Federal University of Catalão, Catalao, Goias, Brazil
e-mail: luiz_ferreira@ufcat.edu.br

 
J. O. Fajemiroye (ed.), *Fundamentals of Drug and Non-Drug Interactions*,
https://doi.org/10.1007/978-3-031-80107-5_1

# 1 Introduction

A biological system is made of organs, tissues, cells, organelles, micro-, and macromolecules that collectively perform diverse body functions including mechanical support, posture, and locomotion (the musculoskeletal system), physical protection of the body surface, sensory reception, vitamin synthesis (the integumentary system), oxygen and carbon-dioxide exchange (the respiratory system), control of vital body functions (the nervous system), filtration of blood and excretion of metabolic products among others in urine (the urinary system), hormone production and regulation of a series of functions (the endocrine system), excess tissue fluid drainage (the lymphatic system), reproduction process (the reproductive system), transportation of oxygen, nutrients, and hormones throughout the body and elimination of cellular metabolic waste (the cardiovascular system), and mechanical and chemical degradation and absorption of food (the digestive system). The performance of these systems (Fig. 1.1) under healthy, diseased, or pharmacologically intervened states is an outcome of the dynamic roles of diverse targets and intrinsic or extrinsic factors. The integrity of both cardiovascular and digestive systems plays an important role regarding possible interactions arising from the circulation or passage and removal of drugs among other xenobiotics.

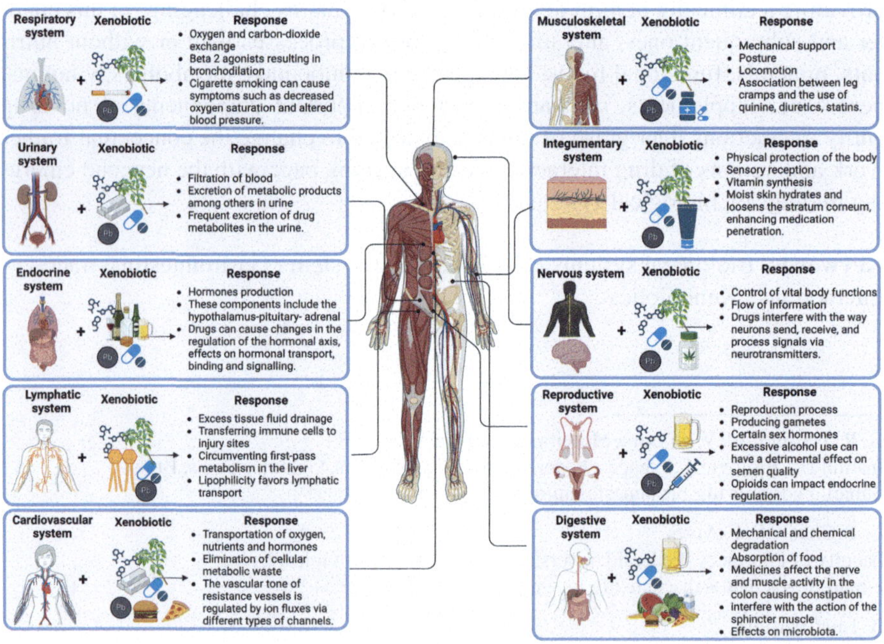


**Fig. 1.1** Major systems (respiratory, urinary, endocrine, lymphatic, cardiovascular, musculoskeletal, integumentary, nervous, reproductive, and digestive systems) and biological responses with or without xenobiotics. (Created with BioRender)

Understanding the function and dysfunction of biological systems, particularly the cardiovascular and gastrointestinal systems, is crucial for understanding how various drugs and nondrug entities interact within the body. By exploring these mechanisms, researchers and healthcare professionals can gain insights into how to manage diseases and avoid or attenuate adverse interactions while optimizing treatments. It is fascinating how interconnected these systems are and how they can impact each other's function in health and disease.

## 2 Cardiovascular System

Homeostasis is the process by which living organisms maintain a stable internal balance despite fluctuations in the external environment. This equilibrium is attained through regulatory mechanisms that continuously monitor and adjust physiological variables, including body temperature, blood glucose levels, blood pressure, and blood pH, to maintain them within normal and healthy ranges. The concept of homeostasis was initially introduced by Claude Bernard, a French physiologist, in the nineteenth century. In his work, Bernard underlined the significance of preserving internal balance for organism survival, and he came up with the term "milieu intérieur" to describe the body's internal environment, which must remain constant despite fluctuations in the external environment (Bernard 1865).

The cardiovascular system (CVS) stands as a cornerstone of human physiology, orchestrating a complex transportation network of organs, tissues, and biochemical processes to sustain life. The heart, which is at the center of this system, is a tireless pump that sends nutrient-rich blood through the arteries, arterioles, and capillaries (Fig. 1.2). Meanwhile, waste-laden blood, depleted of its vital cargo, undergoes a return journey, destined for organs of elimination, such as the lungs and kidneys, where it undergoes purification before being recirculated back into the bloodstream. This organ is the primary force behind blood circulation, propelling blood through a vast network of arteries and veins. Cardiac output, the volume of blood ejected per minute, is a critical determinant of hemodynamic status and a function of ventricular contraction.

The cardiovascular system, which is responsible for transportation, also plays a crucial role in regulating temperature, maintaining acid-base balance, and facilitating hormonal signaling. This system operates through a series of intricate regulatory mechanisms that help preserve the body's internal environment within narrow physiological limits, thereby ensuring optimal conditions for cellular function and survival. Moreover, the cardiovascular system serves as a conduit for hormonal signaling, allowing endocrine glands to distribute chemical messengers to the target tissues and organs throughout the body. Hemodynamics, which is the study of blood flow dynamics within the circulatory system, is central to the functionality of the cardiovascular system. The key principles governing hemodynamics include blood pressure, flow rate, and vascular resistance. Poiseuille's law describes the relationship between pressure, flow, and resistance in a cylindrical vessel, whereas Ohm's

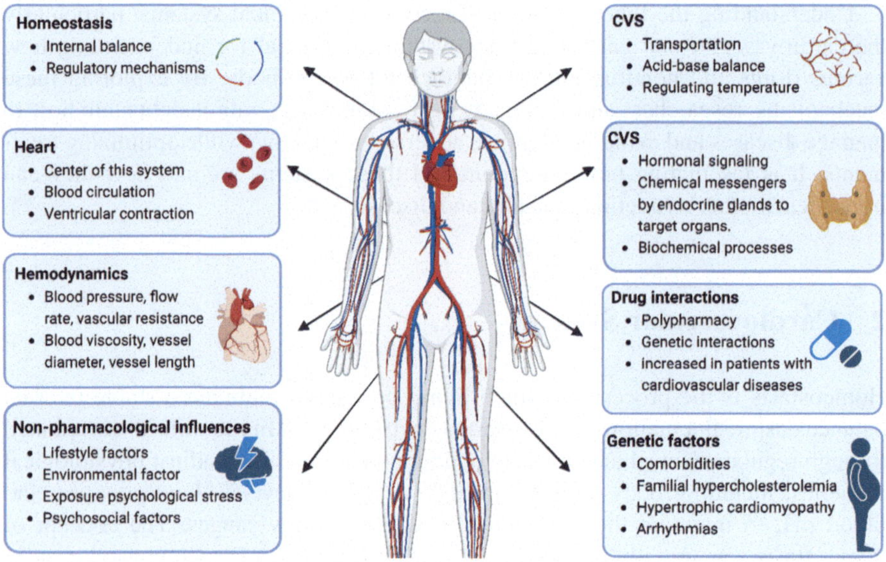


**Fig. 1.2** Cardiovascular system revisited. (Created with BioRender)

law applies analogously to a circulation system. The factors influencing hemodynamics include blood viscosity, vessel diameter, and vessel length. Changes in these parameters can affect cardiovascular function and predispose individuals to various cardiovascular disorders such as hypertension, atherosclerosis, and thrombosis. Therefore, it is crucial to understand how drugs interact with the cardiovascular system to improve health outcomes. Research has shown that noncardiovascular medications can impact cardiovascular health directly and through drug interactions (Schaffer et al. 2022).

The interactions between medical and nutritional elements, including nutrients, minerals, functional foods, herbs, active principles or metabolites, vitamins, beverages, and supplements, have significant implications for cardiovascular health. While these interactions can provide synergistic benefits, they can also pose considerable risks. In addition, polypharmacy, which is common in patients with cardiovascular diseases, can lead to significant drug-drug interactions, especially with medications affecting the cardiovascular system (Assefa et al. 2020). These interactions can also involve genetic interactions, influencing the efficacy and safety of cardiovascular medications (Asiimwe and Pirmohamed 2022).

The complexity of drug interactions is further highlighted by the risks associated with uninformed complementary and alternative supplement use in patients with cardiovascular disease (Bristol et al. 2008). Factors such as the duration of hospital stay, prescribed medications, age, and sex of patients can influence the incidence of potential drug interactions with cardiovascular drugs (Sobhy et al. 2021). Moreover, the risk of potential drug-disease interactions is significantly increased in patients with cardiovascular diseases (Doubova et al. 2007). Understanding the mechanisms of drug interactions is essential for effective co-administration of cardiovascular therapeutics. For example, interactions of cardiovascular drugs such as verapamil and digoxin with P-glycoprotein can lead to drug-drug interactions and associated toxicity (Ledwitch et al. 2016). Additionally, the impact of cardiovascular drugs on the efficacy of local anesthesia in dentistry has broader implications for drug interactions beyond the cardiovascular system (Milosavljević and Janković 2016). In the realm of cardiovascular drug discovery, systems pharmacology approaches have provided valuable insights into the relationships between cardiovascular targets and genes, aiding in the development of new therapies (Fu et al. 2014). It is crucial to consider the cardiovascular safety of medications, as highlighted in studies focusing on cardiovascular risks associated with nonsteroidal anti-inflammatory drugs (Trelle et al. 2011).

Epidemiological studies have demonstrated that the cardiovascular system's susceptibility to nonpharmacological influences, such as lifestyle factors, including unhealthy dietary habits, sedentary behaviors, and chronic stress, further underscores its vulnerability to external perturbations (Khera et al. 2016). For example, a diet rich in saturated fats and sugars can lead to conditions such as obesity, hypertension, and atherosclerosis, all of which are risk factors for cardiovascular diseases (Stanley et al. 2009). Similarly, environmental factors, ranging from air pollution to occupational hazards, can exacerbate underlying cardiovascular conditions and amplify the burden of disease within susceptible populations. Research indicates that prolonged exposure to air pollution is associated with an elevated risk of cardiovascular diseases, such as hypertension, stroke, and heart failure. Similarly, chronic exposure to noise pollution is associated with increased blood pressure, heart rate variability, and a higher likelihood of cardiovascular events (Trelle et al. 2011).

Psychosocial factors, including social support, mental health, and socioeconomic status, can exert an impact on cardiovascular health. Generally, individuals with robust social networks and favorable mental health have more favorable cardiovascular outcomes. Conversely, those experiencing chronic stress, depression, or social isolation face an elevated risk of developing cardiovascular diseases (Murtaza et al. 2016). In addition to taking into account lifestyle, environmental, and psychosocial factors, it is crucial to consider genetic predispositions and underlying medical conditions when assessing cardiovascular health. Genetic factors can make individuals more prone to conditions, such as familial hypercholesterolemia, hypertrophic cardiomyopathy, and arrhythmias, which may increase the likelihood of experiencing cardiovascular events. Furthermore, comorbidities, including diabetes, kidney disease, and autoimmune disorders, can interact with the cardiovascular system and affect treatment outcomes and disease progression (Storelli et al. 2018).

The cardiovascular system is a highly intricate network, with a range of essential functions that support homeostasis and overall health. It is crucial to understand the intricate interplay between the cardiovascular system and various factors such as drugs, lifestyle choices, environmental factors, and psychological stress to promote cardiovascular health and prevent cardiovascular diseases. Healthcare professionals must recognize the potential interactions between medications and the cardiovascular system to optimize therapeutic outcomes and minimize the risk of adverse events.

Modifiable risk factors, such as unhealthy lifestyle behaviors and environmental exposure, are essential to prevent and manage cardiovascular diseases, and comprehensive strategies should be employed to target them. Further research is necessary to elucidate the underlying mechanisms of drug-cardiovascular interactions and develop innovative interventions to mitigate their adverse effects. By integrating the knowledge of cardiovascular physiology, pharmacology, and epidemiology, healthcare professionals can effectively address the complex challenges posed by cardiovascular diseases and improve patient outcomes.

The use of drugs along with various dietary components is becoming more common as patients seek to improve their health through an integrated approach; however, the possibility of harmful interactions requires a full understanding of the pharmacokinetics and pharmacodynamics of these drugs. Medications and dietary components can interact at various levels, including absorption, metabolism, distribution, and excretion, which can potentially lead to a change in therapeutic effects and an increased risk of side effects. One of the most well-documented interactions involves grapefruit juice, which can significantly affect the metabolism of certain drugs. Several studies have demonstrated that grapefruit juice contains furanocoumarins, such as bergamottin and 6′,7′-dihydroxybergamottin, which inhibit the cytochrome P450 3A4 (CYP3A4) enzyme in the small intestine, leading to increased bioavailability of drugs metabolized by this pathway (Lown et al. 1997). Research by Paine and colleagues highlighted that these furanocoumarins can irreversibly inhibit CYP3A4, affecting the metabolism of around 50% of drugs and significantly increasing their systemic exposure (Paine et al. 2006). Moreover, a daily glass of grapefruit juice increases blood levels of simvastatin and lovastatin by about 260% if taken simultaneously, thereby increasing the risk of myopathy and rhabdomyolysis (Lee et al. 2016). Similarly, grapefruit juice can potentiate the effects of calcium channel blockers. Studies have shown that co-administration of grapefruit juice with dihydropyridine calcium channel antagonists felodipine and nifedipine resulted in a large increase in the plasma concentration of these drugs, which can cause serious adverse reactions such as headaches, hypotension, and lightheadedness (Bailey et al. 1991; Lundahl et al. 1995).

Herbal supplements, widely used for their purported health benefits, can also interact with cardiovascular medications. St. John's wort, a popular herbal remedy for depression, induces the CYP3A4 enzyme and the P-glycoprotein transporter, leading to reduced plasma concentrations of drugs such as digoxin, warfarin, and statins (Markowitz 2003; Jiang et al. 2004; Johne et al. 1999). This interaction can compromise the efficacy of these medications, resulting in inadequate therapeutic effects and an increased risk of cardiovascular events. On the other hand, herbs like ginkgo biloba and garlic have antiplatelet properties that can enhance the effects of anticoagulant and antiplatelet medications, such as warfarin and aspirin, increasing the risk of bleeding complications (Ke et al. 2021; Hatfield et al. 2022).

Minerals, essential for various physiological functions, can also influence the effectiveness and safety of cardiovascular drugs. For instance, potassium supplementation can be beneficial in managing hypertension, but excessive intake can lead to hyperkalemia, especially in patients taking potassium-sparing diuretics, angiotensin-converting enzyme (ACE) inhibitors, or mineralocorticoid receptor antagonists (Antoniou 2010; Antoniou et al. 2015). Hyperkalemia can result in life-threatening arrhythmias, necessitating careful monitoring of potassium levels in patients receiving these medications. Similarly, magnesium supplements can interact with calcium channel blockers, enhancing their hypotensive effects and increasing the risk of bradycardia and hypotension. In fact, evidence in the literature collectively demonstrates that magnesium supplements with calcium channel blockers, such as nifedipine and verapamil, enhance their hypotensive effects and increase the risk of bradycardia and hypotension. Houston (2011) highlighted that magnesium supplements can enhance the blood pressure-lowering effects of antihypertensive medications, potentially intensifying the actions of calcium channel blockers. Additionally, Campbell et al. (2014) pointed out that magnesium acts as a calcium channel antagonist, which can lead to bradycardia, hypotension, and other cardiac effects.

The consumption of high-sodium foods has been shown to counteract the effects of antihypertensive medications, leading to uncontrolled hypertension and increased cardiovascular risk. Studies such as those by Zhang et al. (2021) and Gupta et al. (2023) have highlighted the detrimental impact of excessive sodium intake on blood pressure regulation, especially in individuals taking antihypertensive medications. Additionally, research by Surma and Szyndler (2020) emphasized that limiting sodium consumption significantly improves the effects of antihypertensive drugs, underlining the importance of dietary sodium restriction in managing hypertension effectively. Furthermore, a study by Ahmad et al. (2021) has delved into the patterns of high-sodium food consumption among populations, shedding light on the prevalence of this dietary habit and its implications for cardiovascular health.

In conclusion, the cardiovascular risks associated with interactions between medical and nutritional elements, including nutrients, minerals, and supplements, are multi-faceted and significant. These interactions can alter the pharmacokinetics and pharmacodynamics of cardiovascular drugs, leading to suboptimal therapeutic outcomes and an increased risk of adverse effects. Understanding the mechanisms underlying these interactions, coupled with comprehensive patient assessment and education, is essential for optimizing cardiovascular health and preventing adverse outcomes. The studies underscore the importance of holistically considering dietary habits, mineral supplementation, and medication regimens to optimize cardiovascular health outcomes. Moving forward, further research and clinical vigilance are warranted to enhance our understanding of these interactions and improve patient care in managing cardiovascular conditions.

## 3 Gastrointestinal Tract Roles

The gastrointestinal tract (GIT) is one of the most complex and important organ systems in the body. It comprises the alimentary tract, a hollow structure extending from the mouth to the anus, and associated glandular organs that continuously

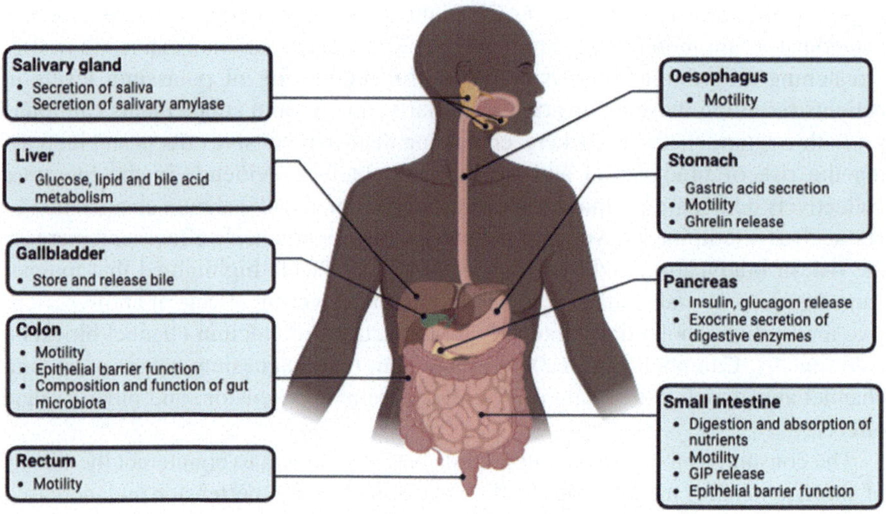


**Fig. 1.3** Overview of the gastrointestinal tract. (Created with BioRender)

supply the body with water, electrolytes, and nutrients (Fig. 1.3). To fulfill this function, the GIT requires the movement of food, the secretion of digestive solutions, the absorption of water, electrolytes, vitamins, and products of digestion, the circulation of blood through the gastrointestinal organs, and control of these functions by the local nervous and hormonal systems (Fitton and Thomas 2009).

Each portion of the GIT is adapted for specific functions, some for the simple passage of food, such as the esophagus, others for temporary storage, such as the stomach, and others for digestion and absorption, such as the small intestine.

The GIT has 7–9 m on average in an adult individual and includes the mouth, esophagus (23–25 cm), stomach, small intestine (6–7 m), large intestine (1.0–1.5 m), rectum, and anus. These regions are separated by three important sphincters (esophageal, pyloric, and anal) that assist in controlling the flow of the food bolus through the GIT.

The accessory glands, which include the salivary glands, the pancreas, and the gallbladder, are connected to the GIT, providing exocrine secretions that play key roles in the digestion process.

The GIT wall is composed of four layers called mucosa, submucosa, muscularis externa, and serosa.

The mucosal layer, the inner lining of the GIT, consists of the mucosal epithelium layer facing the lumen; lamina propria, formed by epithelial connective tissue that stabilizes the epithelium in place; and the mucosal muscle, a thin layer of smooth muscle that can vary in specific regions of the GIT.

The submucosal layer contains connective tissue with large blood and lymphatic vessels and a submucosal plexus (also known as *Meissner's* plexus), one of the two major nerve plexuses of the enteric nervous system that innervates cells of the epithelial layer and smooth muscle of the mucosal muscle to control secretions in the GIT.

The muscularis externa, the outermost layer of the GIT, is composed of an inner circular layer and an outer longitudinal layer of smooth muscles, responsible for the motility of the GIT. The contraction of the circular layer promotes a decrease in the diameter of the lumen, while the contraction of the longitudinal layer shortens the tube. Between the circular and longitudinal layers lies the myenteric nervous plexus, the second nerve plexus of the myenteric nervous system (also known as *Auerbach's* plexus), which regulates and coordinates the motor activity of the muscularis externa.

Finally, the serosa, the outer lining of the entire GIT, is a membrane made up of connective tissues, a continuation of the peritoneal membrane that lines the oral cavity.

Through the physiological processes of digestion, secretion, motility, and absorption, the GIT performs the general function of receiving nutrients, processing them so that they can be used by the body, and finally eliminating the waste.

As the food ingested initially through the mouth contains nonabsorbable macromolecules, the digestion process converts them into products that can be absorbed by mucosal cells. The physiological process of digestion includes physical processes (e.g., chewing and GIT contractions) that fragment food by mixing it with secretions and chemical processes (e.g., digestive enzymes) that degrade the proteins, fats, and polysaccharides present in food.

During the digestion process, large volumes of liquids are secreted into the lumen of the GIT by the exocrine glands. Part of this fluid (salivary, gastric, bile, pancreatic, and intestinal secretions) is absorbed, and the rest is eliminated in the feces.

The different portions of the GIT perform specific functions in the digestive process, acting on the food or the product of its processing so that the secretions and contents of the lumen are moved in the mouth-anus direction and mixed through motility.

The food undergoes, initially in the oral cavity, the process of mechanical digestion by the action of the teeth that cut (incisors) and grind (molars) food to reduce the size of its particles and increase the surface area for later action of the enzymatic secretions of the GIT. It is also in the oral cavity that the process of chemical digestion is initiated by the exocrine secretion of three pairs of salivary glands: the parotid glands (producing aqueous enzymatic solution), the sublingual glands (producing saliva rich in mucus), and the submandibular glands that produce a mixed secretion (enzymes and mucus). The water and mucus present in the saliva soften and lubricate the food, facilitating the swallowing process. Chemical digestion begins in the oral cavity with the secretion of salivary amylase that breaks down starch into maltose. In addition to acting on starch digestion, saliva also contains lysozymes and salivary immunoglobulins that inactivate viruses and bacteria.

After swallowing, the food is conducted through alternating movements of the inner circular and outer longitudinal muscle layers (peristaltic movements) from the pharynx and esophagus to the stomach.

The lumen of the stomach is formed by a mucus-producing epithelium and the gastric pits, which are made up of gastric glands with multiple cell types that produce hydrochloric acid (HCl), enzymes, and hormones.

The presence of amino acids and peptides in the food bolus and gastric distention promote the release of the hormone gastrin by the G cells present in gastric glands. Gastrin, in turn, stimulates the secretion of HCl by parietal cells. In addition to stimulation by gastrin, gastric $H^+$ secretion can occur via a direct route where the vagus nerve, which innervates the parietal cells, directly stimulates $H^+$ secretion or via an indirect route, where the vagus nerve, which innervates the G cell, stimulates it to secrete gastrin. Finally, histamine, released by enterochromaffin-like cells in the gastric mucosa, stimulates histamine $H_2$ receptors, present in the membrane of parietal cells, to secrete $H^+$. Histamine $H_2$-receptor blockers inhibit H+ secretion by blocking the stimulant effect of histamine. HCl has multiple functions such as the release and activation of pepsin, protein denaturation, destruction of swallowed microorganisms, inactivation of salivary amylase, and increase of GIT motility.

Besides secreting HCl, the stomach produces two enzymes: pepsin and a gastric lipase. Pepsin (secreted by gastric chief cells) is responsible for the chemical digestion of proteins that takes place in the stomach, effectively acting on collagen, thus playing an important role in the digestion of meat. Pepsin is secreted in the inactive form "pepsinogen" by the chief cells, and, once in the stomach lumen, pepsinogen is cleaved into pepsin, activated by the action of $H^+$, and starts the digestion of proteins.

Although gastric lipase, co-secreted with pepsin, is responsible for the breakdown of triacylglycerols, only one-third of fat digestion occurs in the stomach.

Peristaltic movements propel the food bolus into the small intestine. Once in the duodenum, the presence of nutrients and HCl stimulates the secretion of the hormones secretin, cholecystokinin (CCK), and gastric inhibitory peptide (GIP).

CCK participates in the chemical digestion that happens in this portion of the GIT as it stimulates the secretion of digestive enzymes by another exocrine gland, the pancreas. Pancreatic secretions are composed of pancreatic amylase that breaks down starch into oligosaccharides, pancreatic lipase that breaks down lipids into fatty acids, and proteinases (trypsin, chymotrypsin, and carboxypeptidases) that break down proteins into oligopeptides. Most pancreatic enzymes are secreted as zymogens, which are to be activated upon arrival in the intestine. This activation process starts when enteropeptidase present on the brush border converts trypsinogen into trypsin. Trypsin is responsible for the conversion of the other zymogens into their active forms.

Secretin stimulates the secretion of bicarbonate, also by the pancreas, and bile salts by the liver. Finally, GIP acts by stimulating insulin secretion by pancreatic beta cells.

Another hormone secreted by endocrine cells present in the stomach is ghrelin, which acts on the central nervous system with orexigenic function. On the other hand, leptin produced by adipocytes acts as an anorexigen.

The final phase of chemical digestion occurs in the brush-border region of the small intestine. In this region, disaccharidases and oligopeptidases cleave oligosaccharides into simple sugars and oligopeptides into amino acids, respectively.

The end products of digestion (glucose, amino acids, fatty acids) enter the body through absorption. The absorbed molecules can pass passively (without energy expenditure) through the epithelial cells lining the lumen (paracellular and transcellular pathways) or by active transport (requiring metabolic energy). Large molecules, such as proteins, are transported by endocytosis and exocytosis.

In animals, glucose is the predominant circulating sugar and, therefore, the main energy source for anaerobic and aerobic cellular metabolism in various body tissues (Dashty 2013; Klip et al. 2024). Glucose is actively absorbed by enterocytes via proteins called glucose transporter 2 (GLUT2) that are expressed in various tissues of the body, including the liver, kidney, and intestines (Biesiekierski 2015; Sala-Rabanal et al. 2018).

In the case of fructose, structurally similar to glucose, the glucose transporter 5 (GLUT5) is the protein used by the intestine, possibly due to its greater affinity, to carry out the transport to enterocytes (Patel et al. 2015; Febbraio and Karin 2021).

The absorption of dietary amino acids involves numerous transporter systems expressed in the brush-border membrane of human enterocytes, which may operate as electroneutral or electrogenic transporters (Kiela and Ghishan 2016).

Dietary fat is predominantly composed of nonpolar lipids, including diacylglycerol (DAG) and triacylglycerol (TAG), as well as cholesterol esters and polar phospholipids. Metabolites of dietary fats are vital sources of energy, and the associated fat-soluble vitamins are essential nutrients for the functioning of body systems. After the action of lipases and bile acids, TAG, the main dietary lipid, is hydrolyzed to monoacylglycerol (MAG) and fatty acids. Fatty acids and MAG are absorbed by enterocytes through passive diffusion, due to the concentration gradient, or by active transport using transporters such as CD36 and fatty acid transport protein-4 (FATP4) (Ko et al. 2020).

Considering that the absorption of water, electrolytes, and nutrients happens at the end of the GIT, disorders of this system mostly affect the nutritional status or the salt and water content of the body.

Gastrointestinal dysfunctions (GID) are characterized by signs and symptoms that affect GIT tissues and organs, which may be manifested as altered food intake (resulting from nausea, vomiting, dysphasia, odynophagia, anorexia), altered bowel movements (constipation or diarrhea), abdominal or thoracic pain, and a wide variety of other systemic diseases that reflect on symptoms in the GIT (Pereira et al. 2020). Gastrointestinal dysfunctions may be limited to the GIT, be manifested as a symptom of a systemic disorder, or be manifested as a systemic disease resulting from a pathological process of the GIT.

In addition to genetic predisposition, several other factors such as the use of some classes of drugs without a medical prescription, inadequate diet, and lifestyle can lead to the onset of GID (Bahramikia and Izadi 2023; Yemula 2024). In general, GID can affect motility, secretion, digestion, and absorption, thus directly influencing the GIT. Furthermore, some systemic diseases, not initiated in the GIT, can produce symptoms in the GIT.

Given that motility results from smooth muscle contraction under neural and hormonal control, disorders in motility may be the result of dysfunctions of the muscles or of the neural or hormonal systems, or even both. Secretory disorders involve dysfunctions in the production of acid, intrinsic factors or mucus by the stomach, digestive enzymes and bicarbonate by the pancreas, and bile by the liver, as well as dysfunctions in the absorption of water and electrolytes by the small intestine in response to secretagogues. Clinically known disorders involving digestion and absorption are concentrated in the small and large intestines, as well as accessory organs such as the pancreas and liver, whose secretions are necessary for digestion and absorption. Several systemic diseases (primary hyperparathyroidism, cancer, systemic lupus erythematosus, rheumatoid arthritis) can cause symptoms such as nausea, vomiting, and gastric ulcers in the GIT.

Altogether, the GIT performs metabolic and physiological functions essential for human survival. In the exercise of these functions that include, among others, absorption of food nutrients, protection against opportunistic pathobionts, and development and maturation of the intestinal epithelium, both the presence and a failure in the metabolism of xenobiotics (drug and nondrug entities) could be impacted on GIT function or dysfunction. These are fundamental for determining general drug interactions in clinical settings.

## 4 Gastrointestinal Microbiota

Trillions of microbes that constitute the gut microbiota live in the GIT, orchestrating human health. This wide range of microbes has often been referred to as bacterial species that reside in the GIT. However, recent studies have shown that nonbacterial microbes, including fungi and bacteriophages, also modulate important host-microbe interactions (Jaswal et al. 2023; Maleki et al. 2024).

The number and diversity of microbes generally increase from proximal to distal regions of the intestine and can be influenced by host characteristics and the dynamics of the microbial community (Martinez-Guryn et al. 2019; Zhang et al. 2022). Other factors such as GIT transit time, nutrient bioavailability, oxygen levels, bile acids, and pH, in addition to immunological factors, are important determinants for microbial selection (Friedman et al. 2018; Lloyd-Price et al. 2019).

In this sense, as shown in Fig. 1.4, a narrowly diverse and lower density, expressed here in terms of colony-forming units (CFU) per milliliter, is observed in the stomach and duodenum ($10^1$–$10^3$ CFU/mL) and jejunum ($10^4$–$10^7$ CFU/mL), while a wide variety and greater abundance of microbes ($10^{11}$–$10^{12}$ CFU/mL) is found in the large intestine (Giuffrè et al. 2020; Afzaal et al. 2022; Jensen et al. 2023).

This diverse microbiota has metabolic activities that are essential for the functioning of enzymes in the intestinal mucosa and for host metabolism (Dasriya et al. 2024).

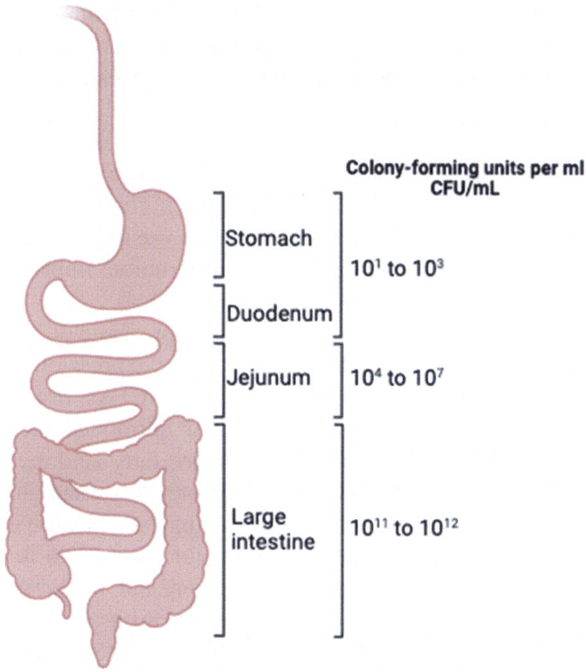


**Fig. 1.4** Distribution of microbial colonization in the gastrointestinal tract. (Created with BioRender)

The gut microbiota reacts efficiently to the main dietary changes. The alternation between a diet rich in vegetables and meat for 10 days, for example, promotes, in just 1 or 2 days, structural changes in the diversity and composition of the microbiome (David et al. 2014). The human gut microbiota is associated with the degradation of dietary fibers, proteins, and peptides by fermentation and anaerobic degradation (Yadav et al. 2018; Pei et al. 2024).

Unlike glucose and fructose, which are rapidly absorbed, polysaccharides such as hemicellulose, pectin, and resistant starch escape digestion in the digestive tract and are fermented by the intestinal microbes to produce mainly short-chain fatty acids (SCFAs). Acetate, propionate, and butyrate, the main SCFAs, perform several essential functions in the body, with a fundamental role in the microbiota-gut-brain crosstalk (Dalile et al. 2019; Blaak et al. 2020; Fusco et al. 2023).

Butyrate is perhaps the most important SCFA for human health, as it is the main energy source of human colonocytes, promoting GIT health by reducing inflammation, preserving the intestinal barrier, and even the ability to participate in the control of colorectal cancer (CRC) progression (Dias et al. 2009; Dalile et al. 2019; Karim et al. 2024). In fact, the energy balance of colonocytes from germ-free mice was restored after recolonization with microbiota derived from conventionally raised mice or with a butyrate-producing bacterial strain, demonstrating the preferential use of butyrate by colonocytes as an energy source (Donohoe et al. 2011). Propionate is an essential energy source for epithelial cells of the liver, playing a vital role in gluconeogenesis, while acetate acts as a key cofactor that contributes to the growth of other bacteria (Cani 2018; Dalile et al. 2019). Along with butyrate, acetate stimulates mucus secretion that acts as a biological lubricant of the GIT (Barcelo et al. 2000). In addition, the intestinal microbiota can also synthesize essential vitamins, such as biotin, folate, and vitamin K, besides neutralizing carcinogenic compounds (Mancabelli et al. 2024).

Metabolites derived from the gut microbiota act as important regulators of the host energy balance (Kong et al. 2021). The absence of the gut microbiota leads to lower expression of enzymes that catalyze important steps of metabolism (Donohoe et al. 2011). The presence of this microbiota improves, for example, the intestinal absorption of glucose (Backhed et al. 2004).

The mechanism through which the presence of intestinal bacteria facilitates glucose uptake involves the increase in the expression of glucose transporters such as GLUT (Yan and Ajuwon 2015). Also, SCFAs, products of bacterial fermentation, modulate the secretion of intestinal hormones such as glucagon-like peptide 1 (GLP-1) through the activation of free fatty acid receptor 2 (FFA2) (Psichas et al. 2015; Yin et al. 2022). GLP-1 promotes increased glucose absorption, metabolism, and glycogen synthesis (Ang and Ding 2016; Dalile et al. 2019).

Studies indicate that a diverse gut microbiota is an important factor for lipid absorption. Some bacterial strains present in the gut microbiota may affect the absorption of lipids and the formation of intracellular lipid droplets in enterocytes, while other strains do not have this effect (Semova et al. 2012). Indeed, conventional mice showed significantly higher levels of fatty acid translocase (CD36) in the small intestine than germ-free mice, suggesting that gut bacteria mediate an increase in gene and protein expression of this lipid transporter (Duca et al. 2012). The gut microbiota and its metabolites also modulate the levels of Angiopoietin-like 4 (ANGPTL4), a lipoprotein lipase (LPL) inhibitor that regulates triglyceride deposition in adipocytes (Sukonina et al. 2006; Korecka et al. 2013). By inhibiting LPL, ANGPTL4 decreases lipolysis of triglyceride-rich lipoproteins, resulting in higher plasma triglyceride levels and subsequently lower fatty acid uptake by body tissues (Lichtenstein and Kersten 2010; Korecka et al. 2013).

The diversity of the gut microbiota and its role in host health and disease is a process entirely dependent on the detection of gut microbes by the host immune system. In this sense, the colic mucosa acts as a physical barrier, preventing entry and promoting the recognition of possible pathogens (Shin et al. 2023; Didriksen et al. 2024).

To accomplish this process, intestinal epithelial cells rely on a wide variety of pattern recognition receptors (PRRs) that detect the pathogen-associated molecular patterns (PAMPs) present in bacteria, fungi, and viruses (Fig. 1.5).

PRRs are classified into five families: Toll-like receptors (TLRs), NOD-like receptors (NLRs), retinoic acid-inducible gene I (RIG-I)-like receptors (RLRs), C-type lectin receptors (CLRs), absence in melanoma 2 (AIM2)-like receptors (ALRs), and cyclic GMP-AMP synthase (cGAS) (Wan et al. 2023).

TLRs include TLR4 that detects bacterial lipopolysaccharide (LPS), TLR2-TLR6 that detects bacterial lipoprotein, TLR5 that detects bacterial flagellin, and TLR9 that detects CpG DNA from both bacteria and viruses (Kawai and Akira

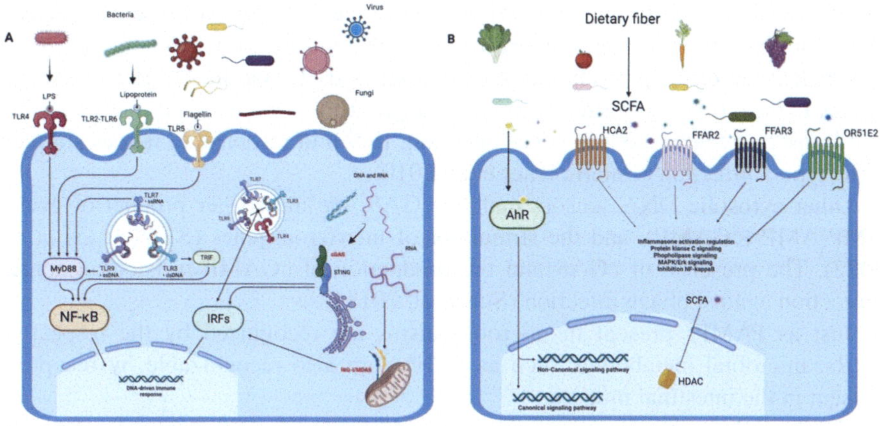


**Fig. 1.5** A. Microorganism-associated molecular patterns (MAMPs) from bacteria, fungi, and viruses are sensed by extensive types of pattern recognition receptors (PRRs). B. Microbiota-processed metabolites could be sensed mainly by G protein-coupled receptors (GPCRs) on the cell membrane or by aryl hydrocarbon receptor (AHR) and nuclear receptors in the cell nucleus and induce downstream signaling pathways. (Created with BioRender)

2010). TLR, the major immune system contributor to inflammation, plays a critical role in the pathogenesis of the inflammatory process that affects the GIT (Esmaealzadeh et al. 2024).

NLRs are another broad family of PRRs. Through their three main subtypes [NODs, NLR family pyrin domain-containing proteins (NLRPs), and NLR family caspase recruitment domain-containing proteins (NRLCs)] NLRs mediate caspase-1 activation, pro-inflammatory cytokine secretion, and cell death in response to diverse cytoplasmic stimuli. NLRP10 can detect a variety of microbial signals to promote intestinal barrier integrity by inducing secretion of IL-1β and IL-18 (Zheng et al. 2023). Genetic deficiency in NOD2 is strongly associated with the development of autoinflammatory diseases (Nomani et al. 2023).

RLRs are cytoplasmic nucleic acid receptors for RNA virions, intermediate RNA replication molecules, and transcription products (Liu et al. 2024). The RLRs family includes three receptors: retinoic acid-inducible gene I (RIG-I), melanoma differentiation factor 5 (MDA5), and laboratory of genetics and physiology 2 (LGP2). Most infections caused by most viruses are recognized by RLRs (Wicherska-Pawłowska et al. 2021).

In turn, CLRs represent a family of transmembrane PRRs, being primarily expressed by myeloid cells. They recognize not only portions of pathogens for host defense but also modified self-antigens such as damage-associated molecular patterns (DAMPs) released by cells that enter the apoptotic process (Sancho and Reis e Sousam 2013). These receptors are linked to sepsis, an immune disorder associated with hyperinflammation (Jondle et al. 2016).

ALRs are involved in the recognition of pathogenic and cytosolic DNA. Currently, two members of the ALR family have been identified: AIM2 and interferon-inducible protein 16 (IFI16) (Liwinski et al. 2020). AIM2 senses dsDNA and forms an inflammasome platform with ASC and caspase-1, leading to enzymatic cleavage of the pro-inflammatory cytokines IL-1β and IL-18 (Hornung et al. 2009). In turn, IFI16 acts directly associated with IFN-β-inducing viral DNA motifs and transcriptionally induces genes encoding IFN-β in response to herpes simplex virus type 1 infection (Unterholzner et al. 2010).

Other cytosolic DNA sensors include cGAS, the messenger product of cyclic GMP-AMP (cGAMP), and the stimulator of interferon genes (STING) (Si et al. 2022). The presence of cGAS and the production of cGAMP in bacteria confer protection against phage infection (Sun et al. 2013).

Just as PAMPs present in microorganisms are recognized by the respective PRRs, microbial metabolites such as SCFAs are also recognizable by receptors present in the intestinal mucosa.

The body has a myriad of receptors that recognize these metabolites. Some of them are the aryl hydrocarbon receptor (AhR), the free fatty acid receptor (FFAR), the hydroxycarboxylic acid receptor 2 (HCA2), and the olfactory receptor family 51, subfamily E, member 2 (OR51E2) that modulate innate immune signaling pathways, impact epithelial integrity, and, mainly, regulate the release of different hormonal signals (Dalile et al. 2019; Wan et al. 2023).

AhRs play a particularly important role in maintaining the integrity of the GIT, and their expression is important for bacterial resistance, inflammation control, and intestinal integrity (Monteleone et al. 2016; Jin et al. 2017). SCFAs, ligands of these receptors, stimulate the production of interleukin 22 (IL-22), a member of the IL-10 family with a central role in protecting the host against inflammatory damage to the colic mucosa (Yang et al. 2020).

FFAR2 and FFAR3 are the most studied SCFA receptors, as they regulate many cellular processes in innate and adaptive immune responses (Dalile et al. 2019; Kimura et al. 2020; Sawabe et al. 2024). FFAR2 is primarily expressed in enteroendocrine L cells and immune cells, including lymphocytes, neutrophils, and monocytes (Kimura et al. 2020). FFAR3 is expressed in the colon, kidneys, sympathetic nervous system, and blood vessels (Nohr et al. 2015; Kimura et al. 2020). These receptors have the most varied physiological functions that regulate energy and immune homeostasis (Milligan et al. 2017; Wu et al. 2021; Shin et al. 2023). The signaling pathway used by SCFAs to promote their anti-inflammatory effects is the FFAR pathway. However, SCFAs may act by inhibiting the histone deacetylases (HDAC) signaling pathway which, in turn, suppresses the expression of important inflammatory transcription factors (Chang et al. 2014; Ahern and Maloy 2020).

The HCA2 receptor is activated by butyrate but not acetate and propionate. It is expressed in epithelial cells of the colon and adipose tissues (Cresci et al. 2010; Feingold et al. 2014). Its activation promotes the suppression of colic inflammation through the modulation of macrophages and dendritic cells in addition to the differentiation of interleukin 10 (IL-10)-producing regulatory T cells (Singh et al. 2014).

OR51E2, expressed in multiple tissues including the intestinal tissue and enteroendocrine cells, can be activated by acetate and propionate (Pluznick et al. 2013; Zhou et al. 2019; Kotlo et al. 2020). Activation of OR51E2, induced by its ligand, inhibits the proliferation and differentiation of cancer cells and inflammation (Rodriguez et al. 2016; Sakellakis 2022).

The interaction between host and microbes also plays a key role in health and disease. Recently, studies have linked the composition and function of the human gut microbiome to numerous pathological processes (Weersma et al. 2020; Simões et al. 2022). Studies have documented that changes in the diversity of intestinal bacteria are implicated not only in GIT dysfunctions, such as inflammatory bowel disease (Lawal et al. 2023) and CRC (Pandey and Bhattacharyya 2024), but also in diseases of other systems and organs, including those that affect the cardiovascular and metabolic systems, autoimmune conditions, and psychiatric disorders (Huang et al. 2021; Romano et al. 2023; Sun et al. 2024; Wang et al. 2024).

In this sense, the presence of microbes and the products of their metabolism can effectively reduce the incidence of enteritis, cardiovascular diseases, CRC, obesity, and diabetes, as well as play an important role in maintaining the balance of energy metabolism (mainly glucose metabolism) and improving insulin tolerance (Yao et al. 2024).

In addition to influencing the absorption of nutrients, protection against pathogens, and modulation of the immune system, gut microbes can also affect the biotransformation of pharmaceuticals (e.g., antihypertensives, proton pump inhibitors (PPI), antidiabetics, antipsychotics, nonsteroidal anti-inflammatory drugs), and other xenobiotics, directly or indirectly influencing their action, fate, and toxicity (Dikeocha et al. 2022; Zhao et al. 2023). This is a bidirectional relationship since drugs can also alter intestinal microenvironments and affect microbial growth. A healthy core native microbiota is important for maintaining host health, and disruption of gut microbiota homeostasis, termed "dysbiosis," besides contributing to the pathogenesis and progression of several diseases, can also influence the outcome of treatment (Weersma et al. 2020; Chen et al. 2022).

The growing field of pharmacomicrobiomics focuses on how inter-individual variation in the microbiome can influence the clinical outcomes (pharmacokinetics and pharmacodynamics) of drugs (Torres-Carrillo et al. 2023). An increasing number of studies have sought to decipher the intricate inter-individual clinical ramifications arising from intestinal microbiota-mediated drug metabolism and vice versa, also exploring the reciprocal impacts of pharmaceutical agents on the composition and functionality of the intestinal microbiome, which can either improve or worsen the effect of the drug (Chen et al. 2022; Heirali et al. 2023; Torres-Carrillo et al. 2023). As the gut ecosystem is easily modifiable, its microbial communities may likely become an important part of personalized treatment in the future, either as a therapeutic target or as a modulator (Wang et al. 2024). In April 2023, the U.S. FDA approved Vowst®, from Seres Therapeutics, the first fecal microbiota product as an oral drug. Vowst® is used for the prevention of recurrence of *Clostridioides difficile* infection in individuals 18 years of age and older (FDA 2023).

Although drug-microbes' interactions are still poorly understood, to date, more than 70 pharmaceuticals are known to be biotransformed by gut microbiota-encoded enzymes (Klünemann et al. 2021). Bacteria generally metabolize medications into more hydrophobic compounds through reactions of reduction, hydrolysis, hydroxylation, dihydroxylation, dealkylation, and, less frequently, oxidation, but they can also remove functional groups (Tsunoda et al. 2021). Microbial enzymes participating in drug metabolism include oxidoreductases, hydrolases, and lyases (Pant et al. 2023). The mechanisms and clinical implications of microbiota-drug interactions are divided into three main types—direct and indirect metabolic transformation as well as bioaccumulation of drugs (Fig. 1.6).

1. Direct metabolic transformation of drugs by the microbiota through oxidation and reduction reactions, which may lead to processes of activation, including conversion of the prodrug to its active form; inactivation, which results in a loss in therapeutic efficacy; or toxicity resulting from the formation of toxic intermediates, leading to adverse drug reactions (Dhurjad et al. 2022; Wang et al. 2024).
2. Indirect biotransformation of drugs by intermeddling with the host metabolism through various mechanisms, such as the modulation of drug-metabolizing enzymes and transporters. Thus, gut microbes can interfere with the absorption of medications in the small intestine by increasing the expression of adhesion proteins between cells, by supporting the mucosal protective layer, and/or directly by confiscating chemicals to prevent their absorption (Collins and Patterson 2020). Also, gut microbes can impact drugs undergoing enterohepatic recycling by altering host enzyme metabolic processes, which can potentially prolong exposure to toxic metabolites (Tsunoda et al. 2021; Wang et al. 2024).
3. Bioaccumulation of drugs by storing the drug intracellularly without chemically modifying it, thereby impeding their absorption or metabolism by the host (Klünemann et al. 2021). Additionally, some drugs can be actively or passively transported into bacterial cells, influencing drug availability and bacterial

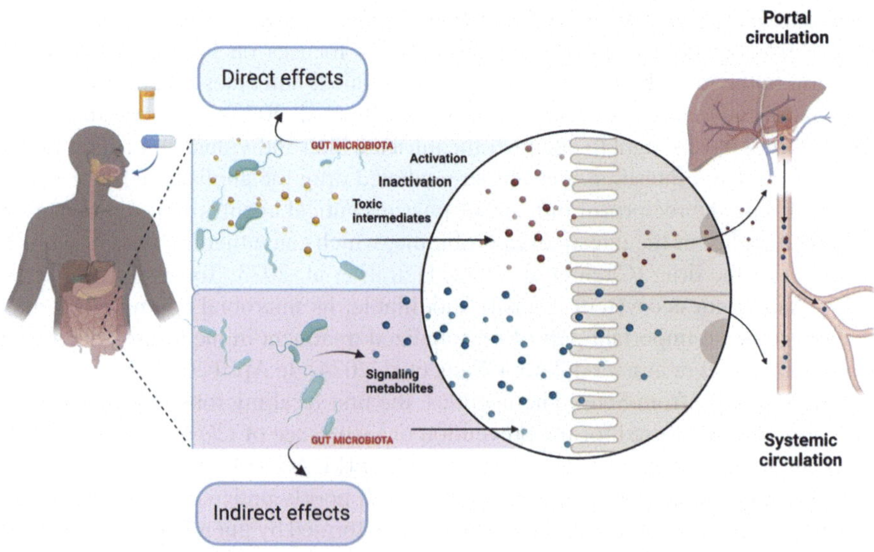


**Fig. 1.6** Influence of gut microbiota on drug and xenobiotic metabolism. (Created with BioRender)

metabolism, which may result in lower plasma concentrations and reduced therapeutic efficacy within the host, to toxicity and overdose symptoms when the accumulated drug is released in the intestinal lumen when bacterial death occurs, which can selectively impact the growth and survival of different bacterial species, in addition to increasing the risk of selecting microorganisms with resistance to antibiotics (Wang et al. 2024).

As previously stated, the interaction between drugs and enteric microbes has a bidirectional nature and can impact microbial fitness and host homeostasis. For instance, it has already been demonstrated in animals that the administration of magnesium can increase the diversity of the intestinal microbiota, while a diet deficient in magnesium results in a decrease in probiotic bacteria such as *Bifidobacterium* sp. (Mahboobi et al. 2022). Furthermore, it has been observed that hypermagnesemia can cause loss of intestinal microbial diversity in rats without magnesium deficiency (García-Legorreta et al. 2020). Indeed, it is well established that other pharmaceuticals can also exert selective pressure on the gut microbiota, such as antibiotics and PPIs, which are associated with dysbiosis that leads to an increased risk of enteric infections or can result in intestinal dysfunctions such as irritable bowel syndrome (IBS) and an increased risk of developing chronic-degenerative diseases such as metabolic syndrome, obesity, and some types of cancer (Mruk-Mazurkiewicz et al. 2023). Moreover, shifts in gut microbial composition and/or function can cause changes in the pharmacodynamics, pharmacokinetics, and toxicity of medications that are metabolized in the digestive system. Intestinal dysbiosis is associated with hepatotoxicity of drugs such as brivudine, an oral antiviral, and acetaminophen, a nonsteroidal anti-inflammatory (Zimmerman et al. 2019; Luo and Zhou 2022).

The interplay between the gut microbiota and vitamins also systematically contributes to human health. Gut microbes are essential for supporting the metabolism of bile acids by influencing the absorption and post-adsorption metabolism of lipids and fat-soluble vitamins and maintaining the function of the intestinal barrier. Gut dysbiosis can lead to the malabsorption of fat-soluble micronutrients, which can increase the risk of developing IBS and colitis-associated colorectal cancer (Stacchiotti et al. 2021). On the other hand, fat-soluble vitamins influence enteric microbes by modulating immunity or by stimulating or inhibiting the proliferation of bacterial species. As it is a bidirectional interplay, gut microbes in turn play a role in the synthesis, metabolism, and transport of fat-soluble vitamins and their bioactive metabolites, originating from the diet or released in the intestine through enterohepatic circulation. Illustrating some of these interactions, *Escherichia coli* lipocalins are involved in the transport of retinoids, while retinaldehyde dehydrogenase expressed by *E. coli* catalyzes the conversion of retinal to retinoic acid. Also, the gut microbiota can produce active metabolites of vitamins A, E, and K that are important for immunomodulation (Stacchiotti et al. 2021).

Gut microbes are also suppliers, by several synthesis pathways, of water-soluble vitamins, including several B vitamins, such as pyridoxine, folates, biotin, cobalamin, pantothenic acid, niacin, riboflavin, and thiamine, to their host (Magnusdottir et al. 2015). Components of the intestinal microbiota assist in the absorption of water-soluble vitamins, such as riboflavin (B2), by altering the rate of gastric emptying. Moreover, the end product of the metabolism of acid intestinal lactic acid bacteria and bifidobacteria is important for the intestinal absorption of thiamine (B1) and helps modulate the expression and activity of cytochrome P450 (CYP) hepatic enzymes, which act through different mechanisms in the metabolism of various drugs (Wang et al. 2024).

Gut symbiont microbes can metabolize plant-derived substances such as phenolic compounds, leading to the formation of a myriad of nutritional metabolites that contribute to metabolic health through their antioxidant and anti-inflammatory properties. Dietary polyphenols exhibit prebiotic action, beneficially influencing gut bacterial composition and function. After ingestion, polyphenols have limited absorption in both the stomach and small intestine. However, upon reaching the large intestine, they are hydrolyzed, demethylated, decarboxylated, dehydroxylated, and ring fissioned by the indigenous microbiota, generating more active and better-absorbed metabolites (Koudoufio et al. 2020). Valuable effects of polyphenols, such as those present in red wine, include an increase in the abundance of beneficial bacteria that improve intestinal barrier health (*Bifidobacterium* and *Lactobacillus*), while inhibiting the growth of less desirable microorganisms (Moreno-Indias et al. 2016; Koudoufio et al. 2020).

Natural methylxanthines, such as caffeine, are associated with an anti-obesity effect through the intestinal microbiota and its metabolites, via mechanisms that include promoting lipid degradation, inhibiting adipogenesis, and inducing changes in serum metabolites (Fortunato et al. 2024). Interestingly, the modification of the intestinal microbiota by the ingestion of methylxanthines present in coffee can affect the pharmacological effects of aspirin/salicylic acid, potentially increasing their absorption (Kim et al. 2022).

Due to their complexity and high molecular weight, saponins have limited bioavailability, depending on interactions with enzymes, nutrients, and intestinal microorganisms for absorption. Metabolization of most saponins by the intestinal microbiota reduces their polarity and increases their solubility in fat, facilitating their absorption (Zhang et al. 2024). Saponins can selectively influence the composition of the intestinal microbiota, increasing beneficial microorganisms, which, in turn, plays a pivotal role in the biotransformation of saponins into sapogenins (He et al. 2019; Zhang et al. 2024). Although knowledge of these mutual interactions is not fully elucidated, the stepwise hydrolysis of sugar is among the main metabolic routes (He et al. 2019). Indeed, after oral administration of ginsenosides, a gradual deglycosylation action occurs in the digestive tract, mediated by gut species of *Bacteroides*, *Bifidobacterium*, *Eubacterium*, and *Fusobacterium* originating secondary ginsenosides with greater pharmacological properties than their original molecules.

Other natural compounds that can also undergo biotransformation reactions generating active metabolites under the action of enteric microbes are alkaloids. These compounds usually contain ether bonds and coordination bonds, which are easily dehydrated and hydrolyzed by nitro- and nitroso-reductases produced by components of the intestinal microbiota (Gong et al. 2020). Some indigenous intestinal microorganisms express carboxylesterases (CEs), which hydrolyze diester diterpenoid alkaloids into monoester diterpenoid alkaloids, which are less toxic (Zhao et al. 2022). Also, selected alkaloids are associated with improving intestinal microbiota and metabolism while decreasing harmful bacteria associated with intestinal dysbiosis, thus alleviating symptoms of chronic-degenerative diseases such as obesity and metabolic syndrome (Zhao et al. 2022; Liu et al. 2023; Peng et al. 2019).

Even though allicin is known to have antimicrobial effects against undesirable microbes, it has been demonstrated that this compound selectively influences the intestinal microbiota, increasing the population of probiotic microbes, in addition to reducing the inflammatory process and increasing intestinal villi (Zhang et al. 2020; Li et al. 2022). The combined use of gliclazide, an antidiabetic, with allicin promoted a more noticeable reduction in blood glucose in diabetic animals than when the medication was used alone (Lagisetty et al. 2018). It is possible to infer that there is a synergism mediated by healthy intestinal microbiota between the two compounds since, according to Đanić et al. (2019), gliclazide-probiotic interactions are important for the biotransformation of the drug, especially via hydroxylation and hydrolysis reactions catalyzed by bacterial enzymes. Allicin also showed therapeutic effects on hepatic steatosis in nonalcoholic fatty liver disease patients (Soleimani et al. 2020).

## 5 Concept of Drug-Nondrug Interactions and Biological Processes

Deciphering the operation of systems and disease biology which is complex can get complicated in the presence of xenobiotics. Insights into the interactions that are associated with the biological responses could ultimately provide useful medical clues. Biological processes in human body systems are readouts of interactions among drugs and nondrug entities involving (1) myriads of endogenous targets/mediators, and (2) xenobiotics—nutrients, minerals, functional foods, herbs, active principles or metabolites, vitamins, beverages, and supplements. The idea that drug molecules interact at specific sites (receptor interactions theory) or other means (e.g., osmotic laxatives) to elicit their effects in the body often defines the concept of drug interactions. The aspect of drug interactions that are associated with alterations in absorption, distribution, metabolism, or excretion and/or effects of one drug by nondrug components and vice versa (see details in Chap. 7). These interactions can lead to important considerations in medical practice vis-a-vis changes in drug effectiveness, safety profile, and limitations in clinical application. According to Zhao et al. (2023), an answer to the reason why a drug with a good response for one patient but not for the next, or in some cases of serious adverse drug reactions, is complicated and multi-faceted. For instance, drug affinity and intrinsic activity of receptors, among other parameters, are crucial factors in understanding how drugs interact with biological systems; they are not the only determinants. Drug interactions can be influenced by a multitude of factors, including the presence of other drugs, displacement from plasma proteins, irreversible combination, and long-lasting interaction through covalent bonding or environmental substances, individual variability, and variations in patient physiology. A mechanistic understanding of drug interactions encompasses biological systems' function and dysfunction with or without nondrug entities. The extension of predominantly drug-drug interaction dogma to DNIs variety of nondrug entities (Fig. 1.7) could potentially redefine pharmacological outcomes (therapeutic, adverse, and toxic effects, etc.), clinically relevant interactions, limitations, and restrictions to drug applications.

The combination of drugs and nondrug therapies in therapeutic settings among patients with diverse physiopathological readouts and medical states (infectious, psychiatric, oncology, cardiovascular, or metabolic diseases) among geriatric or pediatric patients, pregnant and breastfeeding mothers are making cases for a new understanding of drug-related interactions in clinical practice (Fig. 1.8). The drug-disease interactions can occur between certain drugs and specific disease states affecting biological systems (cardiovascular, gastrointestinal, respiratory, endocrine, immune), specific tissues, and organs (liver, heart, kidney, etc.). For instance, severe liver disease can be associated with reduced metabolic clearance and higher plasma levels of drugs extensively metabolized by the liver (Brouwer et al. 1992). Although liver disease reduces drug clearance on average, the change is relatively small and usually not clinically relevant except in patients with near-terminal liver

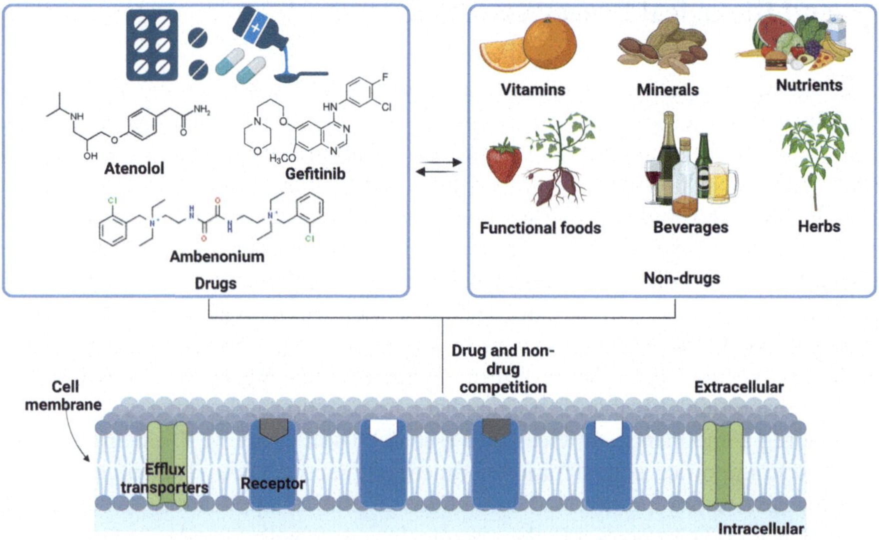


**Fig. 1.7** Drugs and nondrugs entities are represented by nutrients, minerals, functional foods, herbs, active principles or metabolites, vitamins, beverages, and supplements. (Created with BioRender)

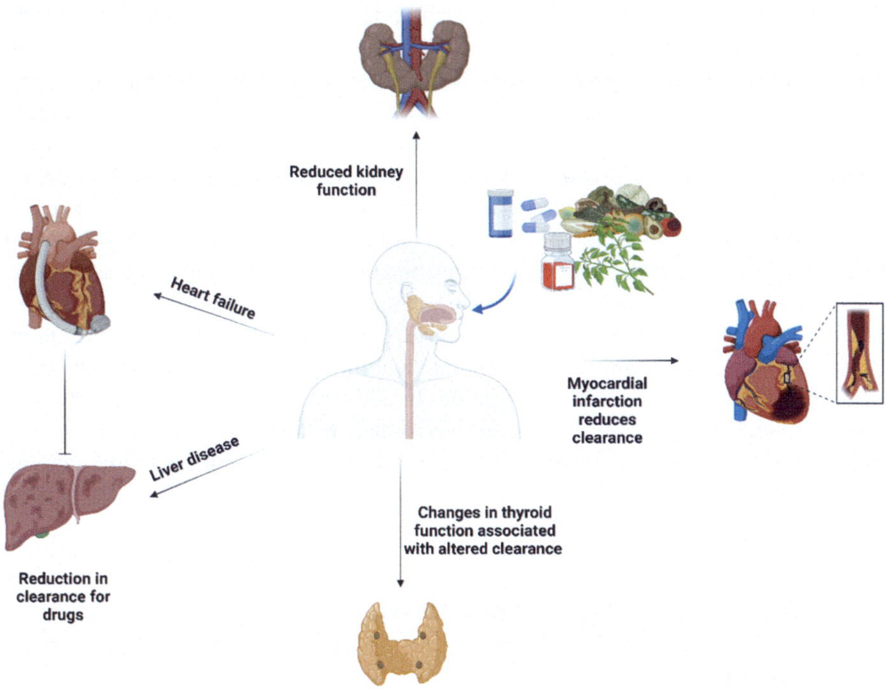


**Fig. 1.8** System, organ, tissue, or cellular physiopathological readouts with or without xenobiotics. (Created with BioRender)

disease. The effects of renal disease on the elimination of drugs that are primarily cleared renally are more predictable (Lam et al. 1997). Heart failure reduces liver blood flow and causes a reduction in clearance for drugs such as lidocaine or propranolol that are usually extensively cleared by the liver (Shammas and Dickstein 1988), and acute myocardial infarction reduces the clearance of some drugs, such as lidocaine, as well (Pieper and Johnson 1992). Acute viral infection and changes in thyroid function have been associated with altered clearance for some drugs, such as theophylline and warfarin (Yamaguchi et al. 2000; Stephens et al. 1989; Pokrajac et al. 1987).

A collection of diverse, interdependent, and interconnected agents (endogenous and exogenous neurochemicals, drugs, foods, beverages, supplements, etc.) that interact with each other could elicit an overall impact on the biological processes, structure, function, or dynamic behavior at the molecular, cellular, tissue, and system level. In this manner, enhancement, suppression, or lack of clinical efficacy of drugs goes beyond the predominantly drug-drug interaction dogma. Recurrence of unanticipated effects and consequences demands an in-depth analysis of the non-drug entities to fully account for drug interaction. For instance, typical nondrug-induced change in the microenvironment could impact the physicochemical properties of drugs or responses from different targets (enzymes, transporters, receptors, etc.), accounting for subsequent pharmacological consequences. The chemical nature of these nondrugs could be analyzed and explored to provide in-depth coverage of unsuspecting significant interactions from physiopathological perspectives to clinical approaches. The dearth of reports on DNIs often jeopardizes our understanding of important interactions that can enhance therapeutic outcomes, safety, and optimization of drug applications. The application of molecular interactions, omics, predictive tools, and artificial intelligence could facilitate DNIs comprehension in a complex biological system and help rationalize drug development programs for the identification, monitoring, and evaluation of pharmaceutical, pharmacokinetic, and pharmacodynamic interactions (Konig et al. 2017; Ramaswami et al. 2018; Prasad and Groop 2019). A grasp of DNIs in the future could enrich clinical understanding toward a tailored medical intervention that observes an individual's variability for precision medicine. In addition, important pharmacological concepts (such as agonism, synergism, resistance, potentiation, and antagonism), and rational decisions on drug choice, profiling, or recommendation can be established.

# 6 Self-Assessment Questions and Answers

## *6.1 Questions*

1. An undergraduate student participates in a clinical research study on intestinal motility. Peristaltic movements of the small intestine:

   A. Involves the simultaneous relaxation of smoo.th muscle throughout the small intestine
   B. Coordinated by the central nervous system
   C. Involves contraction of the smooth muscle behind the food bolus and relaxation of the smooth muscle in front of the food bolus
   D. Mixes the food bolus

2. A patient with a duodenal ulcer is successfully treated with the drug cimetidine. The basis of inhibition of gastric H+ secretion by cimetidine is:

   A. Increases intracellular levels of cyclic adenosine monophosphate (cAMP)
   B. Blocks muscarinic receptors on parietal cells
   C. Intensifies the action of acetylcholine in parietal cells
   D. Blocks histamine H2 receptors on parietal cells

3. The overall outcome of DNIs is determined by

   A. Affinity alone
   B. Efficacy alone
   C. Intrinsic activity of receptor
   D. None of the above

4. Which of the following chemical bonds would most likely create an irreversible combination and long-lasting drug-plant metabolite interaction?

   A. Ionic bond
   B. Hydrogen bond
   C. Van der Waals bond
   D. Covalent bond

## 6.2 Answers

1. C. Peristaltic movements are the contractile activity coordinated by the enteric nervous system, not the central nervous system, that propels intestinal contents forward. Then, for this to happen, there must be simultaneous contraction of the smooth muscle behind the food bolus and relaxed in front of it.
2. D. Cimetidine is a reversible inhibitor of histamine H2 receptors in parietal cells and blocks the secretion of H+.
3. D. Drug and nondrug interactions can have various outcomes, and these outcomes can be influenced by factors such as the specific substances involved, the dosage, individual patient characteristics, and the timing of administration.
4. D. A covalent bond is a strong and stable bond that is essentially irreversibly formed at normal body temperature. The other bonds are much weaker.

# References

Afzaal M, Saeed F, Shah YA, Hussain M, Rabail R, Socol CT, Hassoun A, Pateiro M, Lorenzo JM, Rusu AV, Aadil RM. Human gut microbiota in health and disease: unveiling the relationship. Front Microbiol. 2022;13:999001.

Ahern PP, Maloy KJ. Understanding immune-microbiota interactions in the intestine. Immunology. 2020;159:4–14.

Ahmad MH, Man CS, Othman F, He FJ, Salleh R, Noor NSM, et al. High sodium food consumption pattern among Malaysian population. J Health Popul Nutr. 2021;40(S1):4. https://doi.org/10.1186/s41043-021-00230-5.

Ang Z, Ding JL. GPR41 and GPR43 in obesity and inflammation – protective or causative? Front Immunol. 2016;7:28.

Antoniou T. Trimethoprim-sulfamethoxazole–induced hyperkalemia in patients receiving inhibitors of the renin-angiotensin system. Arch Intern Med. 2010;170(12):1045. https://doi.org/10.1001/archinternmed.2010.142.

Antoniou T, Hollands S, Macdonald EM, Gomes T, Mamdani M, Juurlink DN. Trimethoprim–sulfamethoxazole and risk of sudden death among patients taking spironolactone. Can Med Assoc J. 2015;187(4):E138–43. https://doi.org/10.1503/cmaj.140816.

Asiimwe IG, Pirmohamed M. Drug-drug-gene interactions in cardiovascular medicine. Pharmacogenomics Pers Med. 2022;15:879–911. https://doi.org/10.2147/pgpm.s338601.

Assefa YA, Kedir A, Kahaliw W. Survey on polypharmacy and drug-drug interactions among elderly people with cardiovascular diseases at Yekatit 12 hospital, Addis Ababa, Ethiopia. Integr Pharm Res Pract. 2020;9:1–9. https://doi.org/10.2147/iprp.s231286.

Backhed F, Ding H, Wang T, Hooper LV, Koh GY, Nagy A, Semenkovich CF, Gordon JI. The gut microbiota as an environmental factor that regulates fat storage. Proc Natl Acad Sci USA. 2004;101:15718–23.

Bahramikia S, Izadi R. Plant-based green synthesis of nanoparticles as an effective and safe treatment for gastric ulcer. Inflammopharmacology. 2023;31:2843–55.

Bailey DG, Spence JD, Munoz C, Arnold JM. Interaction of citrus juices with felodipine and nifedipine. Lancet. 1991;337(8736):268–9. https://doi.org/10.1016/0140-6736(91)90872-m.

Barcelo A, Claustre J, Moro F, Chayvialle JA, Cuber JC, Plaisancié P. Mucin secretion is modulated by luminal factors in the isolated vascularly perfused rat colon. Gut. 2000;46:218–24.

Bernard C. Introduction à l'étude de la médecine expérimentale. Paris: J.-B. Baillière et Fils. [French]; 1865.

Biesiekierski JR. Fructose-induced symptoms beyond malabsorption in FGID. United European Gastroenterol J. 2015;2:10–3.

Blaak EE, Canfora EE, Theis S, Frost G, Groen AK, Mithieux G, Nauta A, Scott K, Stahl B, van Harsselaar J, van Tol R, Vaughan EE, Verbeke K. Short chain fatty acids in human gut and metabolic health. Benef Microbes. 2020;11:411–55.

Bristol M, Sonnad SS, Guerra CE. Uninformed complementary and alternative supplement use: a risky behavior for cardiovascular patients. Complement Health Pract Rev. 2008;13(2):100–9. https://doi.org/10.1177/1533210108317281.

Brouwer KLR, Dukes GE, Powell JR. Influence of liver function on drug disposition. In: Evans WE, Schentag JJ, Jusko WJ, editors. Applied pharmacokinetics: principles of therapeutic drug monitoring. Vancouver: Applied Therapeutics, Inc; 1992. p. 6-1–6-59.

Campbell SC, Stockmann C, Balch A, Clark EA, Kamyar M, Varner M, Korgenski EK, Bonkowsky JL, Spigarelli MG, Sherwin CM. Intrapartum magnesium sulfate and the potential for cardiopulmonary drug–drug interactions. Ther Drug Monit. 2014;36(4):544–8. https://doi.org/10.1097/FTD.0000000000000050.

Cani PD. Human gut microbiome: hopes, threats and promises. Gut. 2018;67:1716–25.

Chang PV, Hao L, Offermanns S, Medzhitov R. The microbial metabolite butyrate regulates intestinal macrophage function via histone deacetylase inhibition. Proc Natl Acad Sci USA. 2014;111:2247–52.

Chen HQ, Gong JY, Xing K, Liu MZ, Ren H, Luo JQ. Pharmacomicrobiomics: exploiting the drug-microbiota interactions in antihypertensive treatment. Front Med (Lausanne). 2022;8:742394. https://doi.org/10.3389/fmed.2021.742394.

Collins SL, Patterson AD. The gut microbiome: an orchestrator of xenobiotic metabolism. Acta Pharm Sin B. 2020;10(1):19–32. https://doi.org/10.1016/j.apsb.2019.12.001.

Cresci GA, Thangaraju M, Mellinger JD, Liu K, Ganapathy V. Colonic gene expression in conventional and germ-free mice with a focus on the butyrate receptor GPR109A and the butyrate transporter SLC5A8. J Gastrointest Surg. 2010;14:449–61.

Dalile B, Van Oudenhove L, Vervliet B, Verbeke K. The role of short-chain fatty acids in microbiota-gut-brain communication. Nat Rev Gastroenterol Hepatol. 2019;16:461–78.

Đanić M, Stanimirov B, Pavlović N, Vukmirović S, Lazić J, Al-Salami H, Mikov M. Transport and biotransformation of gliclazide and the effect of deoxycholic acid in a probiotic bacteria model. Front Pharmacol. 2019;10:1083. https://doi.org/10.3389/fphar.2019.01083.

Dashty M. A quick look at biochemistry: carbohydrate metabolism. Clin Biochem. 2013;46:1339–52.

Dasriya VL, Samtiya M, Ranveer S, Dhillon HS, Devi N, Sharma V, Nikam P, Puniya M, Chaudhary P, Chaudhary V, Behare PV, Dhewa T, Vemuri R, Raposo A, Puniya DV, Khedkar GD, Vishweswaraiah RH, Vij S, Alarifi SN, Han H, Puniya AK. Modulation of gut-microbiota through probiotics and dietary interventions to improve host health. J Sci Food Agric. 2024;104:6359–75.

David LA, Maurice CF, Carmody RN, Gootenberg DB, Button JE, Wolfe BE, Ling AV, Devlin AS, Varma Y, Fischbach MA, Biddinger SB, Dutton RJ, Turnbaugh PJ. Diet rapidly and reproducibly alters the human gut microbiome. Nature. 2014;505:559–63.

Dhurjad P, Dhavaliker C, Gupta K, Sonti R. Exploring drug metabolism by the gut microbiota: modes of metabolism and experimental approaches. Drug Metab Dispos. 2022;50(3):224–34. https://doi.org/10.1124/dmd.121.000669.

Dias JC, Suzuki E, de Albuquerque CL, Ferreira AL, Brito AR, Kubota LT. Determination of short-chain fatty acids in dietary fiber extracts using ion-exclusion chromatography with suppressed conductivity detection. J Pharm Biomed Anal. 2009;49:1128–32.

Didriksen BJ, Eshleman EM, Alenghat T. Epithelial regulation of microbiota-immune cell dynamics. Mucosal Immunol. 2024;17:303–13.

Dikeocha IJ, Al-Kabsi AM, Miftahussurur M, Alshawsh MA. Pharmacomicrobiomics: influence of gut microbiota on drug and xenobiotic metabolism. FASEB J. 2022;36(6):e22350. https://doi.org/10.1096/fj.202101986R.

Donohoe DR, Garge N, Zhang X, Sun W, O'Connell TM, Bunger MK, Bultman SJ. The microbiome and butyrate regulate energy metabolism and autophagy in the mammalian colon. Cell Metab. 2011;13:517–26.

Doubova SV, Reyes-Morales H, Torres-Arreola L d P, Suárez-Ortega M. Potential drug-drug and drug-disease interactions in prescriptions for ambulatory patients over 50 years of age in family medicine clinics in Mexico City. BMC Health Serv Res. 2007;7(1):147. https://doi.org/10.1186/1472-6963-7-147.

Duca FA, Swartz TD, Sakar Y, Covasa M. Increased oral detection, but decreased intestinal signaling for fats in mice lacking gut microbiota. PLoS One. 2012;7:202.

Esmaealzadeh N, Ram M, Abdolghaffari A, Marques AM, Bahramsoltani R. Toll-like receptors in inflammatory bowel disease: a review of the role of phytochemicals. Phytomedicine. 2024;123:155178.

Febbraio MA, Karin M. "Sweet death": fructose as a metabolic toxin that targets the gut-liver axis. Cell Metab. 2021;33:2316–28.

Feingold KR, Moser A, Shigenaga JK, Grunfeld C. Inflammation stimulates niacin receptor (GPR109A/HCA2) expression in adipose tissue and macrophages. J Lipid Res. 2014;55:2501–8.

Fitton N, Thomas JS. Gastrointestinal dysfunction. Surgery. 2009;27:492–5.

Fortunato IM, Pereira QC, Oliveira FS, Alvarez MC, Santos TWD, Ribeiro ML. Metabolic insights into caffeine's anti-adipogenic effects: an exploration through intestinal microbiota modulation in obesity. Int J Mol Sci. 2024;25(3):1803. https://doi.org/10.3390/ijms25031803.

Friedman ES, Bittinger K, Esipova TV, Hou L, Chau L, Jiang J, Mesaros C, Lund PJ, Liang X, FitzGerald GA, Goulian M, Lee D, Garcia BA, Blair IA, Vinogradov SA, Wu GD. Microbes vs. chemistry in the origin of the anaerobic gut lumen. Proc Natl Acad Sci USA. 2018;115:4170–5.

Fu Y, Ru J, Huang C, Du J, Zheng C, Chen X, et al. Insights from systems pharmacology into cardiovascular drug discovery and therapy. BMC Syst Biol. 2014;8(1):141. https://doi.org/10.1186/s12918-014-0141-z.

Fusco W, Lorenzo MB, Cintoni M, Porcari S, Rinninella E, Kaitsas F, Lener E, Mele MC, Gasbarrini A, Collado MC, Cammarota G, Ianiro G. Short-chain fatty-acid-producing bacteria: key components of the human gut microbiota. Nutrients. 2023;15:2211.

García-Legorreta A, Soriano-Pérez LA, Flores-Buendía AM, Medina-Campos ON, Noriega LG, Granados-Portillo O, Nambo-Venegas R, Tovar AR, Mendoza-Vargas A, Barrera-Oviedo D, Pedraza-Chaverri J, Palacios-González B. Effect of dietary magnesium content on intestinal microbiota of rats. Nutrients. 2020;12(9):2889. https://doi.org/10.3390/nu12092889.

Giuffrè M, Campigotto M, Campisciano G, Comar M, Crocè LS. A story of liver and gut microbes: how does the intestinal flora affect liver disease? A review of the literature. Am J Physiol Gastrointest Liver Physiol. 2020;318:G889–906.

Gong X, Li X, Bo A, Shi RY, Li QY, Lei LJ, Zhang L, Li MH. The interactions between gut microbiota and bioactive ingredients of traditional Chinese medicines: A review. Pharmacol Res. 2020;157:104824.

Gupta DK, Lewis CE, Varady KA, Su YR, Madhur MS, Lackland DT, et al. Effect of dietary sodium on blood pressure. JAMA. 2023;330(23):2258. https://doi.org/10.1001/jama.2023.23651.

Hatfield J, Saad S, Housewright C. Dietary supplements and bleeding. Baylor Univ Med Cent Proc. 2022;35(6):802–7. https://doi.org/10.1080/08998280.2022.2121575.

He Y, Hu Z, Li A, Zhu Z, Yang N, Ying Z, He J, Wang C, Yin S, Cheng S. Recent advances in biotransformation of saponins. Molecules. 2019;24(13):2365. https://doi.org/10.3390/molecules24132365.

Heirali A, Moossavi S, Arrieta MC, Coburn B. Principles and terminology for host–microbiome–drug interactions. Open Forum. Infect Dis. 2023;10(5):ofad195. https://doi.org/10.1093/ofid/ofad195.

Hornung V, Ablasser A, Charrel-Dennis M, Bauernfeind F, Horvath G, Caffrey DR, Latz E, Fitzgerald KA. AIM2 recognizes cytosolic dsDNA and forms a caspase-1-activating inflammasome with ASC. Nature. 2009;458:514–8.

Houston M. The role of magnesium in hypertension and cardiovascular disease. J Clin Hypertens (Greenwich). 2011;13(11):843–7. https://doi.org/10.1111/j.1751-7176.2011.00538.x.

Huang L, Thonusin C, Chattipakorn N, Chattipakorn SC. Impacts of gut microbiota on gestational diabetes mellitus: a comprehensive review. Eur J Nutr. 2021;60:2343–60.

Jaswal K, Todd OA, Behnsen J. Neglected gut microbiome: interactions of the non-bacterial gut microbiota with enteric pathogens. Gut Microbes. 2023;15:2226916.

Jensen BAH, Heyndrickx M, Jonkers D, Mackie A, Millet S, Naghibi M, Pærregaard SI, Pot B, Saulnier D, Sina C, Sterkman LGW, Van den Abbeele P, Venlet NV, Zoetendal EG, Ouwehand AC. Small intestine vs. colon ecology and physiology: why it matters in probiotic administration. Cell Rep Med. 2023;4:101190.

Jiang X, Williams KM, Liauw W, Ammit AJ, Roufogalis BD, Duke CC, et al. Effect of St John's wort and ginseng on the pharmacokinetics and pharmacodynamics of warfarin in healthy subjects. Br J Clin Pharmacol. 2004;57(5):592–9. https://doi.org/10.1111/j.1365-2125.2003.02051.x.

Jin UH, Cheng Y, Park H, Davidson LA, Callaway ES, Chapkin RS, Jayaraman A, Asante A, Allred C, Weaver EA, Safe S. Short chain fatty acids enhance aryl hydrocarbon (Ah) responsiveness in mouse colonocytes and Caco-2 human colon cancer cells. Sci Rep. 2017;7:10163.

Johne A, Brockmöller J, Bauer S, Maurer A, Langheinrich M, Roots I. Pharmacokinetic interaction of digoxin with an herbal extract from St John's wort (Hypericum perforatum). Clin Pharmacol Ther. 1999;66(4):338–45. https://doi.org/10.1053/cp.1999.v66.a101944.

Jondle CN, Sharma A, Simonson TJ, Larson B, Mishra BB, Sharma J. Macrophage galactose-type lectin-1 deficiency is associated with increased neutrophilia and hyperinflammation in gram-negative pneumonia. J Immunol. 2016;196:3088–96.

Karim MR, Iqbal S, Mohammad S, Morshed MN, Haque MA, Mathiyalagan R, Yang DC, Kim YJ, Song JH, Yang DU. Butyrate's (a short-chain fatty acid) microbial synthesis, absorption, and preventive roles against colorectal and lung cancer. Arch Microbiol. 2024;206:137.

Kawai T, Akira S. The role of pattern-recognition receptors in innate immunity: update on Toll-like receptors. Nat Immunol. 2010;11:373–84.

Ke J, Li MT, Huo YJ, Cheng YQ, Guo SF, Wu Y, Zhang L, Ma J, Liu AJ, Han Y. The synergistic effect of Ginkgo biloba extract 50 and aspirin against platelet aggregation. Drug Des Devel Ther. 2021;15:3543–60. https://doi.org/10.2147/DDDT.S318515.

Khera A, Emdin C, Drake I, Natarajan P, Bick A, Cook N, et al. Genetic risk, adherence to a healthy lifestyle, and coronary disease. N Engl J Med. 2016;375(24):2349–58. https://doi.org/10.1056/nejmoa1605086.

Kiela PR, Ghishan FK. Physiology of intestinal absorption and secretion. Best Pract Res Clin Gastroenterol. 2016;30:145–59.

Kim JK, Choi MS, Yoo HH, Kim DH. The intake of coffee increases the absorption of aspirin in mice by modifying gut microbiome. Pharmaceutics. 2022;14(4):746. https://doi.org/10.3390/pharmaceutics14040746.

Kimura I, Ichimura A, Ohue-Kitano R, Igarashi M. Free fatty acid receptors in health and disease. Physiol Rev. 2020;100:171–210.

Klip A, De Bock K, Bilan PJ, Richter EA. Transcellular barriers to glucose delivery in the body. Annu Rev Physiol. 2024;86:149–73.

Klünemann M, Andrejev S, Blasche S, Mateus A, Phapale P, Devendran S, Vappiani J, Simon B, Scott TA, Kafkia E, Konstantinidis D, Zirngibl K, Mastrorilli E, Banzhaf M, Mackmull MT, Hövelmann F, Nesme L, Brochado AR, Maier L, Bock T, Periwal V, Kumar M, Kim Y, Tramontano M, Schultz C, Beck M, Hennig J, Zimmermann M, Sévin DC, Cabreiro F, Savitski MM, Bork P, Typas A, Patil KR. Bioaccumulation of therapeutic drugs by human gut bacteria. Nature. 2021;597(7877):533–8. https://doi.org/10.1038/s41586-021-03891-8.

Ko CW, Qu J, Black DD, Tso P. Regulation of intestinal lipid metabolism: current concepts and relevance to disease. Nat Rev Gastroenterol Hepatol. 2020;17:169–83.

Kong D, Schipper L, van Dijk G. Distinct effects of short chain fatty acids on host energy balance and fuel homeostasis with focus on route of administration and host species. Front Neurosci. 2021;15:755845.

Konig IR, Fuchs O, Hansen G, von Mutius E, Kopp MV. What is precision medicine? Eur Respir J. 2017;50:1700391.

Koudoufio M, Desjardins Y, Feldman F, Spahis S, Delvin E, Levy E. Insight into polyphenol and gut microbiota crosstalk: are their metabolites the key to understand protective effects against metabolic disorders? Antioxidants (Basel). 2020;9(10):982. https://doi.org/10.3390/antiox9100982.

Korecka A, de Wouters T, Cultrone A, Lapaque N, Pettersson S, Doré J, Blottière HM, Arulampalam V. ANGPTL4 expression induced by butyrate and rosiglitazone in human intestinal epithelial cells utilizes independent pathways. Am J Physiol Gastrointest Liver Physiol. 2013;304:G1025–37.

Kotlo K, Anbazhagan AN, Priyamvada S, Jayawardena D, Kumar A, Chen Y, Xia Y, Finn PW, Perkins DL, Dudeja PK, Layden BT. The olfactory G protein-coupled receptor (Olfr-78/OR51E2) modulates the intestinal response to colitis. Am J Physiol Cell Physiol. 2020;318:C502–13.

Lagisetty U, Mohammed H, Ramaiah S. Effect of capsaicin on pharmacodynamic and pharmacokinetics of gliclazide in animal models with diabetes. Pharmacognosy Research. 2018;10(4):437.

Lam YW, Banerji S, Hatfield C, Talbert RL. Principles of drug administration in renal insufficiency. Clin Pharmacokinet. 1997;32(1):30–57.

Lawal SA, Voisin A, Olof H, Bording-Jorgensen M, Armstrong H. Diversity of the microbiota communities found in the various regions of the intestinal tract in healthy individuals and inflammatory bowel diseases. Front Immunol. 2023;14:1242242.

Ledwitch K, Barnes RW, Roberts AG. Unravelling the complex drug-drug interactions of the cardiovascular drugs, verapamil and digoxin, with P-glycoprotein. Biosci Rep. 2016;36(2):e00309. https://doi.org/10.1042/bsr20150317.

Lee JW, Morris JK, Wald N. Grapefruit juice and statins. Am J Med. 2016;129(1):26–9. https://doi.org/10.1016/j.amjmed.2015.07.036.

Li L, Song Q, Zhang X, Yan Y, Wang X. Allicin alleviates diabetes mellitus by inhibiting the formation of advanced glycation end products. Molecules. 2022;27(24):8793.

Lichtenstein L, Kersten S. Modulation of plasma TG lipolysis by angiopoietin-like proteins and GPIHBP1. Biochim Biophys Acta. 2010;1801:415–20.

Liu D, Ye J, Yan Y, Chen Y, Wang H, Wang M, Feng Y, Li R, Xu X, Jiang Y, Lian C, Yang Y, Meng Y, Liu Y, Jiang W. Ramulus mori (Sangzhi) alkaloids regulates gut microbiota disorder and its metabolism profiles in obese mice induced by a high-fat diet. Front Pharmacol. 2023;14:1166635.

Liu N, Yan X, Lv B, Wu Y, Hu X, Zheng C, Tao S, Deng R, Dou J, Zeng B, Jiang G. A study on the association between gut microbiota, inflammation, and type 2 diabetes. Appl Microbiol Biotechnol. 2024;108:213.

Liwinski T, Zheng D, Elinav E. The microbiome and cytosolic innate immune receptors. Immunol Rev. 2020;297:207–24.

Lloyd-Price J, Arze C, Ananthakrishnan AN, Schirmer M, Avila-Pacheco J, Poon TW, Andrews E, Ajami NJ, Bonham KS, Brislawn CJ, Casero D, Courtney H, Gonzalez A, Graeber TG, Hall AB, Lake K, Landers CJ, Mallick H, Plichta DR, Prasad M, Rahnavard G, Sauk J, Shungin D, Vázquez-Baeza Y, White RA, Braun J, Denson LA, Jansson JK, Knight R, Kugathasan S,

McGovern DPB, Petrosino JF, Stappenbeck TS, Winter HS, Clish CB, Franzosa EA, Vlamakis H, Xavier RJ, Huttenhower C. Multi-omics of the gut microbial ecosystem in inflammatory bowel diseases. Nature. 2019;569:655–62.

Lown KS, Bailey DG, Fontana RJ, Janardan SK, Adair CH, Fortlage LA, et al. Grapefruit juice increases felodipine oral availability in humans by decreasing intestinal CYP3A protein expression. J Clin Invest. 1997;99:2545–53.

Lundahl J, Regårdh CG, Edgar B, Johnsson G. Relationship between time of intake of grapefruit juice and its effect on pharmacokinetics and pharmacodynamics of felodipine in healthy subjects. Eur J Clin Pharmacol. 1995;49(1–2):61–7. https://doi.org/10.1007/BF00192360.

Luo Y, Zhou T. Connecting the dots: Targeting the microbiome in drug toxicity. Med Res Rev. 2022;42(1):83–111. https://doi.org/10.1002/med.21805.

Magnúsdóttir S, Ravcheev D, de Crécy-Lagard V, Thiele I. Systematic genome assessment of B-vitamin biosynthesis suggests co-operation among gut microbes. Front Genet. 2015;6:148. https://doi.org/10.3389/fgene.2015.00148.

Mahboobi S, Ghasvarian M, Ghaem H, Alipour H, Alipour S, Eftekhari MH. Effects of probiotic and magnesium co-supplementation on mood cognition intestinal barrier function and inflammation in individuals with obesity and depressed mood: A randomized double-blind placebo-controlled clinical trial. Front Nutr. 2022;9:1018357. https://doi.org/10.3389/fnut.2022.1018357.

Maleki S, Razavi SH, Yadav H, Letizia Manca M. New horizon to the world of gut microbiome: seeds germination. Crit Rev Food Sci Nutr. 2024;16:1–19.

Mancabelli L, Milani C, De Biase R, Bocchio F, Fontana F, Lugli GA, Alessandri G, Tarracchini C, Viappiani A, De Conto F, Nouvenne A, Ticinesi A, Bussolati O, Meschi T, Cecchi R, Turroni F, Ventura M. Taxonomic and metabolic development of the human gut microbiome across life stages: a worldwide metagenomic investigation. mSystems. 2024;5:e0129423.

Markowitz J. Effect of St John's wort on drug metabolism by induction of cytochrome P450 3A4 enzyme. JAMA. 2003;290(11):1500. https://doi.org/10.1001/jama.290.11.1500.

Martinez-Guryn K, Leone V, Chang EB. Regional diversity of the gastrointestinal microbiome. Cell Host Microbe. 2019;26:314–24.

Milligan G, Shimpukade B, Ulven T, Hudson BD. Complex pharmacology of free fatty acid receptors. Chem Rev. 2017;117:67–110.

Milosavljević M, Janković SM. The impact of cardiovascular drugs on the efficacy of local anesthesia in dentistry. Biomed Pap. 2016;160(4):571–7. https://doi.org/10.5507/bp.2016.040.

Monteleone I, Marafini I, Zorzi F, Di Fusco D, Dinallo V, Rizzo A, Sileri P, Sica G, Monteleone G. Smad7 knockdown restores aryl hydrocarbon receptor-mediated protective signals in the gut. J Crohns Colitis. 2016;10:670–7.

Moreno-Indias I, Sánchez-Alcoholado L, Sánchez-Garrido MÁ, Martín-Núñez GM, Pérez-Jiménez F, Tena-Sempere M, Tinahones FJ, Queipo-Ortuño MI. Neonatal androgen exposure causes persistent gut microbiota dysbiosis related to metabolic disease in adult female rats. Endocrinology. 2016;157(12):4888–98. https://doi.org/10.1210/en.2016-1317.

Mruk-Mazurkiewicz H, Kulaszyńska M, Jakubczyk K, Janda-Milczarek K, Czarnecka W, Rębacz-Maron E, Zacha S, Sieńko J, Zeair S, Dalewski B, Marlicz W, Łoniewski I, Skonieczna-Żydecka K. Clinical relevance of gut microbiota alterations under the influence of selected drugs—updated review. Biomedicines. 2023;11(3):952. https://doi.org/10.3390/biomedicines11030952.

Murtaza G, Khan M, Azhar S, Khan S, Khan T. Assessment of potential drug-drug interactions and its associated factors in the hospitalized cardiac patients. Saudi Pharm J. 2016;24(2):220–5. https://doi.org/10.1016/j.jsps.2015.03.009.

Nohr MK, Egerod KL, Christiansen SH, Gille A, Offermanns S, Schwartz TW, Møller M. Expression of the short chain fatty acid receptor GPR41/FFAR3 in autonomic and somatic sensory ganglia. Neuroscience. 2015;290:126–37.

Nomani H, Deng Z, Navetta-Modrov B, Yang J, Yun M, Aroniadis O, Gorevic P, Aksentijevich I, Yao Q. Implications of combined NOD2 and other gene mutations in autoinflammatory diseases. Front Immunol. 2023;14:1265404.

Paine MF, Widmer WW, Hart HL, Pusek SN, Beavers KL, Criss AB, et al. A furanocoumarin-free grapefruit juice establishes furanocoumarins as the mediators of the grapefruit juice-felodipine interaction. Am J Clin Nutr. 2006;83:1097–105.

Pandey M, Bhattacharyya J. Gut microbiota and epigenetics in colorectal cancer: implications for carcinogenesis and therapeutic intervention. Epigenomics. 2024;16:403–18.

Pant A, Maiti TK, Mahajan D, Das B. Human Gut microbiota and drug metabolism. Microb Ecol. 2023;86(1):97–111. https://doi.org/10.1007/s00248-022-02081-x.

Patel C, Douard V, Yu S, Tharabenjasin P, Gao N, Ferraris RP. Fructose-induced increases in expression of intestinal fructolytic and gluconeogenic genes are regulated by GLUT5 and KHK. Am J Physiol Regul Integr Comp Physiol. 2015;309:R499–509.

Pei Z, Liu Y, Chen Y, Pan T, Sun X, Wang H, Ross RP, Lu W, Chen W. A universe of human gut-derived bacterial prophages: unveiling the hidden viral players in intestinal microecology. Gut Microbes. 2024;16:1–21.

Peng J, Zheng TT, Li X, Liang Y, Wang LJ, Huang YC, Xiao HT. Plant-Derived Alkaloids: The Promising Disease-Modifying Agents for Inflammatory Bowel Disease. Front Pharmacol. 2019;10:351.

Pereira LB, Gonçalves AM, Estancial Fernandes CS, Fontanella AT, Francisco PM, Mengue SS, Borges RB, Pizzol TSD, Costa KS. Use of drugs for gastrointestinal disorders: evidence from National Survey on Access, Use and Promotion of Rational Use of Medicines. Einstein. 2020;18:eAO5314.

Pieper JA, Johnson KE. Lidocaine. In: Evans WE, Schentag JJ, Jusko WJ, editors. Applied pharmacokinetics: principles of therapeutic drug monitoring. Vancouver: Applied Therapeutics Inc; 1992. p. 21-1–21-37.

Pluznick JL, Protzko RJ, Gevorgyan H, Peterlin Z, Sipos A, Han J, Brunet I, Wan LX, Rey F, Wang T, Firestein SJ, Yanagisawa M, Gordon JI, Eichmann A, Peti-Peterdi J, Caplan MJ. Olfactory receptor responding to gut microbiota-derived signals plays a role in renin secretion and blood pressure regulation. Proc Natl Acad Sci USA. 2013;110:4410–5.

Pokrajac M, Simic D, Varagic VM. Pharmacokinetics of theophylline in hyperthyroid and hypothyroid patients with chronic obstructive pulmonary disease. Eur J Clin Pharmacol. 1987;33(5):483–6.

Prasad RB, Groop L. Precision medicine in type 2 diabetes. J Intern Med. 2019;285:40–8.

Psichas A, Sleeth ML, Murphy KG, Brooks L, Bewick GA, Hanyaloglu AC, Ghatei MA, Bloom SR, Frost G. The short chain fatty acid propionate stimulates GLP-1 and PYY secretion via free fatty acid receptor 2 in rodents. Int J Obes. 2015;39:424–9.

Ramaswami R, Bayer R, Galea S. Precision medicine from a public health perspective. Annu Rev Public Health. 2018;39:153–68.

Rodriguez M, Siwko S, Liu M. Prostate-specific G-protein coupled receptor, an emerging biomarker regulating inflammation and prostate cancer invasion. Curr Mol Med. 2016;16:526–32.

Romano KA, Nemet I, Prasad Saha P, Haghikia A, Li XS, Mohan ML, Lovano B, Castel L, Witkowski M, Buffa JA, Sun Y, Li L, Menge CM, Demuth I, König M, Steinhagen-Thiessen E, DiDonato JA, Deb A, Bäckhed F, Tang WHW, Naga Prasad SV, Landmesser U, Van Wagoner DR, Hazen SL. Gut microbiota-generated phenylacetylglutamine and heart failure. Circ Heart Fail. 2023;16:e009972.

Sakellakis M. Orphan receptors in prostate cancer. Prostate. 2022;82:1016–24.

Sala-Rabanal M, Ghezzi C, Hirayama BA, Kepe V, Liu J, Barrio JR, Wright EM, et al. Intestinal absorption of glucose in mice as determined by positron emission tomography. J Physiol. 2018;596:2473–89.

Sancho D, Reis e Sousam C. Sensing of cell death by myeloid C-type lectin receptors. Curr Opin Immunol. 2013;25:46–52.

Sawabe A, Okazaki S, Nakamura A, Goitsuka R, Kaifu T. The orphan G protein-coupled receptor 141 expressed in myeloid cells functions as an inflammation suppressor. J Leukoc Biol. 2024;16:qiae009.

Schaffer AL, Chia J, Brett J, Pearson S, Falster MO. A nationwide study of multimedicine use in people treated with cardiovascular medicines in Australia. Pharmacotherapy. 2022;42(11):828–36. https://doi.org/10.1002/phar.2735.

Semova I, Carten JD, Stombaugh J, Mackey LC, Knight R, Farber SA, Rawls JF. Microbiota regulate intestinal absorption and metabolism of fatty acids in the zebrafish. Cell Host Microbe. 2012;12:277–88.

Shammas FV, Dickstein K. Clinical pharmacokinetics in heart failure. An updated review. Clin Pharmacokinet. 1988;15(2):94–113.

Shin Y, Han S, Kwon J, Ju S, Choi TG, Kang I, Kim SS. Roles of short-chain fatty acids in inflammatory bowel disease. Nutrients. 2023;15:4466.

Si W, Liang H, Bugno J, Xu Q, Ding X, Yang K, Fu Y, Weichselbaum RR, Zhao X, Wang L. Lactobacillus rhamnosus GG induces cGAS/STING-dependent type I interferon and improves response to immune checkpoint blockade. Gut. 2022;71:521–33.

Simões CD, Maganinho M, Sousa AS. FODMAPs, inflammatory bowel disease and gut microbiota: updated overview on the current evidence. Eur J Nutr. 2022;61:1187–98.

Singh N, Gurav A, Sivaprakasam S, Brady E, Padia R, Shi H, Thangaraju M, Prasad PD, Manicassamy S, Munn DH, Lee JR, Offermanns S, Ganapathy V. Activation of Gpr109a, receptor for niacin and the commensal metabolite butyrate, suppresses colonic inflammation and carcinogenesis. Immunity. 2014;40:128–39.

Sobhy KE, Abdelmagged O, Abdelaty K, Khalil D, Abdelgaied M. Potential cardiovascular drug interactions in Egypt: incidence, outcomes, mechanism, and management. Kafrelsheikh Vet Med J. 2021;19(2):22–7. https://doi.org/10.21608/kvmj.2021.97432.1025.

Soleimani D, Paknahad Z, Rouhani MH. Therapeutic effects of garlic on hepatic steatosis in nonalcoholic fatty liver disease patients: a randomized clinical trial. Diabetes Metab Syndr Obes. 2020;13:2389–97. https://doi.org/10.2147/DMSO.S254555.

Stacchiotti V, Rezzi S, Eggersdorfer M, Galli F. Metabolic and functional interplay between gut microbiota and fat-soluble vitamins C. Crit Rev Food Sci Nutr. 2021;61(19):3211–32. https://doi.org/10.1080/10408398.2020.1793728.

Stanley WC, Shah KB, Essop MF. Does junk food lead to heart failure? Hypertension. 2009;54(6):1209–10. https://doi.org/10.1161/hypertensionaha.109.128660.

Stephens MA, Self TH, Lancaster D, Nash T. Hypothyroidism: effect on warfarin anticoagulation. South Med J. 1989;82(12):1585–6.

Storelli F, Samer C, Reny J, Desmeules J, Daali Y. Complex drug-drug-gene-disease interactions involving cytochromes p450: systematic review of published case reports and clinical perspectives. Clin Pharmacokinet. 2018;57(10):1267–93. https://doi.org/10.1007/s40262-018-0650-9.

Sukonina V, Lookene A, Olivecrona T, Olivecrona G. Angiopoietin-like protein 4 converts lipoprotein lipase to inactive monomers and modulates lipase activity in adipose tissue. Proc Natl Acad Sci USA. 2006;103:17450–5.

Sun L, Wu J, Du F, Chen X, Chen ZJ. Cyclic GMP-AMP synthase is a cytosolic DNA sensor that activates the type I interferon pathway. Science. 2013;339(786):791.

Sun C, Zhu D, Zhu Q, He Z, Lou Y, Chen D. The significance of gut microbiota in the etiology of autoimmune hepatitis: a narrative review. Front Cell Infect Microbiol. 2024;14:1337223.

Surma S, Szyndler A. Salt and arterial hypertension—epidemiological, pathophysiological and preventive aspects. Arterial Hypertens. 2020;24(4):148–58. https://doi.org/10.5603/ah.a2020.0012.

Trelle S, Reichenbach S, Wandel S, Hildebrand P, Tschannen B, Villiger PM, et al. Cardiovascular safety of non-steroidal anti-inflammatory drugs: network meta-analysis. BMJ. 2011;342:c7086. https://doi.org/10.1136/bmj.c7086.

Torres-Carrillo N, Martínez-López E, Torres-Carrillo NM, López-Quintero A, Moreno-Ortiz JM, González-Mercado A, Gutiérrez-Hurtado IA. Pharmacomicrobiomics and drug–infection interactions: the impact of commensal symbiotic and pathogenic microorganisms on a host response to drug therapy. Int J Mol Sci. 2023;24(23):17100. https://doi.org/10.3390/ijms242317100.

Tsunoda SM, Gonzales C, Jarmusch AK, Momper JD, Ma JD. Contribution of the gut microbiome to drug disposition pharmacokinetic and pharmacodynamic variability. Clin Pharmacokinet. 2021;60(8):971–84. https://doi.org/10.1007/s40262-021-01032-y.

Unterholzner L, Keating SE, Baran M, Horan KA, Jensen SB, Sharma S, Sirois CM, Jin T, Latz E, Xiao TS, Fitzgerald KA, Paludan SR, Bowie AG. IFI16 is an innate immune sensor for intracellular DNA. Nat Immunol. 2010;11:997–1004.

U.S. Food and Drug Administration, FDA approves first orally administered Fecal microbiota product for the prevention of recurrence of Clostridioides difficile infection, https://www.fda.gov/news-events/press-announcements/fda-approves-first-orally-administered-fecal-microbiota-product-prevention-recurrence-clostridioides (2023).

Wan T, Wang Y, He K, Zhu S. Microbial sensing in the intestine. Protein Cell. 2023;14:824–60.

Wang M, Song Z, Lai S, Tang F, Dou L, Yang F. Depression-associated gut microbes, metabolites and clinical trials. Front Microbiol. 2024;15:1292004.

Weersma RK, Zhernakova A, Fu J. Interaction between drugs and the gut microbiome. Gut. 2020;69:1510–9.

Wicherska-Pawłowska K, Wróbel T, Rybka J. Toll-like receptors (TLRs), NOD-like receptors (NLRs), and RIG-I-like receptors (RLRs) in innate immunity. TLRs, NLRs, and RLRs ligands as immunotherapeutic agents for hematopoietic diseases. Int J Mol Sci. 2021;22:13397.

Wu M, Li Q, Mai K, Ai Q. Regulation of free fatty acid receptor 4 on inflammatory gene induced by LPS in large yellow croaker (Larimichthys crocea). Front Immunol. 2021;12:703914.

Yadav M, Verma MK, Chauhan NS. A review of metabolic potential of human gut microbiome in human nutrition. Arch Microbiol. 2018;200:203–17.

Yamaguchi A, Tateishi T, Okano Y, Matuda T, Akimoto Y, Miyoshi T, et al. Higher incidence of elevated body temperature or increased C-reactive protein level in asthmatic children showing transient reduction of theophylline metabolism. J Clin Pharmacol. 2000;40(3):284–9.

Yan H, Ajuwon KM. Mechanism of butyrate stimulation of triglyceride storage and adipokine expression during adipogenic differentiation of porcine stromovascular cells. PLoS One. 2015;10:e0145940.

Yang W, Yu T, Huang X, Bilotta AJ, Xu L, Lu Y, Sun J, Pan F, Zhou J, Zhang W, Yao S, Maynard CL, Singh N, Dann SM, Liu Z, Cong Y. Intestinal microbiota-derived short-chain fatty acids regulation of immune cell IL-22 production and gut immunity. Nat Commun. 2020;11:4457.

Yao Y, Liu Y, Xu Q, Mao L. Short chain fatty acids: essential weapons of traditional medicine in treating inflammatory bowel disease. Molecules. 2024;29:1–27.

Yemula N. Gut microbiota in celiac disease. Ann Gastroenterol. 2024;37:125–32.

Yin Y, Guo Q, Zhou X, Duan Y, Yang Y, Gong S, Han M, Liu Y, Yang Z, Chen Q, Li F. Role of brain-gut-muscle axis in human health and energy homeostasis. Front Nutr. 2022;9:947033.

Zhang C, He X, Sheng Y, Yang C, Xu J, Zheng A, Liu J, Xu W, Luo Y, Huang K. Allicin-induced host-gut microbe interactions improves energy homeostasis. FASEB J. 2020;34(8):10682–98. https://doi.org/10.1096/fsb2.v34.8. https://doi.org/10.1096/fj.202001007R.

Zhang X, Chen B, Jia P, Han J. Locked on salt? Excessive consumption of high-sodium foods during covid-19 presents an underappreciated public health risk: a review. Environ Chem Lett. 2021;19(5):3583–95. https://doi.org/10.1007/s10311-021-01257-0.

Zhang ZJ, Lehmann CJ, Cole CG, Pamer EG. Translating microbiome research from and to the clinic. Ann Rev Microbiol. 2022;76:435–60.

Zhang Y, Hao R, Chen J, Li S, Huang K, Cao H, Farag MA, Battino M, Daglia M, Capanoglu E, Zhang F, Sun Q, Xiao J, Sun Z, Guan X. Health benefits of saponins and its mechanisms: perspectives from absorption metabolism and interaction with gut. Crit Rev Food Sci Nutr. 2024;64(25):9311–32. https://doi.org/10.1080/10408398.2023.2212063.

Zhao Y, Zhong X, Yan J, Sun C, Zhao X, Wang X. Potential roles of gut microbes in biotransformation of natural products: An overview. Front Microbiol. 2022;13:956378. https://doi.org/10.3389/fmicb.2022.956378.

Zhao Q, Chen Y, Huang W, et al. Drug-microbiota interactions: an emerging priority for precision medicine. Signal Transduct Target Ther. 2023;8:386. https://doi.org/10.1038/s41392-023-01619-w.

Zheng D, Mohapatra G, Kern L, He Y, Shmueli MD, Valdés-Mas R, Kolodziejczyk AA, Próchnicki T, Vasconcelos MB, Schorr L, Hertel F, Lee YS, Rufino MC, Ceddaha E, Shimshy S, Hodgetts RJ, Dori-Bachash M, Kleimeyer C, Goldenberg K, Heinemann M, Stettner N, Harmelin A, Shapiro H, Puschhof J, Chen M, Flavell RA, Latz E, Merbl Y, Abdeen SK, Elinav E. Epithelial Nlrp10 inflammasome mediates protection against intestinal autoinflammation. Nat Immunol. 2023;24:585–94.

Zhou Q, Yang D, Wu M, Guo Y, Guo W, Zhong L, Cai X, Dai A, Jang W, Shakhnovich EI, Liu ZJ, Stevens RC, Lambert NA, Babu MM, Wang MW, Zhao S. Common activation mechanism of class A GPCRs. elife. 2019;8:e50279.

Zimmermann M, Zimmermann-Kogadeeva M, Wegmann R, Goodman AL. Separating host and microbiome contributions to drug pharmacokinetics and toxicity Off-target drug metabolism. Science. 2019;363(6427):eaat9931. https://doi.org/10.1126/science.aat9931.

# Chapter 2
# Nutritional Elements I: Nutrients, Proteins, Carbohydrates, and Lipids

**Barbara Martins Vieira, Giovanna Nascimento de Mello e Silva, and Mariana Isaac Silva**

**Abstract** This chapter addresses macronutrients essential for human nutrition. Nutrients are present in foods that are essential for the body's functioning, providing energy and raw materials for maintaining and repairing body tissues, as well as supporting growth. They are classified into macronutrients, such as carbohydrates, lipids, and proteins, and micronutrients, such as vitamins and minerals. Adequate nutrition requires the intake of essential nutrients that are not produced by the body in sufficient quantities and must be obtained through diet. Carbohydrates, one of the macronutrients, are specific, abundant molecules that account for up to 80% of all energy consumed by humans. They are classified as monosaccharides, disaccharides, oligosaccharides, and polysaccharides. The main function of carbohydrates is to provide energy. Carbohydrate digestion begins in the mouth with the action of salivary amylase and continues in the small intestine with pancreatic amylase. Proteins, composed of amino acids, perform a variety of functions in the body, including the formation of tissues, enzymes, hormones, antibodies, and the provision of energy. Protein digestion begins in the stomach with the action of pepsin and continues in the small intestine with pancreatic enzymes such as trypsin. Lipids, another essential macronutrient, are water-insoluble organic compounds that perform crucial functions, such as energy storage, structural components of cell membranes, and acting as hormones. Lipid digestion begins in the mouth and stomach but occurs primarily in the small intestine with the help of bile salts and pancreatic lipase.

**Keywords** Biochemistry · Nutrients bioavailability · Nutritional requirements · Digestive process · Human nutrition

B. M. Vieira (✉) · G. N. de Mello e Silva
Curso de Nutrição, Universidade Evangélica, Anápolis, Goias, Brazil
e-mail: barbara.vieira@docente.unievangelica.edu.br; giovanna.silva@docente.unievangelica.edu.br

M. I. Silva
Fundação de Ensino e Pesquisa em Ciências da Saúde, Brasília, Distrito Federal, Brazil
e-mail: marianasilva@escs.edu.br

 
J. O. Fajemiroye (ed.), *Fundamentals of Drug and Non-Drug Interactions*,
https://doi.org/10.1007/978-3-031-80107-5_2

# 1 Introduction

Nutrition is a complex and multidisciplinary science that studies the relationship between food and the human body, encompassing much more than the simple ingestion of food. It is essential for maintaining health and preventing and treating diseases (Nelson et al. 2022). According to Krause (2022), nutrition is defined as the science that studies the combination of processes by which living organisms receive and utilize the food necessary for maintaining their functions, growth, and renewal of their components, while feeding refers to the processes by which an organism ingests, digests, absorbs, transports, utilizes, and excretes nutrients.

Humans are capable of extracting the energy and other necessary nutrients for life from the food they consume. One of the main functions of a healthy diet is to provide the body with the nutrients to maintain an adequate state of health and prevent diseases, thus contributing to an improved quality of life. Pedro Escudero, an important nutrition science scholar from the past century, emphasized that through the balanced consumption of food (healthy eating), the body receives in quantity, quality, adequacy, and harmony to remain functional.

Nutrients provide structures that constitute food and are essential for body function. The word nutrient, from Latin "nutriens," means "to nourish." These are substances whose absence from the diet or their reduction below the minimum limit produces a deficiency disease over a certain period. They provide energy, serve as raw materials, maintain and repair body parts, and sustain growth.

Nutrients can be classified into macronutrients, such as carbohydrates, lipids, and proteins, and micronutrients, such as vitamins and minerals (Fig. 2.1). Adequate nutrition requires individuals to receive essential nutrients. These nutrients are indispensable for growth and proper functioning of the body. An essential nutrient

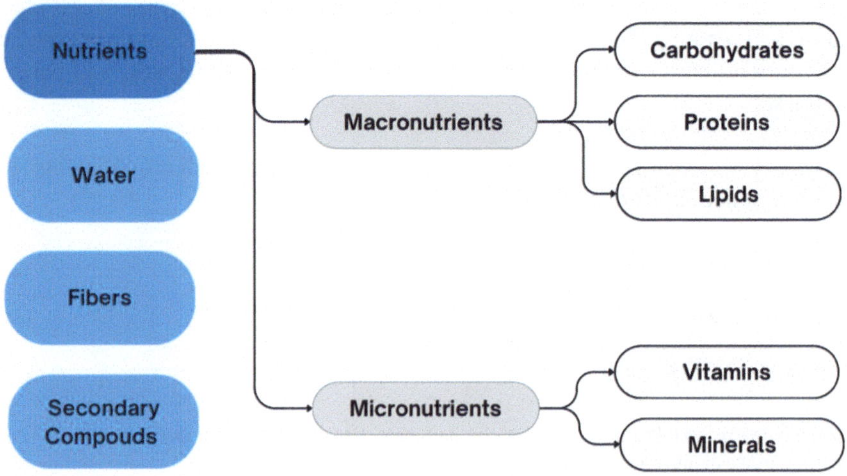


**Fig. 2.1** Classification of nutrients

is defined as one that must be obtained from the diet because it is not produced in sufficient quantities by the body. A non-essential nutrient does not need to be ingested through the diet because it is produced in sufficient quantity by the body.

Each nutrient has various specific functions and is related to others. They play vital roles in maintaining health and well-being, providing energy, building and maintaining body tissues, and controlling metabolic processes such as enzyme production and body temperature regulation. In this chapter, we will discuss macronutrients in more detail.

# 2 Carbohydrates

## *2.1 Classification*

Carbohydrates are the most abundant molecules in life and constitute up to 80% of all the energy consumed by humans. Glucose is the form in which most dietary carbohydrates are absorbed and converted into energy. The liver often converts other sugars into glucose, and all carbohydrates in the body are formed from glucose (Liberato and Oliveira 2019).

They are classified into monosaccharides, disaccharides, oligosaccharides, and polysaccharides (Table 2.1). The word "carbohydrate" originally referred to compounds with the general formula Cn(H2O)n. However, it is now known that only simple sugars or monosaccharides fit exactly into this formula. Other types of carbohydrates are based on units of monosaccharides and have partially distinct general formulas (Campbell and Farrell 2008).

Disaccharides and monosaccharides are not absorbed in this form in the small intestine; they need to undergo hydrolysis and be converted into monosaccharides to be rapidly absorbed in the gastrointestinal tract and pass into the bloodstream. Oligosaccharides and non-hydrolyzed polysaccharides pass intact through the small

**Table 2.1** Subdivision of carbohydrates

| Class | Subgroup | Types |
|---|---|---|
| *Sugars* | *Monosaccharides* | Glucose, fructose, galactose |
| (1–2 monomers) | Disaccharides | Sucrose, lactose, maltose |
| | Polyols | Sorbitol, mannitol |
| *Oligosaccharides* | *Malt oligosaccharides* | Raffinose, stachyose, fructooligosaccharides |
| (3–9 monomers) | Other oligosaccharides | Maltodextrins |
| *Polysaccharides* | *Starch* | Amylose, amylopectin, resistant starch |
| (>9 monomers) | Non-starch Polysaccharides | Cellulose, hemicellulose, fructans (inulin), gums, mucilages, and pectin |

Source: FAO/WHO (1998)

intestine to the large intestine, where they act as fibers. Additionally, oligosaccharides can be broken down by intestinal microflora, producing metabolic products such as acetic and lactic acids, which in large quantities have a laxative effect and can cause diarrhea (Mahan et al. 2013).

To implement these characteristics into food science and nutrition, it was necessary to expand the biochemical classification of carbohydrates, organizing carbohydrate molecules according to their degree of polymerization (number of monosaccharide units), and then subdividing them according to their types of glycosidic linkages and specific chemical characteristics (Cummings and Englyst 1996).

## 2.2 Chemical Composition

Carbohydrates are macromolecules composed of carbon, hydrogen, and oxygen atoms, generally in the proportion of 1:2:1, respectively. They are chemically defined as polyhydroxyketones (ketoses) or polyhydroxyaldehydes (aldoses), which are organic compounds with at least three carbons, possessing a hydroxyl group, except for the one that has a primary carbonyl group (aldehyde group) or a secondary carbonyl group (ketone group). Based on the number of carbon atoms in their structure, monosaccharides can be classified as triose (three carbon atoms), tetrose (four carbon atoms), pentose (five carbon atoms), and so on (Cozzolino 2016).

In an aqueous solution, aldotetroses and all other monosaccharides with more than five carbon atoms predominantly occur in a cyclic (ring) form. This structure is formed by the covalent bond between the carbon of the carbonyl group (anomeric carbon) and the oxygen of one of the hydroxyl groups in the monosaccharide chain. Depending on the position of the hydroxyl group formed during this bond, two structures with the same molecular formula but different molecular arrangements can be formed, resulting in different properties (isomers). In the case of glucose and fructose, α and β isomers are formed (Campbell 2000).

Monosaccharides, or simple sugars, are the most abundant carbohydrates in the diet of living beings. For example, glucose is one of the main energy sources for living organisms. It is an essential carbohydrate, chemically classified as a hexose due to its six carbon atoms, and more specifically as an aldohexose because it contains an aldehyde group. It can exist in two isomeric forms, D-glucose and L-glucose, which are mirror images of each other. D-glucose is the biologically active form found in nature and, in solution, can form two anomers: α-D-glucose and β-D-glucose, which differ in the position of the hydroxyl group on the anomeric carbon (the first carbon of the ring). Additionally, glucose is stored as glycogen, mainly in the liver and muscles, and as starch in plants (FAO/WHO 1998).

Fructose is a sugar that occurs naturally in fruits. It is a hexose, containing six carbon atoms and a ketone group in its structure. It can exist in two isomeric forms: D-fructose and L-fructose, which are mirror images of each other. D-fructose is a naturally occurring and biologically active form. It functions as an energy source, being converted into glucose in the liver before entering the metabolic pathways of

ATP production, and serves as a natural sweetener (Dutra de Oliveira and Marchini 2008).

Galactose is one of the main energy sources for living organisms and plays a crucial role in various biological functions. It is a hexose, composed of six carbon atoms and an aldehyde group. Galactose is not generally found in large quantities in nature but is combined with glucose to form lactose, which is the main sugar found in milk. Lactose is formed by a molecule of galactose and a molecule of glucose linked by a glycosidic bond (Cozzolino 2016).

Disaccharides are formed from two monosaccharides. The condensation reaction occurs when two monosaccharides are linked and a molecule of water is released. This bond is called a glycosidic bond and can be of the alpha or beta type. The three main disaccharides are sucrose (glucose + fructose), lactose (glucose + galactose), and maltose (glucose + glucose). Simple sugars (monomers or dimers) generally impart a sweet taste to foods, making them pleasant to the human palate, and are widely used by the food industry as preservatives, giving characteristic texture and flavor to baked goods and cookies (Campbell and Farrell 2008).

Oligosaccharides are between the simple molecules of mono- and disaccharides and the long spatial arrangements of polysaccharides. Carbohydrates with chains that have these characteristics include tri- and tetrasaccharides such as raffinose and stachyose, maltodextrins, pyrodextrins, fructo-, and galactooligosaccharides, which are highly soluble in water. It should be noted that oligosaccharides are formed by the linkage of monosaccharide units with the consequent loss of a water molecule through the so-called O-glycosidic bond. These bonds are named according to the numbering of the carbon atoms involved in the bond, as well as the configuration ($\alpha$ or $\beta$) of the anomeric carbon, for example, $\alpha(1 \rightarrow 4)$ bond and $\beta(1 \rightarrow 6)$ bond. Generally, oligosaccharides resulting from the hydrolysis of starch are partially digested and absorbed in the human gastrointestinal tract (Cozzolino 2016; Nelson and Cox 2019).

Inulin is a storage oligosaccharide, soluble in water, and belongs to a group of non-digestible carbohydrates called fructans. They are natural food ingredients commonly found in various components of a normal diet. They are part of the storage carbohydrate fraction of plants, which is why they are found in more than 36,000 plant species (Liberato and Oliveira 2019).

Polysaccharides are polymers formed by a large number of monomeric units linked together through glycosidic bonds, and consequently, when hydrolyzed, they release a large number of monosaccharides or oligosaccharides. General characteristics include being insoluble in water, not having a sweet taste, and having a high molecular weight. Most carbohydrates found in nature are in the form of polysaccharides, possessing different functions such as structural, energy storage, and forming the extracellular matrix. Polysaccharides can be classified into homopolysaccharides (formed by a single type of monomeric unit) and heteropolysaccharides (formed by more than one type of monomeric unit) (Liberato and Oliveira 2019).

Starch is the most important dietary source of carbohydrates. It consists of two main substances: amylose, with a helical non-branched structure ($\alpha(1 \rightarrow 4)$), and amylopectin, composed of branched chains formed by 20–25 glucose residues linked by $\alpha(1 \rightarrow 4)$ bonds in the chains and by $\alpha(1 \rightarrow 6)$ bonds at the branching points. When hydrolyzed, starch forms dextrins, then maltose, and finally glucose (Lehninger et al. 2016).

Starch is found only in plants, in the form of amylose, which has straight, long chains of glucose units, and amylopectin, which has branched chains of glucose units. Starch granules of various sizes and shapes are enclosed within the plant cells by cellulose walls. Starches have common characteristics such as insoluble granules in cold water, the ability to form a gel upon cooking, which causes the granules to swell, and the rupture of the structure to make the starch available for enzymatic digestive processes (Chandel 2021).

Glycogen is a form of carbohydrate storage found in humans and animals. It consists of branched chains of glucose units similar to those of plant starch and differs from amylopectin (a form of starch) precisely in this characteristic, that is, in the degree of branching of the chain. It is the first and most readily available source of glucose and energy. In this sense, the small amounts of glycogen in animal foods are converted to lactic acid before being available for consumption (Maughan 2009).

Dietary fiber is a complex fraction composed of a set of components present in plant foods, represented by the sum of lignin and polysaccharides (cellulose, hemicellulose, pectin, mucilage, and gum), classified according to their water solubility as soluble or insoluble. They are composed of cell wall components that are resistant to the enzymes secreted by animals. Both types of fiber have shown beneficial physiological functions in the gastrointestinal tract and in reducing the risk of certain diseases (Suharoschi et al. 2019).

Cellulose and hemicellulose are polysaccharides that constitute the cell structure of plants. Cellulose is similar to starch as it contains many glucose molecules in a non-branched form like amylose, but linked in a way that is not hydrolyzed by the enzymes that hydrolyze starch. Hemicelluloses are non-cellulosic polysaccharides, as they differ from cellulose in structure, having fewer glucose units. They can consist of hexoses and pentoses and form acids from these compounds (Suharoschi et al. 2019).

Pectin is a non-cellulosic polysaccharide composed of units of a galactose derivative. Gums and mucilages are similar to pectin, except that the galactose units are combined with other sugars (glucose) and polysaccharides (Suharoschi et al. 2019).

Prebiotics are non-digestible food substances that selectively stimulate the growth or activity of beneficial bacteria in the colon (probiotics) for the host. Various prebiotics, including inulin, inulin-type fructans, and FOS, stimulate the growth of intestinal bacteria, mainly bifidobacteria. Fructans (synthesized or extracted) have prebiotic properties and are considered functional fibers. Functional fiber is commonly added to liquid nutritional supplements and tube-feeding formulas (Chandel 2021).

Fructans include fructooligosaccharides (FOS), inulin, inulin-type fructans, and oligofructose and are composed of fructose polymers, often linked to an initial glucose. Inulin encompasses a varied group of fructose polymers widely distributed in plants as a storage carbohydrate. Oligofructose is a subgroup of inulin with fewer than 10 fructose units. All are poorly digested in the upper GI tract and thus provide only 1 kcal/g (Suharoschi et al. 2019).

## 2.3 Functions

Carbohydrates have three main functions: providing energy, energy storage, and structural function. They are considered the fuel of life. The carbohydrates from the diet are hydrolyzed by enzymes into smaller sugar units until glucose is produced. In cells, energy in the form of ATP is obtained from glucose, which is used in biological actions that require energy (Lehninger et al. 2016).

Additionally, energy in living beings is stored in the form of the polysaccharides starch and glycogen. Starch is the energy reserve of plants, and glycogen is the energy reserve of animals (Lehninger et al. 2016). Regarding structure, polysaccharides play an important structural role, providing shape, protection, and support to molecular structures and tissues, such as cellulose (found in the cell wall of plants), peptidoglycan (found in the cell wall of bacteria), and chitin (in the exoskeleton of arthropods) (Lehninger et al. 2016).

## 2.4 Dietary Sources

Among the sources of simple carbohydrates, sucrose is represented by crystal sugar, brown sugar, and demerara sugar, which are extracted from sugar cane and are easily found in cooking, sweets, beverages, and industrialized products. Fructose is naturally found in fruits. Lactose is the main representative of disaccharides and is present in dairy products and their derivatives (Cominetti and Cozzolino 2020).

Among complex carbohydrates are starches and fibers. Starches are composed of amylose and amylopectin, which are transformed into glucose by digestive enzymes. They are found in cereals: rice, wheat, corn, flour, tubers, and roots like potatoes, cassava, yam, and legumes such as beans, peas, soybeans, and chickpeas. Fibers are carbohydrates that pass intact through the body and are eliminated in the feces. They are essential for digestion and intestinal function. Additionally, they help control dyslipidemia, diabetes, and weight, prevent intestinal problems like irritable bowel syndrome, and play a role in preventing some types of cancer, such as colon cancer. They are found in plant-based foods: vegetables, fruits, legumes, greens, whole grains (oats, granola, and psyllium), whole bread, pasta, and rice, seeds in general, and nuts (Cominetti and Cozzolino 2020).

## 2.5 Digestion, Absorption, and Metabolism

Since carbohydrates represent more than 60% of the available energy in the human diet, it can be considered that there are efficient digestion and absorption mechanisms in the body to maximize the utilization of carbohydrates. Thus, the digestion and absorption of carbohydrates mainly occur along the small intestine, specifically at the enterocyte membrane and the unstirred water layer, which establishes an interface between the lumen and the apical region of the intestinal mucosa cells. Two classes of proteins are primarily responsible for carbohydrate digestion and absorption: glycolytic enzymes and transporter proteins (Cozzolino 2016).

In the mouth, carbohydrates undergo mechanical action through chewing and enzymatic action by ptyalin (salivary alpha-amylase), an enzyme responsible for initiating the hydrolysis of starch molecules. At this moment, the long polysaccharide chains are broken down, leaving only carbohydrates with a lower degree of polymerization. Upon reaching the stomach, ptyalin is neutralized by the gastric juice, and the food bolus helps disperse the soluble carbohydrates in the acidic content. Peristaltic movements direct the liquid components toward the gastric antrum and determine the fluidity of the chyme about to reach the duodenum. Gastric emptying is controlled by the regulation of the osmolarity and acidity of the chyme, with very low pH levels and hypertonic solutions reducing the rate of stomach emptying (Goodman 2010).

Upon reaching the first portion of the small intestine, the chyme, with larger fragments from starch digestion (di- and oligosaccharides), is acted upon by pancreatic α-amylase and glycosidases synthesized by enterocytes and anchored in the brush border villi. Monosaccharides such as glucose, fructose, and low-polymerization residues like maltose, isomaltose, and dextrins are the resulting products of different enzymes in the small intestine (Goodman 2010). The "membrane contact" effect allows glycosidases to be strategically positioned near transporter systems in the enterocyte membrane, reducing the final digestion process and immediate intracellular uptake of substrates of interest to a short time frame, a process characterized by a significant degree of selectivity. Besides pancreatic α-amylase, three other enzymes show significant activity at the brush border: lactase, sucrase-isomaltase, and maltase-glucoamylase. Transporter proteins are synthesized by enterocytes according to the availability of specific monosaccharides in the brush border region. Hence, the higher the sugar levels in the enterocytes, the greater the synthesis and inclusion of transporter proteins in the membrane (Goodman 2010).

Glucose and galactose rely on the same transporter protein, SGLT1 (sodium-glucose transporter 1), which is responsible for the active transport of these two monosaccharides, along with equimolar amounts of sodium, against a concentration gradient, into the cytoplasm of enterocytes, with consequent ATP expenditure. Once inside the enterocytes, sodium ions are exchanged for potassium ions through the classic mechanism of the sodium-potassium pump (Na+—K+—ATPase), located on the basolateral membrane. Glucose is then pumped into the intracellular space by GLUT 2 (glucose transporter 2) located on the basolateral membrane (Southgate 2002).

GLUT 5 is a transporter located on the apical membrane of enterocytes and assists in the absorption of fructose. Being a type of facilitated transport, it does not involve energy expenditure (ATP). In this context, the quality of fructose absorption is higher in the presence of other carbohydrates in the intestinal lumen, such as glucose, galactose, sucrose, or starch; its absorption is significantly greater than when administered alone. There is likely a complex network involving other processes in addition to what happens with the exclusive fructose transporter (GLUT

5), such as the kinetics of glucose and sucrose absorption and the ion exchanges involved, which generate an osmotic gradient favoring fructose absorption (Lehninger et al. 2016).

The absorption process involving transporter proteins and enterocytes can adapt regarding modifications in the mechanism of protein synthesis and membrane transport capacity. This can generate an important factor for modulating the nutrient uptake process. In this sense, intestinal cells can increase the number of transporters due to increased protein synthesis and the expression of transporters in the enterocyte membrane. The increase in the turnover rate will allow these transporters to have a faster pace (higher turnover). Hypertrophy (larger area per cell) and hyperplasia (greater number of cells) are described as non-specific transport strategies and contribute to the increased glucose transport rates (Southgate 2002).

When the body needs glucose, stored polysaccharides are broken down and simultaneously phosphorylated by the enzyme phosphorylase. Animals possess glycogen phosphorylase, and plants possess starch phosphorylase. Both convert the polysaccharide into glucose-1-phosphate with ATP expenditure. Glycolysis is the first metabolic pathway for the glucose molecule and other hexoses. All living beings (except viruses) perform glycolysis under aerobic or anaerobic conditions, with glycolytic enzymes present in the cytoplasm. In this sense, in glycolysis as an anaerobic process, the formation of lactate, a stable final product in the cell cytoplasm, can be observed. Under aerobic conditions, glucose metabolism continues with other energy-producing pathways (Krebs cycle and respiratory chain) if the cell has functional mitochondria (Lehninger et al. 2016).

Glycolysis occurs in an enzymatic sequence of 11 reactions divided into two phases. The first phase occurs up to the formation of two molecules of glyceraldehyde-3-phosphate, characterized as an energy-consuming phase with 2 ATPs being used in the two phosphorylations that occur in this phase. The second phase is characterized by the energy production of 4 ATPs in oxygen-independent oxidative enzymatic reactions, using NADH as a hydrogen carrier from the dehydrogenation reaction. The net final energy yield of anaerobic glucose metabolism is 2 free ATPs. The first phase of glycolysis is energy-consuming, where the products formed are more energetic than glucose. The second phase recovers the invested energy and releases part of the energy contained in the glucose molecule (Shils et al. 2003).

## 3 Proteins

### *3.1 Chemical Composition*

Proteins are macromolecules composed of carbon (C), hydrogen (H), oxygen (O), and nitrogen (N) and are present in all cells of living organisms. They are formed by various combinations of 20 amino acids, which are connected through peptide bonds. Each amino acid has the same general structure, with the methyl group being different for each one (Fig. 2.2).

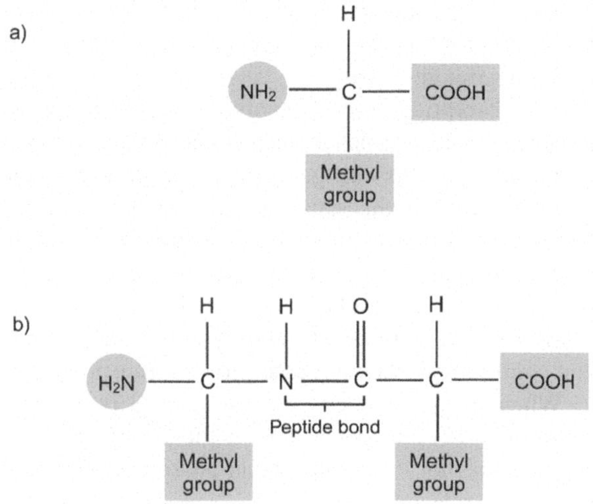


**Fig. 2.2** Structure of an amino acid (**a**) and a dipeptide (**b**). (Source: Adapted from Dutra de Oliveira and Marchini (2008))

In terms of structure, proteins have four levels:

1. *Primary Structure*: Peptide bonds are formed between sequential amino acids according to the instructions in the mRNA. The complete protein is a linear chain of amino acids.
2. *Secondary Structure*: Attractions between the methyl groups of amino acids create helices and pleated sheet structures.
3. *Tertiary Structure*: The helices and pleated sheets are folded into compact domains. Small proteins have one domain, while large proteins have multiple domains.
4. *Quaternary Structure*: Individual polypeptides can serve as subunits in the formation of larger groups or complexes. The subunits are linked by a large number of weak, non-covalent interactions; sometimes, they are stabilized by disulfide bonds. For example, four hemoglobin monomers are joined to form the tetrameric hemoglobin molecule (Fig. 2.3).

## 3.2 Classification

Amino acids can be classified according to their physicochemical and nutritional properties. Based on the ability of the side chains to interact with water, amino acids have hydrophobic potential, such as nonpolar ones with aliphatic or aromatic side chains and uncharged polar ones; in addition, those with hydrophilic characteristics, such as charged polar amino acids. Thus, the behavior and accessibility to interact with components present in the gastrointestinal tract and foods vary according to the lateral polarity of each amino acid.

Regarding nutritional characteristics, they can be grouped into essential (which cannot be synthesized by the body), non-essential (which can be produced by the

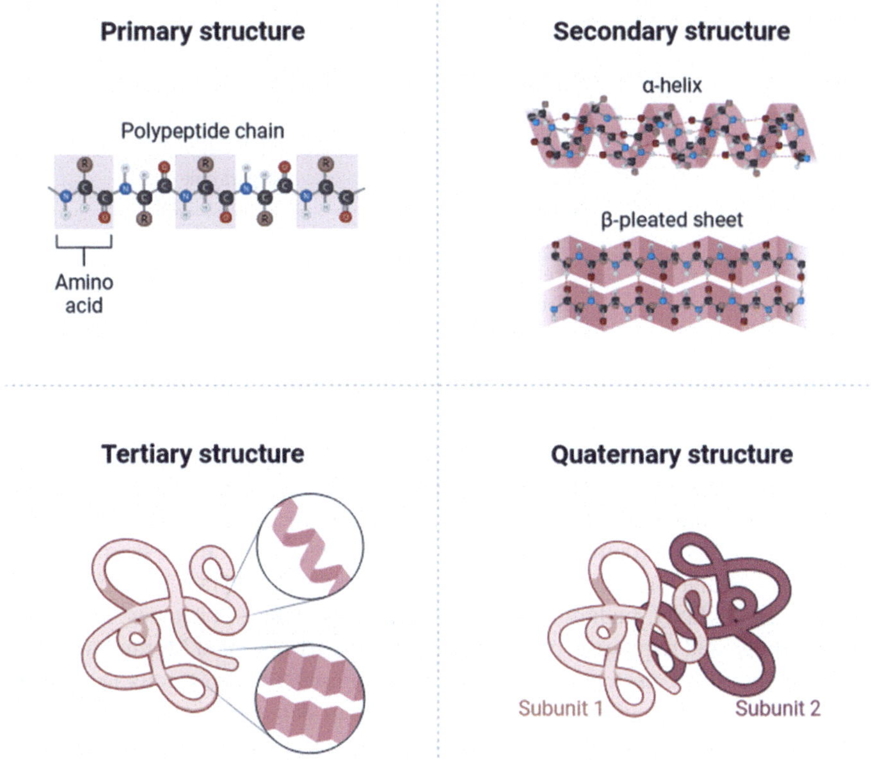


**Fig. 2.3** Protein structure. (Source: Biorender, 2020)

body from other amino acids or other nitrogen-containing metabolites), and conditionally essential (amino acids whose synthesis is impaired under special physiological conditions), as shown on Table 2.2.

Chemically, amino acids are carboxylic acids with an amino group attached to the α-carbon. All amino acids share this same general structure; the side chain is also attached to the α-carbon (the methyl group), which determines the identity and function of each amino acid. The α-carbon is a chiral carbon, and isomers can be formed. The L-isomer is functional in the human body. Many amino acids can be synthesized from carbon skeletons produced as intermediates in major metabolic pathways through a process called transamination, which adds an amino group to another amino acid without producing a free amino group (Lehninger et al. 2016).

Transamination is an important process because it allows the production of nonessential amino acids from metabolic intermediates while using free amino groups, thereby not being available for producing toxic ammonia. For example, pyruvate formed during glycolysis is easily converted into the amino acid alanine by adding the amino group through the enzyme alanine aminotransaminase. On the other hand, essential amino acids have carbon skeletons that humans cannot synthesize (or cannot synthesize enough of) and therefore must be obtained through the diet.

**Table 2.2** Amino acids and their classifications

| Essential | Conditionally essential | Non-essential |
|---|---|---|
| Phenylalanine | Glycine | Alanine |
| Tryptophan | Proline | Aspartic acid |
| Valine | Tyrosine | Glutamic acid |
| Leucine | Serine | Asparagine |
| Isoleucine | Cysteine and cystine | |
| Methionine | Taurine | |
| Threonine | Arginine | |
| Lysine | Histidine | |
| | Glutamine | |

## *3.3 Functions*

The sequence of amino acids determines the final structure and function of the protein, being determined by the genetic code stored in the cell nucleus as deoxyribonucleic acid (DNA). Proteins can be obtained endogenously from the degradation of the body's cellular proteins or exogenously from the diet. They perform various functions in the body, including tissue formation, enzymes, hormones, antibodies, energy supply, and regulation of metabolic processes (Lopez and Mohiuddin 2024).

The functions of proteins are specific to each one and allow cells to maintain their integrity, defend against external agents, repair damage, control, and regulate cellular functions, etc. All proteins perform their function in the same way: by selective binding of molecules. Some proteins, such as structural ones, aggregate with other molecules of the same protein to form a larger structure. Other proteins aggregate with different molecules: antibodies bind to specific antigens, hemoglobin to oxygen, enzymes to their substrates, gene expression regulators to DNA, and hormones to their specific receptors. The main functions are structural, hormonal, gene expression regulation, defense, transport, contractile, storage, and enzymatic.

Some proteins constitute cellular structures, such as certain glycoproteins that are part of cell membranes and act as receptors or facilitate the transport of substances. Histones are part of chromosomes, spatially assisting in the winding of DNA (genes). Other proteins provide elasticity and strength to organs and tissues: collagen in fibrous connective tissue; elastin in elastic connective tissue; and keratin in the epidermis (Wu 2009).

There are protein-based hormones such as insulin and glucagon that regulate blood glucose levels, hormones secreted by the pituitary gland such as growth hormone, and calcitonin, which regulates calcium metabolism. Additionally, some proteins regulate the expression of certain genes, such as transcription and translation factors, while others regulate cell division, such as cyclin, playing a role in gene expression (Wu 2009).

In terms of defense, immunoglobulins act as antibodies against potential antigens; thrombin and fibrinogen contribute to blood clot formation to prevent hemorrhages; and mucins have germicidal effects and protect mucous membranes. Some bacterial toxins, such as botulinum toxin (produced by *Clostridium botulinum*, causing botulism) or snake venoms, are proteins produced for defense purposes (Li et al. 2007).

Transport proteins include hemoglobin, which transports oxygen in the blood of vertebrates; hemocyanin, which transports oxygen in the blood of invertebrates; myoglobin, which transports oxygen in muscles; lipoproteins, which transport lipids in the blood; and cytochromes, which transport electrons. Actin and myosin constitute myofibrils responsible for muscle contraction, along with dynein, a protein related to the movement of cilia and flagella (Wu 2009).

For storage, there are proteins like ovalbumin in egg white, gliadin in wheat grain, and hordein in barley, which constitute amino acid reserves for embryo development, and lactoalbumin in milk, which is a source of amino acids for mammals. Regarding their enzymatic function, proteins are the most numerous and specialized, acting as biological catalysts of chemical reactions in cellular metabolism (Cominetti and Cozzolino 2020).

In addition to the well-recognized biological functions, proteins have been studied for their health properties, notably their hypocholesterolemic effect. The National Health Surveillance Agency and the Joint Health Claims Initiative (2002) approved the claim that including at least 25 g of soy protein in the diet can help reduce blood cholesterol. Both agencies recommend the association of soy consumption with a balanced diet, healthy lifestyle habits, and a diet low in saturated fatty acids.

The most accepted proposal to elucidate how soy works in this system is that there is an alteration in hepatic metabolism, with an increase in the removal of low-density lipoproteins (LDL) and very low-density lipoproteins (VLDL) by hepatocytes, because of increased transcription of LDL receptors (Cho et al. 2007). Additionally, some evidence shows that the isoflavones present in soy stimulate the action of proteins in reducing blood cholesterol, although isoflavones alone do not have this reducing function (Joint Health Claims Initiative 2002).

### *3.4 Dietary Sources*

The World Health Organization (WHO) and the U.S. Food and Drug Administration (FDA) have adopted the method known as the Protein Digestibility-Corrected Amino Acid Score as the official assay for evaluating protein quality in humans. Digestibility is the main factor affecting protein quality and can be influenced by various factors. Meat preparation procedures often involve marinating with wine or vinegar and moist heat to tenderize tough meat cuts through denaturation. Proteins are maintained in their functional three-dimensional structure by hydrogen and ionic interactions; these bonds are loosened in the presence of acid, salt, and heat. As these methods denature proteins, they also soften cartilage or connective tissue proteins and release muscle proteins from their bonds, making all proteins more available to digestive enzymes.

Vegetable protein is less efficient than animal protein because it is embedded in carbohydrates and is less available to digestive enzymes. Some vegetables also contain enzymes that interfere with protein digestion and require heat inactivation before consumption. For example, soy contains trypsinase, which inactivates trypsin, the main enzyme responsible for protein digestion in the intestine (Cominetti and Cozzolino 2020).

The quality of dietary protein can be improved by combining protein sources with different limiting amino acids. Diets based on a single plant-based food do not promote adequate growth because they lack the proper amount of limiting amino acids needed to provide substrates for protein synthesis. If another plant protein, which contains an excess of the limiting amino acid, is added to the diet, the protein combination is complemented; essential amino acids are adequate to support human protein synthesis (Cominetti and Cozzolino 2020).

The concept of complementary proteins is important for populations that do not consume animal protein or are at risk of insufficient dietary diversity. Complementary foods, when consumed together, provide all essential amino acids. It is considered unnecessary to consume complementary amino acids in a single meal, but they should be consumed on the same day (Havala and Dwyer 1993). Children, pregnant women, and breastfeeding mothers who follow vegan diets need to plan their diets carefully to include a mixture of foods containing amino acids. The main sources of protein are animal-based foods, not only for their quantity but primarily for their nutritional quality. Examples of animal proteins include meat, eggs, milk, and dairy products; plant protein sources include cereals and legumes.

## 3.5 Digestion, Absorption, and Metabolism

Proteins are present in virtually all foods and undergo digestion processes. An important factor in protein absorption is its digestibility. The digestion of proteins begins in the stomach by the action of the proteolytic enzyme pepsin, secreted along with gastric juice, and continues in the small intestine with other proteolytic enzymes produced by the pancreas, such as trypsin and chymotrypsin. These proteolytic enzymes are not produced in their active form but as proenzymes or zymogens, which, after partial hydrolysis, lose part of the polypeptide chain and are activated (Shils et al. 2003).

Pepsinogen, by the action of hydrochloric acid present in gastric juice, is hydrolyzed into pepsin, an enzyme that cleaves peptide bonds at specific points in proteins involving a pair of amino acids, one of which is aromatic (phenylalanine, tryptophan, and tyrosine) and the other is leucine. Pancreatic proenzymes are activated by enteropeptidase, a peptidase present in intestinal juice, which hydrolyzes trypsinogen into trypsin. This, through cascade activation, continues the hydrolysis process of other pancreatic proenzymes. Pancreatic enzymes have high specificity: trypsin acts on bonds adjacent to lysine and arginine, chymotrypsin acts on aromatic amino acids, and elastase acts on neutral amino acids. The intestinal mucosa still has the property of producing enzymes that complete peptide digestion into amino acids. Carboxypeptidases release amino acids with a free carboxyl group from the polypeptide chain, while aminopeptidases release amino acids that contain free amine groups (Shils et al. 2003).

After protein digestion, free amino acids and some small peptides are directly absorbed by enterocytes, a process controlled by gastrointestinal hormones such as gastrin, secretin, and cholecystokinin (Shils et al. 2003). Peptides resulting from incomplete protein digestion, when absorbed, can play regulatory roles. The proteins constituting organisms are continuously being renewed (protein turnover). Amino acids present in animal cells originate from dietary proteins (1/4) and endogenous proteins (3/4), forming a pool. This amino acid pool is used for the synthesis of endogenous proteins and other nitrogen-containing molecules (nitrogenous bases, amines, and their derivatives, such as adrenaline and histamine).

In animals, amino acids can undergo oxidative degradation under three different metabolic circumstances: during the normal synthesis and degradation of cellular proteins (protein turnover), some of the amino acids released during protein breakdown undergo oxidative degradation if they are not needed for the synthesis of new proteins; when, due to a protein-rich diet, amino acids are ingested in excess relative to the body's biosynthetic protein needs, the excess is catabolized because free amino acids cannot be stored; during severe fasting or diabetes mellitus, when carbohydrates are inaccessible or not properly utilized, body proteins will be hydrolyzed, and their amino acids will be used as fuel (Lehninger et al. 2016).

The absorption of protein digestion products, AA (amino acids), dipeptides, and tripeptides occurs through complementary processes and can be transported by three mechanisms: passive transfer by simple diffusion, passive transfer by facilitated diffusion, and active transfer by co-transport. Passive transfer can be through cellular or paracellular pathways, while active transfer occurs through cellular pathways. Passive transfer by simple diffusion mainly occurs with free AAs and is inversely proportional to hydrophilicity and directly proportional to the AA concentration gradient. In other words, the more hydrophobic the AA (most of the neutral ones) and the higher its concentration gradient across the membrane, the more significant the simple diffusion transfer. Dipeptides and tripeptides can also be transported by a passive paracellular mechanism, although it seems to be of minor importance as it occurs in small quantities (Friedman 2018).

Active transfer by co-transport occurs with free AAs, dipeptides, and tripeptides. It is mediated by carriers but can "pump" a substance against its concentration gradient (or electrochemical potential), thus requiring energy expenditure. The energy is obtained indirectly through the electrochemical gradient of an ion actively transported. Therefore, they are called co-transports. Free AAs are co-transported along with Na+, and their transport depends on the electrochemical gradient of Na+ generated by the primary active transport Na+-K+-ATPase in the basolateral membrane, known as secondary active transport. Dipeptides and tripeptides are co-transported along with protons (H+), and their transport depends on the electrochemical gradient of H+ generated and maintained by the Na+-proton countertransport in the brush border membrane in response to the Na+ gradient, known as tertiary active transport (Lehninger et al. 2016).

# 4 Lipids

## 4.1 Introduction to Lipids

Lipids are a diverse group of chemical compounds essential for life. Defined as organic substances insoluble in water but soluble in nonpolar solvents such as ether, benzene, and chloroform, these compounds play a variety of crucial biological functions (Philippi 2014). For example, fats and oils primarily act as energy reservoirs, while phospholipids and sterols are fundamental components of cell membranes. Additionally, some lipids, even in small amounts, have vital functions as enzyme cofactors, electron carriers, photosensitive pigments, hydrophobic anchors for proteins, emulsifying agents in the digestive system, hormones, and intracellular messengers (Nelson et al. 2022).

Lipids are often classified as oils (when liquid) or fats (when solid), depending on their physical state at room temperature. This state is determined by the characteristics of the fatty acids they contain. In the human diet, fats and oils account for about 34% of the total energy consumed, providing 9 kcal/g. Due to their insolubility in water, lipids require specialized processes for their digestion, absorption, transport, storage, and utilization (Melo et al. 2020).

## 4.2 Classification and Chemical Composition of Lipids

Lipids are primarily composed of carbon (C), hydrogen (H), and oxygen (O). The proportion of these elements in lipids results in a higher energy density compared to carbohydrates, providing a high caloric value of 9 kcal/g (Galisa et al. 2008). They can be classified into three main categories: simple lipids, compound lipids, and derived lipids.

**Simple Lipids**

- Fatty Acids: Carboxylic acids with hydrocarbon chains ranging from 4 to 36 carbons. They can be saturated (no double bonds) or unsaturated (with one or more double bonds).
- Neutral Fats: Esters of fatty acids with glycerol, such as monoacylglycerols, diacylglycerols, and triacylglycerols.
- Waxes: Esters of fatty acids with high molecular weight alcohols, such as cholesterol ester or retinol palmitate (a vitamin A ester).

**Compound Lipids**

- Phospholipids: Compounds of phosphoric acid, fatty acids, and a nitrogenous base, including glycerophospholipids (lecithins, cephalins, and plasmalogens).
- Sphingolipids: Lipids containing a sphingosine base, such as sphingomyelin, ceramide, cerebrosides, and gangliosides.
- Lipoproteins: Particles that combine lipids and proteins for lipid transport in the blood.

**Derived Lipids**

- Sterols: Such as cholesterol and bile salts.
- Terpenes and Other Compounds: Including sesquiterpenes, chlorophyll, carotenoids, and vitamins A, D, E, and K.

**Fatty Acids**

Fatty acids, essential components of fats and oils, vary in chain length and degree of saturation. They are categorized based on the length of the hydrocarbon chain (short, medium, or long), the presence of branches (unbranched or branched), and the presence of unsaturations (saturated or unsaturated) (Nelson et al. 2022). Unsaturated fatty acids are further classified by the position of the first double bond relative to the terminal methyl group, using the omega (n- or Ω) nomenclature.

## 4.3 Chemical and Physical Properties

Unsaturated fatty acids tend to be liquid at room temperature due to their double bonds, which create bends in the chain, making it difficult for the molecules to pack efficiently. In contrast, saturated fatty acids, especially long-chain ones, are solid. The cis and trans configurations of the double bonds also influence the melting point, with trans fatty acids having higher melting points due to their linearity (Melo et al. 2020).

Hydrogenation, a process that adds hydrogen to unsaturated oils, can increase the melting point and chemical stability of lipids but can also reduce nutritional value by decreasing the content of essential fatty acids. On the other hand, interesterification rearranges fatty acids in the ester bonds of glycerol, resulting in triacylglycerols with different melting and crystallization profiles without altering the fatty acid composition.

## 4.4 Digestion, Absorption, and Metabolism of Lipids

The digestion of lipids begins in the mouth with chewing and salivation. Small amounts of fats are broken down by lingual lipase, an enzyme released by the serous glands of the tongue along with saliva, which cleaves fatty acids at the sn-3 position, preferably short-chain fatty acids. Chewing increases the surface area on which lingual lipase acts. Digestion continues in the stomach with gastric lipase, also called tributyrinase, which hydrolyzes short- and medium-chain triacylglycerols into fatty acids and diacylglycerols (Oliveira and Gazzola 2002).

When fat reaches the duodenum, it consists of about 70% triacylglycerols and the remainder partially digested hydrolysis products. Most fat digestion occurs in the small intestine with the help of bile salts and pancreatic lipase. The presence of fat in the intestine stimulates the release of enterogastrone, which inhibits gastric secretion and motility, slowing down the release of lipids. Additionally, fat in the small intestine stimulates the secretion of cholecystokinin, which in turn stimulates bile and pancreatic secretions (Gallagher 2010).

Bile salts, phospholipids, and sterols are the main lipid components of bile, an emulsifying liquid produced by the liver and secreted by the bile ducts, forming micelles that incorporate lipids. Pancreatic lipase, the main enzyme in triacylglycerol digestion, hydrolyzes ester bonds at the sn-1 and sn-3 positions of glycerol. Fatty acids at the sn-2 position of monoacylglycerols are resistant to hydrolysis by lipase. Lipolysis by pancreatic lipase is rapid, producing monoacylglycerols and free fatty acids faster than their incorporation into micelles (Berg et al. 2008).

The final products of digestion, free fatty acids, and beta-monoacylglycerols are incorporated into micelles that transport these molecules to the enterocytes in the intestine. Near the enterocytes, micelles dissociate, and lipid molecules are absorbed by monomolecular diffusion in the proximal jejunum, while bile acids are absorbed in the terminal ileum (Carpentier and Sobotka 2008). Inside the enterocytes, free fatty acids migrate to the smooth endoplasmic reticulum and are reesterified to glycerol and monoacylglycerols to form triacylglycerols. These triacylglycerols, along with cholesterol, cholesterol esters, phospholipids, and fat-soluble vitamins, form particles that, after the insertion of apoproteins (apoB-48 and apoA-1), are called chylomicrons. Chylomicrons are then released into the lymphatic vessels and reach the systemic venous circulation through the thoracic duct (Ratnayak and Galli 2009).

The metabolism of lipids involves important steps such as lipolysis, fatty acid synthesis, and oxidation. Lipolysis is the process by which triacylglycerols stored in lipid droplets, mainly in adipose tissue, are broken down into fatty acids and glycerol, making these fatty acids available to various tissues in the body, including the liver, adipose tissue, and muscles. During feeding, excess free fatty acids are converted into inert triacylglycerols and stored in adipocytes. When there is an increased energy demand, stored triacylglycerols are mobilized through hydrolytic cleavage, and unesterified fatty acids are released into the circulation, being transported to peripheral tissues for beta-oxidation and production of adenosine triphosphate (ATP).

## 4.5 Functions of Lipids

Lipids perform various essential functions in organisms and foods. They act as energy sources stored in the form of triacylglycerols (neutral fats). Fatty acids released from triacylglycerols are oxidized to meet the energy needs of a cell or organism. Dietary fat is stored in adipose cells present in the human body, allowing humans to remain without food for weeks if necessary (FAO 2010).

Lipids are also important structural elements, especially amphipathic molecules (phospholipids and glycolipids) that are fundamental in the composition of biological membranes. Some lipid derivatives function as hormones and intracellular messengers (Berg et al. 2008). Fat deposits, such as structural fat, are not mobilized during fasting and help protect organs and nerves against traumatic injuries and shocks. The subcutaneous fat layer in humans insulates the body, preserves body heat, and maintains temperature. Dietary fat is crucial for the digestion, absorption, and transport of fat-soluble vitamins, as well as for slowing gastric emptying and stimulating bile and pancreatic flow, facilitating the digestive process (Gallagher 2010). Lipids also significantly contribute to food quality, influencing texture, flavor, nutritional aspects, and caloric density (McClements and Decker 2008).

### *4.6 Dietary Sources of Lipids*

Triacylglycerols are the main source of energy from lipids in the diet. Various oils and fats used in human food come from fruits (such as palm and olive), seeds (such as corn and soy), as well as animal fats (pork, beef, and poultry), dairy products (milk, cheese, and butter), and marine sources (fish, seal, and whale oils). These fats have complex fatty acid profiles with different chain lengths and degrees of unsaturation (Mu and Porsgaard 2005).

Vegetable oils, such as corn and olive oil, are mainly triacylglycerols with unsaturated fatty acids, making them liquid at room temperature. Industrially, these oils can be converted into solid fats through catalytic hydrogenation, which alters their double bonds. Coconut oil, for example, is an exception among vegetable oils as it is rich in saturated fatty acids. Animal fats are predominantly saturated fatty acids, making them solid at room temperature.

Long-chain polyunsaturated fatty acids n-3, such as EPA and DHA, are essential for the body and are primarily found in cold-water fish. The intake of these nutrients is recommended as their synthesis in the body is limited.

Although dietary lipids are associated with health problems such as obesity and cardiovascular diseases due to their high caloric density and the presence of saturated and trans fatty acids, they are essential for various bodily functions. Consumed in moderation, lipids are important for growth, development, and overall health. Studies indicate that bioactive lipids can reduce abdominal fat, improve blood lipid levels and blood pressure, and regulate genes involved in metabolism (Nagao and Yanagita 2008).

Short-chain fatty acids resulting from the fermentation of dietary fibers in the colon are important for energy metabolism and anabolic processes. They contribute about 10% of the total caloric intake in humans (Schönfeld and Wojtczak 2016). Trans fatty acids are mainly found in processed foods made with partially hydrogenated vegetable fat and in small amounts in products derived from ruminant animals.

## 5 Self-Assessment Questions and Answers

### *5.1 Questions*

1. What is the main function of carbohydrates in the human body?

   A. Energy supply
   B. Tissue construction

   C. Oxygen transport
   D. Body temperature regulation

2. What is the human body's main site of protein digestion?

   A. Mouth
   B. Esophagus
   C. Stomach
   D. Small intestine

3. What is the main type of lipid found in the cell membrane?

   A. Triglycerides
   B. Phospholipids
   C. Cholesterol
   D. Free fatty acids

4. Which of the following proteins is responsible for the transport of lipids in the blood?

   A. Hemoglobin
   B. Albumin
   C. Lipoprotein
   D. Ferritin

5. Which of the following compounds is a precursor of steroid hormones and is also found in cell membranes?

   A. Arachidonic acid
   B. Glycerol
   C. Cholesterol
   D. Phosphatidylserine

## 5.2 *Answers*

1. The correct answer is (A). Carbohydrates are the main source of energy for the human body. When consumed, they are metabolized to produce glucose, which is used by cells to generate ATP (adenosine triphosphate), the energy molecule that fuels various cellular activities and essential physiological functions.
2. The correct answer is (D). Although protein digestion begins in the stomach with the action of pepsin, most of the digestion and absorption of proteins occurs in the small intestine. Pancreatic enzymes, such as trypsin and chymotrypsin, and brush border enzymes of enterocytes complete the breakdown of proteins into amino acids and small peptides, which are then absorbed by intestinal cells.
3. The correct answer is (B). Cell membranes are mainly composed of a phospholipid bilayer. These lipids have a hydrophilic head and two hydrophobic tails, which organize into a bilayer, creating a selective barrier that controls the entry and exit of substances in the cell.

4. The correct answer is (C). Lipoproteins are complexes of lipids and proteins that transport lipids through the blood. There are several classes of lipoproteins, such as chylomicrons, VLDL, LDL, and HDL, each with a specific function in transporting different types of lipids, including cholesterol and triglycerides, to tissues and organs.
5. The correct answer is (C). Cholesterol is an essential component of cell membranes, where it contributes to membrane fluidity and stability. Additionally, it is the precursor of steroid hormones, such as cortisol, aldosterone, and sex hormones (testosterone, estrogen, and progesterone). The synthesis of steroid hormones from cholesterol occurs in the adrenal glands and gonads.

## References

Berg JM, Tymoczko JL, Stryer L. Biochemistry. New York: W.H. Freeman and Company; 2008.

Campbell MK. Biochemistry. Porto Alegre: Artmed Editora; 2000. p. 752.

Campbell MK, Farrell SO. Biochemistry. São Paulo: Thomson Learning; 2008. 182 p.

Carpentier YA, Sobotka L. Nutrition and metabolism in intensive care patients. Cambridge: Cambridge University Press; 2008.

Chandel NS. Carbohydrate metabolism. Cold Spring Harb Perspect Biol. 2021;13(1):1–15.

Cho JH, Min BJ, Chen YJ, Yoo JS, Wang Q, Kim JD, Kim IH. Evaluation of FSP (Fermented Soy Protein) to replace soybean meal in weaned pigs: growth performance, blood urea nitrogen and total protein concentrations in serum and nutrient digestibility. Asian Australas J Anim Sci. 2007;20(12):1874–9.

Cominetti C, Cozzolino SMF. Bases bioquímicas e fisiológicas da nutrição: nas diferentes fases da vida, na saúde e na doença. 2nd ed. Editora Manole; 2020.

Cozzolino SMF. Biodisponibilidade de Nutrientes. Barueri: Manole; 2016. 1443 p.

Cummings JH, Englyst HN. Fermentation in the human large intestine and the available substrates. Am J Clin Nutr. 1996;45(5):1243–55.

Dutra de Oliveira JE, Marchini JS. Ciências nutricionais: aprendendo a aprender. São Paulo: Sarvier; 2008.

Food and Agriculture Organization of the United Nations. Fat and fatty acid terminology, methods of analysis and fat digestion and metabolism. In: FAO - Food and Agriculture Organization. Fats and fatty acids in human nutrition: report of an expert consultation. Rome: FAO; 2010. pp. 21–42.

Food and Agriculture Organization/World Health Organization (FAO/WHO). Carbohydrates in human nutrition. Food and nutrition. Rome: FAO; 1998. p. 140.

Friedman M. Absorption and utilization of amino acids, vol. 1. Boca Raton: CRC Press; 2018.

Galisa MS, Esperança MN, Sá LC. Bioquímica dos lipídios: fundamentos e aplicações. São Paulo: Editora Universitária; 2008.

Gallagher M. Human Nutrition. São Paulo: Editora Atheneu; 2010.

Goodman BE. Insights into digestion and absorption of major nutrients in humans. Adv Physiol Educ. 2010;34:44–53.

Havala S, Dwyer J. Position of the American Dietetic Association: vegetarian diets. J Am Diet Assoc. 1993;93(11):1317–9.

Joint Health Claims Initiative. Final technical report, 2002.

Krause MV. Alimentos, nutrição e dietoterapia, 15th ed. Rio de Janeiro: Elsevier; 2022.

Lehninger TM, Nelson DL, Cox MM. Principles of biochemistry. 6th ed. Porto Alegre: Artmed; 2016.

Li P, et al. Amino acids and immune function. Br J Nutr. 2007;98:237–52.

Liberato M, Oliveira M. Biochemistry. Fortaleza: EdUECE; 2019. p. 205.
Lopez MJ, Mohiuddin SS. Biochemistry, essential amino acids. National Library of Medicine; 2024. From: https://www.ncbi.nlm.nih.gov/books/NBK557845/?report=classic
Mahan LK, Escott-Stump S, Raymond JL. Krause's food, nutrition & diet therapy. 13th ed. Rio de Janeiro: Elsevier; 2013. p. 1228.
Maughan R. Carbohydrate metabolism. Surgery. 2009;27(1):6–10.
McClements DJ, Decker EA. Lipid oxidation in foods. Boca Raton: CRC Press; 2008.
Melo AP, Silva R, Mancini Filho J. Impacto das configurações cis e trans nas propriedades físicas dos ácidos graxos. Revista Brasileira de Ciências da Nutrição. 2020;15(2):123–35.
Mu H, Porsgaard T. The metabolism of structured triacylglycerols. Prog Lipid Res. 2005;44(6):430–48.
Nagao K, Yanagita T. Bioactive lipids in metabolic syndrome. Prog Lipid Res. 2008;47(2):127–46.
Nelson DL, Cox MM. Princípios de bioquímica de Lehninger, 7th ed. Porto Alegre: Artmed; 2019. Available from: https://integrada.minhabiblioteca.com.br/#/books/9788582715345/
Nelson DL, Cox MM, Hoskins AA. Lehninger principles of biochemistry, 8th ed. New York: W. H. Freeman; 2022.
Oliveira G, Gazzola J. Nutrition and metabolism. Rio de Janeiro: Editora Revinter; 2002.
Philippi ST. Nutrição e metabolismo dos lipídeos, 4th ed. Barueri: Manole; 2014.
Ratnayak S, Galli G. Lipid metabolism. Berlin: Springer; 2009.
Schönfeld P, Wojtczak L. Short- and medium-chain fatty acids in energy metabolism: the cellular perspective. J Lipid Res. 2016;57(6):943–54.
Shils ME, Olson JA, Shike M, Ross AC. Modern nutrition in health and disease, vol. 1. 9th ed. Barueri: Manole; 2003.
Southgate DA. Data quality in sampling, analysis and compilation. J Food Compos Anal. 2002;15(4):507–13.
Suharoschi R, et al. Dietary fiber and metabolism. In: Dietary fiber: properties, recovery, and applications. Academic Press; 2019. p. 59–77. https://doi.org/10.1016/B978-0-12-816495-2.00003-4.
Wu G. Amino acids: metabolism, functions, and nutrition. Amino Acids. 2009;37:1–17.

# Chapter 3
# Nutritional Elements II: Vitamins and Minerals

**Barbara Martins Vieira, Giovanna Nascimento de Mello e Silva, and Mariana Isaac Silva**

**Abstract** This chapter presents an analysis of the role of vitamins, minerals, and bioactive compounds in human nutrition, emphasizing their importance in maintaining health and preventing diseases. Although these micronutrients are required in small quantities, they play essential roles in physiological processes, such as energy metabolism, tissue synthesis, immune function, and protection against oxidative stress. Vitamins are classified into two major groups: fat-soluble (A, D, E, and K) and water-soluble (B-complex and vitamin C). Each group has unique characteristics in terms of dietary sources, absorption mechanisms, metabolism, and impact on health. Vitamin A, for example, is essential for ocular health, immune function, and epithelial integrity. Deficiency of this vitamin can lead to night blindness and weakened immune function, while its excess can result in toxicity, with symptoms such as nausea, vomiting, and liver damage. B-complex vitamins are crucial for cellular metabolism and energy production, acting as coenzymes in various enzymatic reactions. Deficiency of these vitamins can lead to anemia and severe neurological disorders. Vitamin D is vital for calcium absorption and bone health, while vitamin E possesses antioxidant properties, protecting cells from oxidative damage. Vitamin K is important for blood clotting and bone health. In addition to vitamins, this chapter explores the importance of minerals, such as calcium, iron, magnesium, zinc, sodium, and potassium. Calcium is the most abundant mineral in the human body and is essential for the formation and maintenance of bones and teeth, as well as participating in blood clotting and nerve signal transmission. Its deficiency can lead to osteoporosis and rickets, while excess calcium can cause kidney stones and interfere with the absorption of other minerals. Iron is vital for

B. M. Vieira · G. N. de Mello e Silva (✉)
Curso de Nutrição, Universidade Evangélica, Anápolis, Goias, Brazil
e-mail: barbara.vieira@docente.unievangelica.edu.br; giovanna.silva@docente.unievangelica.edu.br

M. I. Silva
Fundação de Ensino e Pesquisa em Ciências da Saúde, Brasília, Distrito Federal, Brazil
e-mail: marianasilva@escs.edu.br

 
J. O. Fajemiroye (ed.), *Fundamentals of Drug and Non-Drug Interactions*,
https://doi.org/10.1007/978-3-031-80107-5_3

oxygen transport in the blood and energy metabolism, and iron deficiency is one of the most common nutritional deficiencies globally, leading to iron deficiency anemia. This chapter also highlights the role of bioactive compounds found in foods, such as carotenoids, polyphenols, and flavonoids. Although not considered essential nutrients, these compounds have antioxidant, anti-inflammatory, and chemopreventive properties, contributing to the protection against chronic diseases such as cancer, cardiovascular diseases, and neurodegenerative disorders. The bioavailability and impact of these compounds on health can vary depending on factors, such as the food matrix, interactions with other nutrients, and the individual's physiological state.

**Keywords** Metabolism · Absorption · Fat-soluble vitamins · Water-soluble vitamins · Minerals

## 1 Introduction

Vitamins, minerals, and bioactive compounds are fundamental components of human nutrition, playing critical roles in maintaining health and preventing disease. These micronutrients required in small amounts are essential for numerous physiological functions, including the regulation of metabolism and immune function (Shils et al. 2003). The balance of these nutrients is crucial for the optimal functioning of the body, and their deficiency or excess can lead to a variety of health issues (Nelson and Cox 2019).

Vitamins, minerals, and bioactive compounds in foods are the elements responsible for the antioxidant activity of foods. However, each class of compounds acts differently in counteracting radical attacks. While vitamins and bioactive compounds directly neutralize free radicals by donating electrons, which reduce reactive oxygen species, minerals act as cofactors in the production of endogenous antioxidant enzymes, such as superoxide dismutase (SOD) and catalase (CAT) (Catania et al. 2009; Kumar and Goel 2019; Remely et al. 2017).

In this chapter, we will explore the intricate roles that vitamins and minerals play in the human body, from their involvement in energy production and bone health to their role in protecting against oxidative stress and supporting the immune system. We will also examine the importance of bioactive compounds, which, despite not being classified as essential nutrients, have significant health benefits, including antioxidant, anti-inflammatory, and cardioprotective properties. Understanding these compounds' dietary sources, absorption, metabolism, and interactions with other nutrients is vital for comprehending their impact on overall health.

# 2 Vitamins

## *2.1 Vitamin A*

Vitamin A is a generic term that encompasses compounds with biological activity related to retinol, including both carotenoids (also known as provitamin A) and retinoids, which are referred to as preformed vitamin A. Carotenoids and retinoids exhibit differences in their chemical structures. For example, provitamin A is characterized by containing eight isoprenoid units, each with five carbon atoms (C5), totaling 40 carbon atoms in its chain, and it has one or two β-ionone rings at the ends, as illustrated in Fig. 3.1a (Mesquita et al. 2017). Retinoids, on the other hand, are defined by the presence of a trimethylated cyclohexene ring, a conjugated tetraene side chain, and a polar carbon-oxygen functional group, as represented in Fig. 3.1b.

Carotenoids constitute a diverse class, with over 600 identified compounds, including β-carotene, α-carotene, lutein, lycopene, and β-cryptoxanthin. These pigments are responsible for the yellow, red, and orange colors of vegetables such as carrots, bell peppers, and tomatoes. However, only β-carotene, α-carotene, and β-cryptoxanthin are classified as provitamin A, as this function requires the carotenoid to possess an unsubstituted β-ionone ring and an unsaturated hydrocarbon chain. This class of compounds is recognized for its antioxidant and antiproliferative properties, and regular intake of these nutrients is essential, as carotenoids are synthesized only in plant cells, algae, and some bacteria through light stimulation. Frequent consumption of these compounds has been associated with cancer chemoprevention due to their ability to suppress cell growth factors related to cancer pathophysiology (Rowles and Erdman 2020).

Particularly, studies have highlighted the chemopreventive activity of lycopene in prostate cancer. Mirahmadi et al. (2020), in a literature review, evaluated in vitro

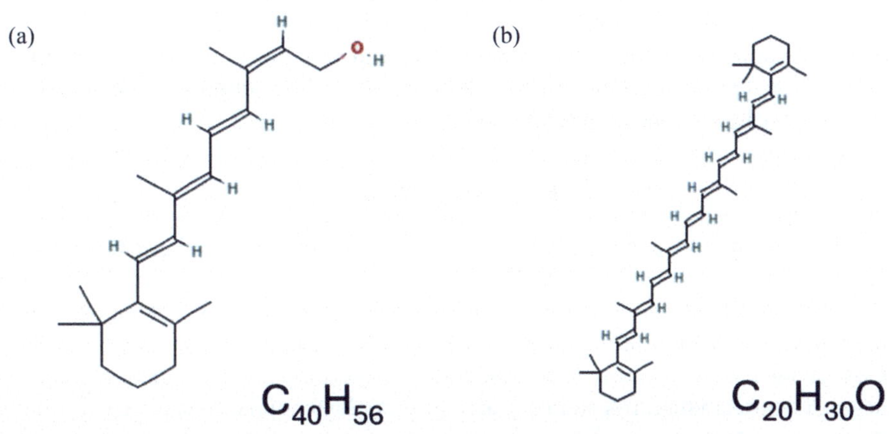


**Fig. 3.1** Structural and molecular formulas of β-carotene (**a**) and 13-cis-retinol (**b**)

and in vivo studies and concluded that, in addition to its ability to neutralize reactive oxygen species, reduce cellular oxidative stress, and protect lipids, proteins, and DNA from damage, lycopene also modulates gene expression, enhances immune response, and regulates the cell cycle. These findings support epidemiological studies that associate regular lycopene consumption with reduced tumor size and PSA levels (Kucuk et al. 2002).

β-Carotene, on the other hand, is a reddish-orange pigment widely found in plants and fruits and is the most active and important provitamin A carotenoid. In addition to acting as a biological pigment, β-carotene functions as a cofactor in cellular production, a ferroptosis inhibitor, and an antioxidant. As an antioxidant, it inhibits free radical damage to DNA, induces cell differentiation and apoptosis in certain types of tumor cells, particularly in the early stages of tumorigenesis, and enhances immune system activity by stimulating the release of natural killer cells, lymphocytes, and monocytes (Kordiak et al. 2022).

Retinoids include compounds such as 13-cis-retinoic acid (13-cRA), 9-cis-retinoic acid (9-cRA), and all-trans-retinoic acid (ATRA). These compounds are synthesized from retinol through the action of the enzyme retinol dehydrogenase, which is zinc-dependent (Maden 2000). Retinol can be obtained from animal-based foods such as liver, meats, eggs, and dairy products.

Retinoic acid plays a crucial role in physiological regulation and maintaining homeostasis. Deficiency of this compound can impair the growth and differentiation of epidermal cells and the activity of sebaceous glands. Furthermore, retinoic acid influences the immune system due to its immunomodulatory and anti-inflammatory properties (Zerres and Stahl 2020; Mukherjee et al. 2006; Singh and Lippman 1998).

Due to the structural diversity of vitamin A, it is measured not in International Units (IU) but in Retinol Activity Equivalents (RAE), which standardizes the equivalence of compounds relative to retinol. Thus, 1 μg of fully transesterified retinol corresponds to 1 μg of RAE or 2 μg of β-carotene (IOM 2002).

As fat-soluble compounds, the absorption of carotenoids and retinoids is optimized when consumed in meals containing some amount of lipids (Tso et al. 2001). Although there is no established upper limit for carotenoid intake, as they do not exhibit toxicity, excessive consumption can lead to carotenodermia, a condition characterized by the accumulation of these compounds in adipose tissue, resulting in a yellowish skin tone (Kapsetaki 2024).

In contrast, excessive intake of retinoids can cause severe adverse effects related to toxicity, such as nausea, vomiting, headaches, dizziness, blurred vision, lack of muscle coordination, liver dysfunction, and bone and joint pain. Hypervitaminosis A is associated with high doses of retinoids, typically from dietary supplements and self-medication (Carazo et al. 2021).

The bioavailability of vitamin A can be influenced by various factors, including the nutritional content of the meal, proper liver function (essential for the metabolism of fat-soluble vitamins), the individual's nutritional status (catabolic states negatively affect vitamin A storage), and the complexity of the food matrix to which the carotenoids are bound. For example, leafy green vegetables contain carotenoids complexed with chloroplast proteins, which reduce their bioavailability. Therefore, regular consumption of small portions of carotenoid- and/or retinoid-rich foods is recommended to ensure adequate nutritional intake (Mourão et al. 2005).

## 2.2 Complex-B Vitamins

The B-complex vitamins constitute a group of eight water-soluble vitamins that play essential roles in cellular metabolism, energy production, and the maintenance of neurological and immune functions. Although often considered together, each of these vitamins has distinct and interrelated functions in the human body. Deficiency or excess of these vitamins can lead to a wide range of health issues, from anemia to neurological disorders (Kennedy 2016; Green et al. 2017). This chapter explores each of the B-complex vitamins, covering their dietary sources, absorption mechanisms, and the clinical consequences of deficiency or overdose.

The B-complex is composed of the following vitamins (Table 3.1):

These vitamins primarily function as coenzymes in a series of enzymatic reactions involved in the metabolism of carbohydrates, proteins, and lipids, making them essential for energy production and biomolecule synthesis (Said 2015; Kennedy 2016). Each B-complex vitamin can be found in a wide variety of foods, with numerous sources derived from both animal and plant products. Below, the main dietary sources of each B-complex vitamin, as well as specific absorption mechanisms, will be explored.

**Vitamin B1 (Thiamine)** Thiamine is primarily found in whole grains, meats (especially pork), seeds, and legumes. Its absorption occurs in the small intestine through both active transport and passive diffusion at higher concentrations (Smith et al. 2021; Thomson 2020). Once absorbed, thiamine is converted into its biologically active form, thiamine pyrophosphate (TPP), which acts as a cofactor in alpha-keto acid decarboxylation reactions and the pentose phosphate pathway (Kennedy 2016).

**Vitamin B2 (Riboflavin)** Riboflavin is abundant in dairy products, eggs, lean meats, and green leafy vegetables. Riboflavin absorption occurs in the small intestine via an energy-dependent active transport process (Thakur et al. 2017). After absorption, riboflavin is converted into its coenzyme forms, FMN (flavin mononucleotide) and FAD (flavin adenine dinucleotide), which play crucial roles in redox reactions within energy metabolism.

**Vitamin B3 (Niacin)** Niacin is found in meats, fish, poultry, nuts, and seeds, and it can also be synthesized by the body from tryptophan, an amino acid present in protein-rich foods. Its absorption occurs in the small intestine through passive diffusion. In the body, niacin is converted into NAD (nicotinamide adenine dinucleotide) and NADP (nicotinamide adenine dinucleotide phosphate), coenzymes essential for oxidation–reduction reactions in cellular metabolism (MacKay et al. 2012).

**Table 3.1** Names and structural formulas of B-complex vitamins

| | |
|---|---|
| Thiamine | 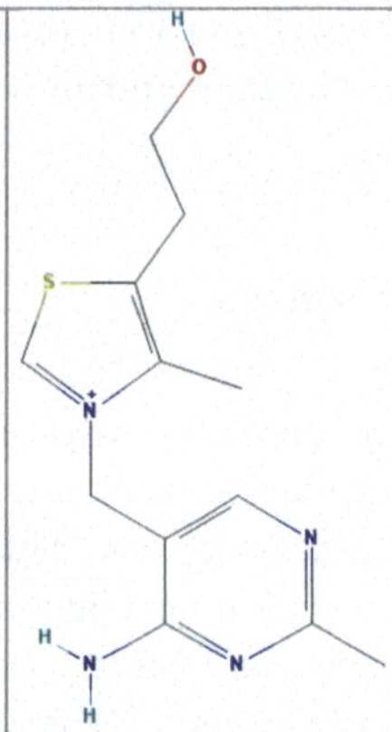 |
| Riboflavin | 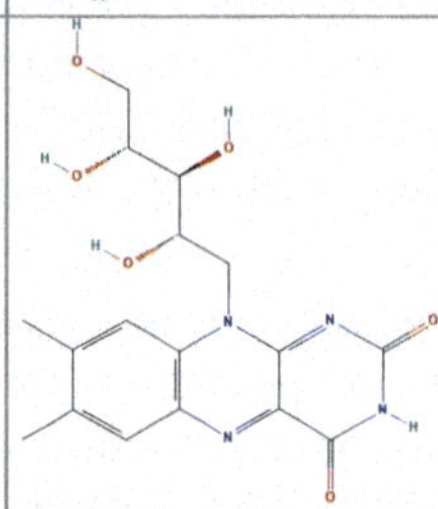 |
| Niacin | 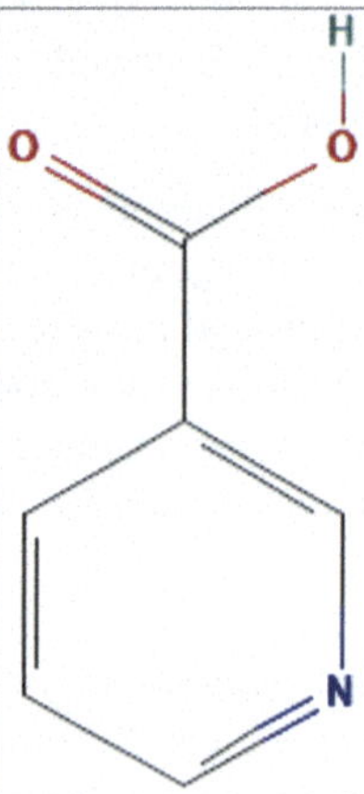 |
| Pantothenic acid | 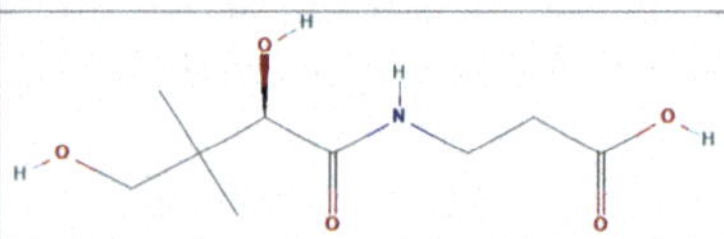 |
| Pyridoxine | 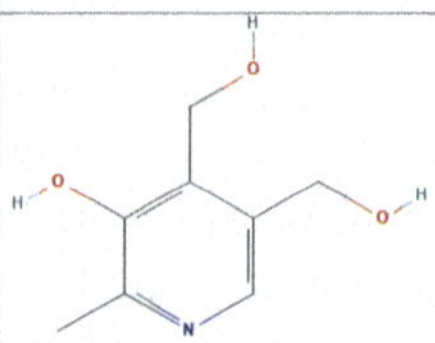 |

(continued)

**Table 3.1** (continued)

| | |
|---|---|
| Biotin | 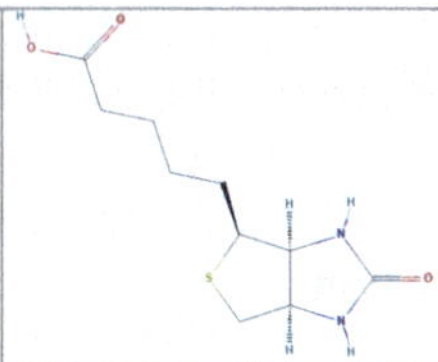 |
| Folic acid | 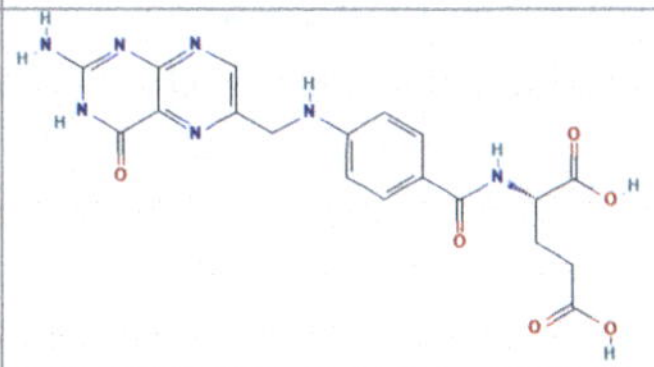 |
| Cobalamin | 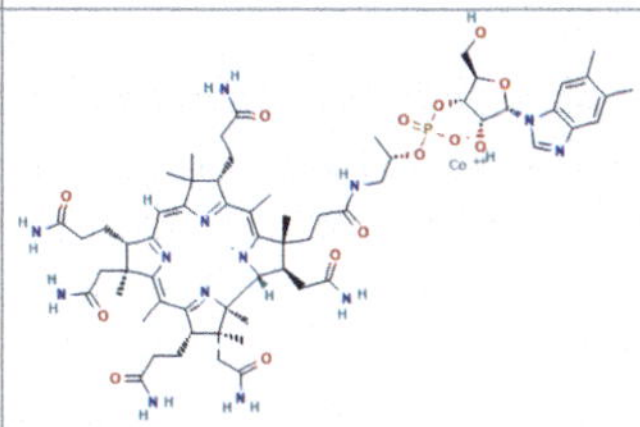 |

**Vitamin B5 (Pantothenic Acid)** Pantothenic acid is widely distributed in foods such as meats, eggs, whole grains, and vegetables. Its absorption in the small intestine involves both active transport and passive diffusion (Miller and Rucker 2012). After absorption, pantothenic acid is incorporated into coenzyme A (CoA), which is essential for fatty acid and carbohydrate metabolism.

**Vitamin B6 (Pyridoxine)** Vitamin B6 is found in poultry, fish, potatoes, whole grains, and bananas. It is absorbed in the small intestine through passive diffusion and is converted into its active form, pyridoxal 5′-phosphate (PLP), which acts as a cofactor in various enzymatic reactions, including amino acid transamination and neurotransmitter synthesis (Wilson et al. 2019).

**Vitamin B7 (Biotin)** Biotin is found in eggs, nuts, seeds, and green leafy vegetables. Its absorption occurs in the small intestine through sodium-mediated active transport (León-Del-Río 2019). Biotin functions as a cofactor for carboxylase enzymes, which are essential for fatty acid synthesis, gluconeogenesis, and amino acid metabolism.

**Vitamin B9 (Folic Acid)** Folic acid is found in leafy vegetables, legumes, liver, and citrus fruits. It is absorbed in the small intestine as monoglutamate, the form available in foods and supplements, through specific folate transporters (Antony 2007). After absorption, folic acid is converted into its active form, tetrahydrofolate (THF), which participates in DNA and RNA synthesis, as well as in the methylation of homocysteine to form methionine.

**Vitamin B12 (Cobalamin)** Cobalamin is found exclusively in animal-derived foods, such as meats, fish, eggs, and dairy products. Its absorption is a complex process that involves an intrinsic factor, a glycoprotein produced in the stomach that binds to vitamin B12, allowing its absorption in the terminal ileum (Green et al. 2017). Vitamin B12 is essential for DNA synthesis, proper neurological function, and red blood cell formation.

The bioavailability of B-complex vitamins can be affected by various factors, including an individual's health status, dietary interactions, and medication use. For example, alcohol consumption can impair the absorption of thiamine, riboflavin, and folic acid, increasing the risk of deficiencies (Miyazaki et al. 2012). Additionally, certain medications, such as anticonvulsants and oral contraceptives, can interfere with the absorption or metabolism of vitamins like B6 and folic acid (Sener et al. 2006). Gastrointestinal disorders, such as celiac disease or short bowel syndrome, can impair the absorption of several B-complex vitamins, leading to multiple deficiencies. Furthermore, certain compounds in foods can interfere with vitamin absorption. Avidin, a protein found in raw egg whites, binds to biotin and prevents its absorption, while phytic acid in grains and legumes can reduce thiamine absorption (Watanabe 2007).

Deficiency in B-complex vitamins can result in a wide range of symptoms, varying depending on the specific vitamin:

- *Vitamin B1 (Thiamine)*: Thiamine deficiency leads to the development of beriberi, which can manifest in two main forms: "dry" (characterized by peripheral neuropathy and muscle weakness) and "wet" (congestive heart failure). Additionally, Wernicke's encephalopathy and Korsakoff's syndrome, severe neurological conditions, are associated with thiamine deficiency, particularly in individuals with chronic alcoholism (Smith et al. 2021).
- *Vitamin B2 (Riboflavin)*: Riboflavin deficiency can result in cheilosis (cracking at the corners of the mouth), glossitis (inflammation of the tongue), and seborrheic dermatitis. Riboflavin deficiency may also impair the metabolism of other B-complex vitamins, such as vitamin B6 and niacin (Thakur et al. 2017).
- *Vitamin B3 (Niacin)*: Niacin deficiency causes pellagra, a condition characterized by the "4 Ds": dermatitis, diarrhea, dementia, and, if untreated, death. Pellagra is more prevalent in populations with maize-based diets, as maize contains niacin in a non-bioavailable form (MacKay et al. 2012).
- *Vitamin B5 (Pantothenic Acid)*: Although rare, pantothenic acid deficiency can cause symptoms such as fatigue, irritability, and neurological disturbances due to its essential role in energy metabolism (Miller and Rucker 2012).
- *Vitamin B6 (Pyridoxine)*: Vitamin B6 deficiency can lead to peripheral neuropathy, seizures, depression, and microcytic anemia. Additionally, B6 deficiency is associated with elevated homocysteine levels, a risk factor for cardiovascular diseases (Wilson et al. 2019).
- *Vitamin B7 (Biotin)*: Biotin deficiency is rare but can cause alopecia (hair loss), dermatitis, conjunctivitis, and neurological disorders (León-Del-Río 2019).

- *Vitamin B9 (Folic Acid)*: Folic acid deficiency results in megaloblastic anemia, characterized by large and immature red blood cells. Additionally, folate deficiency during pregnancy is associated with neural tube defects in the fetus (Antony 2007).
- *Vitamin B12 (Cobalamin)*: Vitamin B12 deficiency can cause pernicious anemia, characterized by megaloblastic anemia and neurological damage, including peripheral neuropathy and dementia (Green et al. 2017).

Regarding excessive intake, B-complex vitamins, being water-soluble, are generally excreted in the urine when consumed in excess, reducing the risk of toxicity. However, very high doses of certain vitamins can cause adverse effects. For example, excessive niacin intake can cause skin flushing, hepatotoxicity, and gastrointestinal discomfort (MacKay et al. 2012). High doses of pyridoxine can lead to sensory neuropathy, while excessive folic acid intake can mask vitamin B12 deficiency, delaying proper diagnosis and treatment (Antony 2007).

B-complex vitamins interact with various nutrients and medications, such as *vitamin B6 and magnesium*. Magnesium is necessary for the conversion of vitamin B6 into its active form, pyridoxal 5′-phosphate. Therefore, magnesium deficiency can impair vitamin B6 function in metabolism (Wilson et al. 2019). Vitamin B9 and Vitamin B12: Folic acid and vitamin B12 have a metabolic interdependence. Both are necessary for the methylation of homocysteine to methionine. Vitamin B12 deficiency can lead to the accumulation of folate in the form of 5-methyltetrahydrofolate, a phenomenon known as the "folate trap" (Green et al. 2017). Folic acid and anticonvulsant medications: anticonvulsant medications, such as phenytoin and phenobarbital, can reduce the levels of folic acid and vitamin B6 in the body, increasing the risk of deficiencies (Sener et al. 2006).

## 2.3 Vitamin C

Vitamin C, or ascorbic acid, is a water-soluble compound essential for the treatment and prevention of scurvy due to its crucial role in collagen cross-linking, promoting connective tissue health, and maintaining bones, teeth, and cartilage in the human body. Additionally, ascorbic acid is widely recognized for its antioxidant properties and its critical role in iron absorption. The antioxidant effect of vitamin C is mediated by its ability to donate electrons, neutralize free radicals, and prevent oxidative stress, which can contribute to the development of chronic diseases (Doseděl et al. 2021).

The molecular formula of vitamin C is C6H8O6, and its nomenclature is L-ascorbic acid. As the human body does not synthesize this vitamin, it must be obtained exclusively through the diet, particularly by consuming citrus or acidic fruits such as oranges, lemons, acerola cherries, and strawberries, as well as vegetables like bell peppers (Abdullah et al. 2024).

Vitamin C absorption occurs predominantly in the small intestine, where it is transported into epithelial cells by two main transporters: SVCT1 and SVCT2 (Sodium-Dependent Vitamin C Transporters). SVCT1 is responsible for intestinal absorption, while SVCT2 distributes vitamin C to various tissues throughout the body, ensuring adequate levels in different tissues. This efficient distribution is vital, as vitamin C is not synthesized by the human body (Doseděl et al. 2021). The regulation of vitamin C absorption also involves its excretion by the kidneys. In situations of excess, unabsorbed vitamin C is readily eliminated to prevent toxicity, which explains why megadoses of this vitamin can be ineffective and, in some cases, potentially harmful (Kaźmierczak-Barańska et al. 2020).

The bioavailability of vitamin C can be influenced by various factors, including the presence of other nutrients and the individual's health status. Inflammatory conditions and the use of certain medications can reduce vitamin C absorption. Additionally, smoking increases the body's need for this vitamin, as the oxidative stress generated by tobacco consumption accelerates the depletion of antioxidant vitamins (Carr and Rowe 2020).

Vitamin C plays a fundamental role in several metabolic processes. It acts as a cofactor in the production of neurotransmitters such as norepinephrine and dopamine and is essential for collagen formation through the hydroxylation of proline and lysine residues, a process crucial for stabilizing the three-dimensional structure of collagen. Studies also indicate that vitamin C increases the expression of genes involved in collagen production in fibroblasts, underscoring its importance in maintaining connective tissue integrity and in wound healing processes (Potekaev et al. 2021).

Acute vitamin C deficiency can lead to scurvy, with symptoms potentially appearing within a month of low or no vitamin C intake (less than 10 mg/day). Early signs include fatigue, malaise, and gum inflammation (Franco et al. 2020).

Furthermore, vitamin C enhances the absorption of non-heme iron, the type of iron found in plant-based foods such as legumes and dark leafy greens. The presence of vitamin C reduces ferric iron to the ferrous state, a form more easily absorbed by the body, making it especially relevant for individuals with iron deficiency or those following vegetarian and vegan diets (Franco et al. 2020). Positive interactions also exist between vitamin C and vitamin E, as vitamin C can regenerate oxidized vitamin E, contributing to its maintenance (Podszun and Frank 2014).

On the other hand, prolonged use of aspirin, a drug commonly used to reduce fever, pain, and inflammation, in high doses, can increase the excretion of vitamin C in the urine and decrease its absorption. This may affect the body's ability to maintain adequate levels of this vitamin (Konturek et al. 2006).

## 2.4 Vitamin D

The term "vitamin D" refers to a group of steroid and fat-soluble compounds that play a crucial role in bone health, particularly in the prevention of rickets. The two most well-known compounds in this group are vitamin D2 (ergocalciferol) and vitamin D3 (cholecalciferol), both derived from the cyclopentanoperhydrophenanthrene ring, sharing the same basic structure as cholesterol. Vitamin D2 is derived from a common plant steroid, ergosterol, and is synthesized through sun exposure in plants. In contrast, vitamin D3 is produced from 7-dehydrocholesterol after ultraviolet (UV)-B irradiation in human skin, a unique characteristic among vitamins. Since the human body can produce vitamin D3, vitamin D does not fit the classic definition of a vitamin, and it is more accurately described as a prohormone (Janoušek et al. 2022).

The production of vitamin D3 in the skin occurs through the conversion of 7-dehydrocholesterol, catalyzed by UV-B rays. The highest concentrations of this lipid compound in the skin are found in the basal and spinous layers of the epidermis. The production of vitamin D is determined by the intensity and appropriate wavelength of UV-B radiation that reaches the epidermis. Individuals living in regions with high sun exposure throughout the year tend to have more efficient vitamin D production. 7-dehydrocholesterol absorbs UV light most efficiently at wavelengths between 270 and 290 nm, so only UV light in this range can produce vitamin D3. Vitamin D3 production can also be reduced when cholesterol levels in the epidermis are low. Melanin, the pigment responsible for skin color, also influences the production of vitamin D3, with higher amounts of melanin reducing UV-B penetration and thereby decreasing vitamin D production (Neville et al. 2021).

Although sun exposure is an important source of vitamin D, dietary intake becomes essential in situations of low sun exposure. However, few foods contain significant amounts of vitamin D. Milk, which contains 25-hydroxycholecalciferol (25-OH D), is an exception, as this metabolite is absorbed more rapidly than other forms of vitamin D. In pasteurized milk, 75% of the vitamin D present is in the form of 25-OH D (Polzonetti et al. 2020). Vitamin D obtained through the diet is absorbed in the small intestine with the help of bile salts, through the lymphatic system and chylomicrons. Only about 50% of ingested vitamin D is effectively absorbed (Ramasamy 2020). Foods rich in vitamin D include fish (5–25 μg/100 g), mushrooms (21.1–58.7 μg/100 g), beef liver, and eggs (1.3–2.9 μg/100 g) (Benedik 2022).

The main circulating metabolite of vitamin D3 is 25-OH D, but the bioactive form is the steroid hormone 1α,25(OH)2D3. After being converted in the liver to 25-OH D3, vitamin D3 is transported to the kidneys, where the enzyme 25(OH) D3-1-α-hydroxylase adds a second hydroxyl group at carbon 1, producing the bioactive form. Along with parathyroid hormone and calcitonin, 1α,25(OH)2D3 regulates serum calcium and phosphate levels by acting on the intestines, kidneys, bones, and parathyroid gland. The antirachitic effect of 1α,25(OH)2D3 results from its ability to stimulate bone remodeling to adjust blood calcium and phosphate levels (Janoušek et al. 2022).

In addition to its antirachitic effects, 1α,25(OH)2D3 influences the immune system. It reduces the proliferation of HL-60 cells and induces their differentiation into monocytes and macrophages, improves monocyte function, and reduces the levels of HLA-DR and CD4+ class II antigens in monocytes or macrophages without affecting class I antigen expression. Recent studies have shown that vitamin D deficiency is associated with worse outcomes in patients infected with the SARS-CoV-2 virus, which causes COVID-19, suggesting that vitamin D deficiency may worsen the course of the disease (Pereira et al. 2022).

Vitamin D deficiency results in inadequate intestinal absorption and renal reabsorption of calcium and phosphate, leading to decreased serum levels of these minerals and increased serum alkaline phosphatase activity. In response to low calcium levels, hyperparathyroidism occurs, and elevated PTH levels, along with the remaining 1α,25(OH)2D3, cause bone demineralization. Endemic vitamin D deficiency can result in a range of clinical problems. However, interlaboratory variation in the measurement of serum 25(OH)D3 levels complicates the accurate diagnosis of vitamin D insufficiency or deficiency (Yang et al. 2020).

Vitamin D deficiency and excess body fat are negatively interrelated. Metabolic changes associated with obesity can result in the accumulation of inactive forms of vitamin D and a reduction in its bioavailability. Additionally, reduced tissue secretion and insulin resistance can impair vitamin D synthesis and expression. It is believed that vitamin D levels are generally lower in obese individuals due to its fat-soluble nature, which facilitates its absorption into adipose tissue (Fiamenghi and Mello 2021).

## *2.5 Vitamin E*

Vitamin E refers to a family of structurally related compounds, with α-tocopherol and α-tocotrienol being the most biologically active forms. Major dietary sources of vitamin E include cereal grains and their oils, meats, eggs, fish, and dairy products, with γ-tocopherol being the most abundant form in these foods. In 2000, the Food and Nutrition Board (FNB) identified alpha-tocopherol as the only form of vitamin E that meets human nutritional needs, as it is the only one capable of reversing vitamin E deficiency symptoms in humans. Vitamin E supplements often contain alpha-tocopherol esters, which, while not antioxidants themselves, have a long shelf life. However, when ingested, these esters are easily hydrolyzed in the intestine, releasing alpha-tocopherol for absorption.

Vitamin E absorption follows the lipid metabolism process, relying on bile salts, micelle formation, and incorporation into chylomicrons within enterocytes for subsequent transport through the lymphatic system. Absorption efficiency appears to be higher when vitamin E is solubilized in micelles containing medium-chain fatty acid triglycerides, compared to long-chain triglycerides (Reboul 2017).

Alpha-tocopherol plays an important role in modulating various cellular functions, including inhibiting protein kinase C (PKC), which results in the inhibition of smooth muscle cell proliferation and a reduction in platelet aggregation and adhesion. Additionally, vitamin E is widely recognized for its antioxidant activity, which is based on its ability to donate electrons from the phenolic hydrogens in its molecule (Jiang 2022).

Both in animals and humans, an increased intake of unsaturated lipids, especially polyunsaturated fatty acids (PUFAs), accelerates vitamin E depletion and raises its requirements. This occurs because PUFAs tend to concentrate in cell membranes, where they capture a significant amount of vitamin E to preserve the oxidative stability of the membranes. Furthermore, a high intake of vitamin A, wheat bran, and pectin has been associated with reduced vitamin E bioavailability. However, the influence of dietary fiber on vitamin E bioavailability is still debated, as studies typically use fiber amounts much higher than those normally consumed in daily diets (Mourão et al. 2005).

Vitamin E deficiency can manifest as neurological symptoms, often described as progressive sensory neuropathy, which includes spinocerebellar ataxia, loss of motor control, and damage to peripheral muscles and nerves. Ocular disorders, such as retinitis pigmentosa and macular degeneration, are also common, especially in conditions such as ataxia with isolated vitamin E deficiency (AVED) and abetalipoproteinemia. Additionally, vitamin E deficiency can lead to anemia due to the increased susceptibility of erythrocytes to hemolysis (Traber 2024).

Patients who use electronic cigarettes have shown an increased risk of developing pulmonary and extrapulmonary complications. This may be related to the increased need for antioxidant nutrients in tissues due to the elevated oxidative stress associated with the use of these devices (Cardoso et al. 2022).

## *2.6 Vitamin K*

Vitamin K is represented by compounds with anti-hemorrhagic activity and a lipophilic chain characterized by the presence of a 2-methyl-1,4-naphthoquinone ring. There are two naturally occurring groups of vitamin K: 2-methyl-3-phytyl-1,4-naphthoquinone, produced in green plants and commonly known as vitamin K1 (phylloquinone), and vitamin K2 (menaquinone), or 2-methyl-3-farnesylgeranylgeranyl-1,4-naphthoquinone, which can be synthesized by bacteria in the gastrointestinal tract. Phylloquinone, the most common form of vitamin K, is found in green leafy vegetables and vegetable oils, while vitamin K2 (menaquinone) is found in animal products such as meats, eggs, and cheeses (Mladěnka et al. 2022).

The biological activity of vitamin K analogs is determined by their absorption, transport, metabolism, and binding efficacy at active sites. The absorption process of vitamin K1 is well understood, but there are still gaps in the understanding of the molecular mechanism of intestinal absorption of menaquinone. It is known that phylloquinone is absorbed mainly in the duodenum with the aid of bile acids and mediated by transporters also involved in the absorption of cholesterol and vitamin E. Due to this interaction, these vitamins are considered antagonistic and should not be administered simultaneously to avoid competitive inhibition (Bus and Szterk 2021).

The most well-known role of vitamin K is as a cofactor in the carboxylation of proteins, an essential process that converts the amino acid glutamate (Glu) into gamma-carboxyglutamate (Gla). This modification enables proteins involved in blood clotting to bind calcium, which is crucial for the interaction between membrane phospholipids, platelets, and endothelial cells. This process is fundamental for normal blood coagulation (Popa et al. 2021).

Although vitamin K1 is the most abundant form, its absorption is not very efficient. Vitamin K absorption is facilitated by bile salts, which form mixed micelles that promote the absorption of lipophilic natural forms of vitamin K by enterocytes. After absorption, the vitamin is encapsulated in chylomicrons and transported through the lymphatic system into the systemic circulation. Phylloquinone is tightly bound to the cellular structures of plants, making its absorption difficult, and the low lipid content of vegetables also interferes with the absorption of this fat-soluble compound (Mladěnka et al. 2022).

Symptoms of vitamin K deficiency include spontaneous skin purpura, nosebleeds, and various types of bleeding. Although rare, vitamin K deficiency can occur in cases of insufficient intake of foods rich in this vitamin, liver diseases, cholestasis, cystic fibrosis, alcoholism, and malabsorption syndromes, and it is most frequently associated with the use of coumarin anticoagulant medications (Simes et al. 2020).

In addition to its role in blood clotting, vitamin K is also essential for bone health due to its influence on the protein osteocalcin, which is produced by mature osteoblasts. Elevated plasma osteocalcin levels are used as an indicator of bone formation and remodeling. The production of this protein is regulated by 1,25-dihydroxyvitamin D, and its ability to bind to minerals depends on carboxylation mediated by vitamin K. Insufficient plasma phylloquinone levels can result in inadequate bone mineralization, linking vitamin K to the prevention of bone diseases such as osteoporosis (Tsugawa and Shiraki 2020).

## 3 Minerals

### *3.1 Calcium*

Calcium is one of the most important and abundant minerals in the human body, classified as an alkaline earth metal and predominantly found in the form of salts such as carbonates and phosphates. Approximately 99% of the body's calcium is present in bones and teeth, while the remaining 1% is distributed throughout bodily fluids and tissues, where it plays crucial physiological roles (Gramlich et al. 2023).

Adequate calcium intake is essential for maintaining bone health and supporting various physiological functions. Calcium-rich foods include dairy products, sardines, almonds, sesame seeds, chia seeds, legumes, and dark leafy green vegetables such as spinach, kale, and broccoli, as well as fortified products (Singh and Prasad 2023).

The intestinal absorption of calcium is a complex process involving multiple factors. About 30% of dietary calcium is absorbed in the small intestine, although this rate can vary. Vitamin D plays a critical role in this process by increasing the expression of calcium transport proteins in the intestine, thereby facilitating its absorption (Hollander and Truscott 1976).

Certain compounds in foods can interfere with calcium bioavailability. Oxalic and phytic acids, found in legumes and some vegetables, can bind to calcium and inhibit its absorption. Additionally, high intakes of protein and sodium can increase urinary calcium excretion, reducing serum calcium levels. Conditions that compromise the integrity of the intestinal barrier, such as celiac disease and irritable bowel syndrome, can also impair calcium absorption (Stephensen 2023).

Calcium performs several essential functions in the human body, including the formation and maintenance of bones and teeth, muscle contraction, blood clotting, nerve impulse transmission, and hormone secretion, such as insulin. Calcium deficiency can lead to various health conditions, including osteoporosis, rickets, osteomalacia, and hypocalcemia (Albahrani and Greaves 2016).

On the other hand, excessive calcium intake can lead to health problems such as hypercalcemia, kidney stone formation, and deficiencies in iron, zinc, and magnesium, as excess calcium can impair the intestinal absorption of these minerals. Calcium also interacts with other nutrients, such as vitamin D, optimizing its absorption in enterocytes. However, medications such as corticosteroids and aluminum-containing antacids can affect calcium absorption and metabolism, leading to deficiencies. Additionally, calcium can reduce the effectiveness of certain antibiotics and thyroid hormone-containing medications when taken simultaneously (Reboul and Borel 2011).

## *3.2 Iron*

Iron is an essential mineral for the proper functioning of the human body, playing a vital role in various biological functions. One of its primary roles is in the formation of hemoglobin, a protein found in red blood cells responsible for transporting oxygen from the lungs to tissues and organs. Additionally, iron is a component of myoglobin, a protein found in muscles that store oxygen and release it during muscle contraction, making it crucial for muscle energy and endurance. Iron also participates in the production of several enzymes, such as cytochromes, which play a fundamental role in cellular respiration and energy metabolism, and aconitase, involved in the Krebs cycle. Furthermore, iron is essential for the immune system, being necessary for the proliferation and maturation of immune cells, including lymphocytes and macrophages, which help protect the body against infections. Iron homeostasis is also important for cognitive function and neurological development, particularly in children, where iron deficiency can lead to mental and behavioral developmental problems (McLean et al. 2009).

Iron is a transition metal whose molecular structure in biological systems varies, mainly existing as heme iron and non-heme iron. Heme iron, found in hemoglobin and myoglobin, is vital for oxygen transport in the blood and muscles, respectively. Hemoglobin consists of four globin subunits, each containing a heme group, which is a porphyrin structure with a central iron atom capable of binding oxygen. In contrast, non-heme iron is present in enzymes and proteins, playing important catalytic and structural roles. Heme iron is found in animal-based foods such as red meat, poultry, fish, and seafood and is more easily absorbed by the body compared to non-heme iron, which is predominantly found in plant-based foods such as legumes, whole grains, nuts, seeds, dark leafy green vegetables, and fortified foods (Pawlak et al. 2017).

Iron absorption in the intestine is a complex process dependent on various factors. Heme iron is efficiently absorbed, with absorption rates ranging from 15% to 35%. In contrast, non-heme iron has a significantly lower absorption rate, ranging from 2% to 20%, depending on the presence of dietary inhibitors and facilitators. Vitamin C, for example, facilitates non-heme iron absorption by reducing ferric iron to ferrous iron, a more easily absorbable form. Animal proteins also enhance non-heme iron absorption. However, inhibitors of absorption include phytates found in grains and legumes, polyphenols present in teas and coffees, and calcium, which can reduce non-heme iron absorption efficiency (Zimmermann and Hurrell 2007).

Iron deficiency is one of the most common nutritional deficiencies worldwide, particularly affecting children, women of reproductive age, and pregnant women. The consequences of iron deficiency can be severe and include iron deficiency anemia, characterized by a reduction in the number of red blood cells or an insufficient amount of hemoglobin, leading to fatigue, weakness, pallor, and shortness of breath; cognitive deficits, more common in children, resulting in learning difficulties and behavioral problems; and lowered immunity, increasing susceptibility to infections (WHO 2021).

While iron is essential, excessive amounts of this mineral can be harmful. Hemochromatosis, a genetic condition that leads to excessive iron accumulation in the body, is the most common form of iron overload. The consequences of iron excess include damage to the liver, heart, and pancreas, potentially leading to liver cirrhosis, heart failure, and diabetes. Excess iron can also increase oxidative stress, resulting in cellular damage, and heighten the risk of infections, as excess iron can promote the growth of pathogens (Fraser et al. 2019).

Iron can also interact with other nutrients and medications, influencing their absorption and effectiveness. High amounts of calcium and zinc can negatively impact iron levels, as calcium can compete for iron absorption in the intestines, and zinc competes for the same intestinal transporters. Medications that reduce stomach acidity, such as proton pump inhibitors, can decrease iron absorption since gastric acidity helps maintain iron in its ferrous form, which is more easily absorbed. Additionally, some antibiotics, such as tetracyclines, can form insoluble complexes with iron, reducing their effectiveness. High doses of iron supplements can also interfere with the absorption of medications like levodopa (used in Parkinson's disease treatment) and levothyroxine (used in hypothyroidism treatment), reducing their efficacy (Singh and Prasad 2023).

## 3.3 Magnesium

Magnesium (Mg) is an essential mineral for the human body, playing a fundamental role in nearly all cellular functions. Its biological importance is evidenced by its central position in energy metabolism and the maintenance of the functional integrity of various systems. Approximately 60% of the body's magnesium is located in the bones, 20% in the muscles, and the remainder is distributed across soft tissues and extracellular fluid (IOM 1997).

As a divalent cation ($Mg^{2+}$), magnesium is essential for human physiology, acting as a cofactor in over 300 enzymatic reactions, many of which are related to energy metabolism and the synthesis of proteins and nucleic acids (Schwalfenberg and Genuis 2017). In addition to its structural role in bones, magnesium plays a regulatory role in processes such as muscle contraction, nerve transmission, heart rhythm control, and blood pressure regulation (Dominguez et al. 2021).

Magnesium is primarily found in plant-based foods, such as leafy green vegetables, nuts, and whole grains (Ross et al. 2014). Its absorption occurs in the small intestine through both passive diffusion and active transport, mediated by proteins such as TRPM6 (Blaine et al. 2015). Approximately 30–40% of ingested magnesium is absorbed, although this rate can vary based on individual factors.

Magnesium bioavailability is influenced by several factors, such as the presence of phytates in plant foods, which can reduce its absorption (Santos et al. 2018). Conversely, proteins and carbohydrates like inulin can enhance magnesium absorption (Miao et al. 2022). Additionally, physiological states such as pregnancy can alter magnesium requirements (Fudge and Kovacs 2010).

Magnesium deficiency, known as hypomagnesemia, can result from inadequate intake or increased urinary excretion (Tong et al. 2017). Symptoms include muscle cramps, fatigue, and, in more severe cases, cardiac arrhythmias (Long and Romani 2014). Excess magnesium, or hypermagnesemia, is rare in individuals with normal renal function but can occur in cases of renal failure or excessive supplementation (Al Alawi et al. 2018).

Magnesium interacts with other nutrients, such as calcium, which competes for the same active absorption sites, and vitamin D, which promotes intestinal magnesium absorption (Plante et al. 2016). Additionally, medications like thiazide diuretics can affect magnesium levels in the body by increasing its excretion (Blaine et al. 2015).

### 3.4 Zinc

Zinc (Zn) is an essential micronutrient for human health, playing a crucial role in various biological processes, including immune function, protein synthesis, and wound healing. Although required in small amounts, zinc deficiency or inadequacy can have significant health consequences (Maares and Haase 2020).

As a transition metal, zinc exists in the body as a divalent cation ($Zn^{2+}$). Unlike other minerals, the body does not have a specialized storage system for zinc, making continuous dietary intake essential to maintaining adequate physiological concentrations (Chen et al. 2024). Approximately 85% of the body's zinc is found in skeletal muscles and bones, with the remainder distributed across tissues such as the liver and skin (Roohani et al. 2013).

The primary dietary sources of zinc include animal-based foods such as red meat and seafood. However, plant-based foods, such as whole grains and legumes, have lower zinc bioavailability due to the presence of phytates, which bind to the mineral and interfere with its absorption (Castro-Alba et al. 2019). Zinc absorption occurs in the small intestine and is mediated by specific transporters, such as ZIP4 (Maares and Haase 2020).

Zinc bioavailability can be influenced by various factors. For example, phytates inhibit its absorption, while excessive iron and calcium intake can compete for the same intestinal transporters, reducing zinc absorption efficiency (Hess and Brown 2017). In contrast, the consumption of animal proteins tends to increase zinc bioavailability (Cunnane 2012).

Zinc deficiency is a public health concern, particularly in populations that follow predominantly plant-based diets. Deficiency can result in impaired immune function, delayed wound healing, and other complications such as growth and developmental delays in children (Roohani et al. 2013). On the other hand, excessive zinc intake can lead to toxicity, causing symptoms such as nausea, vomiting, and abdominal pain, and can interfere with the absorption of other minerals, such as copper, potentially resulting in copper deficiency (Wani et al. 2021).

Zinc also interacts with various nutrients and medications. For instance, proton pump inhibitors, commonly used to treat conditions such as gastroesophageal reflux disease, can reduce zinc levels in the body over time, increasing the need for supplementation (Farrell et al. 2011). Therefore, monitoring and adjusting zinc intake is crucial to maintaining nutritional balance and preventing deficiencies or excesses.

### 3.5 Sodium

Sodium (Na) is an essential mineral for the human body, playing a fundamental role in fluid balance regulation, muscle function, and nerve impulse conduction. Although vital for life, excessive sodium consumption is strongly associated with health issues such as cardiovascular diseases and hypertension. This chapter explores sodium's functions in the body, its absorption, and the impacts of inadequate intake.

Chemically, sodium is an alkali metal, identified by the symbol Na and atomic number 11. In the human body, sodium primarily exists in the form of an ion ($Na^+$), playing crucial roles in maintaining cellular homeostasis, regulating blood volume and blood pressure, as well as contributing to nerve impulse conduction and muscle contraction (Mattes and Donnelly 1991). Sodium levels in the body are primarily regulated by the kidneys, which adjust urinary excretion based on dietary intake. The balance between sodium and potassium is vital for proper cellular function, particularly in excitable tissues such as nerves and muscles (WHO 2023).

Sodium is found in a wide variety of foods, either naturally or as an additive, primarily in the form of sodium chloride (NaCl) or table salt. While fresh foods such as fruits and vegetables contain low levels of sodium, most dietary sodium comes from processed foods like bread, cheese, cured meats, canned goods, and ready-to-eat meals (Brown et al. 2009). It is estimated that over 70% of sodium consumed in the Western diet comes from processed foods, with added salt contributing a smaller portion. In Brazil, industrialized bread, ready-made seasonings, and fast food are significant sources of dietary sodium (Sarno et al. 2009).

Sodium absorption primarily occurs in the small intestine through mechanisms such as sodium-glucose cotransport (SGLT1), which allows for the simultaneous absorption of sodium and glucose (Wright and Loo 2000). Sodium can also be absorbed via specific channels in intestinal epithelial cells. Sodium bioavailability is high, with most of the ingested sodium being efficiently absorbed by the body. Factors influencing sodium bioavailability include overall diet composition and the presence of nutrients that can affect electrolyte balance, such as potassium and magnesium (He and MacGregor 2009).

Sodium bioavailability can also be influenced by dietary and physiological factors. For example, potassium-rich diets increase urinary sodium excretion, helping to mitigate its effects on blood pressure (Morris et al. 2006). Additionally, dehydration or prolonged use of diuretics can alter sodium balance in the body, affecting its bioavailability and plasma levels. The sodium content of drinking water can also influence overall sodium balance, particularly in regions with high water salinity (WHO 2017).

Sodium deficiency (hyponatremia) is rare in individuals with a balanced diet but can occur in cases of excessive sodium loss, such as during diarrhea, vomiting, or heavy sweating. Symptoms of hyponatremia include nausea, headaches, confusion, fatigue, and, in severe cases, seizures and coma (Adrogué and Madias 2000). Conversely, excessive sodium intake (hypernatremia) is a common concern, particularly in populations that consume large amounts of processed foods. Excess sodium intake is associated with the development of hypertension, increasing the risk of cardiovascular diseases such as myocardial infarction and stroke (Stamler et al. 1996). Additionally, high sodium intake can contribute to the development of chronic kidney disease and osteoporosis due to increased urinary calcium excretion (Heaney 2006).

The World Health Organization (WHO) recommends that daily sodium intake should be less than 2000 mg, equivalent to approximately 5 g of salt per day (WHO 2012). However, the average sodium intake in many regions exceeds this recommendation. In the United States, for example, the average sodium intake is estimated to be around 3400 mg/day (Soh et al. 2024). In Brazil, data from the 2008–2009 Household Budget Survey (POF) indicated that the average sodium intake among the population was approximately 4700 mg/day, with processed foods and added salt being the main contributors (Sarno et al. 2009).

Sodium interacts significantly with other nutrients, particularly potassium, and a proper balance between these minerals is essential for cardiovascular health. Potassium-rich diets, derived from fruits and vegetables, are associated with reduced risk of hypertension and heart disease (He and MacGregor 2009). Additionally, sodium interacts with various medications. Diuretics, commonly prescribed to treat hypertension, increase sodium excretion, but prolonged use can lead to sodium depletion and other electrolyte imbalances, requiring careful monitoring (Brater 2000).

## *3.6 Potassium*

Potassium (K) is an essential mineral for the human body, playing fundamental roles in fluid balance regulation, muscle contraction, nerve function, and blood pressure control. As the primary intracellular cation, potassium is crucial for maintaining cellular homeostasis and executing vital physiological functions. Despite being widely available in the diet, inadequate potassium intake is a global public health issue with significant implications for cardiovascular diseases and other metabolic disorders.

Potassium is an alkali metal element, identified by the symbol K and atomic number 19. In the human body, potassium primarily exists in ion form ($K^+$), with approximately 98% of total potassium found within cells. The intracellular concentration of potassium is about 30 times higher than in the extracellular fluid, a distribution maintained primarily by the sodium-potassium pump ($Na^+/K^+$-ATPase), an enzyme essential for maintaining membrane potential and cellular function (Aburto et al. 2013).

Potassium is widely found in plant-based foods, with fruits, vegetables, legumes, and tubers being some of the richest sources. Foods like bananas, oranges, spinach, potatoes, and avocados are well known for their high potassium content. The mineral is also present in meats, fish, and dairy products, though in smaller amounts (He and MacGregor 2008). Due to its presence in a wide variety of natural foods, potassium is easily obtained through a balanced diet based on fresh, minimally processed foods.

Potassium absorption occurs predominantly in the small intestine through a passive transport process, which does not require energy expenditure. The absorption is highly efficient, with about 85–90% of ingested potassium being absorbed (Cotran et al. 2015). Once absorbed, potassium is rapidly distributed to the intracellular compartment, where it performs its physiological functions.

Potassium bioavailability can be influenced by dietary and physiological factors. Different chemical forms of potassium in foods (e.g., potassium from salts versus naturally occurring potassium in foods) may have varying absorption rates. Additionally, an individual's health status, particularly kidney function, plays a crucial role in regulating potassium levels in the body. Patients with kidney failure, for instance, may struggle to excrete potassium efficiently, leading to hyperkalemia (excess potassium in the blood) (Weaver 2013).

The interaction of potassium with other nutrients, such as sodium, also affects its bioavailability. A high sodium intake can increase potassium excretion, creating an electrolyte imbalance that can negatively impact health, particularly in terms of blood pressure control (He and MacGregor 2008).

Potassium deficiency, known as hypokalemia, is relatively common and can result from inadequate intake or excessive losses due to vomiting, diarrhea, prolonged use of diuretics, or medical conditions that affect electrolyte balance (Weaver 2013). Symptoms of hypokalemia range from fatigue, muscle weakness, and cramps to severe cardiac arrhythmias and, in extreme cases, paralysis. Hypokalemia is particularly dangerous for individuals with heart disease, as potassium is vital for normal cardiac function.

Conversely, excess potassium, or hyperkalemia, typically occurs in individuals with impaired kidney function, as the kidneys are responsible for excreting excess potassium. Hyperkalemia can cause muscle weakness, nausea, cardiac arrhythmias, and, in severe cases, cardiac arrest (Adrogué and Madias 2000). While excessive potassium intake through food is rare in healthy individuals, using potassium supplements without medical supervision can pose significant risks.

Potassium interacts in complex ways with other nutrients, particularly sodium. The balance between sodium and potassium is crucial for maintaining blood pressure and cardiovascular health. Diets high in sodium and low in potassium are associated with a higher risk of hypertension, while increasing potassium intake can help mitigate sodium's effects on blood pressure (Houston 2011). For this reason, increasing potassium intake is often recommended as part of dietary interventions to prevent and treat hypertension.

Moreover, potassium can interact with certain medications, such as diuretics and angiotensin-converting enzyme inhibitors (ACE inhibitors), used to treat hypertension and heart failure. Thiazide diuretics, for example, increase urinary potassium excretion, raising the risk of hypokalemia, while potassium-sparing diuretics and ACE inhibitors can reduce potassium excretion, leading to a risk of hyperkalemia (Sfairopoulos et al. 2021).

## 3.7 *Phosphorus*

Phosphorus (P) is one of the most abundant minerals in the human body, playing essential roles in bone health, energy metabolism, and acid–base balance regulation. This mineral is crucial for life, being present in all cells and forming part of various molecules fundamental to cellular function. This chapter explores the key characteristics of phosphorus, including its definition, dietary sources, absorption, bioavailability, and health impacts.

Phosphorus is a chemical element with atomic number 15, predominantly found in the body as phosphate ions ($PO_4^{3-}$). Approximately 85% of the body's total phosphorus is located in bones and teeth, in the form of hydroxyapatite, a mineral matrix that provides rigidity to the skeleton. The remaining phosphorus is distributed in the blood, muscles, and other tissues, where it participates in various essential biochemical reactions (Institute of Medicine 1997). Among its most important functions are the production of ATP (adenosine triphosphate), the molecule that stores and transports energy within cells, and the formation of nucleic acids (DNA and RNA), which are critical for storing and transmitting genetic information (Ross et al. 2014).

Phosphorus is widely available in a variety of foods. The primary dietary sources include animal products such as meat, fish, poultry, eggs, and dairy, which contain highly bioavailable phosphorus. Plant-based foods, such as vegetables, whole grains, and nuts, are also good sources of phosphorus; however, the phosphorus in these foods is often bound to phytate, which reduces its bioavailability (Shils et al. 2006). Additionally, phosphorus-containing food additives are commonly used in processed foods, further increasing dietary phosphorus intake in modern diets.

Phosphorus absorption occurs predominantly in the small intestine through both passive and active processes. Passive absorption depends on the concentration of phosphorus in the intestinal lumen, while active absorption is regulated by vitamin D, which enhances phosphorus absorption efficiency during periods of low intake or increased physiological need (Kumar et al. 2012). On average, about 60–70% of dietary phosphorus is absorbed by the body.

Phosphorus bioavailability can be influenced by various dietary and physiological factors. The type of phosphorus consumed is a key determinant: phosphorus from animal sources is more easily absorbed than phosphorus from plant sources, due to the presence of phytates in plants, which bind to phosphorus and reduce its absorption (Wasserman 2004). Additionally, the presence of other nutrients, such as calcium and magnesium, can interfere with phosphorus absorption, as they compete for the same transporters in the intestine (Heaney and Weaver 2005).

Vitamin D plays an essential role in regulating phosphorus absorption. Vitamin D deficiency can impair the absorption of both phosphorus and calcium, leading to potential deficits in bone mineralization and, in extreme cases, conditions such as rickets in children and osteomalacia in adults (Hill Gallant and Spiegel 2017).

Although phosphorus deficiency is rare due to its wide availability in the diet, it can occur in cases of severe malnutrition, chronic alcoholism, or conditions that affect intestinal absorption, such as malabsorption syndrome (Bird and Eskin 2021). Symptoms of phosphorus deficiency include muscle weakness, bone pain, mental confusion, and, in severe cases, respiratory failure and seizures.

On the other hand, excessive phosphorus consumption, particularly in the form of phosphorus-containing food additives in processed foods, has become a growing concern. Excessive phosphorus intake can lead to an imbalance in calcium metabolism, resulting in decreased calcium absorption and increased calcium mobilization from bones, which can contribute to the development of osteoporosis over time (Calvo and Uribarri 2013). Additionally, hyperphosphatemia (excess phosphorus in the blood) is associated with an increased risk of cardiovascular diseases, especially in patients with chronic kidney disease, due to the kidneys' reduced ability to excrete phosphorus (Vervloet et al. 2017).

Phosphorus interacts significantly with other nutrients, particularly calcium, with which it shares a close relationship in bone metabolism. A proper balance between calcium and phosphorus is crucial for bone health. Diets with an excessively high phosphorus-to-calcium ratio can impair bone mineralization and increase the risk of osteoporosis (Heaney and Weaver 2005).

Furthermore, certain medications can affect phosphorus levels in the body. For example, aluminum-containing antacids can bind to phosphorus in the gastrointestinal tract, reducing its absorption and potentially leading to a deficiency in individuals who use these medications chronically (Shils et al. 2006). Conversely, medications used to treat chronic kidney disease, such as phosphorus binders, are prescribed to reduce phosphorus absorption and prevent hyperphosphatemia in patients with this condition (Vervloet et al. 2017).

## 4 Self-Assessment Questions and Answers

### *4.1 Questions*

1. What is the primary function of vitamin A in the human body?

    A. Regulates calcium absorption
    B. Supports immune function
    C. Acts as an antioxidant and is crucial for vision
    D. Facilitates iron absorption

2. Which vitamin is essential for the formation of collagen and is also known for its antioxidant properties?

    A. Vitamin A
    B. Vitamin B12
    C. Vitamin C
    D. Vitamin E

3. Which vitamin deficiency is associated with neural tube defects in the fetus?
   A. Vitamin A
   B. Vitamin B9 (folic acid)
   C. Vitamin C
   D. Vitamin D
4. Which mineral is most abundant in the human body and is essential for bone health?
   A. Magnesium
   B. Sodium
   C. Iron
   D. Calcium
5. What is the primary source of vitamin K1 (phylloquinone) in the diet?
   A. Leafy green vegetables
   B. Red meat
   C. Dairy products
   D. Fish oils

## *4.2 Answers*

1. Correct answer is (C). Vitamin A is essential for maintaining healthy vision, particularly in low-light conditions, as it is a key component of the visual pigments in the retina. It also plays a role in immune function, reproduction, and cellular communication. Additionally, vitamin A has antioxidant properties that help protect cells from damage by free radicals.
2. Correct answer is (C). Vitamin C, or ascorbic acid, is essential for the synthesis of collagen, a protein that helps maintain the integrity of skin, blood vessels, bones, and connective tissues. It also acts as an antioxidant, protecting cells from oxidative damage by neutralizing free radicals.
3. Correct answer is (B). Folic acid (vitamin B9) is crucial during pregnancy, especially in the early stages of fetal development. Adequate intake of folic acid before conception and during early pregnancy reduces the risk of neural tube defects (NTDs), such as spina bifida, in the developing fetus.
4. Correct answer is (D). Calcium is the most abundant mineral in the human body, with the majority stored in bones and teeth, where it provides structural support. It is also vital for muscle function, nerve transmission, and blood clotting. Adequate calcium intake is crucial for preventing bone diseases like osteoporosis.
5. Correct answer is (A). Vitamin K1, or phylloquinone, is primarily found in leafy green vegetables like spinach, kale, and broccoli. It is essential for blood clotting and bone metabolism. The body uses vitamin K1 to produce proteins involved in coagulation and bone formation.

## References


Abdullah M, Jamil RT, Attia FN. Vitamin C (Ascorbic acid) [Updated 2023 May 1]. In: StatPearls [Internet]. Treasure Island: StatPearls Publishing; 2024. Available from: https://www.ncbi.nlm.nih.gov/books/NBK499877/.

Aburto NJ, Hanson S, Gutierrez H, Hooper L, Elliott P, Cappuccio FP. Effect of increased potassium intake on cardiovascular risk factors and disease: systematic review and meta-analyses. BMJ. 2013;346:f1378. https://doi.org/10.1136/bmj.f1378.

Adrogué, H. J., Madias, N. E. Sodium deficiency (hyponatremia) is rare in individuals with a balanced diet but can occur in cases of excessive sodium loss, such as during diarrhea, vomiting, or heavy sweating. N Engl J Med. 2000;342(21):1581–1589. https://doi.org/10.1056/NEJM200005253422107.

Al Alawi AM, Majoni SW, Falhammar H. Magnesium and human health: perspectives and research directions. Int J Endocrinol. 2018;2018:9041694. https://doi.org/10.1155/2018/9041694.

Albahrani AA, Greaves RF. Fat-soluble vitamins: clinical indications and current challenges for chromatographic measurement. Clin Biochem Rev. 2016;37(1):27–47.

Antony AC. In utero physiology: role of folic acid in nutrient delivery and fetal development. Am J Clin Nutr. 2007;85(2):598S–603S. https://doi.org/10.1093/ajcn/85.2.598S.

Benedik E. Sources of vitamin D for humans. Int J Vitam Nutr Res. 2022;92(2):93–102. https://doi.org/10.1024/0300-9831/a000733.

Bird RP, Eskin NAM. The emerging role of phosphorus in human health. Adv Food Nutr Res. 2021;96:27–88. https://doi.org/10.1016/bs.afnr.2020.10.004.

Blaine J, Chonchol M, Levi M. Renal control of calcium, phosphate, and magnesium homeostasis. Clin J Am Soc Nephrol. 2015;10(7):1257–72. https://doi.org/10.2215/CJN.09750913.

Brater DC. Diuretic therapy in congestive heart failure. Congest Heart Fail. 2000;6(4):197–201. PMID: 12147953.

Brown IJ, Tzoulaki I, Candeias V, Elliott P. Salt intakes around the world: implications for public health. Int J Epidemiol. 2009;38(3):791–813. https://doi.org/10.1093/ije/dyp139.

Bus K, Szterk A. Relationship between structure and biological activity of various vitamin K forms. Foods. 2021;10(12):2907. https://doi.org/10.3390/foods10122907.

Calvo MS, Uribarri J. Public health impact of dietary phosphorus excess on bone and cardiovascular health in the general population. Am J Clin Nutr. 2013;98(1):6–15. https://doi.org/10.3945/ajcn.112.053934.

Carazo A, Fernández-Delgado A, Blanco-Navarro I, Pérez-Sacristán B, García-López JP. Vitamin A update: forms, sources, kinetics, detection, function, deficiency, therapeutic use and toxicity. Nutrients. 2021;13(5):1706. https://doi.org/10.3390/nu13051706.

Cardoso BEM, Costa IF, Santana AN, Silva IL, Rezende AM. Complicações pulmonares e extrapulmonares associadas ao uso de cigarros eletrônicos: uma revisão integrativa. Res Soc Dev. 2022;11(15):e280111536348. https://doi.org/10.33448/rsd-v11i15.36348.

Carr AC, Rowe S. Factors affecting vitamin C status and prevalence of deficiency: a global health perspective. Nutrients. 2020;12(7):1963. https://doi.org/10.3390/nu12071963.

Castro-Alba V, Lazarte CE, Bergenståhl B, Granfeldt Y. Phytate, iron, zinc, and calcium content of common Bolivian foods and their estimated mineral bioavailability. Food Sci Nutr. 2019;7(9):2854–65. https://doi.org/10.1002/fsn3.1127.

Catania AS, Barros CR, Ferreira SRG. Vitaminas e minerais com propriedades antioxidantes e risco cardiometabólico: controvérsias e perspectivas. Arq Bras Endocrinol Metab. 2009;53(5):550–9. https://doi.org/10.1590/S0004-27302009000500008.

Chen B, Liang Y, Song S, Zhang J, Zhuo Y, Huo S, et al. Cellular zinc metabolism and zinc signaling: from biological functions to diseases and therapeutic targets. Signal Transduct Target Ther. 2024;9(1):4. https://doi.org/10.1038/s41392-024-01056-y.

Cotran RS, Kumar V, Collins T. Robbins and Cotran pathologic basis of disease. 9th ed. Philadelphia: Elsevier; 2015.

Cunnane SC. Zinc: clinical and biochemical significance. Prog Food Nutr Sci. 2012;7:197–212.

Dominguez LJ, Veronese N, Barbagallo M. Magnesium and hypertension in old age. Nutrients. 2021;13(1):139. https://doi.org/10.3390/nu13010139.

Doseděl M, Jirkovský E, Macáková K, Mladěnka P, Hrdina R, Patočka J. Vitamin C – sources, physiological role, kinetics, deficiency, use, toxicity, and determination. Nutrients. 2021;13(2):615. https://doi.org/10.3390/nu13020615.

Farrell CP, Morgan M, Rudolph DS, Hwang A, Albert NE, Valenzano MC, et al. Proton pump inhibitors interfere with zinc absorption and zinc body stores. Gastroenterology Res. 2011;4(6):243–51. https://doi.org/10.4021/gr379w.

Fiamenghi VI, Mello ED. Vitamin D deficiency in children and adolescents with obesity: a meta-analysis. J Pediatr. 2021;97(3):273–9. https://doi.org/10.1016/j.jped.2020.07.004.

Franco LP, Rocha AV, Cominetti C. Vitamina C. In: Cominetti C, Cozzolino SMF, editors. Bases bioquímicas e fisiológicas da nutrição: nas diferentes fases da vida, na saúde e na doença. Barueri: Editora Manole; 2020.

Fraser GM, Marx N, Elias SM, Heyman SN. Hereditary hemochromatosis and the excess risk of hepatic and extrahepatic diseases: a hospital-based study. J Clin Gastroenterol. 2019;53(6):e243–8. Available from: https://journals.lww.com/jcge/Abstract/2019/07000/Hereditary_Hemochromatosis_and_the_Excess_Risk_of.3.aspx.

Fudge NJ, Kovacs CS. Pregnancy up-regulates intestinal calcium absorption and skeletal mineralization independently of the vitamin D receptor. Endocrinology. 2010;151(3):886–95. https://doi.org/10.1210/en.2009-1010.

Gramlich L, Meddings L, Alberda C. Essential fatty acid deficiency in 2015: the impact of novel intravenous lipid emulsions. JPEN J Parenter Enteral Nutr. 2023;39(1 Suppl):61S–6S.

Green R, Allen LH, Bjørke-Monsen AL, Brito A, Guéant JL, Miller JW, et al. Vitamin B12 deficiency. Nat Rev Dis Primers. 2017;3:17040. https://doi.org/10.1038/nrdp.2017.40.

He FJ, MacGregor GA. Beneficial effects of potassium on human health. Physiol Plant. 2008;133(4):725–35. https://doi.org/10.1111/j.1399-3054.2008.01113.x.

He FJ, MacGregor GA. Reducing population salt intake worldwide: from evidence to implementation. Prog Cardiovasc Dis. 2009;52(5):363–82. https://doi.org/10.1016/j.pcad.2009.12.006.

Heaney RP. Role of dietary sodium in osteoporosis. J Am Coll Nutr. 2006;25(3):271S–6S. https://doi.org/10.1080/07315724.2006.10719580.

Heaney RP, Weaver CM. Calcium and phosphorus. In: Shils ME, Shike M, Ross AC, Caballero B, Cousins RJ, editors. Modern nutrition in health and disease. 10th ed. Philadelphia: Lippincott Williams & Wilkins; 2005. p. 153–76.

Hess SY, Brown KH. Impact of zinc fortification on zinc nutrition. J Nutr. 2017;147(6):1071–7. https://doi.org/10.3945/jn.116.242420.

Hill Gallant KM, Spiegel DM. Calcium balance in chronic kidney disease. Curr Osteoporos Rep. 2017;15(3):198–205. https://doi.org/10.1007/s11914-017-0357-8.

Hollander D, Truscott TC. Mechanism and site of small intestinal uptake of vitamin D3 in pharmacological concentrations. Am J Clin Nutr. 1976;29(9):970–5. https://doi.org/10.1093/ajcn/29.9.970.

Houston MC. The importance of potassium in managing hypertension. Curr Hypertens Rep. 2011;13(4):309–17. https://doi.org/10.1007/s11906-011-0197-8.

Institute of Medicine. Dietary reference intakes for calcium, phosphorus, magnesium, vitamin D, and fluoride. Washington, DC: National Academies Press; 1997.

Institute of Medicine. Dietary reference intakes for vitamin A, vitamin K, arsenic, boron, chromium, copper, iodine, iron, manganese, molybdenum, nickel, silicon, vanadium, and zinc. Washington, DC: National Academies Press; 2002. p. 8–9.

Janoušek J, Pilařová V, Macáková K, Nomura A, Veiga-Matos J, Silva DDD, et al. Vitamin D: sources, physiological role, biokinetics, deficiency, therapeutic use, toxicity, and overview of analytical methods for detection of vitamin D and its metabolites. Crit Rev Clin Lab Sci. 2022;59(8):517–54. https://doi.org/10.1080/10408363.2022.2070595.

Jiang Q. Metabolism of natural forms of vitamin E and biological actions of vitamin E metabolites. Free Radic Biol Med. 2022;179:375–87. https://doi.org/10.1016/j.freeradbiomed.2021.11.012.

Kapsetaki ME. Diet-induced carotenodermia: a literature review. Int J Dermatol. 2024;63(2):161–8. https://doi.org/10.1111/ijd.16891.

Kaźmierczak-Barańska J, Boguszewska K, Adamus-Grabicka A, Karwowski BT. Two faces of vitamin C – antioxidative and pro-oxidative agent. Nutrients. 2020;12(5):1501. https://doi.org/10.3390/nu12051501.

Kennedy DO. B vitamins and the brain: mechanisms, dose and efficacy—a review. Nutrients. 2016;8(2):68. https://doi.org/10.3390/nu8020068.

Konturek PC, Kania J, Hahn EG, Konturek JW. Ascorbic acid attenuates aspirin-induced gastric damage: role of inducible nitric oxide synthase. J Physiol Pharmacol. 2006;57(Suppl 5):125–36.

Kordiak J, Słomiński JM, Kozłowska B, Antosiewicz J, Lango R, Woźniak P, et al. Role of beta-carotene in lung cancer primary chemoprevention: a systematic review with meta-analysis and meta-regression. Nutrients. 2022;14(4):660. https://doi.org/10.3390/nu14030660.

Kucuk O, Sarkar FH, Djuric Z, Sakr W, Pollak MN, Khachik F, et al. Effects of lycopene supplementation in patients with localized prostate cancer. Exp Biol Med (Maywood). 2002;227(10):881–5.

Kumar N, Goel N. Phenolic acids: natural versatile molecules with promising therapeutic applications. Biotechnol Rep (Amst). 2019;24:e00370. https://doi.org/10.1016/j.btre.2019.e00370.

Kumar R, Cooney R, Saif MW. Vitamin D, calcium, and the kidney: physiological and molecular insights. Am J Physiol Renal Physiol. 2012;302(12):F1697–704. https://doi.org/10.1152/ajprenal.00326.2011.

León-Del-Río A. Biotin in metabolism, gene expression, and human disease. J Inherit Metab Dis. 2019;42(4):596–604. https://doi.org/10.1002/jimd.12037.

Long S, Romani AM. Role of cellular magnesium in human diseases. Austin J Nutr Food Sci. 2014;2(10):1051.

Maares M, Haase H. A guide to human zinc absorption: general overview and recent advances of in vitro intestinal models. Nutrients. 2020;12(3):762. https://doi.org/10.3390/nu12030762.

MacKay D, Hathcock J, Guarneri E. Niacin: chemical forms, bioavailability, and health effects. Nutr Rev. 2012;70(6):357–66. https://doi.org/10.1111/j.1753-4887.2012.00479.x.

Maden M. The role of retinoic acid in embryonic and post-embryonic development. Proc Nutr Soc. 2000;59(1):65–73.

Mattes RD, Donnelly D. Relative contributions of dietary sodium sources. J Am Coll Nutr. 1991;10(4):383–93. https://doi.org/10.1080/07315724.1991.10718167.

McLean E, Cogswell M, Egli I, Wojdyla D, de Benoist B. Worldwide prevalence of anaemia, WHO Vitamin and Mineral Nutrition Information System, 1993–2005. Public Health Nutr. 2009;12(4):444–54. https://doi.org/10.1017/S1368980008002401.

Mesquita SDS, Teixeira CMLL, Servulo EFC. Carotenoids: properties, applications and market. Rev Virtual Quim. 2017;9(2):672–88.

Miao M, Dai H, Zhang Y, Chen J, Wu H, Gu X, et al. The protective effects of inulin-type fructans against high-fat/sucrose diet-induced gestational diabetes mice in association with gut microbiota regulation. Front Microbiol. 2022;13:863446. https://doi.org/10.3389/fmicb.2022.863446.

Miller J, Rucker R. Pantothenic acid. In: Erdman JW, MacDonald IA, Zeisel SH, editors. Present knowledge in nutrition. 10th ed. Oxford: Wiley-Blackwell; 2012. p. 357–66.

Mirahmadi M, Hosseinpour-Moghaddam Z, Nayeri H, Asemi Z. Potential inhibitory effect of lycopene on prostate cancer. Biomed Pharmacother. 2020;128:110304. https://doi.org/10.1016/j.biopha.2020.110304.

Miyazaki A, Sano M, Fukuwatari T, Shibata K. Effects of ethanol consumption on the B-group vitamin contents of liver, blood and urine in rats. Br J Nutr. 2012;108(6):1034–41. https://doi.org/10.1017/S0007114511006192.

Mladěnka P, Pilařová V, Macáková K, Saso L, Remião F, Silva DDD, et al. Vitamin K – sources, physiological role, kinetics, deficiency, detection, therapeutic use, and toxicity. Nutr Rev. 2022;80(4):677–98. https://doi.org/10.1093/nutrit/nuaa134.

Morris RC, Schmidlin O, Frassetto LA, Sebastian A. Relationship and interaction between sodium and potassium. J Am Coll Nutr. 2006;25(3 Suppl):262S–70S. https://doi.org/10.1080/07315724.2006.10719576.

Mourão DMC, Botelho RBA, Akutsu RD, Rocha AC, Silva DM. Bioavailability of fat-soluble vitamins. Rev Nutr. 2005;18(4):529–39.

Mukherjee S, Date A, Patravale V, Korting HC, Roeder A, Weindl G. Retinoids in the treatment of skin aging: an overview of clinical efficacy and safety. Clin Interv Aging. 2006;1(4):327–48.

Nelson DL, Cox MM. Princípios de bioquímica de Lehninger. 7th ed. Porto Alegre: Artmed; 2019. Available from: https://integrada.minhabiblioteca.com.br/#/books/9788582715345/.

Neville JJ, Palmieri T, Young AR. Physical determinants of vitamin D photosynthesis: a review. JBMR Plus. 2021;5(1):e10492. https://doi.org/10.1002/jbm4.10492.

Pawlak R, Berger J, Hines I. Iron status of vegetarian adults: a review of literature. Am J Lifestyle Med. 2017;11(5):439–51. https://doi.org/10.1177/1559827616682933.

Pereira M, Damascena AD, Azevedo LM, Oliveira TA, Santana JM. Vitamin D deficiency aggravates COVID-19: systematic review and meta-analysis. Crit Rev Food Sci Nutr. 2022;62(5):1308–16. https://doi.org/10.1080/10408398.2020.1841090.

Plante L, Phan N, Han A, Brahim A, Glorieux F, Rauch F. Effect of high-dose vitamin D supplementation on bone density in youth with osteogenesis imperfecta: a randomized controlled trial. Bone. 2016;86:36–42. https://doi.org/10.1016/j.bone.2016.02.005.

Podszun M, Frank J. Vitamin E-drug interactions: molecular basis and clinical relevance. Nutr Res Rev. 2014;27(2):215–31. https://doi.org/10.1017/S0954422414000122.

Polzonetti V, Caré V, Ferri B, Pucciarelli S, Vincenzetti S, Polidori P, et al. Dietary intake of vitamin D from dairy products reduces the risk of osteoporosis. Nutrients. 2020;12(6):1741. https://doi.org/10.3390/nu12061741.

Popa DS, Bigman G, Rusu ME. The role of vitamin K in humans: implication in aging and age-associated diseases. Antioxidants. 2021;10(4):566. https://doi.org/10.3390/antiox10040566.

Potekaev NN, Nenasheva NM, Demina TN, Churina EG, Savva NN, Smirnova EY. Genetic and epigenetic aspects of skin collagen fiber turnover and functioning. Cosmetics. 2021;8(4):115. https://doi.org/10.3390/cosmetics8040115.

Ramasamy I. Vitamin D metabolism and guidelines for vitamin D supplementation. Clin Biochem Rev. 2020;41(3):103–26.

Reboul E. Vitamin E bioavailability: mechanisms of intestinal absorption in the spotlight. Antioxidants. 2017;6(4):95. https://doi.org/10.3390/antiox6040095.

Reboul E, Borel P. Proteins involved in uptake, intracellular transport and basolateral secretion of fat-soluble vitamins and carotenoids by mammalian enterocytes. Prog Lipid Res. 2011;50(4):388–402. https://doi.org/10.1016/j.plipres.2011.04.002.

Remely M, Ferk F, Sterneder S, Setayesh T, Kepcija T, Roth S, et al. Vitamin E modifies high-fat diet-induced increase of DNA strand breaks, and changes in expression and DNA methylation of DNMT1 and MLH1 in C57BL/6J male mice. Nutrients. 2017;9(6):607. https://doi.org/10.3390/nu9060607.

Roohani N, Hurrell R, Kelishadi R, Schulin R. Zinc and its importance for human health: an integrative review. J Res Med Sci. 2013;18(2):144–57.

Ross AC, Caballero B, Cousins RJ, Tucker KL, Ziegler TR. Modern nutrition in health and disease. 11th ed. Philadelphia: Lippincott Williams & Wilkins; 2014.

Rowles JL, Erdman JW. Carotenoids and their role in cancer prevention. Biochim Biophys Acta Mol Cell Biol Lipids. 2020;1865(11):158613. https://doi.org/10.1016/j.bbalip.2020.158613.

Said HM. Water-soluble vitamins. World Rev Nutr Diet. 2015;111:30–7. https://doi.org/10.1159/000362294.

Santos, M. A., et al. Factors influencing magnesium absorption and bioavailability. J Nutr Biochem. 2018;55(2):56–62.

Sarno F, Claro RM, Levy RB, Bandoni DH, Monteiro CA. Estimativa de consumo de sódio pela população brasileira, 2008-2009. Rev Saude Publica. 2009;43(2):219–25. https://doi.org/10.1590/S0034-89102009000200015.

Schwalfenberg GK, Genuis SJ. The importance of magnesium in clinical healthcare. Scientifica. 2017;2017:4179326. https://doi.org/10.1155/2017/4179326.

Sener U, Zorlu Y, Karagöz Y, Oztürk E, Akıncı A, Tunca Y, et al. Effects of common anti-epileptic drug monotherapy on serum levels of homocysteine, vitamin B12, folic acid and vitamin B6. Seizure. 2006;15(2):79–85. https://doi.org/10.1016/j.seizure.2005.11.006.

Sfairopoulos D, Arseniou A, Korantzopoulos P. Serum potassium and heart failure: association, causation, and clinical implications. Heart Fail Rev. 2021;26(4):479–86. https://doi.org/10.1007/s10741-020-10039-9.

Shils, M. E.; Olson, J. A.; Shike, M.; Ross, A. C. Tratado de nutrição moderna na saúde e na doença. 9ª ed. v.1. Barueri: Manole, 2003.

Shils ME, Shike M, Ross AC, Caballero B, Cousins RJ. Modern nutrition in health and disease. 10th ed. Philadelphia: Lippincott Williams & Wilkins; 2006.

Simes DC, Viegas CSB, Araújo N, Marreiros C, Simes MR. Vitamin K as a diet supplement with impact in human health: current evidence in age-related diseases. Nutrients. 2020;12(1):138. https://doi.org/10.3390/nu12010138.

Singh DK, Lippman SM. Cancer chemoprevention. Part 1: Retinoids and carotenoids and other classic antioxidants. Oncology (Williston Park). 1998;12(11):1643–53.

Singh P, Prasad S. A review on iron, zinc and calcium biological significance and factors affecting their absorption and bioavailability. J Food Compos Anal. 2023;123:105529. https://doi.org/10.1016/j.jfca.2023.105529.

Smith TJ, Johnson CR, Koshy R, Hess SY, Qureshi UA, Mynak ML, et al. Thiamine deficiency disorders: a clinical perspective. Ann N Y Acad Sci. 2021;1498(1):9–28. https://doi.org/10.1111/nyas.14536.

Soh YC, Fairley A, Alawad M, Lee SS, Su TT, Stephan BCM, et al. Assessing sodium intake in middle-aged and older adults with elevated blood pressure: validation of spot urine excretion and dietary survey-derived estimates. Nutrients. 2024;16(10):1461. https://doi.org/10.3390/nu16101461.

Stamler J, Rose G, Stamler R, Elliott P, Dyer A, Marmot M. Worldwide differences in hypertension: an overview of the INTERSALT study. J Hum Hypertens. 1996;10(Suppl 3):S3–9.

Stephensen CB. Vitamin A, infection, and immune function. Annu Rev Nutr. 2023;21:167–92.

Thakur K, Tomar SK, Singh AK, Mandal S, Arora S. Riboflavin and health: a review of recent human research. Crit Rev Food Sci Nutr. 2017;57(17):3650–60. https://doi.org/10.1080/10408398.2016.1145104.

Thomson, A. D. Tiamina: absorção, metabolismo e implicações na saúde humana. Journal of Nutritional Science, v. 9, p. e56, 2020. Disponível em: https://doi.org/10.1017/jns.2020.48. Acesso em: 24 dez. 2024.

Tong, G. M., et al. Clinical implications of hypomagnesemia in chronic disease. Am J Med. 2017;130(7):742–751.

Traber MG. Human vitamin E deficiency, and what is and is not vitamin E? Free Radic Biol Med. 2024;213:285–92. https://doi.org/10.1016/j.freeradbiomed.2024.01.024.

Tso P, Lee T, DeMichele SJ. Structured triglycerides increase lymphatic absorption of tocopherol and retinol compared with the equivalent physical mixture in a rat model of fat malabsorption. J Nutr. 2001;131(8):2157–63. https://doi.org/10.1093/jn/131.8.2157.

Tsugawa N, Shiraki M. Vitamin K nutrition and bone health. Nutrients. 2020;12(7):2275. https://doi.org/10.3390/nu12072275.

Vervloet M, Sezer S, Massy Z, Jahnen-Dechent W, Cozzolino M, Brandenburg V, et al. The role of phosphate in kidney disease. Nat Rev Nephrol. 2017;13(1):27–38. https://doi.org/10.1038/nrneph.2016.164.

Wani AL, Hammad AAS, Afzal M. Lead and zinc interactions – an influence of zinc over lead related toxic manifestations. J Trace Elem Med Biol. 2021;64:126702. https://doi.org/10.1016/j.jtemb.2020.126702.

Wasserman RH. Vitamin D and the dual processes of intestinal calcium absorption. J Nutr. 2004;134(11):3137S–9S. https://doi.org/10.1093/jn/134.11.3137S.

Watanabe F. Vitamin B12 sources and bioavailability. Exp Biol Med (Maywood). 2007;232(10):1266–74. https://doi.org/10.3181/0703-MR-67.
Weaver CM. Potassium and health. Adv Nutr. 2013;4(3):368S–77S. https://doi.org/10.3945/an.112.003533.
Wilson MP, Plecko B, Mills PB, Clayton PT. Disorders affecting vitamin B6 metabolism. J Inherit Metab Dis. 2019;42(4):629–46. https://doi.org/10.1002/jimd.12060.
World Health Organization. Sodium intake for adults and children: guideline. Geneva: WHO; 2012.
World Health Organization. Guidelines for drinking-water quality. 4th ed. Geneva: WHO; 2017.
World Health Organization. WHO guideline on use of ferritin concentrations to assess iron status in individuals and populations. Geneva: WHO; 2021. Available from: https://www.who.int/publications/i/item/9789240000124.
World Health Organization. WHO global report on sodium intake reduction. Geneva: WHO; 2023. Available from: https://www.who.int/publications/i/item/9789240064690.
Wright EM, Loo DD. Coupling between Na+, sugar, and water transport across the intestine. Ann N Y Acad Sci. 2000;915:54–66. https://doi.org/10.1111/j.1749-6632.2000.tb05278.x.
Yang C, Chen Y, Xue Y, Zhang J, Wang M, Yang M, et al. Prevalence of vitamin D deficiency and insufficiency among 460,537 children in 825 hospitals from 18 provinces in mainland China. Medicine (Baltimore). 2020;99(44):e23058. https://doi.org/10.1097/MD.0000000000023058.
Zerres S, Stahl W. Carotenoids in human skin. Biochim Biophys Acta Mol Cell Biol Lipids. 2020;1865(11):158588. https://doi.org/10.1016/j.bbalip.2019.158588.
Zimmermann, M. B., Hurrell, R. F. Nutritional iron deficiency. The Lancet, v. 370, n. 9586, p. 511-520, 2007. Disponível em: https://www.sciencedirect.com/science/article/pii/S0140673607606179. Acesso em: 14 ago. 2024.

# Chapter 4
# Herb-Drug Interactions

**James O. Fajemiroye, Anderson Luiz-Ferreira, and Roberto Saavedra-Rodríguez**

**Abstract** The use of herbs or botanicals is markedly increasing worldwide. Herbs are frequently administered in combination with medications, potentially increasing herb-drug interactions and overall effects. These effects can occur due to incorrect use, lack of knowledge, and awareness regarding the active ingredients involved. As herbal products are perceived as safe, patients usually self-administer herb products without informing their doctors, pharmacists, or healthcare providers. This is particularly true for elderly patients, who typically take a lot of medications alongside herbal products with a combination of poorly defined ingredients. Herb-drug combinations can elicit adverse effects when herb compounds regulate drug-metabolizing enzymes, transporters, and others. A systematic approach and regulatory measures are necessary against unwanted consequences. This chapter explores herbs, active compounds, and drugs with mechanistic targets that are relevant to interactions.

**Keywords** Alternative medicine · Herbs · Active principle · Drugs · Toxicity

## 1 Introduction

Medicinal plants offer alternative sources for the treatment of various diseases through their innumerous metabolites (Thembane et al. 2024). The use of medicinal plants has increased over the last three decades, with approximately 80% of the

J. O. Fajemiroye (✉)
Institute of Biological Sciences, Federal University of Goias, Goiania, Goias, Brazil
e-mail: jamesfajemiroye@ufg.br

A. Luiz-Ferreira
Institute of Biotechnology, Federal University of Catalão, Catalão, Goias, Brazil
e-mail: luiz_ferreira@ufcat.edu.br

R. Saavedra-Rodríguez
Department of Basic Science, Universidad Santo Tomas, Campus Puerto Montt, Puerto Montt, Chile



J. O. Fajemiroye (ed.), *Fundamentals of Drug and Non-Drug Interactions*,
https://doi.org/10.1007/978-3-031-80107-5_4

world's population dependent on natural products, including medicinal plants, for primary healthcare (Périco et al. 2020; Luiz-Ferreira et al. 2023; Fajemiroye et al. 2016). The international medicinal plants market was estimated at $138,350 million in 2020 and is predicted to reach $218,940 million by the end of 2026 (Hossain et al. 2022).

Most recently, the resurgence of herbal medicine has increased considerably as seen by traditional or folk medicine present in different regions of the world. Herbals have become very popular in the general population and among patients and physicians due to their global acceptance as a complementary and alternative remedy (Parvez and Rishi 2019). Approximately, one-third of adults in the Western World use alternative therapies, including herbs (Zhou et al. 2004a). Meanwhile, synthetic drugs are commercially available through laboratory validations, clinical trials, and approval from drug regulatory authorities (Parvez and Rishi 2019).

Complementary and alternative medicine (CAM) is popular in some regions. Some cultures still use traditional medicine such as Chinese, Tibetan, Japanese, Ayurvedic, or Yunani, where their medical healthcare systems include botanicals and nutritional products such as herbs and supplements, which contain bioactive principles that can interact with drugs. In addition, a culture of polypharmacy with herbs, supplements, and food increases the chance of possible interaction with drugs (Martins et al. 2022). Many herbs that have been used traditionally for therapeutic and culinary applications are generally safe it is also important to know that some plants are highly toxic and can even be deadly (Gouws and Hamman 2020; Fajemiroye et al. 2014).

In general, documentation of herb-drug interactions (HDIs) is limited. Despite herbal usage, herbal products in the market lack scientific studies supporting their efficacy and safety. Many case reports and clinical observations show in the scientific literature associations of herb-drug interactions with clinical significance (Hu et al. 2005; Gouws and Hamman 2020; Oyanna and Clarke 2024). Consequently, adverse effects of HDIs, supplement-drug interactions (SDIs), or drug-drug interactions (DDIs) have been reported in the literature. Unfortunately, there is a significant lack of large-scale clinical studies that address these questions. Therefore, there is a need to alert health system communities, health authorities, clinicians, pharmaceutical industries, as well as consumers about the dangers of combining CAM with conventional drugs. Thus, emerging public health issues related to herb-drug interactions are becoming a serious problem (Fugh-Berman 2000; Chavez et al. 2006; Parvez and Rishi 2019; Martins et al. 2022). Furthermore, healthcare professionals should ask their patients about previous use of dietary supplements, herbs, and synthetic drugs as well as warn them about the possible interactions of concomitant use (Fig. 4.1).

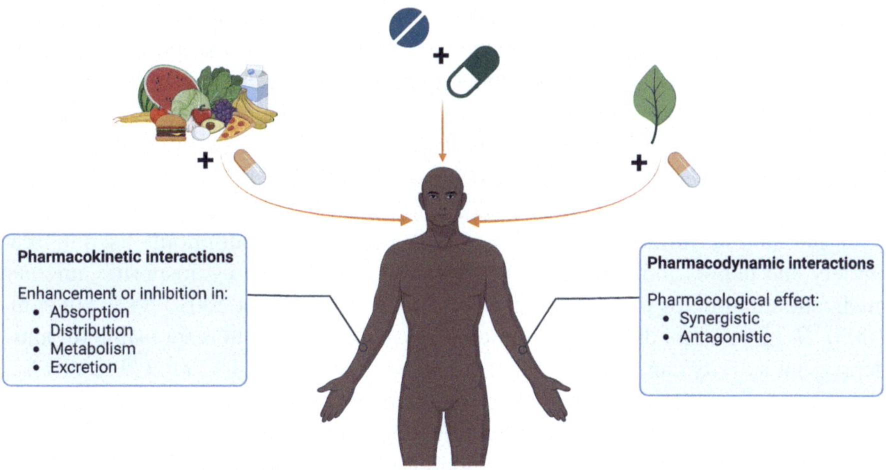


**Fig. 4.1** Overview of HDIs, SDIs, and DDIs interactions. (Created with BioRender)

## 2 Herb-Drug Interactions

It is known that prescribed drugs along with herbal medicines can interact leading to both beneficial aspects and serious clinical consequences. HDIs or SDIs interactions can be insignificant, with no inconvenience for the patient, up to treatment failure and potentially fatal effects or even death (Campos et al. 2024). Herbal products (such as green tea, *Allium sativum*, *Aloe vera*, *Ginkgo biloba*, berberine, ginseng, and cinnamon) are included in the composition of several highly affordable dietary supplements that have various biological effects (Nyulas et al. 2024). Drug metabolism is largely mediated by cytochrome P450 (CYP) metabolizing enzymes, with an important contribution from the CYP3A subfamily, as well as by drug transporters such as P-glycoprotein (P-gp) and organic anion-transporting polypeptides (OATP) (Chang 2009; Tan et al. 2018; Egeland et al. 2020; Takiyama et al. 2022).

Parvez and Rishi (2019) also discuss herb-drug interactions and potential risks related to hepatotoxicity, which can lead to an adverse side effect, and their underlying mechanisms, including drug-metabolizing enzymes and their regulation (Hermann and von Richter 2012; Posadzki et al. 2013). Globally, patients are attracted to the use of herbal remedies with prescription drugs with the misconception that these are completely safe. Still, the toxicological effect of a drug or herb is due to the inhibition of drug-metabolizing enzymes (e.g., cytochrome P450), including interactions with certain prescription drugs through various mechanisms.

The adverse effects of herb-drug interactions may be caused by increased plasma drug levels beyond minimal toxic concentrations (Gouws and Hamman 2020; Oyanna and Clarke 2024). This fact becomes clinically important for patients who use medication with a narrow therapeutic window, such as tacrolimus. Tacrolimus, a calcineurin inhibitor, is one of the most prescribed immunosuppressant drugs after solid organ transplantation (Henkel et al. 2023). Tacrolimus undergoes pre-systemic metabolism via CYP3A5 and P-gp in the intestine (Henkel et al. 2023; Füessl et al. 2024). Although tacrolimus has variable bioavailability with absorption affected by the fat content of foods, well-documented clinical and preclinical studies reported interactions with herbs (Woon et al. 2024). These studies demonstrated that the coadministration of *Schisandra sphenanthera* increased the bioavailability of tacrolimus and reduced the dosage required to maintain therapeutic concentration (Yan et al. 2019; Qin et al. 2020; Qu et al. 2022).

*Hypericum perforatum* (St. John's wort), a plant species commonly used to treat anxiety and depression, decreased plasma concentrations of cyclosporine, another medication that helps prevent organ rejection (Izzo and Ernst 2001, 2009; Hu et al. 2005). *H. perforatum* through the action of its main component responsible for antidepressant activity can also modulate the activity of CYP3A4, CYP2C9, CYP1A2, and P-gp isoforms (Sato et al. 2015; Soleymani et al. 2017). Hypericum can also interact with several medications, affecting their distribution, increasing their levels or enhancing their effects. These include anticancer medications, oral contraceptives (ethinyl estradiol, norethindrone, ketodesogestrel), cardiovascular medications (anticoagulants, statins, beta-blockers), antimicrobials (including anti-HIV, voriconazole), antidepressants and anxiolytics (benzodiazepines and buspirone), anticonvulsant agents, oral hypoglycemic agents (tolbutamide, gliclazide), anesthetics, respiratory (fexofenadine) and gastrointestinal (omeprazole) agents, anti-migraine medicines (eletriptan), muscle-relaxing medicines, and medications (e.g., methadone) used in drug users (Borrelli and Izzo 2009; Soleymani et al. 2017).

Green tea, which contains higher levels of endogenous polyphenols than other teas, is one of the most popular beverages worldwide (Khan and Mukhtar 2013). Green tea catechins have been reported to inhibit drug transporters such as OATP affecting the bioavailability of many drugs (Roth et al. 2011; Abdelkawy et al. 2020). A randomized, double-blind study in healthy volunteers found that green tea extract decreases the absorption of atorvastatin, due to OATP deficiency. This work demonstrates that the coadministration of green tea extract with atorvastatin should be monitored in clinical practice (Abdelkawy et al. 2020).

According to Hu et al. (2005), many case reports showed herb-drug interactions of clinical importance and limited clinical observations. These therapeutic cases reported enhanced anticoagulation and bleeding in patients who took *Salvia miltiorrhiza* (danshen) with long-term warfarin therapy. In volunteers, *Allium sativum* diminished the area under the plasma concentration and maximum plasma concentration of saquinavir, but not ritonavir and acetaminophen. Additionally, *A. sativum* also potentiated hypoglycemia when administered with chlorpropamide while increasing clotting time and the international normalized ratio of warfarin (Hu et al. 2005).

*Aloe vera* contains a variety of specialized metabolites that have diuretic, anti-inflammatory, and immunomodulatory effects (Catalano et al. 2024). The mucilaginous extract obtained from the meaty leaves has anti-inflammatory effects and is used by certain physicians to treat patients with ulcerative colitis (Bendjedid et al. 2021). Cui and coworkers (2017) showed that aloe vera polysaccharide supplementation significantly downregulated the gene expression, protein synthesis, and enzyme activities of hepatic CYP1A2 and 3A4 proteins. There is therapeutic significance to this mechanistic suppression of CYP3A4 and CYP2D6 enzymes, as CYP3A4 and CYP2D6 are the most important enzymes in drug responses, representing approximately 60–75% of all Phase I drug responses (Lynch and Price 2007). These interferences may have consequences for the medications metabolized through these pathways (Djuv and Nilsen 2012).

*Ginkgo biloba* is used in traditional Chinese medicine and throughout the world as both a medicine and a dietary supplement to treat cardiovascular disorders, improve memory, and manage age-related dementia (Hashiguchi et al. 2015; Crawford et al. 2020; Chang et al. 2021; Luo et al. 2023). *G. biloba,* when administered in combination with acetylsalicylic acid or warfarin, promoted bleeding. Flavonoids and tripertenic lactones (gingkoloids and bilobalids) can reduce platelet aggregation by acting on CPY2C9 and CYP3A4 isoenzymes. Therefore, patients taking antiplatelet or anticoagulant medications should avoid Ginkgo preparations (Ke et al. 2021). Furthermore, when co-administered with thiazide diuretic and trazodone, it raised blood pressure and induced coma, respectively (Izzo and Ernst 2001, 2009; Hu et al. 2005). Ginkgo extract can also inhibit cytochrome P450 isoforms, responsible for other interactions involving the calcium channel blocker, nifedipine, and proton pump inhibitor omeprazole. Its flavonoids have an agonistic action on the GABA system and a serotonergic effect, interacting with benzodiazepines and selective serotonin reuptake inhibitors, respectively (Yoshitake et al. 2010; D'Alessandro et al. 2022). Li et al. (2021) demonstrated that *G. biloba* can alter OATP activity and increase plasma exposure to pitavastatin, a medication for hypercholesterolemia and dyslipidemia.

*Panax ginseng* is widely used as a supplement throughout the world and, due to its therapeutic actions, including hepatoprotective effects, prevention of aging, effects on the immune system, and central nervous system, it is one of the top 40 best-selling herbal supplements in the world (Shim et al. 2024). Yang et al. (2020) reported that *P. ginseng* extract alters the oral bioavailability of selegiline, an inhibitor of monoamine oxidase B, prescribed during the early stages of Parkinson's disease. Its ginsenosides can strongly inhibit cytochrome-dependent metabolism and transmembrane transporters (Ramanathan and Penzak 2017). *P. ginseng* reduced warfarin blood concentrations and induced mania when administered with phenelzine. Interestingly, *P. ginseng* increased the effectiveness of influenza vaccination (Hu et al. 2005). *P. ginseng* should not be consumed with some anticoagulants, phenelzine, chemotherapy drugs, oral hypoglycemic agents and insulin, digoxin, anticonvulsants, and anti-estrogenic therapies (Choi and Song 2019).

Curcumin is a compound isolated from the rhizome of *Turmeric longa*, a plant known for its medicinal use for inflammatory and neoplastic diseases (Vollono et al. 2019). Curcumin inhibits the activity of both the transmembrane transporters belonging to the ATP-binding cassette (ABC) family and CYP isoforms. It may modify the pharmacokinetics of cardiovascular agents and drugs acting on the central nervous system.

*Glycyrrhiza glabra*, which belongs to the Fabaceae family, has been recognized since ancient times for its ethnopharmacological values (Pastorino et al. 2018;

El-Saber Batiha et al. 2020). High consumption of liquorice from *G. glabra* roots can increase arterial blood pressure and hypokalemia (Kwon et al. 2020). Its metabolite glycyrrhizin acid inhibits 11-beta-hydroxysteroid-dehydrogenase, the hepatic and renal metabolism of corticosteroids. This effect increases the aldosterone-like activity of cortisol leading to primary hyperaldosteronism, hypertension, hypokalemia, metabolic alkalosis, and hydro-saline retention. Therefore, patients with hypertension or hypokalemia should avoid these herbal preparations (Pastorino et al. 2018).

Traditionally, kava is a drink prepared from the root of *Piper methysticum* originated and was cultivated predominantly in the islands of the South Pacific (Volgin et al. 2020). Kava has been consumed by indigenous peoples for centuries as a religious and celebratory drink due to its relaxing properties (Minh et al. 2022; Freeman et al. 2023). A recent study demonstrated that *P. methysticum* inhibited the intestinal metabolism of midazolam. Moreover, kava-kava extract increased the absorption of fexofenadine. These changes were attributed to changes in CYP3A and P-gp (Nascimento et al. 2024). In addition, *P. methysticum* suspended "off" periods in patients with parkinsonism when administered together with levodopa and induced a semicomatose state when administered concomitantly with alprazolam (Izzo and Ernst 2001; Hu et al. 2005).

Ginger from the rhizome of *Zingiber officinalis* has long been used as an herbal medicine for its antiemetic effect (Fahimi et al. 2011). Ginger has anticoagulant activity, reduces blood pressure, and modulates glucose serum levels. Because of that, people with diabetes or who use antiplatelet drugs or anticoagulants should avoid ginger ingestion (Xu et al. 2017; Fugh-Berman 2000). Altogether, CYP and P-gp play a significant role in drug and/or herb plasma concentrations (Fig. 4.2) and the ensuing interactions.

## *2.1 Natural Active Compounds Interaction with Organic Anion Transporter*

Drug transporters in the human body can be divided into two categories according to their functions (Lin et al. 2023). The first is the solute carrier (SLC) superfamily, which mediates the absorption of drugs into cells and is widely distributed throughout the tissues and organs of the human body. The other is the ATP-binding cassette (ABC) superfamily, which can utilize energy by hydrolyzing ATP to transport drugs out of cells, thereby reducing intracellular drug concentration and producing drug resistance (Tatrai et al. 2022). Either influx or efflux of these substances from the blood to the epithelium is intervened by SLC and ABC (Nigam 2015; Tatrai et al. 2022; Lin et al. 2023). The SLC22 family is covered by six main subfamilies: organic anion transporter (OAT), OAT-like, OAT-related, organic cation transporter (OCT), organic cation and carnitine transporter (OCTN), and OCT/OCTN-related (Nigam 2018; Zou et al. 2021).

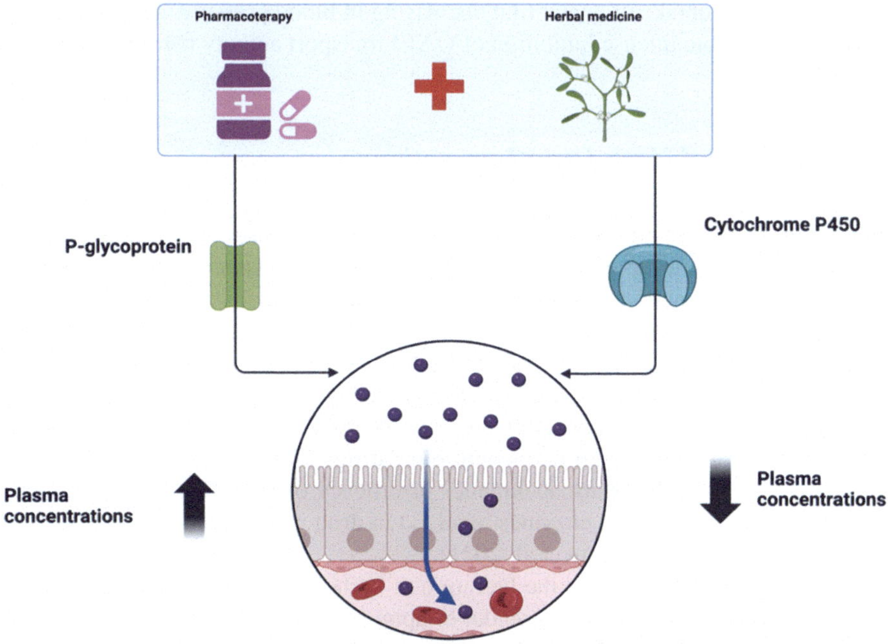


**Fig. 4.2** Overview of the herb-drug interactions. (Created with BioRender)

OAT3 (encoded by SLC22A8) prime property is its broad substrate specificity, which places it as the main transporter in OAT. Its distribution covers various organs such as the kidney, liver, choroid plexus, olfactory mucosa, brain, retina, and placenta (Nigam et al. 2015; Lin et al. 2023). However, it is mainly manifested on the basolateral membrane of renal tubular cells (Bush et al. 2017; Lin et al. 2023). The driving force of OAT3 is related to the concentration gradient when dicarboxylate (e.g., α-Ketoglutarate) exchanges organic anions. First, it carries organic anions (OAs) from the blood to the proximal tubules through the basolateral membrane and leaves OAs into the urine through the apical membrane of tubular epithelial cells. This allows reabsorption of specific compounds in the glomerular filtrate back into the internal circulation (Otani et al. 2017; Chen et al. 2023).

OAT3 is crucial for the uptake, distribution, and excretion of several endogenous and exogenous substances and is also relevant as a competitive target for DDI (Yang et al. 2022). This is visualized when more than two drugs are ingested simultaneously and change in their toxicity, pharmacokinetics, and functions is expressed (Chung and Kim 2021; Zhang et al. 2021). Specifically, when herbs and drugs are taken simultaneously, severe HDIs take place. Effects manifested may be a liver injury and increased clotting time, headache, insomnia, and possibly coma or death (Parvez and Rishi 2019). The previous statement is not negative, because OAT3 effects along with DDIs or HDIs can have either positive or negative results. For instance, Ni et al. (2020) argue that the coadministration of enalaprilat, the active metabolite of the antihypertensive drug enalapril maleate, with flavonoids, inhibits

OAT3-mediated uptake of enalaprilat decreasing in blood pressure. This is because flavonoids promote intense inhibition of OAT3 transport activity (Ma et al. 2015).

## 2.2 OAT3 and Herb-Drug Interactions

Natural products and prescription drugs are commonly consumed over the counter, basically at the same time, by the elderly population. Polyphenols and flavonoids are some of the functional components that natural medicines contain, and these compounds are utilized in clinical therapies for different illnesses, such as hypertension, hyperlipidemia, and hyperuricemia (Mota et al. 2009; Chen et al. 2023). As described by Agbabiaka et al. (2018), 78% of elderly survey respondents take both prescription drugs and dietary supplements, and 32.6% of them experience adverse drug effects when they ingest herbs with other drugs. Some herbs can cause direct organ toxicity, alter the pharmacokinetics or efficacy of OAT3-mediated prescription medications, or inhibit enzymes involved in drug metabolism (Ma et al. 2015; Parvez and Rishi 2019; Zhang et al. 2021).

Many examples reported in the literature of herb-drug interaction effects can be seen with puerarin, steviol glucuronide, or apigenin when combined with different drugs administered. The first, puerarin, an isoflavone extracted from *Pueraria lobata*co administered with methotrexate (MTX), inhibited OAT3 and increased MTX exposure when MTX was administered by both oral and intravenous routes (Liu et al. 2014). The steviol glycosides have gained popularity as natural sweeteners due to their intense sweetness and low caloric intake (Orellana-Paucar 2023). The steviol acyl glucuronide, a major metabolite of steviol glycoside, is uptake transport by OAT3 (Wang et al. 2015). Zhou et al. (2020) demonstrated that probenecid and glimepiride, medications used to manage hyperuricemia or hyperglycemia, displayed potential inhibition toward OAT3-mediated steviol acyl glucuronide transport, altering its pharmacokinetic characteristics. Thus, people with impaired renal function should consider not taking steviol glycoside products.

Apigenin (4′,5,7-trihydroxyflavone) is one of the most widespread flavonoids in plants and formally belongs to the flavone sub-class. Plants belonging to the Asteraceae, such as those belonging to Artemisia, Achillea, Matricaria, and Tanacetum genera, are the main sources of this compound (Ornano et al. 2016; Venditti et al. 2016, 2018; Sharifi-Rad et al. 2018). Apigenin in combination with imipenem (a broad-spectrum antibiotic) can result in OAT3 inhibition and protect against renal toxicity and renal adverse reaction of imipenem in the clinic as well as galangin, which can exhibit a high inhibitory effect on the activity of OAT3 when administered with prescribed drugs (Huo et al. 2020; Ni et al. 2020). In vitro studies in HeLa cells and additional in vivo studies showed that the other flavonoid naringin, a constituent of grapefruit juice, inhibits the uptake of fexofenadine mediated by OATP1A2 (Bailey et al. 2007; Glaeser et al. 2007).

The dichloromethane extract from *Juncus effusus* in combination with over-the-counter drugs may also exhibit an interaction exerting effects on the

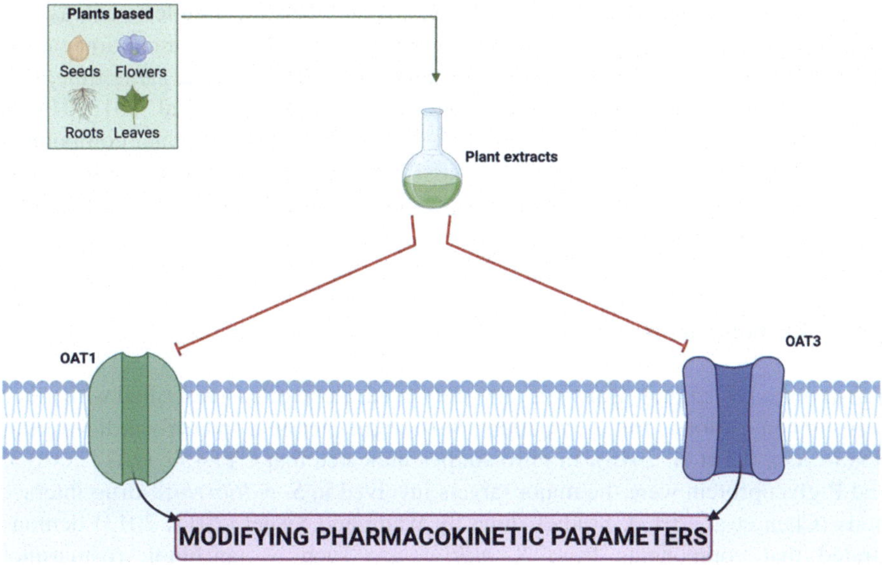


**Fig. 4.3** Interactions of medicinal plants with organic anion transporters OAT1 and OAT3. (Created with BioRender)

pharmacokinetics of furosemide (Lu et al. 2017; Li et al. 2019; Ni et al. 2020). The influence of flavonoids on intestinal drug metabolism and transport is of clinical importance because flavonoids can reach high concentrations in the gut lumen. The inhibition of OATP-mediated drug uptake in the intestine may decrease the plasma concentration of a substrate of OATPs due to reduced absorption (Mandery et al. 2010). Interactions of medicinal plants with organic anion transporters OAT1 and OAT3 (Fig. 4.3) have continued to gain attention.

## 2.3 *Natural Bioactive Inhibitors of OAT*

With increasing public interest in complementary and alternative medicines, several herbal products are available without a prescription in the form of phytopharmaceuticals, and the use of herbal products has become more popular for health maintenance or as a complement to conventional medical care (Ustuner Top et al. 2021). Consequently, there is an increasing risk of pharmacokinetic interactions mediated by herbal ingredients with conventional medicines, as demonstrated by numerous studies (Parvez and Rishi 2019; Gouws and Hamman 2020; Oyanna and Clarke 2024). Different plant polyphenolic compounds have exhibited, in addition to diverse biological activities, actions on the absorption and disposition of drugs through modulation of drug-metabolizing enzymes, as well as drug transporters (Hong et al. 2007; Nguyen et al. 2014; Cao et al. 2020; Wang et al. 2024).

As elucidating the effects on how flavonoids and their metabolites, such as sulfates and glucuronides, act as inhibitors or substrates of OAT3, a prediction can be estimated to prevent potential OAT3-mediated DDIs/HDIs. At the same time, this prediction may have beneficial effects on renal disorders (Xu et al. 2013; Hagos et al. 2017; Parvez et al. 2017; Xiao et al. 2023). The most important compounds identified as OAT3 inhibitors can be classified into nine groups: phenolic acids, flavonoids, alkaloids, anthraquinones, phenols, phenylpropanoids, terpenoids, phenanthrenoids, and others.

### 2.3.1 Phenolic Acids

*Salvia miltiorrhiza* root (Danshen) is widely used, alone or co-prescribed with other herbs or conventional drugs, by patients with angina pectoris and myocardial infarction in Asia (Li et al. 2018). In vitro studies indicated that CYP450, OAT1, OAT3, and P-glycoprotein were the major targets involved in *S. miltiorrhiza*-drug interactions (Chen et al. 2017). Studies done by Wang and Sweet (2012, 2013) demonstrated that components from *S. miltiorrhiza* such as tanshinol, rosmarinic, lithospermic acid, salvianolic acid A, salvianolic acid B, and protocatechuic acid promoted significant inhibitory effect on substrate uptake mediated by OAT3. In addition, caffeic acid, 1,3-dicaffeoylquinic acid, and ginkgolic acid significantly inhibited the transport of OAT3 (Wang and Sweet 2013). The molecular structures of representative OAT-inhibiting phenolic acids are shown in Fig. 4.4.

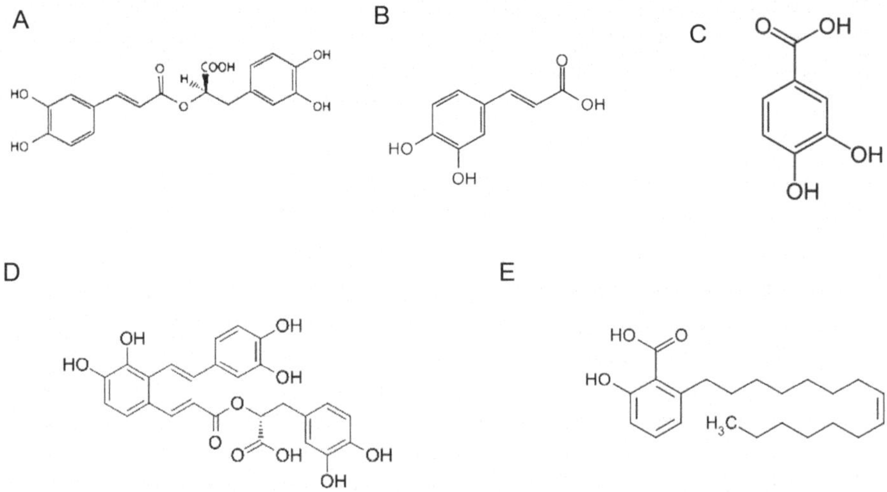


**Fig. 4.4** Molecular structures of some OAT-inhibiting phenolic acids. (**a**) Rosmarinic acid. (**b**) Caffeic acid. (**c**) Protocatechuic acid. (**d**) Salvianolic acid A. (**e**) Ginkgolic acid

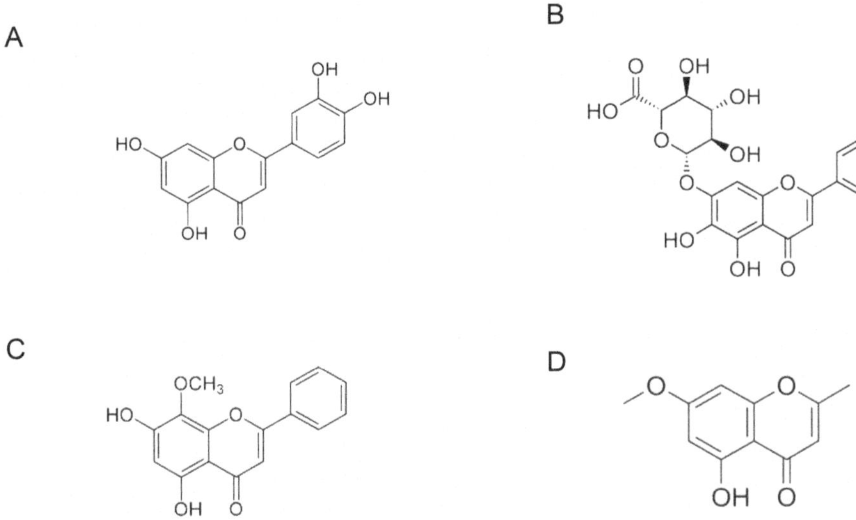


**Fig. 4.5** Molecular structures of some OAT-inhibiting flavonoids. (**a**) Luteolin. (**b**) Baicalin. (**c**) Wogonin. (**d**) Eugenin

### 2.3.2 Flavonoids

*Scutellaria baicalensis* is a plant of the genus Lamiaceae, and its root is the main part used as a medicine to treat respiratory tract infections, pneumonia, and colitis (Liao et al. 2021; Fang et al. 2023). According to Xu et al. (2013), baicalin, baicalein, and wogonin, the main chemical components of *S. baicalensis*, significantly inhibit the OAT1, OAT3-mediated uptake. Li et al. (2020) found that the flavonoids eugenin, calicosin, wogonin, luteolin, quercetol, and scullcapflavone II had non-competitive inhibitory effects on OAT3 while oroxylin A, viscidulin III, and 3,5,7,4′-tetra-O-methyl-kaempferol inhibited competitively OAT3. The molecular structures of representative OAT-inhibiting flavonoids are shown in Fig. 4.5.

### 2.3.3 Alkaloids

The alkaloids are substances of a very diverse class of plant-specialized metabolites with an extensive list of biological activities, such as being anticholinergic, antitumor, and anti-inflammatory (Alves de Almeida et al. 2017). Zheng et al. (2016) stated that camptothecin, 10-hydroxycamptothecin, 10-methoxycamptothecin, and 9-nitrocamptothecin inhibit OAT3-mediated substrate uptake. Also, Shams et al. (2018) assayed five clinically relevant alkaloids (dendrobine, matrine, oxymatrine, tryptanthrin, and chelerythrine) on the substrate transport through OAT and found that tryptanthrin and chelerythrine acted as an inhibitor to the substrate transport mediated through OAT3 and OATP1A2, respectively. Recently, Huang et al. (2020) reported that vindesine, a semisynthetic vinca alkaloid used to treat acute

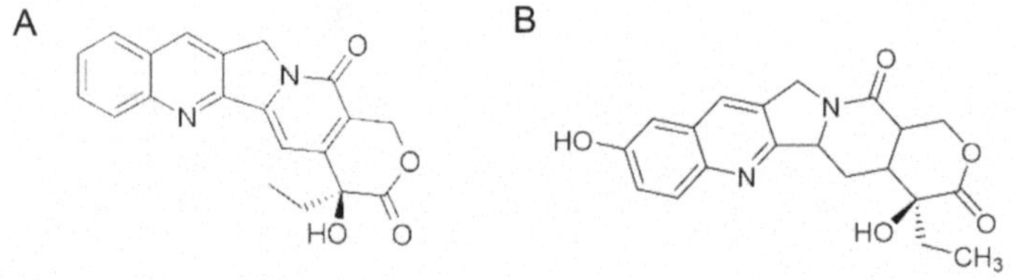


**Fig. 4.6** Molecular structures of some OAT-inhibiting alkaloids. (**a**) Camptothecin. (**b**) 10-hidroxycamptothecin

lymphoblastic leukemia, increased the exposure to and decreased the excretion of MTX by inhibiting OAT1 and OAT3. The molecular structures of representative OAT-inhibiting alkaloids are shown in Fig. 4.6.

### 2.3.4 Anthraquinones

Anthraquinones are bioactive natural products, often found in medicinal herbs that exert pharmacological actions including anti-inflammation, neuroprotective, and anticancer effects (Zhao and Zheng 2023). Some studies have been done on the interaction of anthraquinones with the ABC family (Ye et al. 2013). Anthraquinones, extracted from Rhubarb, are used in traditional Chinese medicine to treat chronic kidney diseases and inhibit the uptake of 6-carboxyl fluorescein via OAT3 (Ma et al. 2014; Zhang et al. 2023). Wang et al. (2018) evaluated 22 substances isolated from *Semen cassiae* and found that six anthraquinones significantly inhibited OAT3-mediated transport. Newly, Pang et al. (2019) demonstrated that a new anthraquinone analog isolated from *Senna obtusifolia* exhibited a strongly specific inhibitory effect on OAT1. The molecular structures of representative OAT-inhibiting anthraquinones are shown in Fig. 4.7.

### 2.3.5 Phenols

Green tea is rich in phenolic compounds, such as (-)-epigallocatechin-3-gallate, which has antioxidant and anticancer activities (Cheng et al. 2020). Results from Kawasaki et al. (2021) showed the inhibition of human and rat OAT1 and OAT3 and the reduction of the renal elimination of fluorescent organic anion 6-carboxyfluorescein. Jia et al. (2016) reported resveratrol enhanced MTX absorption in the intestine and decreased MTX renal elimination by inhibiting P-gp, OAT1, and OAT3 in vivo and in vitro. The molecular structures of representative OAT-inhibiting phenols are shown in Fig. 4.8.

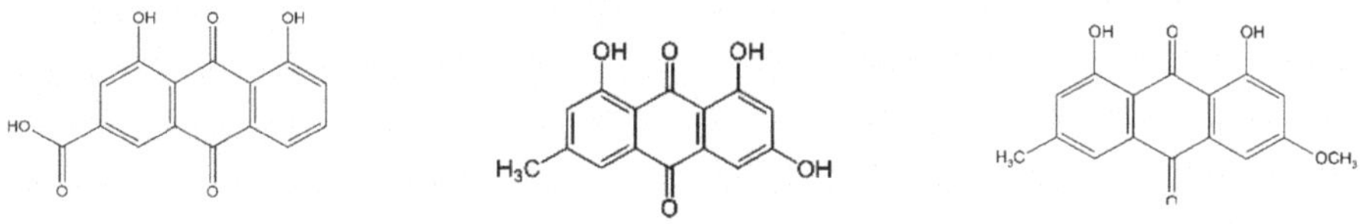


**Fig. 4.7** Molecular structures of some OAT-inhibiting anthraquinones. (**a**) Rhein. (**b**) Emodin. (**c**) Physcion

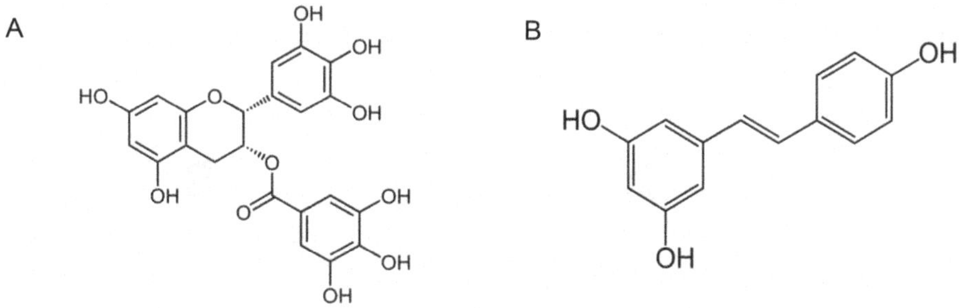


**Fig. 4.8** Molecular structures of some OAT-inhibiting phenols. (**a**) (-)-epigallocatechin-3-gallate. (**b**) Resveratrol

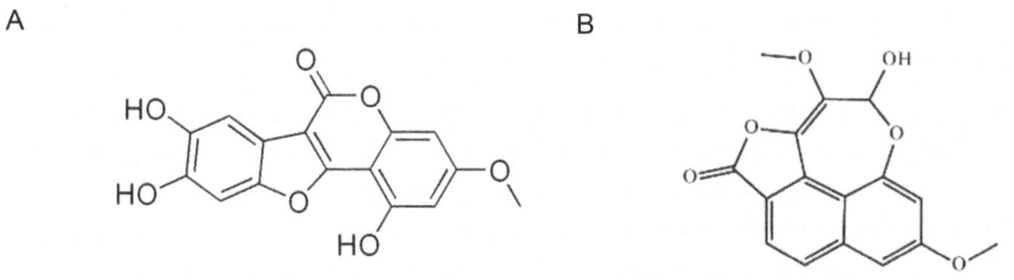


**Fig. 4.9** Molecular structures of some OAT-inhibiting phenylpropanoids. (**a**) Wedelolactone. (**b**) Dioscorealide B

### 2.3.6 Phenylpropanoids

Kawasaki et al. (2017) found that the phenylpropanoids, 3-(4′-isopentenyloxyphenyl)-benzoic acid, 3-(4′-isopentenyloxy-3′-methoxyphenyl)-benzoic acid, and 3-(4′-geranyloxy-3′-methoxy phenyl)-benzoic acid inhibit OAT1 and OAT3. In addition, Dioscorealide B, a phenylpropanoid compound extracted from *Dioscorea esculenta*, is an effective noncompetitive inhibitor of OAT3. Another phenylpropanoid compound, wedelolactone, extracted from *Eclipta prostrata*, is also an effective inhibitor of OAT3 (Li et al. 2020). The molecular structures of representative OAT-inhibiting phenylpropanoids are shown in Fig. 4.9.

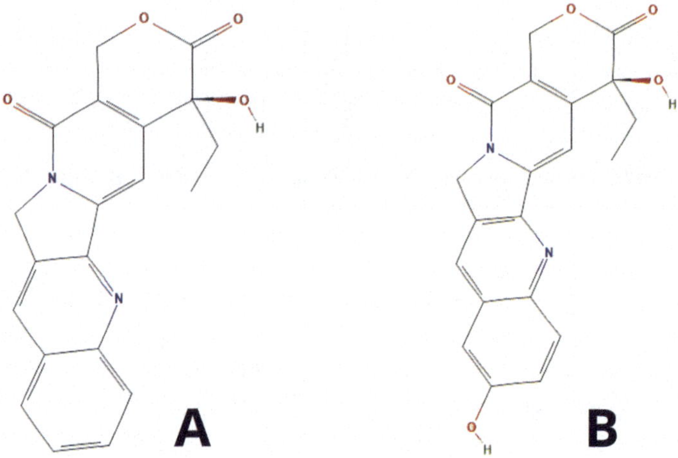


**Fig. 4.10** Molecular structures of some OAT-inhibiting terpenes. (**a**) Ursolic acid. (**b**) 3-oxo-16α-hydroxy-olean-12-en-28β-oic acid

### 2.3.7 Terpenes

*Punica granatum*, pomegranate, is an ancient and mystical fruit that has been widely consumed by various cultural systems for centuries to treat acidosis, hemorrhage, diarrhea, helminthiasis, and microbial infections (Maphetu et al. 2022). Coadministration of ursolic acid, the main pentacyclic triterpenoid constituent of *P. granatum* with other drugs, could result in altered drug pharmacokinetics, possibly due to competing for transporter proteins. Li et al. (2014) showed that ursolic acid from pomegranate significantly inhibits transport mediated by OAT3 and OATP2B1. Another triterpenoid from *Eclipta prostrata* L., 3-oxo-16α-hydroxy-olean-12-en-28β-oic acid, can act as an OAT3 competitive inhibitor (Zhu et al. 2020). The molecular structures of representative OAT-inhibiting terpenes are shown in Fig. 4.10.

## 3 Conclusions

Common problems associated with herb-drug interactions are (a) the lack of standardization of herbal drugs, (b) the lack of randomized clinical trials regarding the majority of these products, (c) the unawareness of risks by the patients who buy and use them, (d) the problem of underreporting, (e) commonly used herbal products and slimming aids and their potential hepatotoxicity mechanisms, (f) the causality assessment necessary for a correct diagnosis, and (g) the clinical patterns for which these products seem to be responsible (Licata et al. 2013). Numerous interactions between herbal medicines and conventional drugs have been documented. While the significance of many interactions is uncertain, several interactions, particularly those with St John's wort, may have serious clinical consequences (Izzo 2004; Mouly et al. 2017).

Bioactivation of herbal compounds is of main importance in terms of toxicity induction. Reactive intermediates formed by induction may bind covalently to proteins and DNA affecting toxicity to organs or inhibiting various metabolizing enzymes and drug transporters. These resulting metabolites can modify the heme, apoprotein, or both by covalent binding showing that this mechanism can explain some reported herb-drug interactions (Zhou et al. 2004b; Sharma et al. 2022). Active substances and cellular molecules should be considered as targets that can give evidence of the pharmacological activity of herbal medicines. Genomic approaches such as high-density microarray analysis are powerful tools for predicting and identifying genes that can be repressed or activated by herbal compounds (Zhou et al. 2004b; Wang et al. 2023).

There should be better safety and quality control of herbal medicines that require the following topics: (a) pharmacovigilance, (b) research, (c) regulatory control, and (d) better communication between patients and health professionals. Licata et al. (2013) report that regulations and pharmacovigilance of herbal drugs must be monitored and standardized. In many countries, herbals are sold as food supplements, which cannot be controlled by drug regulatory agencies (Sanghvi et al. 2023). Many herbal drugs are also commonly used in children and women during pregnancy and lactation because they are believed to be harmless (Kaygusuz et al. 2021).

Regarding drug/herb-drug interactions mediated by OAT3 (DDIs/HDIs), it is necessary to mention their possible risks and benefits in clinical practice. First, drug efficacy can be improved; however, nephrotoxicity can appear as greater drug depositions accumulate in rinses (Chen et al. 2023). Second, being able to identify natural active inhibitors of OAT3 can be used to make variations and avoid possible interactions; its regulatory pathways involved in these processes may also lead to the development of new methods to prevent drug-drug interactions/herb-drug interactions (DDIs/HDIs) mediated by OAT3. There are possible interactions with unreported medications. Finally, in vitro studies are the most practiced methods for researching natural OAT3 inhibitors. Therefore, more in vivo studies are needed to analyze the effects of specific natural compounds and their mechanisms of action.

## 4 Self-Assessment Questions and Answers

### *4.1 Questions*

1. Potential consequences of herb-drug interactions for patients include all except?

   A. Reduced drug efficacy.
   B. Increased drug toxicity.
   C. New adverse effects.
   D. Changed drug structure.

2. Which of the following statements is correct?

    A. ATP-binding cassette and solute carrier transporters are responsible for the absorption, distribution, and elimination of various substances.
    B. OAT3 is a specific transporter of alkaloids.
    C. OAT3 transports organic anions from urine through the apical membrane into the blood through the basolateral membrane.
    D. Tryptanthrin is an anthraquinone that potentiates OAT3-mediated substrate uptake.

3. Name an herb known for improving cognitive function but should be avoided with antiplatelet or anticoagulant drugs due to its effects on CYP enzymes.

    A. *Ginkgo biloba.*
    B. *Atropa belladonna.*
    C. *Curcumin longa.*
    D. *Hypericum perforatum.*

4. Why should patients with diabetes or those using anticoagulant drugs avoid consuming *Zingiber officinalis*?

    A. *Zingiber officinalis* reduces the metabolism of corticosteroids by inhibiting the 11β-hydroxysteroid dehydrogenase enzyme.
    B. *Zingiber officinalis* inhibits the activity of the ATP-binding cassette (ABC family).
    C. *Zingiber officinalis has anticoagulant action and modulates serum glucose levels.*
    D. *Zingiber officinalis* inhibits neurotransmitters such as serotonin, norepinephrine, dopamine, glutamate, and γ-aminobutyric acid.

5. The clinical management manual for transplant patients does not recommend using Hypericum by transplant patients. Check the alternative with the drug that can interact with Hypericum.

    A. *Amitriptyline.*
    B. *Indinavir.*
    C. *Dexamethasone.*
    D. *Cyclosporin.*

## 4.2 Answers

1. The correct answer is D. Changed drug structure: This option is incorrect because herb-drug interactions typically do not alter the chemical structure of the drug molecule itself. Instead, they affect how the drug is processed or utilized in the body.
2. The correct statement is A. ATP-binding cassette and solute carrier transporters are responsible for the absorption, distribution, and elimination of various sub-

stances. Option B—OAT3 is not specific to alkaloids; it is a transporter primarily for organic anions, not limited to alkaloids; Option C—OAT3 transports organic anions from urine through the apical membrane into the blood through the basolateral membrane—incorrect. OAT3 transports organic anions from the blood (basolateral side) into the urine (apical side) across renal tubular cells, not vice versa; Option D—Tryptanthrin is an anthraquinone that potentiates OAT3-mediated substrate uptake—incorrect. Tryptanthrin is not an anthraquinone; it is an alkaloid. Additionally, it inhibits OAT3-mediated substrate uptake rather than potentiating it.

3. The correct answer is A. *Ginkgo biloba* is known for improving cognitive function. Still, it should be avoided with antiplatelet or anticoagulant drugs due to its effects on CYP enzymes, specifically CYP2C9 and CYP3A4, which are involved in the metabolism of these medications. This interaction can potentially lead to increased bleeding risk when used together with antiplatelet or anticoagulant drugs. Therefore, it is important for patients using these medications to avoid concurrent use of *Ginkgo biloba* to prevent adverse effects.
4. The correct answer is C. *Zingiber officinalis* is known for its anticoagulant action and ability to modulate serum glucose levels. The interaction between these classes of drugs can impair coagulation and plasma glucose levels.
5. The correct answer is D. Cyclosporine, which is an immunosuppressive agent used to treat organ rejection post-transplant. *Hypericum perforatum* decreases the plasma concentration of cyclosporine, which can lead to organ rejection.

## References

Abdelkawy KS, Abdelaziz RM, Abdelmageed AM, Donia AM, El-Khodary NM. Effects of green tea extract on atorvastatin pharmacokinetics in healthy volunteers. Eur J Drug Metab Pharmacokinet. 2020;45(3):351–60.

Agbabiaka TB, Spencer NH, Khanom S, Goodman C. Prevalence of drug-herb and drug-supplement interactions in older adults: a cross-sectional survey. Br J Gen Pract. 2018;68:e711–7.

Alves de Almeida AC, de-Faria FM, Dunder RJ, Manzo LP, Souza-Brito AR, Luiz-Ferreira A. Recent trends in pharmacological activity of alkaloids in animal colitis: potential use for inflammatory bowel disease. Evid Based Complement Alternat Med. 2017;2017:8528210.

Bailey DG, Dresser GK, Leake BF, Kim RB. Naringin is a major and selective clinical inhibitor of organic anion-transporting polypeptide 1A2 (OATP1A2) in grapefruit juice. Clin Pharmacol Ther. 2007;81:495–502.

Bendjedid S, Lekmine S, Tadjine A, Djelloul R, Bensouici C. Analysis of phytochemical constituents, antibacterial, antioxidant, photoprotective activities and cytotoxic effect of leaves extracts and fractions of *Aloe vera*. Biocatal Agric Biotechnol. 2021;2021:33.

Borrelli F, Izzo AA. Herb-drug interactions with St John's Wort (Hypericum Perforatum): an update on clinical observations. AAPS J. 2009;11:710–27.

Bush KT, Wu W, Lun C, Nigam SK. The drug transporter OAT3 (SLC22A8) and endogenous metabolite communication via the gut-liver-kidney axis. J Biol Chem. 2017;292:15789–803.

Campos MG, China M, Cláudio M, Capinha M, Torres R, Oliveira S, Fortuna A. Drug-cannabinoid interactions in selected therapeutics for symptoms associated with epilepsy, autism spectrum disorder, cancer, multiple sclerosis, and pain. Pharmaceuticals (Basel). 2024;17:613.

Cao Y, Shi Y, Cai Y, Hong Z, Chai Y. The effects of traditional chinese medicine on P-glycoprotein-mediated multidrug resistance and approaches for studying the herb-P-glycoprotein interactions. Drug Metab Dispos. 2020;48(10):972–9.

Catalano A, Ceramella J, Iacopetta D, Marra M, Conforti F, Lupi FR, Gabriele D, Borges F, Sinicropi MS. *Aloe vera*-an extensive review focused on recent studies. Food Secur. 2024;13(13):2155.

Chang TKH. Activation of pregnane X receptor (PXR) and constitutive androstane receptor (CAR) by herbal medicines. AAPS J. 2009;11:590–601.

Chang TT, Chen YA, Li SY, Chen JW. Nrf-2 mediated heme oxygenase-1 activation contributes to the anti-inflammatory and renal protective effects of Ginkgo biloba extract in diabetic nephropathy. J Ethnopharmacol. 2021;266:113474.

Chavez ML, Jordan MA, Chavez PI. Evidence- based drug–herbal interactions. Life Sci. 2006;78(18):2146–57. Natureceuticals (natural products), nutraceuticals, herbal botanicals, and psychoactives: drug discovery and drug-drug interactions.

Chen F, Li L, Tian DD. Salvia miltiorrhiza Roots against Cardiovascular Disease: Consideration of Herb-Drug Interactions BioMed Research Int. 2017; 2017:9868694.

Chen Y, Li H, Wang K, Wang Y. Recent advances in synthetic drugs and natural actives interacting with OAT3. Molecules. 2023;28:4740.

Cheng Z, Zhang Z, Han Y, Wang J, Wang Y, Chen X, Shao Y, Cheng Y, Zhou W, Lu X, Wu Z. A review on anti-cancer effect of green tea catechins. J Funct Foods. 2020;74:104172.

Choi MK, Song IS. Interactions of ginseng with therapeutic drugs. Arch Pharm Res. 2019;42:862–78.

Chung S, Kim GH. Urate transporters in the kidney: what clinicians need to know. Electrolyte Blood Press. 2021;19:1–9.

Crawford C, Boyd C, Avula B, Wang YH, Khan IA, Deuster PA. A public health issue: dietary supplements promoted for brain health and cognitive performance. J Altern Complement Med. 2020;26:265–72.

Cui Y, Cheng Y, & Guo Y, Xie Y & Yao W, & Zhang W, & Qian, H. Evaluating the hepatoprotective efficacy of Aloe vera polysaccharides against subchronic exposure of aflatoxins B1. J Taiwan Inst Chem Eng. 2017;76.

D'Alessandro C, Benedetti A, Di Paolo A, Giannese D, Cupisti A. Interactions between food and drugs, and nutritional status in renal patients: a narrative review. Nutrients. 2022;14:212.

Djuv A, Nilsen OG. *Aloe vera* juice: IC(5)(0) and dual mechanistic inhibition of CYP3A4 and CYP2D6. Phytother Res. 2012;26(3):445–51.

Egeland EJ, Witczak BJ, Zaré HK, Christensen H, Åsberg A, Robertsen I. Chronic inhibition of CYP3A is temporarily reduced by each hemodialysis session in patients with end-stage renal disease. Clin Pharmacol Ther. 2020;108(4):866–73.

El-Saber Batiha G, Magdy Beshbishy A, El-Mleeh A, Abdel-Daim MM, Prasad Devkota H. Traditional uses, bioactive chemical constituents, and pharmacological and toxicological activities of *Glycyrrhiza glabra* L. (Fabaceae). Biomolecules. 2020;10(3):352.

Fahimi F, Khodadad K, Amini S, Naghibi F, Salamzadeh J, Baniasadi S. Evaluating the effect of zingiber officinalis on nausea and vomiting in patients receiving Cisplatin based regimens. Iran J Pharm Res. 2011 Spring;10(2):379–84.

Fajemiroye JO, da Silva EF, de Oliveira TS, de Oliveira LP, Akanmu MA, Ghedini PC, Pedrino GR, Costa EA. Hypotensive and vasorelaxant effects of (E) – methyl isoeugenol: a naturally occurring food flavour. Food Chem Toxicol. 2014;70:214–21. https://doi.org/10.1016/j.fct.2014.05.007. Epub 2014 May 16. PMID: 24842837.

Fajemiroye JO, da Silva DM, de Oliveira DR, Costa EA. Treatment of anxiety and depression: medicinal plants in retrospect. Fundam Clin Pharmacol. 2016;30(3):198–215. https://doi.org/10.1111/fcp.12186. Epub 2016 Mar 4. PMID: 26851117.

Fang DN, Zheng CW, Ma YL. Effectiveness of *Scutellaria baicalensis* Georgi root in pregnancy-related diseases: a review. J Integr Med. 2023;21(1):17–25.

Freeman B, Mamallapalli J, Bian T, Ballas K, Lynch A, Scala A, Huo Z, Fredenburg KM, Bruijnzeel AW, Baglole CJ, Lu J, Salloum RG, Malaty J, Xing C. Opportunities and challenges of Kava in lung cancer prevention. Int J Mol Sci. 2023;24(11):9539.

Füessl L, Kreuzer L, Nierychlewski K, Seibt T, Stangl MJ, Koliogiannis D, Meiser B, Schwarz M, Fischereder M, Kemmner S. The use of extended-release tacrolimus twice a day might be beneficial for selected kidney transplant recipients: a case report. Front Med (Lausanne). 2024;11:1336035.

Fugh-Berman A. Herb-drug interactions. Lancet. 2000;355(9198):134–8.

Glaeser H, Bailey DG, Dresser GK, Gregor JC, Schwarz UI, McGrath JS, Jolicoeur E, Lee W, Leake BF, Tirona RG, Kim RB. Intestinal drug transporter expression and the impact of grapefruit juice in humans. Clin Pharmacol Ther. 2007;81:362–70.

Gouws C, Hamman JH. What are the dangers of drug interactions with herbal medicines? Expert Opin Drug Metab Toxicol. 2020;16:165–7.

Hagos FT, Daood MJ, Ocque JA, Nolin TD, Bayir H, Poloyac SM, Kochanek PM, Clark RS, Empey PE. Probenecid, an organic anion transporter 1 and 3 inhibitor, increases plasma and brain exposure of N-acetylcysteine. Xenobiotica. 2017;47:346–53.

Hashiguchi M, Ohta Y, Shimizu M, Maruyama J, Mochizuki M. Meta-analysis of the efficacy and safety of Ginkgo biloba extract for the treatment of dementia. J Pharm Health Care Sci. 2015;1(1):14.

Henkel L, Jehn U, Thölking G, Reuter S. Tacrolimus-why pharmacokinetics matter in the clinic. Front Transplant. 2023;2:1160752.

Hermann R, von Richter O. Clinical evidence of herbal drugs as perpetrators of pharmacokinetic drug interactions. Planta Med. 2012;78:1458–77.

Hong SS, Seo K, Lim SC, Han HK. Interaction characteristics of flavonoids with human organic anion transporter 1 (hOAT1) and 3 (hOAT3). Pharmacol Res. 2007;56(6):468–73.

Hossain S, Yousaf M, Liu Y, Chang D, Zhou X. An overview of the evidence and mechanism of drug-herb interactions between propolis and pharmaceutical drugs. Front Pharmacol. 2022;13:876183.

Hu Z, Yang X, Ho PC, Chan SY, Heng PW, Chan E, Duan W, Koh HL, Zhou S. Herb-drug interactions: a literature review. Drugs. 2005;65(9):1239–82.

Huang C, Xia F, Xue L, Liu L, Bian Y, Jin Z, Miao L. Coadministration of vindesine with high-dose methotrexate therapy increases acute kidney injury via BCRP, MRP2, and OAT1/OAT3. Cancer Chemother Pharmacol. 2020;85(2):433–41.

Huo X, Meng Q, Wang C, Wu J, Zhu Y, Sun P, Ma X, Sun H, Liu K. Targeting renal OATs to develop renal protective agent from traditional Chinese medicines: protective effect of Apigenin against Imipenem-induced nephrotoxicity. Phytother Res. 2020;34(11):2998–3010.

Izzo AA. Drug interactions with St. John's Wort (*Hypericum perforatum*): a review of the clinical evidence. Int J Clin Pharmacol Ther. 2004;42(3):139–48.

Izzo AA, Ernst E. Interactions between herbal medicines and prescribed drugs: a systematic review. Drugs. 2001;61(15):2163–75. https://doi.org/10.2165/00003495-200161150-00002.

Izzo AA, Ernst E. Interactions between herbal medicines and prescribed drugs: an updated systematic review. Drugs. 2009;69(13):1777–98.

Jia Y, Liu Z, Wang C, Meng Q, Huo X, Liu Q, Sun H, Sun P, Yang X, Ma X, Liu K. P-gp, MRP2 and OAT1/OAT3 mediate the drug-drug interaction between resveratrol and methotrexate. Toxicol Appl Pharmacol. 2016;306:27–35.

Kawasaki T, Takeichi Y, Tomita M, Uwai Y, Epifano F, Fiorito S, Taddeo VA, Genovese S, Nabekura T. Effects of phenylpropanoids on human organic anion transporters hOAT1 and hOAT3. Biochem Biophys Res Commun. 2017;489(4):375–80.

Kawasaki T, Kondo M, Hiramatsu R, Nabekura T. (-)-Epigallocatechin-3-gallate inhibits human and rat renal organic anion transporters. ACS Omega. 2021;6(6):4347–54.

Kaygusuz M, Gümüştakım RŞ, Kuş C, İpek S, Tok A. TCM use in pregnant women and nursing mothers: a study from Turkey. Complement Ther Clin Pract. 2021;42:101300.

Ke J, Li MT, Huo YJ, Cheng YQ, Guo SF, Wu Y, Zhang L, Ma J, Liu AJ, Han Y. The synergistic effect of *Ginkgobiloba* extract 50 and aspirin against platelet aggregation. Drug Des Devel Ther. 2021;14:3543–60.

Khan N, Mukhtar H. Tea and health: studies in humans. Curr Pharm Des. 2013;19(34):6141–7.

Kwon YJ, Son DH, Chung TH, Lee YJ. A review of the pharmacological efficacy and safety of licorice root from corroborative clinical trial findings. J Med Food. 2020;23:12–20.

Li Z, Wang K, Zheng J, Cheung FS, Chan T, Zhu L, Zhou F. Interactions of the active components of Punica granatum (pomegranate) with the essential renal and hepatic human solute carrier transporters. Pharm Biol. 2014;52:1510–7.

Li ZM, Xu SW, Liu PQ. *Salvia miltiorrhiza* Burge (Danshen): a golden herbal medicine in cardiovascular therapeutics. Acta Pharmacol Sin. 2018;39(5):802–24.

Li X, Qiao Y, Wang X, Ma R, Li T, Zhang Y, Borris RP. Dihydrophenanthrenes from *Juncus effusus* as inhibitors of OAT1 and OAT3. J Nat Prod. 2019;82:832–9.

Li C, Wang X, Bi Y, Yu H, Wei J, Zhang Y, Han L, Zhang Y. Potent inhibitors of organic anion transporters 1 and 3 from natural compounds and their protective effect on aristolochic acid nephropathy. Toxicol Sci. 2020;175:279–91.

Li Z, Tian S, Wu Z, Xu X, Lei L, Li Y, Wang B, Huang Y. Pharmacokinetic herb-disease-drug interactions: effect of *ginkgo biloba* extract on the pharmacokinetics of pitavastatin, a substrate of Oatp1b2, in rats with non-alcoholic fatty liver disease. J Ethnopharmacol. 2021;280:114469.

Liao H, Ye J, Gao L, Liu Y. The main bioactive compounds of *Scutellaria baicalensis* Georgi. for alleviation of inflammatory cytokines: a comprehensive review. Biomed Pharmacother. 2021;133:110917.

Licata A, Macaluso FS, Craxi A. Herbal hepatotoxicity: a hidden epidemic. Intern Emerg Med. 2013;8(1):13–22. https://doi.org/10.1007/s11739-012-0777-x.

Lin K, Kong X, Tao X, Zhai X, Lv L, Dong D, Yang S, Zhu Y. Research methods and new advances in drug-drug interactions mediated by renal transporters. Molecules. 2023;28(13):5252.

Liu Q, Wang C, Meng Q, Huo X, Sun H, Peng J, Ma X, Sun P, Liu K. MDR1 and OAT1/OAT3 mediate the drug-drug interaction between puerarin and methotrexate. Pharm Res. 2014;31:1120–32.

Lu H, Lu Z, Li X, Li G, Qiao Y, Borris RP, Zhang Y. Interactions of 172 plant extracts with human organic anion transporter 1 (SLC22A6) and 3 (SLC22A8): a study on herb-drug interactions. PeerJ. 2017;5:e3333.

Luiz-Ferreira A, Pacifico T, Cruz ÁC, Laudisi F, Monteleone G, Stolfi C. TRAIL-sensitizing effects of flavonoids in cancer. Int J Mol Sci. 2023;24:16596.

Luo M, Zheng Y, Tang S, Gu L, Zhu Y, Ying R, Liu Y, Ma J, Guo R, Gao P, Zhang C. Radical oxygen species: an important breakthrough point for botanical drugs to regulate oxidative stress and treat the disorder of glycolipid metabolism. Front Pharmacol. 2023;14:1166178.

Lynch T, Price A. The effect of cytochrome P450 metabolism on drug response, interactions, and adverse effects. Am Fam Physician. 2007;76(3):391–6.

Ma L, Zhao L, Hu H, Qin Y, Bian Y, Jiang H, Zhou H, Yu L, Zeng S. Interaction of five anthraquinones from rhubarb with human organic anion transporter 1 (SLC22A6) and 3 (SLC22A8) and drug-drug interaction in rats. J Ethnopharmacol. 2014;153:864–71.

Ma L, Qin Y, Shen Z, Hu H, Zhou H, Yu L, Jiang H, Zeng S. Time-dependent inhibition of hOAT1 and hOAT3 by anthraquinones. Biol Pharm Bull. 2015;38(7):992–5.

Mandery K, Bujok K, Schmidt I, Keiser M, Siegmund W, Balk B, König J, Fromm MF, Glaeser H. Influence of the flavonoids apigenin, kaempferol, and quercetin on the function of organic anion transporting polypeptides 1A2 and 2B1. Biochem Pharmacol. 2010;80(11):1746–53.

Maphetu N, Unuofin JO, Masuku NP, Olisah C, Lebelo SL. Medicinal uses, pharmacological activities, phytochemistry, and the molecular mechanisms of *Punica granatum* L. (pomegranate) plant extracts: a review. Biomed Pharmacother. 2022;153:113256.

Martins A, Maia E, Praça I. Herb–Drug Interactions: A Holistic Decision Support System in Healthcare. In Proceedings of the 2022 IEEE International Conference on E-health Networking, Application & Services (HealthCom), Genoa, Italy, 17–19 October 2022; pp. 1–6

Minh TN, Van TM, Khanh TD, Xuan TD. Isolation and identification of constituents exhibiting antioxidant, antibacterial, and antihyperuricemia activities in *Piper methysticum* root. Foods. 2022;11(23):3889.

Mota KS, Dias GE, Pinto ME, Luiz-Ferreira A, Souza-Brito AR, Hiruma-Lima CA, Barbosa-Filho JM, Batista LM. Flavonoids with gastroprotective activity. Molecules. 2009;14(3):979–1012.

Mouly S, Lloret-Linares C, Sellier PO, Sene D, Bergmann JF. Is the clinical relevance of drug-food and drug-herb interactions limited to grapefruit juice and Saint-John's Wort? Pharmacol Res. 2017;118:82–92.

Nascimento ML, do Nascimento SB, Lima ESP, de Oliveira FM, Dos Santos RR, Cesar IDC, de Castro WV. Evaluation of the effects of extracts containing *Valeriana officinalis* and *Piper methysticum* on the activities of cytochrome P450 3A and P-Glycoprotein. Planta Med. 2024;90(10):792–800.

Nguyen MA, Staubach P, Wolffram S, Langguth P. Effect of single-dose and short-term administration of quercetin on the pharmacokinetics of talinolol in humans – implications for the evaluation of transporter-mediated flavonoid-drug interactions. Eur J Pharm Sci. 2014;61:54–60.

Ni Y, Duan Z, Zhou D, Liu S, Wan H, Gui C, Zhang H. Identification of structural features for the inhibition of OAT3-mediated uptake of enalaprilat by selected drugs and flavonoids. Front Pharmacol. 2020;11:802.

Nigam SK. What do drug transporters really do? Nat Rev Drug Discov. 2015;14:29–44.

Nigam SK. The SLC22 transporter family: a paradigm for the impact of drug transporters on metabolic pathways, signaling, and disease. Annu Rev Pharmacol Toxicol. 2018;58:663–87.

Nigam SK, Bush KT, Martovetsky G, Ahn SY, Liu HC, Richard E, Bhatnagar V, Wu W. The organic anion transporter (OAT) family: a systems biology perspective. Physiol Rev. 2015;95:83–123.

Nyulas KI, Simon-Szabó Z, Pál S, Fodor MA, Dénes L, Cseh MJ, Barabás-Hajdu E, Csipor B, Szakács J, Preg Z, Germán-Salló M, Nemes-Nagy E. Cardiovascular effects of herbal products and their interaction with antihypertensive drugs-comprehensive review. Int J Mol Sci. 2024;25(12):6388.

Orellana-Paucar AM. Steviol glycosides from *Stevia rebaudiana*: an updated overview of their sweetening activity, pharmacological properties, and safety aspects. Molecules. 2023;28(3):1258.

Ornano L, Venditti A, Donno Y, Sanna C, Ballero M, Bianco A. Phytochemical analysis of non-volatile fraction of *Artemisia caerulescens* subsp. densiflora (Viv.) (Asteraceae), an endemic species of La Maddalena Archipelago (Sardinia–Italy). Nat Prod Res. 2016;30:920–5.

Otani N, Ouchi M, Hayashi K, Jutabha P, Anzai N. Roles of organic anion transporters (OATs) in renal proximal tubules and their localization. Anat Sci Int. 2017;92:200–6.

Oyanna VO, Clarke JD. Mechanisms of intestinal pharmacokinetic natural product-drug interactions. Drug Metab Rev. 2024;12:1–17.

Pang X, Wang LM, Zhang YC, Kang LP, Yu HS, Fan GW, Han LF. New anthraquinone and eurotinone analogue from the seeds of *Senna obtusifolia* and their inhibitory effects on human organic anion transporters 1 and 3. Nat Prod Res. 2019;33(23):3409–16.

Parvez K, Rishi V. Herb-drug interactions and hepatotoxicity. Curr Drug Metab. 2019;20:03.

Parvez MM, Shin HJ, Jung JA, Shin JG. Evaluation of Para-aminosalicylic acid as a substrate of multiple solute carrier uptake transporters and possible drug interactions with nonsteroidal anti-inflammatory drugs in vitro. Antimicrob Agents Chemother. 2017;61:e02392–16.

Pastorino G, Cornara L, Soares S, Rodrigues F, Oliveira MBPP. Liquorice (*Glycyrrhiza glabra*): a phytochemical and pharmacological review. Phytother Res. 2018;32:2323–39.

Périco LL, Emílio-Silva MT, Ohara R, Rodrigues VP, Bueno G, Barbosa-Filho JM, Rocha LRMD, Batista LM, Hiruma-Lima CA. Systematic analysis of monoterpenes: advances and challenges in the treatment of peptic ulcer diseases. Biomol Ther. 2020;10:265.

Posadzki P, Watson LK, Ernst E. Adverse effects of herbal medicines: an overview of systematic reviews. Clin Med. 2013;13:7–12.

Qin XL, Li JL, Wang SH, Chen X, Huang M, Bi HC. Co-administration of Wuzhi tablet (*Schisandra sphenanthera* extract) alters tacrolimus pharmacokinetics in a dose- and time-dependent manner in rats. J Ethnopharmacol. 2020;263:113233.

Qu J, Bian R, Liu B, Chen J, Zhai J, Teng F, Guo W, Wei H. The pharmacokinetic study of tacrolimus and Wuzhi capsule in Chinese liver transplant patients. Front Pharmacol. 2022;13:956166.

Ramanathan MR, Penzak SR. Pharmacokinetic drug interactions with Panax ginseng. Eur J Drug Metab Pharmacokinet. 2017;42:545–57.

Roth M, Timmermann BN, Hagenbuch B. Interactions of green tea catechins with organic anion-transporting polypeptides. Drug Metab Dispos. 2011;39(5):920–6.

Sanghvi K, Subraya CK, Pai V, Nanjundaiah ARHH, Kunhikatta V. Pharmacovigilance in herbal drugs: a challenge. Curr Drug Saf. 2023;18(2):138–42.

Sato Y, Sasaki T, Takahashi S, Kumagai T, Nagata K. Development of a highly reproducible system to evaluate inhibition of cytochrome P450 3A4 activity by natural medicines. J Pharm Pharm Sci. 2015;4:316–27.

Shams T, Lu X, Zhu L, Zhou F. The inhibitory effects of five alkaloids on the substrate transport mediated through human organic anion and cation transporters. Xenobiotica. 2018;48:197–205.

Sharifi-Rad M, Nazaruk J, Polito L, Morais-Braga MFB, Rocha JE, Coutinho HDM, Salehi B, Tabanelli G, Montanari C, Del Mar Contreras M, Yousaf Z, Setzer WN, Verma DR, Martorell M, Sureda A, Sharifi-Rad J. Matricaria genus as a source of antimicrobial agents: from farm to pharmacy and food applications. Microbiol Res. 2018;215:76–88.

Sharma AK, Kapoor VK, Kaur G. Herb-drug interactions: a mechanistic approach. Drug Chem Toxicol. 2022;45(2):594–603.

Shim D, Bak Y, Choi HG, Lee S, Park SC. Effects of Panax species and their bioactive components on allergic airway diseases. J Ginseng Res. 2024;48(4):354–65.

Soleymani S, Bahramsoltani R, Rahimi R, Abdollahi M. Clinical risks of St John's Wort (*Hypericum perforatum*) co-administration. Expert Opin Drug Metab Toxicol. 2017;13(10):1047–62.

Takiyama M, Matsumoto T, Kaifuchi N, Mizuhara Y, Warabi E, Ohbuchi K, Mizoguchi K. *In vitro* assessment of the inhibitory effect of goreisan extract and its ingredients on the P-glycoprotein drug transporter and cytochrome P-450 metabolic enzymes. Xenobiotica. 2022;52(5):511–9.

Tan ML, Yoshida K, Zhao P, Zhang L, Nolin TD, Piquette-Miller M, Galetin A, Huang SM. Effect of chronic kidney disease on nonrenal elimination pathways: a systematic assessment of CYP1A2, CYP2C8, CYP2C9, CYP2C19, and OATP. Clin Pharmacol Ther. 2018;103(5):854–67.

Tatrai P, Erdo F, Krajcsi P. Role of hepatocyte transporters in drug-induced liver injury (DILI)-*in vitro* testing. Pharmaceutics. 2022;15:29.

Thembane N, Hlatshwayo S, Ngcobo M, Ngubane P, Gqaleni N. Review on the anti-hyperglycemic potential of *Psidium guajava* and *Seriphium plumosum* L. Plants (Basel). 2024;13:1608.

Ustuner Top F, Ekim A, Ozdemir Alkanat H. Use of complementary and alternative medicine in pediatric respiratory diseases. Holist Nurs Pract. 2021;35(2):92–7.

Venditti A, Guarcini L, Bianco A, Rosselli S, Bruno M, Senatore F. Phytochemical analysis of *Achillea ligustica* all. From Lipari Island (Aeolian islands). Nat Prod Res. 2016;30:912–9.

Venditti A, Frezza C, Sciubba F, Serafini M, Bianco A, Cianfaglione K, Maggi F. Volatile components, polar constituents and biological activity of tansy daisy (*Tanacetum macrophyllum* (Waldst. et Kit.) Schultz Bip.). Ind Crop Prod. 2018;118:225–35.

Volgin A, Yang L, Amstislavskaya T, Demin K, Wang D, Yan D, Wang J, Wang M, Alpyshov E, Hu G, et al. DARK classics in chemical neuroscience: Kava. ACS Chem Neurosci. 2020;11:3893–904.

Vollono L, Falconi M, Gaziano R, Iacovelli F, Dika E, Terracciano C, Bianchi L, Campione E. Potential of curcumin in skin disorders. Nutrients. 2019;11(9):2169.

Wang L, Sweet DH. Active hydrophilic components of the medicinal herb *Salvia miltiorrhiza* (Danshen) potently inhibit organic anion transporters 1 (Slc22a6) and 3 (Slc22a8). Evid Based Complement Alternat Med. 2012;2012:872458.

Wang L, Sweet DH. Interaction of natural dietary and herbal anionic compounds and flavonoids with human organic anion transporters 1 (SLC22A6), 3 (SLC22A8), and 4 (SLC22A11). Evid Based Complement Alternat Med. 2013;2013:612527.

Wang MY, Qi HX, Li JJ, Xu YT, Zhang HJ. Transmembrane transport of steviol glucuronide and its potential interaction with selected drugs and natural compounds. Food Chem Toxicol. 2015;86:217–24.

Wang X, Han L, Li G, Peng W, Gao X, Klaassen CD, Fan G, Zhang Y. From the cover: identification of natural products as inhibitors of human organic anion transporters (OAT1 and OAT3) and their protective effect on mercury-induced toxicity. Toxicol Sci. 2018;161:321–34.

Wang X, Zhang X, Li J, Fu J, Zhao M, Zhang W, Weng W, Li Q. Network pharmacology and LC-MS approachs to explore the active compounds and mechanisms of Yuanjiang decoction for treating bradyarrhythmia. Comput Biol Med. 2023;152:106435.

Wang FH, Tan HX, Hu JH, Duan XY, Bai WT, Wang XB, Wang BL, Su Y, Hu JP. Inhibitory interaction of flavonoids with organic anion transporter 3 and their structure-activity relationships for predicting nephroprotective effects. J Asian Nat Prod Res. 2024;26(3):353–71.

Woon TH, Tan MJH, Kwan YH, Fong W. Evidence of the interactions between immunosuppressive drugs used in autoimmune rheumatic diseases and Chinese herbal medicine: a scoping review. Complement Ther Med. 2024;80:103017.

Xiao SQ, Wang W, Liu Y. Research progress on extraction and separation of active components from loquat leaves. Separations. 2023;10:126.

Xu F, Li Z, Zheng J, Gee Cheung FS, Chan T, Zhu L, Zhuge H, Zhou F. The inhibitory effects of the bioactive components isolated from *Scutellaria baicalensis* on the cellular uptake mediated by the essential solute carrier transporters. J Pharm Sci. 2013;102:4205–11.

Xu W, Choi H-K, Huang L. State of Panax ginseng research: a global analysis. Molecules. 2017;22:1518.

Yan L, Yang ZQ, Shi YY, Ren J, Yang CL, Wan ZL, Bai YJ, Luo LM, Wang LL, Li Y. Effects of Wuzhi capsules on blood concentration of tacrolimus in renal transplant recipients. Ann Transplant. 2019;24:594–604.

Yang L, Li CL, Tsai TH. Preclinical herb-drug pharmacokinetic interaction of *Panax ginseng* extract and selegiline in freely moving rats. ACS Omega. 2020;5(9):4682–8.

Yang L, Wang B, Ma L, Fu P. Traditional Chinese herbs and natural products in hyperuricemia-induced chronic kidney disease. Front Pharmacol. 2022;13:971032.

Ye L, Lu L, Li Y, Zeng S, Yang X, Chen W, Feng Q, Liu W, Tang L, Liu Z. Potential role of ATP-binding cassette transporters in the intestinal transport of rhein. Food Chem Toxicol. 2013;58:301–5.

Yoshitake T, Yoshitake S, Kehr J. The Ginkgo biloba extract EGb761(R) and its main constituent flavonoids and ginkgolides increase extracellular dopamine levels in the rat prefrontal cortex. Br J Pharmacol. 2010;159:659–68.

Zhang J, Wang H, Fan Y, Yu Z, You G. Regulation of organic anion transporters: role in physiology, pathophysiology, and drug elimination. Pharmacol Ther. 2021;217:107647.

Zhang F, Wu R, Liu Y, Dai S, Xue X, Li Y, Gong X. Nephroprotective and nephrotoxic effects of Rhubarb and their molecular mechanisms. Biomed Pharmacother. 2023;160:114297.

Zhao L, Zheng L. A review on bioactive anthraquinone and derivatives as the regulators for ROS. Molecules. 2023;28(24):8139.

Zheng J, Chan T, Zhu L, Yan X, Cao Z, Wang Y, Zhou F. The inhibitory effects of camptothecin (CPT) and its derivatives on the substrate uptakes mediated by human solute carrier transporters (SLCs). Xenobiotica. 2016;46:831–40.

Zhou S, Chan E, Pan S-Q, Huang M, Lee E. Pharmacokinetic interactions of drugs with St John's wort. J Psychopharmacol. 2004a;18(2):262–76. https://doi.org/10.1177/0269881104042632.

Zhou S, Koh HL, Gao Y, Gong A, Lee EJD. Herbal bioactivation: the good, the bad and the ugly. Life Sci. 2004b;74:935–68.

Zhou D, Xu Y, Wang Y, Li J, Gui C, Zhang H. Interaction of organic anion transporter 3-mediated uptake of steviol acyl glucuronide, a major metabolite of rebaudioside a, with selected drugs. J Agric Food Chem. 2020;68:1579–87.

Zhu Y, Huo X, Wang C, Meng Q, Liu Z, Sun H, Tan A, Ma X, Peng J, Liu K. Organic anion transporters also mediate the drug-drug interaction between imipenem and cilastatin. Asian J Pharm Sci. 2020;15:252–63.

Zou L, Matsson P, Stecula A, Ngo HX, Zur AA, Giacomini KM. Drug metabolites potently inhibit renal organic anion transporters, OAT1 and OAT3. J Pharm Sci. 2021;110:347–53.

# Chapter 5
# Phytotherapy Products and Active Principles

**Camila Aline Romano, José Realino de Paula, Christianah A. Elusiyan, and James O. Fajemiroye**

**Abstract** Phytotherapy, phytomedicine, or herbal medicine is a medicinal application of plants' repertoire of metabolites to substitute or complement conventional therapy with unsatisfactory therapeutic outcomes or evitable adverse effects. Plants are complex chemical industries that produce different phytochemicals/secondary metabolites with diverse structural functionalities and biological effects. Some of these metabolites often provide a modifiable structural framework to develop a series of derivatives or inspire the synthesis or repurposing of drugs. Extensive clinical evaluations are still required to provide robust evidence of the benefits and quality control of the available therapeutic products. Training of prescribers, awareness education or campaigns among the population, and rational or sustainable use of world plant biodiversity remain crucial for the effective application of phytotherapy.

**Keywords** Phytochemicals · Medicinal plants · Biological activities · Indications

---

C. A. Romano
Natural Products Research Laboratory, Faculty of Pharmacy, Federal University of Goiás, Goiania, Goias, Brazil

Insect Physiology Biology Laboratory, Institute of Tropical Pathology and Public Health, Federal University of Goiás, Goiania, Goias, Brazil
e-mail: camila.aline@ufg.br

J. R. de Paula
Natural Products Research Laboratory, Faculty of Pharmacy, Federal University of Goiás, Goiania, Goias, Brazil
e-mail: jose_realino@ufg.br

C. A. Elusiyan
Drug Research and Production Unit, Faculty of Pharmacy, Obafemi Awolowo University, Ile-Ife, Osun State, Nigeria
e-mail: caelusiyan@oauife.edu.ng

J. O. Fajemiroye (✉)
Institute of Biological Sciences, Federal University of Goias, Goiania, Goias, Brazil
e-mail: jamesfajemiroye@ufg.br

 
J. O. Fajemiroye (ed.), *Fundamentals of Drug and Non-Drug Interactions*,
https://doi.org/10.1007/978-3-031-80107-5_5

# 1 Introduction

Herbal medicine has its origins in ancient cultures and is practiced all around the world. The ancient Egyptian, Indian, and Chinese cultures provide millennial evidence documenting the origins of herbal medicines (Na 2017). Over the decades, phytotherapy has emerged through traditional means of treating various ailments (Fajemiroye et al. 2016). This form of medicine has gained more popularity due to accessibility by the population of less developed countries to traditional practices for their basic health care compared to conventional medicine (Mazzari and Prieto 2014). The World Health Organization (WHO) acknowledges and reports that on average 80% of people in many developing countries still rely heavily on traditional medicine for their primary health care (Mohamed et al. 2012). With this recognition of the therapeutic values of medicinal plants by WHO, scientific investigations were launched into repackaging them to safeguard efficacy, safety, and quality (Ricardo et al. 2015). The use and promotion of phytotherapy have continued to grow globally. To mention a few global examples, traditional practices that intertwine with rich plant and cultural diversity are often explored in Brazilian phytotherapy in the Brazilian Unified Public Health System—a leading and successful primary care provider worldwide. Around 40% and 70% of the German prescriptions and use of herbal medicines, respectively, have been reported (Santos et al. 2011; Jütte et al. 2017). The Chinese and Indians use herbs more frequently to substitute allopathic intervention. In African traditional medicine, the use of natural products to treat diseases is an age-long practice, though not in an organized system such as standardized medicaments or over-the-counter phytomedicinal products as practiced in the developed countries.

The phytotherapy or herbal medicines market has grown significantly. New Zealand's herbal medicines industry accounts for approximately USD 685 million annually (WHO 2023). The values of phytotherapy or herbs are attributable to numerous phytoconstituents in their leaves, bark, roots, flowers, fruits, stems, etc. Plants synthesize active principles through complex biochemical pathways. These pathways are involved in the conversion of simpler molecules into more complex compounds through enzymatic reactions. Some common pathways (Fig. 5.1) involved in the synthesis of active principles include the phenylpropanoid pathway to produce phenolic compounds (flavonoids, coumarins, lignin, tannis, etc.); melvonic acid leading to the formation of terpenes and terpenoidal compounds (saponins, triterpenes, carotenoids, etc.); and shikimic acid pathway for the synthesis of nitrogen-containing compounds (alkaloids, amino acids).

Initially, the function of plant secondary metabolites was unknown. It was believed that these substances were metabolic excretions. Over time, it was observed that these substances were important agents for plant protection. Given the sessile nature of plants, the only way to escape attacks by pathogens and herbivores is to use chemical defense mechanisms. It is believed that the production of these metabolites arose through mutations in the classical pathways of primary metabolism and, through natural selection, which were maintained generation after generation (Taiz

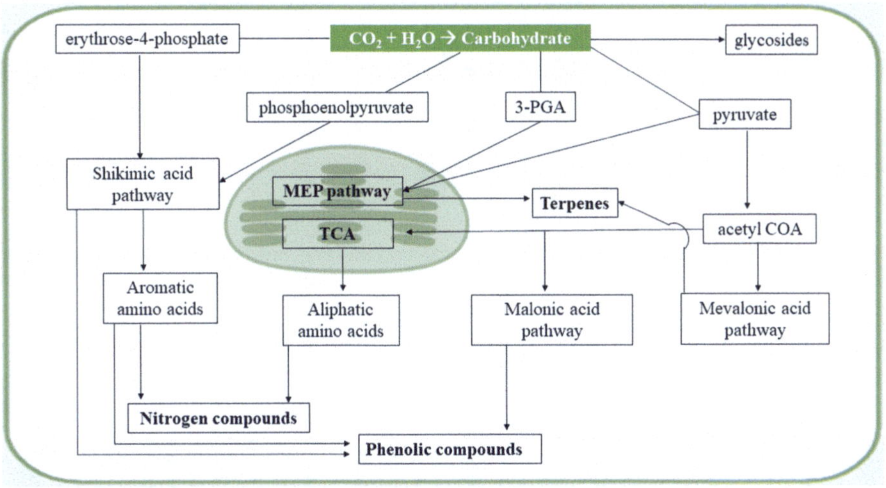


**Fig. 5.1** Representative biosynthetic pathways. Plant secondary metabolites are produced to a greater extent in the cytosol. Two exceptions are products of the tricarboxylic acid cycle and the methyl-D-erythritol phosphate (MEP) pathway, which occur in the chloroplasts. (Source: Adapted from Taiz and Zeiger (2009))

and Zeiger 2009). In addition to their function as plant defense agents against predators, harsh or extreme weather conditions, etc., secondary metabolites can also attract pollinators, and seed dispersers involved in plant-plant interactions (allelopathy). The main groups of secondary metabolites are terpenes, phenolic compounds, and nitrogen compounds. Terpenes are synthesized from acetyl-CoA or glycolytic intermediates (via MEP). They are responsible for numerous biological activities, especially as repellents, deterrents, and insecticides. They are mainly associated with the defense function against herbivory, although some aromas also function as attractants for pollinators. Phenolic compounds originate through the shikimic and malonic acid pathways. Their production is associated with defense against microorganisms. Some classes have their production activated by UV radiation (flavones and flavanols), acting as photoprotectors (Gobbo-Neto and Lopes 2007). Some classes can also confer attractive colors to both pollinators and seed dispersers (anthocyanins). Nitrogen compounds (cyanogenic alkaloids and heterosides) are synthesized from amino acids in the cytosol or vacuoles under basic pH conditions. In nature, these components are palatable. Some insect species have co-evolved closely and are not affected by these substances. Although all alkaloids are toxic in sufficient quantities, in low concentrations they may present interesting pharmacological activities (Taiz and Zeiger 2009). Many phytoconstituents are current drugs in clinical use or prototypes of drugs. Some of the main examples are shown in Table 5.1.

**Table 5.1** Some active principles and their biological activities

| Class | Chemistry | Source | Component | Pharmacological activity | References |
|---|---|---|---|---|---|
| Fatty acids | Carboxylic acids with a long chain | *Borago offIcinalis* | ɣ-linolenic acid | Anti-inflammatory | Erdinest et al. (2012) |
| | | *Oenothera* spp. | | | |
| | | *Ribes nigrum* | | | |
| Phloroglucinol derivatives | Compounds containing phloroglucin (1,3.5-trihydroxybenzene) | *Hypericum perforatum* | Hyperforin | Antidepressant | El Hamdaoui et al. (2022) |
| Purines | Methylxanthines | *Thea sinensis* | Caffeine | CNS stimulant | Fisone et al. (2004) and Reddy et al. (2024) |
| | | *Ilex paraguariensis* | | | |
| | | *Paullinia cupana* | | | |
| Alkaloids | Aporfine | *Peumus boldus* | Boldine | Choleretic and cholagogue | Schulz et al. (2002) |
| | Tropane (derivate of ornithine) | *Atropa belladonna* | Atropine | Antimuscarinic, pupil dilation, and paralysis of visual accommodation | Iribarren et al. (2024) |
| | Morphinans | *Papaver somniferum* | Morphine and codeine | Analgesia | Le Cornec et al. (2024) and Oon et al. (2024) |
| | Quinoline | *Cinchona* spp. | Quinine<br>Quinidine | Antimalarial<br>Antiarrhythmic | |
| | Vinca | *Catharanthus roseus* | Vincristine and vinblastine | Anticancer | Chikezie et al. (2015) |
| Terpenes | Triterpenoid saponins | *Panax ginseng* | Ginsenoside | Antihypertensive, antidiabetic, antistress activity | Christensen (2009) |
| | | *Bacopa monnieri* | Bacoside A | CNS protector | Saunshi et al. (2023) |
| | Terpene trilactones<br>Tetracyclic diterpene<br>Sesquiterpene lactone<br>Monoterpene | *Ginkgo biloba*<br>*Taxus brevifolia*<br>*Artemisia annua*<br>*Chrysanthemum cinerariifolium* | Ginkgolides and bilobalide<br>Paclitaxel (Taxol)<br>Artemisinin<br>Pyrethrins | Memory, learning, and neuroprotective<br>Anticancer<br>Antimalarial<br>Insecticide | Feng et al. (2019) |
| Lignoids | Dibenzylbutyrolactones | *Hydrocotyle umbellata* | Hibalactone | Anxiolytic | De Oliveira et al. (2022) |
| Glycosides | Cardiac glycoside | *Digitalis purpurea* | Digoxin | Cardiotonic | Whayne (2018) |

## 2 Phytochemical Analysis: Extraction, Fractionation, and Isolations

The phytochemical approach consists of a study that elucidates the compounds produced from the metabolic activity of plants (Matos 2009). The production of secondary metabolites in plants is associated, among others, with climate and soil conditions, as well as the availability of water resources (Gobbo-Neto and Lopes 2007) and injuries (Corrêa et al. 2008). Phytoconstituents are not equally available among plants. Their occurrence/abundance is associated with environmental, genetic (lineages and chemotypes), and phenological state factors. These conditions influence the production of certain components or classes, altering their proportions (Schulz et al. 2002).

Studies of the chemistry of plant species allow the confirmation or identification of substances present in species popularly used, for the investigation of bioactive potential and the development of new products (Matos 2009). In phytochemical research, it is essential to respect crucial steps to ensure the robustness of the results. In methodological planning, quality control of the material must consider everything from obtaining to processing the sample when the raw material is obtained directly from nature. The location of the collection must be considered. For example, materials obtained from different places may present differences in their composition (heterogeneity) known as geographical variation. However, material from the same area, or even cultivated, tends to present less significant variations (homogeneity), resulting in extracts with a certain standard in composition (Schulz et al. 2002). For commercially obtained samples, it is necessary to purchase from suppliers that offer a report on the material. In both cases, quality control of the plant drug (determination of impurities, foreign materials, moisture content, swelling, for example) is advisable to ensure that the entire process can be carried out in a way that guarantees reproducibility.

In research, it is very common to obtain plant samples directly (in situ) from known locations where they are cultivated or from the wild. In these cases, it is always important to emphasize the need for taxonomic identification of the specimens evaluated. A sample of the plant with leaves, flowers, or fruits (in some cases) is necessary for a taxonomist botanist to correctly determine the material and ensure that the collected material is the species of interest. This material must be deposited in an indexed herbarium to serve as a witness and reference for future studies (Matos 2009). Once the species and the good quality of the raw material chosen are known, the plant material will be subjected to the stages of extraction, isolation, purification (using various chromatographic techniques), chemical characterization (using the combination of different spectroscopic techniques), and biological evaluation using in vivo and in vitro test systems. In the industry, the raw material is already processed and used directly in the extraction process.

## 2.1 Obtaining Raw Materials and Producing Plant Extracts

When harvesting plant material, certain conditions must be observed so that the material is obtained in a way that is most suitable for obtaining the constituents of interest. The time of day and the season of the year are two factors to be considered. Leaves and shrubs are best collected at the beginning of flowering or during the dry season so that they can regrow during the rainy season. Stem bark is best obtained during the rainy season since it is more easily decomposed due to the high humidity. Underground parts can be obtained during the winter when they are generally dormant. Fruits should be harvested before they are fully mature, but if the interest is in obtaining seeds, the best time is when the fruits are ripe (Sharapin 2000). To obtain essential oils, it is best to collect them in the morning or late afternoon.

After harvesting, the processing of the plant material must be meticulous to guarantee the quality of the material until the extraction process. Different morphological plant parts must be collected separately with the right label and detailed information about the location, time of collection, date, and coordinates. The material must be checked for the presence of foreign material, fungi, or dead parts. After collection, the plant material must be authenticated by an expert or taxonomist. The material should preferably be dried in an oven with air circulation. This interrupts metabolic processes that can compromise the quality of the material (oxidation, hydrolysis, fermentation by microorganisms). The drying temperature depends on the material and the extraction to be performed. Roots and stem barks can withstand up to 60 °C, while samples for obtaining essential oils should not exceed 40 °C. After drying, the material should be pulverized and stored in a dry, dark place, in packaging that allows air circulation and is free from insects, rodents, and other deteriorating agents. The processed material must again be well labeled with all the necessary information for correct identification. Under adequate drying conditions, the material can be stored for an average of 12 months, a value that may vary depending on the sample (Sharapin 2000).

In the production of extracts, the choice of appropriate extracting solvent is closely linked to the constituents that are expected to be obtained. For targeted extraction, the researcher must mimic the way the locals extract the plant material to achieve the desired biological effects. Generally, solvents that are commonly used in the extraction of plant materials range from polar solvents (e.g., water and alcohols) to moderately polar solvents (e.g., acetone, ethyl acetate, dichloromethane) and non-polar solvents (e.g., n-hexane, ether, chloroform). Thus, if a greater diversity of phytoconstituents is desired, more generalist solvents can be chosen, such as alcohols; whereas when a specific substance or class is desired, organic solvents or solvent mixtures can be used, which will result in an extract with a specific composition. The two most important points in extraction are the solvent system used and the consistency in the composition of the final product (Sharapin 2000). An example is the production of the standardized extract of *Ginkgo biloba* (Ginkgoaceae), where the herb/extract ratio is respected in the range of 35:1–67:1 (plant: solvent), to a mixture of water and acetone. Under these conditions,

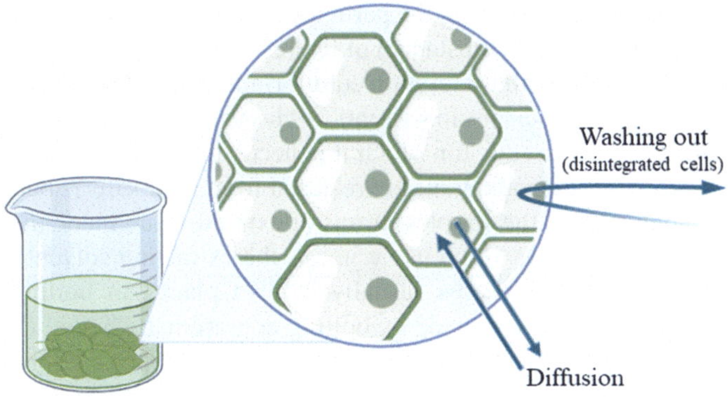


**Fig. 5.2** Penetration of the solvent into fragments of a plant. (Source: List and Schmidt (1989), with adaptation)

undesirable components are eliminated, such as Ginkgolic acids (<5 ppm), for example (Schulz et al. 2002).

The process of producing an extract can be theoretically divided into two stages (Fig. 5.2). In the first, the extracting solvent promotes the washing of the substances present in the plant cells disintegrated by the grinding/shaving process. In the second, the extracting solvent penetrates the whole cells (diffusion). These cells increase in volume, promoting the release of endocellular substances and gelation of the protoplasm (List and Schmidt 1989). The solvent inside the cell generates a dipole in the substances. Therefore, the extraction will depend on the affinity of the substance with the extracting solvent. It is possible to numerically determine the associative capacity of a solvent by the dielectric constant. The greater the polarity of the solvent (or mixture of solvents), the greater the dielectric constant. The opposite is also true. Thus, systems with a high dielectric constant attract a greater quantity of polar substances (Schulz et al. 2002). For example, for an extraction in which the objective is to obtain flavonoids or tannins, which are polar substances, an extraction system based on ethanol:water, in the proportion 7:3, may be preferable. Additionally, if the objective of the extraction is to obtain a fraction rich in tannins, an extraction solution composed of acetone:water, in the proportion 1:1, can be used. On the other hand, waxes and lipids, among other non-polar compounds, can be obtained by means of non-polar solvents, which have a low dielectric constant.

The production of an extract can be influenced by factors associated with the plant drug, such as particle size; the extracting solvent, such as the nature of the solvent and pH; and the process used, such as agitation, temperature, and extraction time. The particle size (granulometry) should be moderately coarse for most extracts. This is because, during percolation, for example, very fine powders can facilitate the formation of channels. During maceration, they can be carried with the extract, requiring additional filtration processes. On the other hand, powders with high particle size can hinder both the penetration of the solvent into the cells and the exit of the extract. Agitation during the extraction process can shift the equilibrium in the solvent saturation, displacing the particles and allowing their contact with the extractant. When the product (substance) of interest is not thermolabile, heating can favor extraction by shifting the saturation equilibrium. pH, on the other hand, influences the solubility of substances in solvents. A classic example is the production of alkaloids with organic solvents, for which it is necessary to alkalize the plant drug before extraction so that the alkaloids are released from their salts (Sharapin 2000).

In traditional practices, the simplest forms of extraction are made in the form of teas (infusion or decoction), which use water as the extracting solvent under heating. Infusions use aromatic herbs (usually at 5%) placed in boiling water for 10–20 minutes. Decoctions are made by boiling non-aromatic herbs, also at a proportion of 5%, for approximately 15 minutes. In industrial production and also in research, the most common ways of producing an extract are maceration and percolation. In maceration, the plant drug (usually at a proportion of 2%) is placed in contact with the extracting solvent for a certain period, at room temperature. The maceration time can take from 24 to 72 hours, or a few hours if the maceration is carried out under agitation. In this case, it is common to replace the extracting solvent after saturation, every 3 hours. Percolation occurs differently. In this method, the extract is obtained by leaching the plant drug using the extracting solvent. It is common to use hydroalcoholic solutions as a solvent. The flow rate is an important factor in percolation. Thus, in slow percolation, the flow rate occurs at 1 mL/min; a moderate flow rate would be between 1 and 3 mL/min. And fast flow, between 3 and 5 mL/min (Brazil 2019).

## 2.2 *Obtaining Essential Oils*

Essential oils can be extracted from aromatic plants, for which extraction methods may vary depending on the arrangement of the secretory cells. The chosen method, extraction time, and initial processing of the raw material can directly impact the chemical constitution and yield of the essential oil. Plants with parenchymatous secretory glands show better performance in the extractive process when dehydrated and pulverized. For plant species where the essential oil is concentrated in glandular trichomes, fresh extraction immediately after harvest offers better results.

The most used methods to obtain essential oils are expression, steam distillation, hydro distillation, cold extraction, solvent-mediated, and supercritical fluid. Among these, steam distillation is the most industrially used extraction method, as in addition to its low cost, it allows large-scale extraction (Da Silva et al. 2014). The method consists of a vast system with a simple boiler. Boiling water in the boiler generates steam that reaches the vegetable drug in the tank. Due to the sensitivity of volatile compounds to heat, the water vapor can extract the volatile compounds and carry them to the condenser, where there is condensation and, due to the physical-chemical characteristics, the essential oil separates from the water in a liquid state, when it is possible to obtain it. Hydro distillation also takes advantage of the principle of volatility and the loading of volatile components by steam; however in this method, the vegetable drug is mixed with water. The application of heat must be sufficient to cause rupture and breakdown of the cellular structure of the plant material and release the essential oil (Tougnuanchan et al. 2012).

The hydro distillation process is usually used in extractions where the amount of raw material is reduced, in pilot tests and laboratory scales. A disadvantage of the process is the occurrence of unidentifiable substances due to the oxidation of substances, for example, due to exposure of the plant drug in contact with water. The most used hydro distillation is that using a Clevenger extractor (Tougnuanchan et al. 2012; Fancello et al. 2016). Over the years, adaptations were incorporated, such as microwave-assisted hydro distillation. In this procedure, the thermal energy dissipation of the process is superior to traditional methods, significantly reducing the extraction time from hours to minutes (on average 30–50 minutes) (Fancello et al. 2016). Despite the temporal advantage, some reports in the literature also show an increase in the percentage of unidentified substances in essential oils obtained through microwave-assisted extraction processes (Aziz et al. 2018).

In addition to distillations, essential oils can be obtained by cold extraction through pressing, especially for *Citrus* fruits (Sun et al. 2014). This method is considered one of the oldest for extraction, it was practiced even before the discovery of the distillation process. The advantage is little or no heat generation during the process, but it offers little yield. In this process, the essential oil is removed from the pressed material with water before decantation or centrifugation. For flowers with reduced amounts of volatiles, extraction can occur by *enfleurage*. In this process, the flowers are deposited on a layer of fat until the extracting fat is saturated. Subsequently, the fat is distilled into ethanol at low temperatures. The essential oil obtained in this process is used in perfumery and has high-added commercial value (Sharapin 2000).

Due to their lipophilic nature, essential oils can also be obtained by extraction with organic solvents. The most used are diethyl ether, petroleum ether, and *n*-hexane. They can be used in simple percolation, making the extraction process longer, or in extraction assisted by microwaves and ultrasound. After extraction, all solvents must be removed from the essential oil (Silveira et al. 2012). During solvent evaporation, substances with low molecular weight can be volatilized with the solvent. Furthermore, other lipophilic materials such as fixed oils and resins can be extracted along with the essential oil, significantly modifying the aroma. Solvent-mediated extractions also have the disadvantage of the persistence of residues of the solvent used, if evaporation is not well conducted. The presence of extraction residues can compromise the essential oil's biopotential, aromatic profile, and even toxicological risks (Sharapin 2000).

Another possible method to extract essential oils is supercritical fluid extraction. This technique has become promising for industrial application, as it is a faster and more selective extraction of compounds. In addition, it is an ecologically correct technique when compared to traditional methods. The supercritical solvent of choice for the extraction of essential oils is carbon dioxide ($CO_2$), as it has low viscosity, low surface tension, high diffusivity, good density, considered non-toxic, non-flammable, low cost, and good availability. It is chemically inert under controlled conditions and gaseous at normal pressure and temperature, which eliminates the solvent evaporation step after extraction (Quitain et al. 2013; Pourmortazavi et al. 2018).

### *2.3 Fractionation and Isolation*

After obtaining the extract, which is a complex mixture of organic compounds belonging to several classes of secondary metabolites, the next phytochemical procedure is fractionation leading to the isolation of the putative phytoconstituents. Fractionation is a process of separating a plant extract into various fractions. It involves an initial selection of appropriate solvents. The selected solvents are added to the extract sequentially in the increasing order of polarity starting with the least polar, followed by the intermediate polar and the most polar solvent. Components of the extracts are distributed into the various solvents based on their polarity. Fractionation can be achieved using different methods including physical and chemical methods. Chemical methods entail the use of appropriate reagents to target the extraction of specific functional groups by chemical reactions. Physical methods are commonly used in the fractionation and eventual isolation of individual compounds. Physical methods used in the separation of plant extract mixture include solvent/solvent partitioning using a separation funnel, simple chromatographic techniques (e.g., paper chromatography, thin-layer chromatography, column chromatography), and the more sophisticated techniques (e.g., high-performance liquid chromatography and gas chromatography) (Abubakar and Haque 2020).

## 3 Phytotherapy for CNS Disorders

Psychiatric disorders such as depression and anxiety are among the leading causes of disability globally (Fajemiroye et al. 2014, 2016). The use of medicinal plants to treat these conditions is widely reported. Triterpenoid-like oleanolic acid from *Pimenta pseudocaryophyllus* exhibits multiple mechanisms of anxiolytic and antidepressant effects (Fajemiroye et al. 2014). The efficacy of *Eschscholtzia* and *Valerian* combination (tablet form) against anxiety and sleep disorders (clinical trial—NCT04812418); saffron, lemon balm, valerian, and tea extract for sleep quality enhancement (clinical trial—NCT05609890); and combination of dronabinol—a synthetic analog of tetrahydrocannabinol from *Cannabis* and palmitoylethanolamide (phase II clinical trial—NCT05239390) being assessed for sleep quality are good indicators of phytotherapy applications in CNS disorders. The preparations and formulations of *Hypericum perforatum, Valeriana officinalis, Piper methysticum, Passiflora incarnata, Humulus lupulus, Melissa officinalis, Lavandula officinalis, and Rhodiola rosea* are being used to treat these psychiatric disorders (Fajemiroye et al. 2014). Several active principles of these species participate in the mechanisms of actions of these species and the modulatory interplay of gut microbiota, neurotransmitters, receptors, enzymes, and transporters among other mediators. The anti-cholinesterase, antioxidant, anti-inflammatory, anti-aggregation, restoration of mitochondrial function, stimulation of autophagy, and anti-apoptotic effects of phytotherapy offer valuable therapeutic options for complex, multifactorial, and fatal diseases.

In the treatment of neurodegenerative diseases (Alzheimer's disease—AD, Parkinson's disease—PD, Huntington's disease, Amyotrophic lateral sclerosis, and stroke, etc.), plant formulations with chemopreventive and neuroprotective properties against inflammation, oxidative stress, protein aggregation, mitochondrial dysfunction, and apoptosis are of interest. Accumulation of misfolded proteins in the aged brain, high levels of oxidative stress-induced damage involving 8-hydroxydeoxyguanosine (DNA oxidation products in mitochondria and nuclei), metabolic loss, and synapse malfunction were reported in AD (John et al. 2022). In this neurodegenerative disease, senile plaques and neurofibrillary tangles, Aβ peptides, glutamatergic, and cholinergic pathways play a vital role in AD pathogenesis. Lower concentrations and functions of ACh as a result of cholinergic neuron degeneration often disrupt cognitive performance in AD (Francis 2005; Shoaib et al. 2023). Phytotherapy treatments of AD with *Ginkgo biloba, Centella Asiatica, Withania somnifera, Salvia officinalis, Tinospora cordifolia, Lepidum meyenii, Curcuma longa, Glycyrrhiza glabra, Magnolia officinalis, Convolvulus pluricaulis* among others often target the neurotransmitters, enzymes, receptors, and oxidative processes. The isolates of these species like resveratrol (potent antioxidant) restore neuronal cell death by mimicking neurotrophic factors and enhancing mitochondrial biogenesis. The presence of stigmasterol, β-sitosterol, and linoleic acid in *Panax ginseng* was attributed to the inhibition of the NF-κB and MAPK pathway-mediated apoptosis and inflammation. From this species, ginsenosides inhibit aggregation of Aβ and acetylcholinesterase (AChE) activities, increase the release of neurotrophic factor, and restore mitochondrial function (Shin et al. 2021; Li et al. 2022) as gintonin reduces the formation of Aβ, increases the expression of choline acetyltransferase, activates phosphatidic acid receptors against hemolysis, and regulates the G protein-coupled lysophosphatidic acid receptors with overall cholinergic and neurotrophic actions (Shin et al. 2021; Fasae et al. 2021; Kim et al. 2015). Gintonin induces the autophagic flux in astrocytes through the activation of the AMPK-mTOR regulatory pathways and an increase in intracellular [$Ca^{2+}$] regulating the release of neurotransmitters to improve cognition (Rahman et al. 2020).

A plant-derived extract ZT-1 that blocks the activity of AChE and restores ACh levels is under clinical trial (NCT00423228) for its safety and efficacy in AD. The flavonoids, alkaloids, saponins, tannins, and isothiocyanates-rich *Moringa olifera* decreased tau hyperphosphorylation and improved cognitive capacity by reducing the level of lipid peroxides (Mahaman et al. 2018; Ganguly 2005). The extracts of *Ginkgo biloba* with flavone glycosides (quercetin, kaempferol, and isorhamnetin and terpene lactones consist of A, B, and C ginkgolides and bilobalide), terpene lactones, and organic acid could improve cognitive function in AD patients. They inhibit apoptosis, reactive oxygen species (ROS) accumulation, glucose uptake, mitochondrial dysfunction, and extracellular signal-regulated kinase and c-JUN N-terminal kinase pathways through the regulation of glutathione peroxidase, catalase, and SOD activity (Essa et al. 2012; Chen et al. 2019). *Glycyrrhiza glabra* that contains glabridin, coumarins, isoflavonoids, saponins, licochalcone A, liquiritigenin, flavonoids, and stilbenoids inhibits Aβ aggregation, reactive oxygen species (ROS) release, decreased oxidative stress, mitochondrial biogenesis, GSH, and radical-scavenging activities (Tapia-Rojas et al. 2019; Anandhan et al. 2012). *Curcuma longa* with anti-amyloidogenic, antioxidant, and anti-inflammatory activities inhibits NF-kB-induced nitric oxide synthase (iNOS), cyclooxygenase (COX −2), and the formation of fibrils and Aβ aggregation (Cole et al. 2007; Ono et al. 2004). A *Pistacia* spp. inhibits Aβ aggregation, neuroinflammatory processes, and AChE activity (Moeini et al. 2019). The *Phyllanthus acidus* reduces lipid peroxidation and AChE activity and increases catalase, nicotinamide adenine dinucleotide (NAD) + hydrogen (H) (NADH) dehydrogenase, and superoxide dismutase activities (Uddin et al. 2016; Alagan et al. 2019). *Salvia lavandulaefolia* with terpenes, alpha-, and beta-pinene and 1,8-cineol, geraniol exhibits free oxygen radical neutralizing action, estrogenic activity, inflammation-reducing properties, and anti-butyryl and anti-AChE activities (Kennedy and Scholey 2006).

*Trifolium pratense* comprises many compounds of polycyclic phenols, genistein, formononetin, biochanin A, daidzein, glycitein, isoflavone segregated pratensein. The extract of this species increased the level of estrogen-inducible gene presenelin-2 (pS2) in S30 cells. Besides a reduction in the activity of the enzyme ACh breakdown, pratensein reduced the MDA, nitric oxide (NO), neuronal nitric oxide synthase (nNOS), IL-1β, and TNF-α and attenuate the manifestation of amyloid precursor protein (APP), and cathepsin B (CatB), the Aβ-synthesizing enzyme β-secretase (BACE1), the insulin-degrading enzyme, and augments synaptic plasticity (Wei et al. 2015). *Sophora flavescens* with prenylflavonoids (sophoflavescenol) lavandulyl flavonoids (Jung et al. 2011) blocks progressive glycation endproducts (AGE), BACE1, inhibits AChE, and decreases Aβ release by human embryonic kidney (HEK-293) cells. Estrogen exhibits antioxidant properties, protects brain integrity (neuroprotective mechanisms), intensifies the action of cholinergic neurons, and accelerates non-amyloidogenic metabolism of the amyloid precursor protein (Behl et al. 1995). A reduction in estrogen level could heighten the risk of brain damage to the brain, and accumulation of amyloid protein and plaques (Yare and Woodward 2020). In addition to the estrogen-progestogen hormone replacement therapy, phytoestrogen formulation Phyto-β-SERM, which elicits central estrogenic neuroprotective effects, provides evidence of alternate therapeutic options in AD treatment (Szego et al. 2011).

Preclinical and clinical trials have continued to provide evidence of phytotherapy and active isolates. Genistein, resveratrol, quercetin, and catechins enhance neuronal survival cell-signaling pathways with inhibition of misfolded protein aggregates. A lower risk of dementia by activating sirtuin proteins was reported (phase II clinical trial—NCT01504854) on resveratrol. A naturally occurring small compound NIC5-15 in soy and other plants inhibits β-amyloid accumulation in AD (phase II clinical trial—NCT01928420) while the dietary strawberry (clinical trial—NCT02051140) and blueberry (clinical trial—NCT01888848) supplementation with antioxidant and anti-inflammatory properties improved cognition and mobility. Sulforaphane from cruciferous plants which enhances dopaminergic neuron survival and acts as an epigenetic modulator, antioxidant, and anti-inflammatory agent is being assessed (phase II clinical trial—NCT05084365) in PD patients for the safety, tolerability, cognitive functions, and motor symptoms. Cannabidiol from *Cannabis sativa* elicits neuroprotective and neuromodulatory effects with potential in the treatment of CNS disorders (Patricio et al. 2020). The tolerability and efficacy of cannabidiol were studied in a randomized, double-blinded, placebo-controlled phase II clinical trial (NCT02818777) on tremors in PD. A plant-derived active component ethnodyne and vitamin B2 improved motor and non-motor signs of PD in a clinical trial (NCT02815800). Anti-inflammation, delayed onset, and progression of multiple sclerosis were reported with gum and fiber-rich Guar beans (Robinson 2019). The fact that the currently available drugs in the market offer symptomatic relief will continue to spur investigations toward the discovery of phytotherapy that can restore the daily functional capacity of the patients alongside the symptoms.

## 4 Phytotherapy in the Management of Diabetes

Diabetes mellitus (categorized into type I, II, and III) remains one of the hallmarks of metabolic anomalies and syndromes. Concurrence of hypertension, obesity, dyslipidemia, and altered glycemia with a high risk of diabetes and cardiovascular diseases often define metabolic diseases (Stern et al. 2004; O'Neill and O'Driscoll 2015). Complex molecular mechanistic pathways of diabetes include defects in glucose metabolism, insulin action, oxidative stress, and inflammation, among others. With some of the clinical limitations and adverse effects of the available synthetic drug classes (sulfonylureas, sodium-glucose co-transporter-2 inhibitors, thiazolidinediones, dipeptidyl peptidase-4 inhibitors, biguanides, glucagon-like peptide 1 analogs), diabetes are increasingly being managed by the use of phytotherapy in patients (Governa et al. 2018). *Allium cepa* (quercetin, rutin, l-cysteine sulfoxides allyl-propyl disulphide), *Nigella sativa (high polyphenol content), Azadirachta indica* (azadirachtins), *Moringa oleifera* (chlorogenic acid, quercetin, kaempferol, and glucosinolates), *Momordica charantia* (cucurbitane triterpenoids, momordin), *Punica granatum (*ellagic acid, punicalagin, flavonoids, anthocyanins), *Panax* spp. (protopanaxidiols, ginsenoside-Rg1, 20(S)-ginsenoside-Rg3, ginsenoside-Rh2), *Rehmannia glutinosa* (polysaccharides, catalpol), *Trigonella foenum-graecum* (trigonelline, diosgenin, 4-hydroxyisoleucine), *Gymnema sylvestre* (gymnemic acids; gymnemasaponins gurmarin), and *Cinnamomum* spp. (Cinnamaldehyde). Curcumin, tannins, lignans resveratrol (nonflavonoid polyphenols) as well as anthocyanins, epigallocatechin gallate, quercetin, naringin, rutin, and kaempferol (flavonoids) are among the active ingredients with antidiabetic potential (Mutha et al. 2021; Jubaidi et al. 2021).

Some of the mechanisms of action reported on phytotherapy with hypoglycemic properties include a reduction of blood intestinal/ blood glucose uptake and insulin resistance, inhibition of pancreatic and intestinal α-amylase and α-glucosidase, increase, and stimulation of glucose transporter-4 translocation and glucose uptake, increase of glucagon-like peptide-1, modulation of insulin secretion and stimulation of insulin action, regulation of insulin receptor insulin receptor kinase/substrate-1, free radical scavenging, improvement of redox homeostasis, regulation of protein-tyrosine phosphatase 1B, stimulation of antioxidant enzymes, an increase of superoxide dismutase and catalase activity, increase of 5′-adenosine monophosphate-activated protein kinase phosphorylation, reduction of tumor necrosis factor-alpha, Interleukin 6 release, cyclooxygenase-2, monocyte chemoattractant protein-1, inducible protein-10, nitric oxide synthase and vascular endothelial growth factor expression, nuclear factor-κB translocation, modulation of insulin secretion, modulation of PPARγ mRNA expression, reduction of malondialdehyde activity, phosphoenolpyruvate carboxykinase and glucose-6-phosphatase, an increase of insulin receptor substrate-1, cell signal transduction involving protein kinase C, phosphatidylinositol 3-kinase/protein kinase B, p38 mitogen-activated protein kinase, and extracellular signal-related kinase, inhibition of N-methyl-D-aspartate receptor-mediated nitrosative stress, stimulation of nucleic acid and energy metabolism, increase β-endorphin secretion, reduction of hepatic glucose-6-phosphatase activity, increase of hepatic glycogen level, reduction of reactive oxygen species production, suppression of adenine dinucleotide phosphate-oxidase activity, induction of β-cells regeneration, promotion of adipocyte differentiation, and inhibition of lipid- and carbohydrate-hydrolyzing enzymes (Balbaa et al. 2021; Governa et al. 2018). The toxicity and efficacy of phytotherapy in managing diabetes still require additional preclinical and clinical studies.

## 5 Phytotherapy in the Management of Cancer

The rising cases of cancers, limited knowledge of the underlying mechanisms of cancer, a wide range of cancer types, complications of antineoplastic agents, and increasing need for drug combinations, invasiveness, and long-lasting adverse effects of conventional therapeutic approaches continued to stimulate interest in phytotherapy and their innumerous active constituents. The plants and active principles that are responsible for the anticancer properties are provided. Solubility and adverse effects, which constitute clinical limitations to the applications of the available ones, are being explored in continuous preclinical studies and clinical trials. Extensive studies have revealed a wide range of mechanisms underlying promising anticancer property of phytochemicals including berberine (benzyl-tetra isoquinoline alkaloid from *Berberis* sp.), catechins (lycopene, quercetin, resveratrol, and sulforaphane), 6-Shogaol (phenylpropanoid from *Zingiber officinale*) elicits anticancer through Akt and STAT signaling pathways (Kim et al. 2014; Saha et al. 2014), curcumin (polyphenolic pigment from *Curcuma longa*) modulates cell-signaling and gene expression regulatory pathways (Kunnumakkara et al. 2017), allicin (organosulfurs from *Allium sativum*) with anticancer mediated through STAT3 signaling pathway (Huang et al. 2017; Chen et al. 2018), andrographolide (diterpenoid from *Andrographis paniculata*)—HIF-1α, VEGF, and PI3K pathway (Li et al. 2015), apigenin (flavonoid from *Petroselinum crispum*)—intrinsic apoptosis pathway (Chang et al. 2018; Yan et al. 2018), baicalein and baicalin (flavonoid from *Scutellaria baicalensis*)- MAPK, ERK, and p38 signaling pathways (Dou et al. 2018; Tao et al. 2018), dicumarol from *Melilotus officinalis*—Intrinsic apoptosis pathway (Zhang et al. 2017b), epigallocatechin (flavonoids from *Camellia sinensis*)—inhibit cell proliferation and apoptosis (Xu et al. 1992; Thangapazham et al. 2007), decursin and decursinol (coumarin from *Angelica gigas*) (Wu et al. 2017a), nimbolide (triterpene from *Azadirachta indica*)—PI3K/AKT/mTOR and ERK signaling (Subramani et al. 2016), emodin (resin from *Rheum palmatum*)—PI3K/AKT and MAPK signaling pathways (Iwanowycz et al. 2016; Lin et al. 2016; Su et al. 2017), genistein (isoflavonoid from *Glycine max*)—WNT/β-catenin and Akt signaling pathway (Zhang et al. 2013; Hsiao et al. 2019), Gingerol (polyphenol) from *Zingiber officinale*)—Intrinsic apoptosis pathway (Joo et al. 2016; Martin et al. 2017), hispidulin (flavone from *Salvia involucrate*) -Intrinsic apoptosis pathway (Gao et al. 2017; Han et al. 2018), HS-1793 (stilbenoid from *Polygonum cuspidatum*)—HIF-1α, VEGF, Ki-67, and CD31 (Kim et al. 2020), Licochalcone A (chalcone from *Glycyrrhiza glabra*)—Cyclins and CDKs (Lu et al. 2018), withaferin-A (phytosterols from *Withania somnifera*)—AKT signaling FOX03a-Par-4 cell death pathway, ERK, and p38 pathway (Choi and Kim 2015; Suman et al. 2016; Kuppusamy et al. 2017), physapubescin B (steroid from *Physalis pubescens*)—Ki-67, Cdc25C, and PARP (Ding et al. 2015), glycyrrhizin (triterpenes from *Glycyrrhiza glabra*)—TxA2 and JAK/STAT signaling pathway (Deng et al. 2017), pterostilbene (polyphenol from *Polygonum cuspidatum*)—Mitochondrial mediated apoptosis; ERK and STAT3 signaling (Feng et al. 2016; Kong et al. 2016; Wen et al. 2017), sulforaphane (organosulfur from *Brassica oleracea*)—Cell cycle arrest and apoptosis. Targets: caspase 8, p21, hsp90 (Qazi et al. 2010), Thymol (monoterpenoids from *Thymus vulgaris*) Mitochondrial mediated apoptosis (De La Chapa et al. 2018), Thymoquinone (quinone from *Nigella sativa*)—STAT3 and associated protein (Zhu et al. 2016; Odeh et al. 2018), resveratrol (phenol from *Polygonum cuspidatum*)—Regulating cell cycle and apoptosis pathways (Banerjee et al. 2002), ursolic acid (triterpenoids from *Oldenlandia diffusa*)—Ki-67, CD31, and miR-29a (Prasad et al. 2012; Zhang et al. 2018a, b), alpinumisoflavone (pyranoisoflavone from *Derris eriocarpa*)—Nrf2, NQO-1, HO-1, miR-101, and Akt signaling (Wang et al. 2017; Zhang et al. 2017a).

Vinca alkaloids, taxane diterpenoids, camptothecin derivatives, epipodophyllotoxin, combretastatins, homoharringtonine (omacetaxine mepesuccinate, cephalotaxine alkaloid), and ingenol mebutate are among the anticancer compounds already approved for clinical applications. Vinca alkaloids like vinblastine and vincristine from *Catharanthus roseus* (Apocynaceae) elicit cytotoxicity, metaphase arrest, and cell death by binding to a distinct site of β-tubulin to inhibit polymerization and microtubule assemblage. Vinorelbine, vindesine, and vinflunine are semi-synthetic approved analogs for the treatment of leukemia, Hodgkin and non-Hodgkin lymphomas, advanced testicular carcinoma, breast and lung cancers, Kaposi's sarcoma, transitional cell carcinoma of the urothelium, etc. (Martino et al. 2018). Taxanes (Paclitaxel and its semi-synthetic derivatives—docetaxel, cabazitaxel, larotaxel, milataxel, ortataxel, and tesetaxel) isolated from *Taxus brevifolia*—Yew tree stabilizes microtubules to induce cell cycle arrest and aberrant mitosis in the treatment of breast, ovarian, pancreas, prostate, and lung cancer therapies (Kotsakis et al. 2016; Oudard et al. 2017). Quinolone alkaloid camptothecin isolated irinotecan and topotecan (semi-synthetic) from *Camptotheca acuminata* complexes with type I DNA topoisomerase to inhibit DNA cleavage and religation (Hertzberg et al. 1989). Irinotecan is indicated for advanced cancers of the large intestine and rectum as topotecan is for recurring ovarian, small cell lung cancer, and cervical cancer. Podophyllotoxin from *Podophyllum peltatum* and *Podophyllum emodi* reversibly bind to tubulin, while etoposide and teniposide (semi-synthetic) target topoisomerase II. A podophyllotoxin derivative CIP-36 failed in clinical trials possesses anti-multidrug resistance of adriamycin-resistant human leukemic cell line K562/ADR by regulating the activity of topoisomerase-IIa (Cao et al. 2015). A hydrophobic ester of the diterpene ingenol (Ingenol mebutate—IM) from Euphorbia peplus induces rapid induction of cell death in the treated area or activates an inflammatory response, capable of eliminating the residual cells (Skroza et al. 2018). It is indicated for actinic keratosis caused by chronic ultraviolet radiation exposure. Ester of the alkaloid cephalotaxine homoharringtonine from *Cephalotaxus* genus binds to the A-site cleft in the large ribosomal subunit and interferes with chain elongation and protein synthesis. It is indicated for chronic myeloid leukemia while omacetaxine mepesuccinate (semi-synthetic) showed efficacy for myelodysplastic syndromes and chronic myelomonocytic leukemia in patients with azacitidine and decitabine resistance and intolerance (Short et al. 2019). The combretastatins (Combretastatin A1, A4, and its phosphate prodrug—CA4P) from *Combretum caffrum* act to inhibit tubulin polymerization disrupting the tumor endothelial cells lining the tumor vasculature, inducing rapid vascular collapse in solid tumors (Tozer et al. 2002).

Phytochemicals that are currently under clinical trials against various cancers include berberine that has been established in cancers such as colon, breast, gastrointestinal, oral, liver, pancreas, prostate, and cervical (Mao et al. 2018; Zhao et al. 2017; Hesari et al. 2018; Lin et al. 2017; Tsang et al. 2015; Abrams et al. 2019; Youn et al. 2018) and ovarian (Hou et al. 2017; Mahata et al. 2011) cancers. A clinical trial is ongoing to determine the efficacy of berberine hydrochloride against the occurrence of new colorectal adenomas among patients with a history of colorectal cancer (NCT03281096). Curcumin plays chemotherapeutic role in different cancer cells including blood (Taverna et al. 2015), breast (Mock et al. 2015), head and neck (Wilken et al. 2011), liver (Darvesh et al. 2012), prostate (Nakamura et al. 2002), ovary (Yallapu et al. 2010), and skin cancers (Huang et al. 1997). Phase I clinical trials have shown the safety and tolerability of curcumin at high doses (8 g/day) but exhibited poor bioavailability in humans (Sharma et al. 2004; Kanai et al. 2013). A phase 2/3 trial is ongoing to determine the efficacy of curcumin along with Paclitaxel against advanced and metastatic breast cancer patients (NCT03072992).

Epigallocatechin elicits anticancer activity in different cell lines (Wang and Bachrach 2002; Fujiki et al. 2015). The use of epigallocatechin in combination with indole-3-carbinol could improve treatment outcomes in advanced ovarian cancer (Kiselev et al. 2018). Currently, a phase 1 trial is ongoing to evaluate the chemopreventive effects of Teavigo™ (highly purified and refined green tea extract providing 94% epigallocatechin) in colorectal cancer patients with curative resections (NCT03072992). The intake of lycopene/tomato was associated with a relatively minor reduction in the risk of prostate cancer (Chen et al. 2013). Phase 2 trial is ongoing to assess the effectiveness of lycopene in reducing skin toxicity in metastasis colorectal carcinoma patients treated with panitumumab (NCT03167268). A delayed recurrence of prostate cancer was reported with pulverized muscadine grape skin extract which contains resveratrol (Paller et al. 2015). RASSF-1a methylation and PGE2 expression in the breast were reduced by trans-resveratrol (Zhu et al. 2012).

## 6 Other Potential and Clinical Applications of Phytotherapy

Phytotherapy has shown promising outcomes in managing microbial infections, and cardiovascular, gastrointestinal, and respiratory diseases. Antibacterial, antiprotozoal, antiviral, and antifungal among other spectrum of antimicrobial activities of some phytotherapy are widely reported. The inhibition of efflux pumps and numerous virulence factors, nucleic acid synthesis, cell division protein, disruption of microbial homeostasis, outer membrane, and the integrity of the cytoplasmic bacterial membrane are among the mechanisms of antimicrobial. The activity of *Aristolochia cymbifera* extracts against promastigotes of *Leishmania chagasi* and tripomastigotes of *Trypanossoma cruzi*; *Aristolochia paucinervis* against the growth of *Clostridium perfringens*, *Clostridium difficile*, *Enterococcus faecalis*, *Micrococcus lutens*, and *Bacillus subtilis; Pistacia lentiscus* bactericidal effect on *S. aureus*; *Myrospermum erythroxylon* against the growth of *Staphylococcus aureus* and *Pseudomonas aeruginosa* have been reported elsewhere (Tempone et al. 2008; Mezni et al. 2015; Machado Barros et al. 2005). In immunocompromised patients, some of these phytotherapy could boost resistance to infections. The antimicrobial properties are attributed to the phytoconstituents acting on different targets. Berberine and jatrorrhizine from *Mahonia aquifolium* against *Cutibacterium acnes* (Slobodníková et al. 2004), 8-hydroxylated benzo[c]phenanthridine from *Chelidonium majus* against methicillin-resistant *Staphylococcus aureus* (Zuo et al. 2009), Cucurbitacin B against *S. aureus*, and synergistic activity with tetracycline and oxacillin (Mezni et al. 2015) together exhibited great antibacterial potentials of phytotherapy and widened therapeutic approach in managing the antibiotic resistance that limits the effectiveness of pharmacological interventions.

The increasing cases of cardiovascular diseases (CVDs) including myocardial infarction, hypertension, peripheral vascular diseases, coronary heart disease, cardiomyopathies, and dyslipidemias demand varying therapeutic approaches that are safe, effective, and affordable. Evidence demonstrating phytotherapy-induced modulations of molecular, metabolic, and cellular targets that are relevant to CVDs provide support for their application in treating these pathological conditions—the most common cause of health burden and death globally. Free radical generation and subsequent free radical-antioxidant imbalance contribute to the evolution of CVDs. Historically, aspirin from the *Salix alba*, digoxin (cardiac glycoside) from Digitalis purpurea, ephedrine from *Ephedra sinica*, lovastatin from *Monascus purpureus*, taxol from *Taxus brevifolia*, reserpine (effective treatment for hypertension) from *Rauvolfia serpentina*, digitalis (potent inhibitor of $Na^+/K^+$-ATPase capable of triggering depolarization, smooth muscle contraction and vasoconstriction) from *Digitalis purpurea* strengthens muscle heart contractions (Liu et al. 2016; Harvey 2000; Frishman et al. 2009; Cragg and Newman 2013), papaverine (base for making the verapamil—antihypertensive drug) from *Papaver somniferum* (Fabricant and Farnsworth 2001) reinforces evidence of phytotherapy relevance in cardiovascular diseases. *Ginseng, Ginkgo biloba, and Ganoderma lucidum* are among the commonly used or screened for phytotherapy potential. The treatment of coronary heart disease, myocardial infarction, atherosclerosis, and angina pectoris with *Salvia miltiorrhiza* is effective. Its isolate salvianolic acid B relieves fibrosis and ischemia-reperfusion injury (Lay et al. 2003). *Astragalus membranaceus* possesses antioxidants (decreasing the levels of the free radicals and malondialdehyde), and antiischemia (Zhang et al. 2006) while protecting against myocardial cell injury (Ma et al. 2013). In patients with congestive heart failure, its main bioactive astragaloside IV improves left ventricular ejection by eliciting a positive inotropic effect (Luo et al. 1995). The *Allium sativum* therapeutic potential in atherosclerosis and hyperlipidemia conditions is associated with reducing cholesterol, LDL, lipid content of arterial cells, and proliferation of vascular smooth muscle cells (Sun et al. 2018). The blockade of $Ca^{2+}$ channels via endothelium-independent mechanisms by Crocus sativus confers vasodilatory properties on this species (Razavi et al. 2016). *Hibiscus sabdariffa* inhibits ACE to reduce blood pressure (Ojeda et al. 2010).

Ginseng or its ginsenoside constituents could increase eNOS expression and NO production, and elicit hypotensive, vasorelaxation, anti-oxidation, and anti-inflammation activities (Kim 2012; Choi et al. 2013; Valli and Giardina 2002; Jang et al. 2011; Hong et al. 2012; Pan et al. 2013; Lee et al. 2016). They inhibit arginase activity before increasing NO production, cGMP levels, activation of $Ca^{2+}$-gated potassium channels, inhibition of ACE activity, and blockade of $Ca^{2+}$-gated channels (Shin et al. 2013). Hence, the regulation of $Ca^{2+}$ channels and anti-hyperlipidemic, anti-inflammatory, and antioxidant effects have been attributed to the underlying mechanism of cardioprotection.

*Ginkgo biloba* and its constituents (ginkgo-flavone glycosides, ginkgolides, and bilobalide) with antioxidant, anti-atherothrombotic, and anti-inflammatory effects provide health benefits in CVDs. The ACE inhibition, lipid reduction, cholinergic stimulations, vasodilatory, antihypertensive, and restoration of oxidative balance altogether contribute to the cardioprotective benefits of *G. biloba* and its isolates (Mansour et al. 2011; Liou et al. 2015; Huang et al. 2018; Perez-Vizcaino et al. 2009). Vascular actions and mechanisms of this species decrease the activity of the

ROS-producing enzyme, production of the enzyme involved in the rupture of atherosclerotic plaques, MMP-1, and phosphorylation of mitogen-activated protein kinases, which in turn suppresses aortic smooth muscle cells toll-like-receptor-4 expression (Lin et al. 2007; Akiba et al. 2007). Ginkgolide B-induced inhibition of monocyte chemotactic protein-1, intercellular adhesion molecule-1, and vascular cell adhesion molecule-1 facilitate the attenuation of endothelial dysfunction. Ginkgolide C can decrease the expression of PPAR adipogenesis-related transcription factors and adipogenesis and enhance the Sirt1/AMPK pathway, stimulating the production of adipose triglyceride lipase and hormone-sensitive lipase with increased lipolysis leading to suppression of lipid accumulation with a decrease in acetyl-CoA carboxylase and fatty acid synthesis activity (Liou et al. 2015).

*Ganoderma lucidum* has been found to protect human lymphocyte DNA from hydrogen peroxide-induced oxidative damage (Shi et al. 2002) and regulates the expression of Nrf2, which in turn regulates antioxidant genes such as HO-1, GST, and NQO-1. This antioxidant property of *G. lucidum* supports its antiatherosclerosis. Peptides QLVP, QDVL, and QLDL from *G. lucidum* with ACE inhibitory activity, enhanced Angiotensin 1-mediated phosphorylation of eNOS, and reduced expression of peptide endothelin-1 (vasoconstrictor) can be used to treat hypertension (Wu et al. 2017b). The ganoderic acid A from *G. lucidum* extracts activates the PI3K/AKT signaling pathway causing increased proliferation and decreased apoptosis of rat H9c2 cardiomyocytes exposed to hypoxic injury suggesting cardioprotective potential (Zhang et al. 2018a, b). In all these activities, antioxidants and anti-inflammatory actions are important when it comes to the prevention of CVDs. Despite all these activities of phytotherapy, reports are showing their ineffectiveness in treating some aspects of CVDs (Klupp et al. 2015).

In gastrointestinal diseases (gastritis, ulcers, spastic constipation, irritable bowel syndrome, diarrhea, gastralgias, pylorospasm, intestinal spasms, and dyspepsia), phytotherapy has always played important roles. The spasmolytic, prokinetic influence on gastrointestinal motilities, anti-inflammatory action, beneficial influence on intestinal microbiota, and inhibitory or enhanced gastric secretion are among the mechanisms supporting the clinical application of phytotherapy. According to Kelber et al. (2017), *Cnicus benedictus* with bitter cnicin stimulates gastric secretion; *Atropa belladonna* with scopolamine, apoatropine, and belladonine lowers the tone, excitation, gastric and intestinal motilities; *Gentiana lutea* with bitter amarogentin and gentiopicroside stimulates saliva and gastric acid secretion, increases gastric blood flow; *Acorus calamus* with asarone, acorin, and tannins *reliefs* hypoacidic stomach diseases, chronic dyspepsia; *Carum carvi* with (S)-(+)-carvon and (R)-(+)-limonene stimulates the gastrointestinal smooth muscle tone and increases mucus secretion with spasmolytic and carminative properties; *Matricaria chamomilla* with apigenin-7-glycoside exhibits spasmolytic and antiperistalsis activities; *Melissa officinalis with* citral, citronellal, linalool, geraniol, and β-caryophyllene possess spasmolytic and carminative property.

Proinflammatory mediators, cyclooxygenases, and lipoxygenases often raise the risk of bronchial or pulmonary inflammation leading to respiratory disorders. This supports the use of phytotherapy with anti-inflammatory properties for lung diseases. A comprehensive review on the potential of extracts, isolates, and mechanisms of *Psoralea corylifolia* (papain-like protease inhibition), *Herba paederiae* (MAPK and NF-kB pathways modulation), *Scutellaria baicalensis* (activation of ACE2/Ang-(1–7)/Mas), *Calophyllum brasiliense and Tripterygium wilfordii* (inhibit transcription and replication), *Camellia sinensis* (inhibit MIP-2 and TNF-α production, activate ERK1/2 and JNK signaling pathway), *Curcuma longa* (enhancement of p66Shc and p-p66Shc proteins inhibition), *Polygonum cuspidatum and Rheum officinale* (inhibits the spike protein and ACE-2 interactions), *Azadirachta indica* (trigger histone deacetylases enzyme, TNF-α regulated NF-kB), *Picrorhiza scrophulariiflora* (reduction in ROS release and complement activation), *Baccharis retusa* (reduction in collagen, elastic fiber accumulation, alveolar size, MMP-9 and MMP-12 cells), etc., for lung diseases was reported by Rahman et al. (2022). Baicalein, ginkogolic acid, resveratrol, and shiraiachrome A are important isolates against coronavirus and lung cancer progression; nimbolide inhibits storm of inflammatory cytokine, iNOS, nitrotyrosine levels, nitrosative-oxidative stress, myeloperoxidase; luteolin seems to attenuate neutrophilic inflammation, lung hemorrhage, and interstitial edema; Ellagic acid reduces IL-6 and increases IL-10 level in bronchoalveolar lavage fluid; kaempferol and quercetin lower serum cytokines TNF-α, IL1-β, IL-6, or nitric oxide; and cepharanthine with nelfinavir, and linoleic acid with remdesivir alleviate lung cancer; curcumin and cisplatin, and celastrol with gefitinib. Further studies will reveal the clinical utility of phytotherapy either as monotherapy or adjunct therapy in the treatment of chronic obstructive pulmonary disease, coronavirus infections, acute respiratory distress syndrome, asthma, pneumonia, and lung cancer.

# 7 Final Remarks

Understanding phytotherapy products and active principles could help screen through their usefulness in patients with a wide range of medical profiles including age, frailty, gender, comorbidities and genetic disposition, high medication burden, recent changes in medicines, exposure to the drugs of a low therapeutic index, limited elimination pathways, long duration of actions and interactions with multiple targets. There is still no straightforward approach to understanding of phytotherapy mechanism, possible toxicity, formulation doses, and side effects. Hence, the phytocomplexity of these phytotherapy products offers challenges and opportunities for the discovery of new drugs. However, as cases of adverse reactions are being reported, the labeling of phytotherapy products as "safe" because it is from "natural" sources is rather misleading. Characterization (phytochemical studies, toxicity, resistance tendency, solubility, and crossing into the CNS, etc.), isolation, and screening of relevant active principles or derivatives are expected to shape chemical modification strategies toward the establishment of their clinical utility and optimization.

# 8 Self-Assessment Questions and Answers

## *8.1 Questions*

1. Why is quality control important in the context of phytotherapy?
   A. To ensure that only synthetic drugs are used.
   B. To verify the effectiveness and safety of therapeutic products derived from plants.
   C. To eliminate the need for clinical research.
   D. To standardize the flavor of herbal products.
2. Which of the following pathways is responsible for the production of phenolic compounds like flavonoids and tannins?
   A. Shikimic acid pathway
   B. Melvonic acid pathway
   C. Phenylpropanoid pathway
   D. Glycolysis pathway
3. What are the main groups of secondary metabolites?
   A. Carbohydrates, proteins, and lipids
   B. Terpenes, phenolic compounds, and nitrogen compounds
   C. Nucleic acids, minerals, and vitamins
   D. Sugars, starches, and fibers
4. What is the primary mechanism by which resveratrol is thought to be effective in neuroprotection?
   A. Inhibiting AChE activity
   B. Inhibiting mitochondrial biogenesis
   C. Reducing oxidative stress as a potent antioxidant
   D. Suppressing brain-derived neurotrophic factors
5. Which of the following is NOT a mechanism through which phytochemicals exhibit anticancer properties?
   A. Modulating cell-signaling and gene expression pathways
   B. Inhibiting DNA cleavage and religation
   C. Stabilizing microtubules to induce cell cycle arrest
   D. Enhancing oxidative stress in human cells

## *8.2 Answers*

1. B. Quality control ensures that phytotherapeutic products are effective, safe, and consistent in their composition. This is crucial for ensuring that the products provide the intended therapeutic benefits and do not pose health risks due to contamination or variability in active ingredient concentrations.

2. C. The phenylpropanoid pathway is the primary metabolic route responsible for producing phenolic compounds, including flavonoids, tannins, and other related substances. This pathway starts with the amino acid phenylalanine and involves several enzymatic steps leading to the formation of phenolic compounds.
3. B. The main groups of secondary metabolites include terpenes (e.g., essential oils, resins), phenolic compounds (e.g., flavonoids, tannins), and nitrogen-containing compounds (e.g., alkaloids, glycosides).
4. B. Resveratrol is widely recognized for its ability to reduce oxidative stress, which helps protect neurons from damage and contributes to its neuroprotective effects. While it can also influence other mechanisms such as mitochondrial biogenesis and brain-derived neurotrophic factors, its role in reducing oxidative stress is considered a primary mechanism for its neuroprotective properties.
5. D. Phytochemicals typically exhibit anticancer properties by modulating cell-signaling and gene expression pathways, inhibiting DNA cleavage and religation, and stabilizing microtubules to induce cell cycle arrest. Enhancing oxidative stress in human cells is not a common or desired mechanism for anticancer activity, as increased oxidative stress generally contributes to cellular damage and can be detrimental rather than therapeutic.

## References


Abrams SL, Follo MY, Steelman LS, Lertpiriyapong K, Cocco L, Ratti S, et al. Abilities of berberine and chemically modified berberines to inhibit proliferation of pancreatic cancer cells. Adv Biol Regul. 2019;71:172–82. https://doi.org/10.1016/j.jbior.2018.10.003.

Abubakar AR, Haque M. Preparation of medicinal plants: basic extraction and fractionation procedures for experimental purposes. J Pharm Bioallied Sci. 2020;12(1):1–10.

Akiba S, Yamaguchi H, Kumazawa S, Oka M, Sato T. Suppression of oxidized LDL-induced PDGF receptor beta activation by ginkgo biloba extract reduces MMP-1 production in coronary smooth muscle cells. J Atheroscler Thromb. 2007;14(5):219–25. https://doi.org/10.5551/jat.e516.

Alagan A, Jantan I, Kumolosasi E, Azmi N. Phyllanthus amarus protects against spatial memory impairment induced by lipopolysaccharide in mice. Bioinformation. 2019;15(8):535–41. https://doi.org/10.6026/97320630015535.

Anandhan A, Tamilselvam K, Radhiga T, Rao S, Essa MM, Manivasagam T. Theaflavin, a black tea polyphenol, protects nigral dopaminergic neurons against chronic MPTP/probenecid induced Parkinson's disease. Brain Res. 2012;1433:104–13. https://doi.org/10.1016/j.brainres.2011.11.021. Epub 2011 Nov 13. PMID: 22138428.

Aziz Z, Ahmad A, Setapar S, Karakucuk A, Azim MM, Lokhat D, Rafatullah M, Ganash M, Kamal MA, Ashraf GM. Essential oils: extraction techniques, pharmaceutical and therapeutic potential – a review. Curr Drug Metab. 2018;19(13):1100–10.

Balbaa M, El-Zeftawy M, Abdulmalek SA. Therapeutic screening of herbal remedies for the management of diabetes. Molecules. 2021;26:6836. https://doi.org/10.3390/molecules26226836.

Banerjee S, Bueso-Ramos C, Aggarwal BB. Suppression of 7,12-dimethylbenz(a)anthracene-induced mammary carcinogenesis in rats by resveratrol: role of nuclear factor-κB, cyclooxygenase 2, and matrix metalloprotease 9. Cancer Res. 2002;62:4945–54.

Behl C, Widmann M, Trapp T, Holsboer F. 17-beta estradiol protects neurons from oxidative stress-induced cell death in vitro. Biochem Biophys Res Commun. 1995;216(2):473–82. https://doi.org/10.1006/bbrc.1995.2647. PMID: 7488136.
Brazil. Brazilian Pharmacopoeia 6th edition. Brazilian Health Regulatory Agency, 2019.
Cao B, Chen H, Gao Y, Niu C, Zhang Y, Li L. CIP-36, a novel topoisomerase II-targeting agent, induces the apoptosis of multidrug-resistant cancer cells in vitro. Int J Mol Med. 2015;35(3):771–6. https://doi.org/10.3892/ijmm.2015.2068.
Chang JH, Cheng CW, Yang YC, Chen WS, Hung WY, Chow JM, et al. Downregulating CD26/DPPIV by apigenin modulates the interplay between Akt and Snail/Slug signaling to restrain metastasis of lung cancer with multiple EGFR statuses. J Exp Clin Cancer Res. 2018;37(1):199. https://doi.org/10.1186/s13046-018-0869-1.
Chen J, Song Y, Zhang L. Lycopene/tomato consumption and the risk of prostate cancer: a systematic review and meta-analysis of prospective studies. J Nutr Sci Vitaminol (Tokyo). 2013;59(3):213–23. https://doi.org/10.3177/jnsv.59.213.
Chen H, Zhu B, Zhao L, Liu Y, Zhao F, Feng J, et al. Allicin inhibits proliferation and invasion in vitro and in vivo via SHP-1-mediated STAT3 signaling in cholangiocarcinoma. Cell Physiol Biochem. 2018;47(2):641–53. https://doi.org/10.1159/000490019.
Chen L, Zhang C, Han Y, Meng X, Zhang Y, Chu H, Ma H. Gingko biloba extract (EGb) inhibits oxidative stress in neuro 2A cells overexpressing APPsw. Biomed Res Int. 2019;2019:7034983. https://doi.org/10.1155/2019/7034983. PMID: 31380435; PMCID: PMC6657629.
Chikezie PC, Ibegbulem CO, Mbagwu FN. Bioactive principles from medicinal plants. Res J Phytochem. 2015;9(3):88–115.
Choi BY, Kim BW. Withaferin-a inhibits colon cancer cell growth by blocking STAT3 transcriptional activity. J Cancer Prev. 2015;20(3):185–92. https://doi.org/10.15430/JCP.2015.20.3.185.
Choi H, Zhao T, Shin K, Kim S, Hwang B, Lee C, et al. Anxiolytic effects of herbal ethanol extract from Gynostemma pentaphyllum in mice after exposure to chronic stress. Molecules. 2013;18(4):4342–56. https://doi.org/10.3390/molecules18044342.
Christensen LP. Ginsenosides chemistry, biosynthesis, analysis, and potential health effects. Adv Food Nutr Res. 2009;55:1–99. https://doi.org/10.1016/S1043-4526(08)00401-4.
Cole GM, Teter B, Frautschy SA. Neuroprotective effects of curcumin. Adv Exp Med Biol. 2007;595:197–212. https://doi.org/10.1007/978-0-387-46401-5_8. PMID: 17569212; PMCID: PMC2527619.
Corrêa PG, Pimentel RMM, Cortez JSA, Xavier HS. Herbivoria e anatomia foliar em plantas tropicais brasileiras. Ciên Cult. 2008;60(3):54–7.
Cragg GM, Newman DJ. Natural products: a continuing source of novel drug leads. Biochim Biophys Acta. 2013;1830(6):3670–95. https://doi.org/10.1016/j.bbagen.2013.02.008.
Da Silva A, Bergamo L, De Camargo L, Fernandes C, Mussato D, Canazart D, De Abreu Filho B. Atividade microbiológica de óleos essenciais obtidos por arraste a vapor. Rev Uningá Rev. 2014;20(3):33–9.
Darvesh AS, Aggarwal BB, Bishayee A. Curcumin and liver cancer: a review. Curr Pharm Biotechnol. 2012;13(1):218–28. https://doi.org/10.2174/138920112798868791.
De La Chapa JJ, Singha PK, Lee DR, Gonzales CB. Thymol inhibits oral squamous cell carcinoma growth via mitochondria-mediated apoptosis. J Pathol Med. 2018;47(7):674–82. https://doi.org/10.1111/jop.12735.
Deng QP, Wang MJ, Zeng X, Chen GG, Huang RY. Effects of glycyrrhizin in a mouse model of lung adenocarcinoma. Cell Physiol Biochem. 2017;41(4):1383–92. https://doi.org/10.1159/000467897.
de Oliveira MG, Moreira LKS, Turones LC, et al. Mechanism of action involved in the anxiolytic-like effects of Hibalactone isolated from Hydrocotyle umbellate L. J. Tradit. Med. Complement. 2022;12(4):318–29. https://doi.org/10.1016/j.jtcme.2021.08.012.
Ding W, Hu Z, Zhang Z, Ma Q, Tang H, Ma Z. Physapubescin B exhibits potent activity against human prostate cancer in vitro and in vivo. J Agric Food Chem. 2015;63(43):9504–12. https://doi.org/10.1021/acs.jafc.5b03045.

Dou J, Wang Z, Ma L, Peng B, Mao K, Li C, et al. Baicalein and baicalin inhibit colon cancer using two distinct fashions of apoptosis and senescence. Oncotarget. 2018;9(28):20089–102. https://doi.org/10.18632/oncotarget.24015.

Erdinest N, Shmueli O, Grossman Y, Ovadia H, Solomon A. Anti-Inflammatory effects of alpha linolenic acid on human corneal epithelial cells. Invest Ophthalmol Vis Sci. 2012;53(8):4396–406. https://doi.org/10.1167/iovs.12-9724.

El Hamdaoui Y, Zheng F, Fritz N, Ye L, Tran MA, Schwickert K, Schirmeister T, Braeuning A, Lichtenstein D, Hellmich UA, Weikert D, Heinrich M, Treccani G, Schäfer MKE, Nowak G, Nürnberg B, Alzheimer C, Müller CP, Friedland K. Analysis of hyperforin (St. John's wort) action at TRPC6 channel leads to the development of a new class of antidepressant drugs. Mol Psychiatry. 2022;27(12):5070–85. https://doi.org/10.1038/s41380-022-01804-3.

Essa MM, Vijayan RK, Castellano-Gonzalez G, Memon MA, Braidy N, Guillemin GJ. Neuroprotective effect of natural products against Alzheimer's disease. Neurochem Res. 2012;37(9):1829–42. https://doi.org/10.1007/s11064-012-0799-9. Epub 2012 May 22. Erratum in: Neurochem Res. 2012;37(10):2293. PMID: 22614926.

Fabricant DS, Farnsworth NR. The value of plants used in traditional medicine for drug discovery. Environ Health Perspect. 2001;109(Suppl 1):69–75. https://doi.org/10.1289/ehp.01109s169.

Fajemiroye JO, Galdino PM, Florentino IF, Da Rocha FF, Ghedini PC, Polepally PR, Zjawiony JK, Costa EA. Plurality of anxiety and depression alteration mechanism by oleanolic acid. J Psychopharmacol. 2014;28(10):923–34. https://doi.org/10.1177/0269881114536789. Epub 2014 Jun 11. PMID: 24920136.

Fajemiroye JO, da Silva DM, de Oliveira DR, Costa EA. Treatment of anxiety and depression: medicinal plants in retrospect. Fundam Clin Pharmacol. 2016;30(3):198–215. https://doi.org/10.1111/fcp.12186. PMID: 26851117.

Fancello F, Petretto GL, Zara S, Sanna ML, Addis R, Maldini M, et al. Chemical characterization, antioxidant capacity and antimicrobial activity against food related microorganisms of Citrus limon var. pompia leaf essential oil. Food Sci Technol. 2016;69:579–85.

Fasae KD, Abolaji AO, Faloye TR, Odunsi AY, Oyetayo BO, Enya JI, Rotimi JA, Akinyemi RO, Whitworth AJ, Aschner M. Metallobiology and therapeutic chelation of biometals (copper, zinc and iron) in Alzheimer's disease: limitations, and current and future perspectives. J Trace Elem Med Biol. 2021;67:126779. https://doi.org/10.1016/j.jtemb.2021.126779. Epub 2021 May 15. PMID: 34034029.

Feng Y, Yang Y, Fan C, Di S, Hu W, Jiang S, et al. Pterostilbene inhibits the growth of human esophageal cancer cells by regulating endoplasmic reticulum stress. Cell Physiol Biochem. 2016;38(3):1226–44. https://doi.org/10.1159/000443071.

Feng Z, Sun Q, Chen W, et al. The neuroprotective mechanisms of ginkgolides and bilobalide in cerebral ischemic injury: a literature review. Mol Med. 2019;25:57. https://doi.org/10.1186/s10020-019-0125-y.

Fisone G, Borgkvist A, Usiello A. Caffeine as a psychomotor stimulant: mechanism of action. Cell Mol Life Sci. 2004;61(7-8):857–72. https://doi.org/10.1007/s00018-003-3269-3.

Francis PT. The interplay of neurotransmitters in Alzheimer's disease. CNS Spectr. 2005;10:6–9. https://doi.org/10.1017/s1092852900014164.

Frishman WH, Beravol P, Carosella C. Alternative and complementary medicine for preventing and treating cardiovascular disease. Dis Mon. 2009;55(3):121–92. https://doi.org/10.1016/j.disamonth.2008.12.002.

Fujiki H, Sueoka E, Watanabe T, Suganuma M. Primary cancer prevention by green tea, and tertiary cancer prevention by the combination of green tea catechins and anticancer compounds. J Cancer Prev. 2015;20(1):1–4. https://doi.org/10.15430/JCP.2015.20.1.1.

Ganguly R. Effect of moringa oleifera in experimental model of Alzheimer's disease role of antioxidants. Ann Neurosci. 2005;12:33–6.

Gao MQ, Gao H, Han M, Liu KL, Peng JJ, Han YT. Hispidulin suppresses tumor growth and metastasis in renal cell carcinoma by modulating ceramide-sphingosine 1-phosphate rheostat. Am J Cancer Res. 2017;7(7):1501–14.

Gobbo-Neto L, Lopes NP. Plantas medicinais: fatores de influência no conteúdo de metabólitos secundários. Quím Nova. 2007;30(2):374–81.

Governa P, Baini G, Borgonetti V, Cettolin G, Giachetti D, Magnano AR, Miraldi E, Biagi M. Phytotherapy in the management of diabetes: a review. Molecules. 2018;23(1):105. https://doi.org/10.3390/molecules23010105. PMID: 29300317; PMCID: PMC6017385.

Han M, Gao H, Xie J, Yuan YP, Yuan Q, Gao MQ, et al. Hispidulin induces ER stress-mediated apoptosis in human hepatocellular carcinoma cells in vitro and in vivo by activating AMPK signaling pathway. Acta Pharmacol Sin. 2018:666–76. https://doi.org/10.1038/s41401-018-0159-7.

Harvey A. Strategies for discovering drugs from previously unexplored natural products. Drug Discov Today. 2000;5(7):294–300. https://doi.org/10.1016/S1359-6446(00)01511-7.

Hertzberg RP, Caranfa MJ, Hecht SM. On the mechanism of topoisomerase I inhibition by camptothecin: evidence for binding to an enzyme-DNA complex. Biochemistry. 1989;28(11):4629–38. https://doi.org/10.1021/bi00437a018.

Hesari A, Ghasemi F, Cicero AFG, Mohajeri M, Rezaei O, Hayat SMG, et al. Berberine: a potential adjunct for the treatment of gastrointestinal cancers? J Cell Biochem. 2018;119(12):9655–63. https://doi.org/10.1002/jcb.27392.

Hong SY, Kim JY, Ahn HY, Shin JH, Kwon O. Panax ginseng extract rich in ginsenoside protopanaxatriol attenuates blood pressure elevation in spontaneously hypertensive rats by affecting the Akt-dependent phosphorylation of endothelial nitric oxide synthase. J Agric Food Chem. 2012;60(12):3086–91. https://doi.org/10.1021/jf204447y.

Hou D, Xu G, Zhang C, Li B, Qin J, Hao X, et al. Berberine induces oxidative DNA damage and impairs homologous recombination repair in ovarian cancer cells to confer increased sensitivity to PARP inhibition. Cell Death Dis. 2017;8(10):e3070. https://doi.org/10.1038/cddis.2017.471.

Hsiao YC, Peng SF, Lai KC, Liao CL, Huang YP, Lin CC, et al. Genistein induces apoptosis in vitro and has antitumor activity against human leukemia HL-60 cancer cell xenograft growth in vivo. Environ Toxicol. 2019;34(4):443–56. https://doi.org/10.1002/tox.22698.

Huang MT, Newmark HL, Frenkel K. Inhibitory effects of curcumin on tumorigenesis in mice. J Cell Biochem Suppl. 1997;27:26–34. https://doi.org/10.1002/(sici)1097-4644(1997)27+<26::aid-jcb7>3.0.co;2-3.

Huang L, Song Y, Lian J, Wang Z. Allicin inhibits the invasion of lung adenocarcinoma cells by altering tissue inhibitor of metalloproteinase/matrix metalloproteinase balance via reducing the activity of phosphoinositide 3-kinase/AKT signaling. Oncol Lett. 2017;14(1):468–74. https://doi.org/10.3892/ol.2017.6129.

Huang WC, Chen YL, Liu HC, Wu SJ, Liou CJ. Ginkgolide C reduced oleic acid-induced lipid accumulation in HepG2 cells. Saudi Pharm J. 2018;26(8):1178–84. https://doi.org/10.1016/j.jsps.2018.07.006.

Iribarren R, Cunha C, Kaymak H, Grzybowski A. Preservatives and pH in low-dose atropine formulations for clinical trials. Acta Ophthalmol. 2024;102:364–66. Available from: https://doi.org/10.1111/aos.16604.

Iwanowycz S, Wang J, Hodge J, Wang Y, Yu F, Fan D. Emodin inhibits breast cancer growth by blocking the tumor-promoting feedforward loop between cancer cells and macrophages. Mol Cancer Ther. 2016;15(8):1931–42. https://doi.org/10.1158/1535-7163.MCT-15-0987.

Jang SJ, Lim HJ, Lim DY. Inhibitory effects of total ginseng saponin on catecholamine secretion from the perfused adrenal medulla of SHRs. J Ginseng Res. 2011;35(2):176–90. https://doi.org/10.5142/jgr.2011.35.2.176.

John OO, Amarachi IS, Chinazom AP, Adaeze E, Kale MB, Umare MD, Upaganlawar AB. Phytotherapy: a promising approach for the treatment of Alzheimer's disease. Pharmacol Res Mod Chin Med. 2022;2:100030. https://doi.org/10.1016/j.prmcm.2021.100030.

Joo JH, Hong SS, Cho YR, Seo DW. 10-Gingerol inhibits proliferation and invasion of MDA-MB-231 breast cancer cells through suppression of Akt and p38MAPK activity. Oncol Rep. 2016;35(2):779–84. https://doi.org/10.3892/or.2015.4405.

Jubaidi FF, Zainalabidin S, Taib IS, Hamid ZA, Budin SB. The potential role of flavonoids in ameliorating diabetic cardiomyopathy via alleviation of cardiac oxidative stress, inflammation and apoptosis. Int J Mol Sci. 2021;22:5094.

Jung HA, Jin SE, Park J, Choi JS. Antidiabetic complications and anti-Alzheimer activities of sophoflavescenol, a prenylated flavonol from Sophora flavescens, and its structure – activity relationship. Phytother Res. 2011;715:709–15.

Jütte R, Heinrich M, Helmstädter A, Langhorst J, Meng G, Niebling W, Pommerening T, Trampisch HJ. Herbal medicinal products – evidence and tradition from a historical perspective. J Ethnopharmacol. 2017;207:220–5.

Kanai M, Otsuka Y, Otsuka K, Sato M, Nishimura T, Mori Y, et al. A phase I study investigating the safety and pharmacokinetics of highly bioavailable curcumin (Theracurmin) in cancer patients. Cancer Chemother Pharmacol. 2013;71(6):1521–30. https://doi.org/10.1007/s00280-013-2151-8.

Kelber O, Bauer R, Kubelka W. Phytotherapy in functional gastrointestinal disorders. Dig Dis. 2017;35(Suppl 1):36–42. https://doi.org/10.1159/000485489. Epub 2018 Feb 8. PMID: 29421793.

Kennedy DO, Scholey AB. The psychopharmacology of European herbs with cognition-enhancing properties. Curr Pharm Des. 2006;12(35):4613–23. https://doi.org/10.2174/138161206779010387. PMID: 17168769.

Kim JH. Cardiovascular diseases and Panax ginseng: a review on molecular mechanisms and medical applications. J Ginseng Res. 2012;36(1):16–26. https://doi.org/10.5142/jgr.2012.36.1.16.

Kim MO, Lee MH, Oi N, Kim SH, Bae KB, Huang Z, et al. [6]-shogaol inhibits growth and induces apoptosis of non-small cell lung cancer cells by directly regulating Akt1/2. Carcinogenesis. 2014;35(3):683–91. https://doi.org/10.1093/carcin/bgt365.

Kim HJ, Shin EJ, Lee BH, Choi SH, Jung SW, Cho IH, Hwang SH, Kim JY, Han JS, Chung C, Jang CG, Rhim H, Kim HC, Nah SY. Oral administration of gintonin attenuates cholinergic impairments by scopolamine, amyloid-β protein, and mouse model of Alzheimer's disease. Mol Cells. 2015 Sep;38(9):796–805. https://doi.org/10.14348/molcells.2015.0116.

Kim Y-S, Nam Y, Song J, Kim H. Gastroprotective and healing effects of polygonum cuspidatum root on experimentally induced gastric ulcers in rats. Nutrients. 2020;12(8):2241. https://doi.org/10.3390/nu12082241.

Kiselev VI, Ashrafyan LA, Muyzhnek EL, Gerfanova EV, Antonova IB, Aleshikova OI, et al. A new promising way of maintenance therapy in advanced ovarian cancer: a comparative clinical study. BMC Cancer. 2018;18(1):904. https://doi.org/10.1186/s12885-018-4792-9.

Klupp NL, Chang D, Hawke F, Kiat H, Cao H, Grant SJ, Bensoussan A. Ganoderma lucidum mushroom for the treatment of cardiovascular risk factors. Cochrane Database Syst Rev. 2015;2015(2):CD007259. https://doi.org/10.1002/14651858.

Kong Y, Chen G, Xu Z, Yang G, Li B, Wu X, et al. Pterostilbene induces apoptosis and cell cycle arrest in diffuse large B-cell lymphoma cells. Sci Rep. 2016;6:37417. https://doi.org/10.1038/srep37417.

Kotsakis A, Matikas A, Koinis F, Kentepozidis N, Varthalitis II, Karavassilis V, et al. A multicentre phase II trial of cabazitaxel in patients with advanced non-small-cell lung cancer progressing after docetaxel-based chemotherapy. Br J Cancer. 2016;115(7):784–8. https://doi.org/10.1038/bjc.2016.281.

Kunnumakkara AB, Bordoloi D, Harsha C, Banik K, Gupta SC, Aggarwal BB. Curcumin mediates anticancer effects by modulating multiple cell signaling pathways. Clin Sci (Lond). 2017;131(15):1781–99. https://doi.org/10.1042/CS20160935.

Kuppusamy P, Nagalingam A, Muniraj N, Saxena NK, Sharma D. Concomitant activation of ETS-like transcription factor-1 and death Receptor-5 via extracellular signal-regulated kinase in withaferin a-mediated inhibition of hepatocarcinogenesis in mice. Sci Rep. 2017;7(1):17943. https://doi.org/10.1038/s41598-017-18190-4.

Lay S, Chiu J-H, Shiao M-S, Lui W-Y, Wu C-W. Crude extract of Salvia miltiorrhiza and salvianolic acid B enhance in vitro angiogenesis in murine SVR endothelial cell line. Planta Med. 2003;69(01):26–32. https://doi.org/10.1055/s-2003-37034.

Le Cornec C, Le Pottier M, Broch H, et al. Ketamine compared with morphine for out-of-hospital analgesia for patients with traumatic pain: a randomized clinical trial. JAMA Netw Open. 2024;7(1):e2352844. https://doi.org/10.1001/jamanetworkopen.2023.52844.

Lee H, Choi J, Shin SS, Yoon M. Effects of Korean red ginseng (Panax ginseng) on obesity and adipose inflammation in ovariectomized mice. J Ethnopharmacol. 2016;178:229–37. https://doi.org/10.1016/j.jep.2015.12.017.

Li J, Zhang C, Jiang H, Cheng J. Andrographolide inhibits hypoxia-inducible factor-1 through phosphatidylinositol 3-kinase/AKT pathway and suppresses breast cancer growth. Onco Targets Ther. 2015;8:427–35. https://doi.org/10.2147/OTT.S76116.

Li YJ, He FQ, Zhao HH, Li Y, Chen J. Screening and identification of acetylcholinesterase inhibitors from Terminalia chebula fruits by immobilized enzyme on cellulose filter paper coupled with ultra-performance liquid chromatography-quadrupole time-of-flight mass spectrometry and molecular docking. J Chromatogr A. 2022 Jan;25(1663):462784. https://doi.org/10.1016/j.chroma.2021.462784.

Lin FY, Chen YH, Chen YL, Wu TC, Li CY, Chen JW, et al. Ginkgo biloba extract inhibits endotoxin-induced human aortic smooth muscle cell proliferation via suppression of toll-like receptor 4 expression and NADPH oxidase activation. J Agric Food Chem. 2007;55(5):1977–84. https://doi.org/10.1021/jf062945r.

Lin W, Zhong M, Yin H, Chen Y, Cao Q, Wang C, et al. Emodin induces hepatocellular carcinoma cell apoptosis through MAPK and PI3K/AKT signaling pathways in vitro and in vivo. Oncol Rep. 2016;36(2):961–7. https://doi.org/10.3892/or.2016.4861.

Lin CY, Hsieh PL, Liao YW, Peng CY, Lu MY, Yang CH, et al. Berberine-targeted miR-21 chemosensitizes oral carcinomas stem cells. Oncotarget. 2017;8(46):80900–8. https://doi.org/10.18632/oncotarget.20723.

Liou CJ, Lai XY, Chen YL, Wang CL, Wei CH, Huang WC. Ginkgolide C suppresses adipogenesis in 3T3-L1 adipocytes via the AMPK signaling pathway. Evid Based Complement Alternat Med. 2015;2015:298635. https://doi.org/10.1155/2015/298635.

List PH, Schmidt PC. Phytopharmaceutical technology. London: Heyden; 1989.

Liu L, Wu J, Kennedy DJ. Regulation of cardiac remodeling by cardiac Na(+)/K(+)-ATPase isoforms. Front Physiol. 2016;7:382. https://doi.org/10.3389/fphys.2016.00382.

Lu WJ, Wu GJ, Chen RJ, Chang CC, Lien LM, Chiu CC, et al. Licochalcone A attenuates glioma cell growth in vitro and in vivo through cell cycle arrest. Food Funct. 2018;9(8):4500–7. https://doi.org/10.1039/c8fo00728d.

Luo HM, Dai RH, Li Y. Nuclear cardiology study on effective ingredients of Astragalus membranaceus in treating heart failure. Zhongguo Zhong Xi Yi Jie He Za Zhi. 1995;15(12):707–9.

Ma X, Zhang K, Li H, Han S, Ma Z, Tu P. Extracts from Astragalus membranaceus limit myocardial cell death and improve cardiac function in a rat model of myocardial ischemia. J Ethnopharmacol. 2013;149(3):720–8. https://doi.org/10.1016/j.jep.2013.07.036.

Machado Barros T, et al. Brazilian phytopharmaceuticals--evaluation against hospital bacteria. Phytother Res. 2005;19(6):519–25.

Mahaman YAR, Huang F, Wu M, Wang Y, Wei Z, Bao J, Salissou MTM, Ke D, Wang Q, Liu R, Wang JZ, Zhang B, Chen D, Wang X. Moringa oleifera alleviates homocysteine-induced Alzheimer's disease-like pathology and cognitive impairments. J Alzheimers Dis. 2018;63(3):1141–59. https://doi.org/10.3233/JAD-180091. PMID: 29710724; PMCID: PMC6004908.

Mahata S, Bharti AC, Shukla S, Tyagi A, Husain SA, Das BC. Berberine modulates AP-1 activity to suppress HPV transcription and downstream signaling to induce growth arrest and apoptosis in cervical cancer cells. Mol Cancer. 2011;10:39. https://doi.org/10.1186/1476-4598-10-39.

Mansour SM, Bahgat AK, El-Khatib AS, Khayyal MT. Ginkgo biloba extract (EGb 761) normalizes hypertension in 2K, 1C hypertensive rats: role of antioxidant mechanisms, ACE inhibiting activity and improvement of endothelial dysfunction. Phytomedicine. 2011;18(8–9):641–7. https://doi.org/10.1016/j.phymed.2011.01.014.

MATOS FJA. Introdução à fitoquímica experimental. 3ª ed. Fortaleza: Edições UFC, 2009.

Mao L, Chen Q, Gong K, Xu X, Xie Y, Zhang W, et al. Berberine decelerates glucose metabolism via suppression of mTORdependent HIF1alpha protein synthesis in colon cancer cells. Oncol Rep. 2018;39(5):2436–42. https://doi.org/10.3892/or.2018.6318.

Martin A, Fuzer AM, Becceneri AB, da Silva JA, Tomasin R, Denoyer D, et al. [10]-gingerol induces apoptosis and inhibits metastatic dissemination of triple negative breast cancer in vivo. Oncotarget. 2017;8(42):72260–71. https://doi.org/10.18632/oncotarget.20139.

Martino E, Casamassima G, Castiglione S, Cellupica E, Pantalone S, Papagni F, et al. Vinca alkaloids and analogues as anti-cancer agents: looking back, peering ahead. Bioorg Med Chem Lett. 2018;28(17):2816–26. https://doi.org/10.1016/j.bmcl.2018.06.044.

Mazzari AL, Prieto JM. Herbal medicines in Brazil: pharmacokinetic profile and potential herb-drug interactions. Front Pharmacol. 2014;5:162. https://doi.org/10.3389/fphar.2014.00162. Erratum in: Front Pharmacol. 2015;6:23. https://doi.org/10.3389/fphar.2015.00023. PMID: 25071580; PMCID: PMC4087670.

Mezni F, Aouadhi C, Khouja ML, Khaldi A, Maaroufi A. In vitro antimicrobial activity of Pistacia lentiscus L. edible oil and phenolic extract. Nat Prod Res. 2015;29:565–70. https://doi.org/10.1080/14786419.2014.952232.

Mock CD, Jordan BC, Selvam C. Recent advances of curcumin and its analogues in breast cancer prevention and treatment. RSC Adv. 2015;5(92):75575–88. https://doi.org/10.1039/C5RA14925H.

Moeini R, Memariani Z, Asadi F, Bozorgi M, Gorji N. Pistacia genus as a potential source of neuroprotective natural products. Planta Med. 2019;85(17):1326–50. https://doi.org/10.1055/a-1014-1075.

Mohamed I, Shuid A, Borhanuddin B, Fozi N. The application of phytomedicine in modern drug development. Internet J Herb Plant Med. 2012;1(2):1–9.

Mutha RE, Tatiya AU, Surana SJ. Flavonoids as natural phenolic compounds and their role in therapeutics: an overview. Futur J Pharm Sci. 2021;7:25–38.

Na AM. Effects of medicinal plants on Alzheimer's disease and memory deficits. Neural Regen Res. 2017;12:660–70. https://doi.org/10.4103/1673-5374.205108.

Nakamura K, Yasunaga Y, Segawa T, Ko D, Moul JW, Srivastava S, et al. Curcumin down-regulates AR gene expression and activation in prostate cancer cell lines. Int J Oncol. 2002;21(4):825–30. https://doi.org/10.3892/ijo.21.4.825.

O'Neill S, O'Driscoll L. Metabolic syndrome: a closer look at the growing epidemic and its associated pathologies. Obes Rev. 2015;16:1–12. https://doi.org/10.1111/obr.12229.

Odeh LH, Talib WH, Basheti IA. Synergistic effect of thymoquinone and melatonin against breast cancer implanted in mice. J Cancer Res Ther. 2018;14(Supplement):S324–30. https://doi.org/10.4103/0973-1482.235349.

Ojeda D, Jimenez-Ferrer E, Zamilpa A, Herrera-Arellano A, Tortoriello J, Alvarez L. Inhibition of angiotensin convertin enzyme (ACE) activity by the anthocyanins delphinidin- and cyanidin-3-O-sambubiosides from Hibiscus sabdariffa. J Ethnopharmacol. 2010;127(1):7–10. https://doi.org/10.1016/j.jep.2009.09.059.

Ono K, Hasegawa K, Naiki H, Yamada M. Curcumin has potent anti-amyloidogenic effects for Alzheimer's beta-amyloid fibrils in vitro. J Neurosci Res. 2004;75(6):742–50. https://doi.org/10.1002/jnr.20025. PMID: 14994335.

Oon MB, Nik A, Rahman NH, Mohd Noor N, et al. Patient-controlled analgesia morphine for the management of acute pain in the emergency department: a systematic review and meta-analysis. Int J Emerg Med. 2024;17:37. https://doi.org/10.1186/s12245-024-00615-3.

Oudard S, Fizazi K, Sengelov L, Daugaard G, Saad F, Hansen S, et al. Cabazitaxel versus docetaxel as first-line therapy for patients with metastatic castration-resistant prostate cancer: a randomized phase III trial-firstana. J Clin Oncol. 2017;35(28):3189–97. https://doi.org/10.1200/JCO.2016.72.1068.

Paller CJ, Rudek MA, Zhou XC, Wagner WD, Hudson TS, Anders N, et al. A phase I study of muscadine grape skin extract in men with biochemically recurrent prostate cancer: safety, tolerability, and dose determination. Prostate. 2015;75(14):1518–25. https://doi.org/10.1002/pros.23024.

Pan SY, Zhou SF, Gao SH, Yu ZL, Zhang SF, Tang MK, et al. New perspectives on how to discover drugs from herbal medicines: CAM's outstanding contribution to modern therapeutics. Evid Based Complement Alternat Med. 2013;2013:627375. https://doi.org/10.1155/2013/627375.

Patricio F, Morales-Andrade AA, Patricio-Martínez A, Limón ID. Cannabidiol as a therapeutic target: evidence of its neuroprotective and neuromodulatory function in Parkinson's disease. Front Pharmacol. 2020;11:595635.

Perez-Vizcaino F, Duarte J, Jimenez R, Santos-Buelga C, Osuna A. Antihypertensive effects of the flavonoid quercetin. Pharmacol Rep. 2009;61(1):67–75. https://doi.org/10.1016/S1734-1140(09)70008-8.

Pourmortazavi SM, Saghafi Z, Ehsani A, Yousefi M. Application of supercritical fluids in cholesterol extraction from foodstuffs: a review. J Food Sci Technol. 2018;55(8):2813–23.

Prasad S, Yadav VR, Sung B, et al. Ursolic acid inhibits growth and metastasis of human colorectal cancer in an orthotopic nude mouse model by targeting multiple cell signaling pathways: chemosensitization with capecitabine. Clin Cancer Res. 2012;18(18):4942–53. https://doi.org/10.1158/1078-0432.CCR-11-2805.

Qazi A, Pal J, Maitah M, Fulciniti M, Pelluru D, Nanjappa P, et al. Anticancer activity of a broccoli derivative, sulforaphane, in barrett adenocarcinoma: potential use in chemoprevention and as adjuvant in chemotherapy. Trans Oncol. 2010;3(6):389–99. https://doi.org/10.1593/tlo.10235.

Quitain AT, Kai T, Sasaki M, Goto M. Supercritical carbon dioxide extraction of fucoxanthin from Undaria pinnatifida. J Agric Food Chem. 2013;61(24):5792–7.

Rahman MA, Hwang H, Nah SY, Rhim H. Gintonin stimulates autophagic flux in primary cortical astrocytes. J Ginseng Res. 2020;44(1):67–78. https://doi.org/10.1016/j.jgr.2018.08.004. Epub 2018 Aug 27. PMID: 32148391; PMCID: PMC7033340.

Rahman MM, Bibi S, Rahaman MS, Rahman F, Islam F, Khan MS, Hasan MM, Parvez A, Hossain MA, Maeesa SK, Islam MR, Najda A, Al-Malky HS, Mohamed HRH, AlGwaiz HIM, Awaji AA, Germoush MO, Kensara OA, Abdel-Daim MM, Saeed M, Kamal MA. Natural therapeutics and nutraceuticals for lung diseases: traditional significance, phytochemistry, and pharmacology. Biomed Pharmacother. 2022;150:113041. https://doi.org/10.1016/j.biopha.2022.113041.

Razavi BM, Amanloo MA, Imenshahidi M, Hosseinzadeh H. The Relaxant Activity of Safranal in Isolated Rat Aortas is Mediated Predominantly via an Endothelium-Independent Mechanism: -Vasodilatory mechanism of safranal. J Pharmacopuncture. 2016;4:329–35. https://doi.org/10.3831/KPI.2016.19.034.

Reddy VS, Shiva S, Manikantan S, Ramakrishna S. Pharmacology of caffeine and its effects on the human body. Eur J Med Chem. 2024;100138. https://doi.org/10.1016/j.ejmcr.2024.100138.

Ricardo LM, Goulart EMA, Brandão MGL. Plantas medicinais da Bacia do Rio das Velhas: avaliação das condições para produção e uso em saúde pública. Rev Bras Plantas Med. 2015;17:398–406.

Robinson H. The gut brain axis: impact of dietary fiber on a murine model of multiple sclerosis. Ph.D. dissertation. Vancouver, BC, Canada: University of British Columbia; 2019.

Saha A, Blando J, Silver E, Beltran L, Sessler J, DiGiovanni J. 6-Shogaol from dried ginger inhibits growth of prostate cancer cells both in vitro and in vivo through inhibition of STAT3 and NF-kappaB signaling. Cancer Prev Res (Phila). 2014;7(6):627–38. https://doi.org/10.1158/1940-6207.CAPR-13-0420.

Santos RL, Guimaraes GP, Nobre MSC, Portela AS. Análise sobre a fitoterapia como prática integrativa no Sistema Único de Saúde. Rev Bras Plantas Med. 2011;13:486–91.

Saunshi YB, David CG, Pushpadass HA, Emerald Franklin ME, Awachat VB, Kadakol VR. Characterization of withanolides and bacoside A-loaded proniosomes: effect on oxidative stress and survival under hypergravity in rodent model. Drug Dev Ind Pharm. 2023;49(12):748–58. https://doi.org/10.1080/03639045.2023.2286702.

Schulz V, Hänsel R, Tyler VE. Fitoterapia Racional: um guia para as ciências da saúde. 4ª ed. Barueri: Manole; 2002. p. 386.

Sharapin N. Fundamentos de Tecnología de Productos Fitoterapéuticos. Bogotá: Convenio Andres Bello y CYTED; 2000.

Sharma RA, Euden SA, Platton SL, Cooke DN, Shafayat A, Hewitt HR, et al. Phase I clinical trial of oral curcumin: biomarkers of systemic activity and compliance. Clin Cancer Res. 2004;10(20):6847–54. https://doi.org/10.1158/1078-0432.CCR-04-0744.

Shi YL, James AE, Benzie IF, Buswell JA. Mushroom-derived preparations in the prevention of $H_2O_2$-induced oxidative damage to cellular DNA. Teratog Carcinog Mutagen. 2002;22(2):103–11. https://doi.org/10.1002/tcm.10008.

Shin W, Yoon J, Oh GT, Ryoo S. Korean red ginseng inhibits arginase and contributes to endotheliumdependent vasorelaxation through endothelial nitric oxide synthase coupling. J Ginseng Res. 2013;37(1):64–73. https://doi.org/10.5142/jgr.2013.37.64.

Shin SJ, Nam Y, Park YH, Kim MJ, Lee E, Jeon SG, Bae BS, Seo J, Shim SL, Kim JS, Han CK, Kim S, Lee YY, Moon M. Therapeutic effects of non-saponin fraction with rich polysaccharide from Korean red ginseng on aging and Alzheimer's disease. Free Radic Biol Med. 2021;20(164):233–48. https://doi.org/10.1016/j.freeradbiomed.2020.12.454. Epub 2021 Jan 8. PMID: 33422674.

Shoaib S, Ansari MA, Fatease AA, Safhi AY, Hani U, Jahan R, Alomary MN, Ansari MN, Ahmed N, Wahab S, et al. Plant-derived bioactive compounds in the management of neurodegenerative disorders: challenges, future directions and molecular mechanisms involved in neuroprotection. Pharmaceutics. 2023;15:749. https://doi.org/10.3390/pharmaceutics15030749.

Short NJ, Jabbour E, Naqvi K, Patel A, Ning J, Sasaki K, et al. A phase II study of omacetaxine mepesuccinate for patients with higher-risk myelodysplastic syndrome and chronic myelomonocytic leukemia after failure of hypomethylating agents. Am J Hematol. 2019;94(1):74–9. https://doi.org/10.1002/ajh.25318.

Silveira JC, Busato NV, da Costa AOS, Junior EFC. Levantamento e análise de métodos de extração de óleos essenciais. Encicl Biosf, Centro Científico Conhecer – Goiânia. 2012;8(15):2038.

Skroza N, Bernardini N, Proietti I, Potenza C. Clinical utility of ingenol mebutate in the management of actinic keratosis: perspectives from clinical practice. Ther Clin Risk Manag. 2018;14:1879–85. https://doi.org/10.2147/TCRM.S145779.

Slobodníková L, Košálová D, Labudová D, Kotulová D, Kettmann V. Antimicrobial activity of Mahonia aquifolium crude extract and its major isolated alkaloids. Phytother Res. 2004;18:674–6. https://doi.org/10.1002/ptr.1517.

Stern MP, Williams K, Gonzalez-Villalpando C, Hunt KJ, Haffner SM. Does the metabolic syndrome improve identification of individuals at risk of type 2 diabetes and/or cardiovascular disease? Diabetes Care. 2004;27:2676–81. https://doi.org/10.2337/diacare.27.11.2676.

Su J, Yan Y, Qu J, Xue X, Liu Z, Cai H. Emodin induces apoptosis of lung cancer cells through ER stress and the TRIB3/NF-kappaB pathway. Oncol Rep. 2017;37(3):1565–72. https://doi.org/10.3892/or.2017.5428.

Subramani R, Gonzalez E, Arumugam A, Nandy S, Gonzalez V, Medel J, et al. Nimbolide inhibits pancreatic cancer growth and metastasis through ROS-mediated apoptosis and inhibition of epithelial-to-mesenchymal transition. Sci Rep. 2016;6:19819. https://doi.org/10.1038/srep19819.

Suman S, Das TP, Sirimulla S, Alatassi H, Ankem MK, Damodaran C. Withaferin-a suppress AKT induced tumor growth in colorectal cancer cells. Oncotarget. 2016;7(12):13854–64. https://doi.org/10.18632/oncotarget.7351.

Sun H, Ni H, Yang Y, Wu L, Cai HN, Xiao AF, Chen F. Investigation of sunlight-induced deterioration of aroma of pummelo (Citrus maxima) essential oil. J Agric Food Chem. 2014;62(49):11818–30.

Sun M, Chai L, Lu F, Zhao Y, Li Q, Cui B, et al. Efficacy and safety of ginkgo Biloba pills for coronary heart disease with impaired glucose regulation: study protocol for a series of N-of-1 randomized, double-blind, placebo-controlled trials. Evid Based Complement Alternat Med. 2018;2018:7571629. https://doi.org/10.1155/2018/7571629.

Szego EM, Csorba A, Janáky T, Kékesi KA, Abrahám IM, Mórotz GM, Penke B, Palkovits M, Murvai U, Kellermayer MS, Kardos J, Juhász GD. Effects of estrogen on beta-amyloid-induced cholinergic cell death in the nucleus basalis magnocellularis. Neuroendocrinology. 2011;93(2):90–105. https://doi.org/10.1159/000321119. Epub 2010 Oct 9. PMID: 20938166.

Taiz L, Zeiger E. Fisiologia Vegetal. 3ª Ed. Porto Alegre: Artmed. 2009.

Tao Y, Zhan S, Wang Y, Zhou G, Liang H, Chen X, et al. Baicalin, the major component of traditional Chinese medicine Scutellaria baicalensis induces colon cancer cell apoptosis through inhibition of oncomiRNAs. Sci Rep. 2018;8(1):14477. https://doi.org/10.1038/s41598-018-32734-2.

Tapia-Rojas C, Cabezas-Opazo F, Deaton CA, Vergara EH, Johnson GVW, Quintanilla RA. It's all about tau. Prog Neurobiol. 2019;175:54–76. https://doi.org/10.1016/j.pneurobio.2018.12.005. Epub 2018 Dec 31. PMID: 30605723; PMCID: PMC6397676.

Taverna S, Giallombardo M, Pucci M, Flugy A, Manno M, Raccosta S, et al. Curcumin inhibits in vitro and in vivo chronic myelogenous leukemia cells growth: a possible role for exosomal disposal of miR-21. Oncotarget. 2015;6(26):21918–33. https://doi.org/10.18632/oncotarget.4204.

Tempone AG, et al. Brazilian flora extracts as source of novel antileishmanial and antifungal compounds. Mem Inst Oswaldo Cruz. 2008;103(5):443–9.

Thangapazham RL, Singh AK, Sharma A, Warren J, Gaddipati JP, Maheshwari RK. Green tea polyphenols and its constituent epigallocatechin gallate inhibits proliferation of human breast cancer cells in vitro and in vivo. Cancer Lett. 2007;245:232–41. https://doi.org/10.1016/j.canlet.2006.01.027.

Tougnuanchan P, Benjakul S, Prodpran T. Properties and antioxidant activity of fish skin gelatin film incorporated with citrus essential oils. Food Chem. 2012;134(3):1571–9.

Tozer GM, Kanthou C, Parkins CS, Hill SA. The biology of the combretastatins as tumour vascular targeting agents. Int J Exp Pathol. 2002;83(1):21–38. https://doi.org/10.1046/j.1365-2613.2002.00211.x.

Tsang CM, Cheung KC, Cheung YC, Man K, Lui VW, Tsao SW, et al. Berberine suppresses Id-1 expression and inhibits the growth and development of lung metastases in hepatocellular carcinoma. Biochim Biophys Acta. 2015;1852(3):541–51. https://doi.org/10.1016/j.bbadis.2014.12.004.

Uddin M, Mamun A, Hossain M, Ashaduzzaman M, Noor M. Neuroprotective effect of phyllanthus acidus L. on learning and memory impairment in scopolamine-induced animal model of dementia and oxidative stress: natural wonder for regulating the development and progression of alzheimer's disease. Advances in Alzheimer's Disease, 2016;5:53–72. https://doi.org/10.4236/aad.2016.52005.

Valli G, Giardina EG. Benefits, adverse effects and drug interactions of herbal therapies with cardiovascular effects. J Am Coll Cardiol. 2002;39(7):1083–95. https://doi.org/10.1016/s0735-1097(02)01749-7.

Wang YC, Bachrach U. The specific anti-cancer activity of green tea (-)-epigallocatechin-3-gallate (EGCG). Amino Acids. 2002;22(2):131–43. https://doi.org/10.1007/s007260200002.

Wang T, Jiang Y, Chu L, Wu T, You J. Alpinumisoflavone suppresses tumour growth and metastasis of clear-cell renal cell carcinoma. Am J Cancer Res. 2017;7(4):999–1015.

Wei L, Lv S, Huang Q, Wei J, Zhang S, Huang R, Zhongpeng L, Lin X. Pratensein attenuates Aβ-induced cognitive deficits in rats: enhancement of synaptic plasticity and cholinergic function. Fitoterapia. 2015;101:208–17. https://doi.org/10.1016/j.fitote.2015.01.017.

Wen W, Lowe G, Roberts CM, Finlay J, Han ES, Glackin CA, et al. Pterostilbene, a natural phenolic compound, synergizes the antineoplastic effects of megestrol acetate in endometrial cancer. Sci Rep. 2017;7(1):12754. https://doi.org/10.1038/s41598-017-12922-2.

Whayne TF Jr. Clinical use of digitalis: a state of the art review. Am J Cardiovasc Drugs. 2018;18(6):427–40. https://doi.org/10.1007/s40256-018-0292-1.

WHO. WHO traditional medicine strategy 2014–2023; 2023.

Wilken R, Veena MS, Wang MB, Srivatsan ES. Curcumin: a review of anti-cancer properties and therapeutic activity in head and neck squamous cell carcinoma. Mol Cancer. 2011;10:12. https://doi.org/10.1186/1476-4598-10-12.

Wu W, Tang SN, Zhang Y, Puppala M, Cooper TK, Xing C, et al. Prostate cancer xenograft inhibitory activity and pharmacokinetics of decursinol, a metabolite of angelica gigas pyranocoumarins, in mouse models. Am J Chin Med. 2017a;45(8):1773–92. https://doi.org/10.1142/S0192415X17500963.

Wu Q, Zhang H, Wang PG, Chen M. Evaluation of the efficacy and safety of Ganoderma lucidum mycelium-fermented liquid on gut microbiota and its impact on cardiovascular risk factors in human RSC Adv. 2017b;7(71):45093–100. https://doi.org/10.1039/C7RA08087E.

Xu Y, Ho CT, Amin SG, Han C, Chung FL. Inhibition of tobacco-specific nitrosamine-induced lung tumorigenesis in A/J Mice by green tea and its major polyphenol as antioxidants. Cancer Res. 1992;52:3875–9. https://doi.org/10.1016/0169-5002(93)90327-T.

Yallapu MM, Maher DM, Sundram V, Bell MC, Jaggi M, Chauhan SC. Curcumin induces chemo/radio-sensitization in ovarian cancer cells and curcumin nanoparticles inhibit ovarian cancer cell growth. J Ovarian Res. 2010;3:11. https://doi.org/10.1186/1757-2215-3-11.

Yan XB, Xie T, Wang SD, Wang Z, Li HY, Ye ZM. Apigenin inhibits proliferation of human chondrosarcoma cells via cell cycle arrest and mitochondrial apoptosis induced by ROS generation-an in vitro and in vivo study. Int J Clin Exp Med. 2018;11(3):1615–31.

Yare K, Woodward M. Hormone therapy and effects on sporadic Alzheimer's disease in post-menopausal women: importance of nomenclature. J Alzheimer's Dis: JAD. 2020;73(1):23–37. https://doi.org/10.3233/jad-190896. PMID: 31744009.

Youn DH, Park J, Kim HL, Jung Y, Kang J, Lim S, et al. Berberine improves benign prostatic hyperplasia via suppression of 5 alpha reductase and extracellular signal-regulated kinase in vivo and in vitro. Front Pharmacol. 2018;9:773. https://doi.org/10.3389/fphar.2018.00773.

Zhang W-D, Chen H, Zhang C, Liu R-H, Li H-L, Chen H-Z. Astragaloside IV from Astragalus membranaceus shows cardioprotection during myocardial ischemia in vivo and in vitro. Planta Med. 2006;72(01):4–8. https://doi.org/10.1055/s-2005-873126.

Zhang Y, Li Q, Zhou D, Chen H. Genistein, a soya isoflavone, prevents azoxymethane-induced up-regulation of WNT/beta-catenin signalling and reduces colon pre-neoplasia in rats. Br J Nutr. 2013;109(1):33–42. https://doi.org/10.1017/S0007114512000876.

Zhang B, Fan X, Wang Z, Zhu W, Li J. Alpinumisoflavone radiosensitizes esophageal squamous cell carcinoma through inducing apoptosis and cell cycle arrest. Biomed Pharmacother. 2017a;95:199–206. https://doi.org/10.1016/j.biopha.2017.08.048.

Zhang W, Su J, Xu H, Yu S, Liu Y, Zhang Y, et al. Dicumarol inhibits PDK1 and targets multiple malignant behaviors of ovarian cancer cells. PLoS One. 2017b;12(6):e0179672. https://doi.org/10.1371/journal.pone.0179672.

Zhang Y, Huang L, Shi H, Chen H, Tao J, Shen R, et al. Ursolic acid enhances the therapeutic effects of oxaliplatin in colorectal cancer by inhibition of drug resistance. Cancer Sci. 2018a;109(1):94–102. https://doi.org/10.1111/cas.13425.

Zhang X, Xiao C, Liu H. Ganoderic acid a protects rat H9c2 cardiomyocytes from hypoxia-induced injury via up-regulating miR-182-5p. Cell Physiol Biochem. 2018b;50(6):2086–96. https://doi.org/10.1159/000495053.

Zhao Y, Jing Z, Lv J, Zhang Z, Lin J, Cao X, et al. Berberine activates caspase-9/cytochrome c-mediated apoptosis to suppress triple-negative breast cancer cells in vitro and in vivo. Biomed Pharmacother. 2017;95:18–24. https://doi.org/10.1016/j.biopha.2017.08.045.

Zhu W, Qin W, Zhang K, Rottinghaus GE, Chen YC, Kliethermes B, et al. Trans-resveratrol alters mammary promoter hypermethylation in women at increased risk for breast cancer. Nutr Cancer. 2012;64(3):393–400. https://doi.org/10.1080/01635581.2012.654926.

Zhu WQ, Wang J, Guo XF, Liu Z, Dong WG. Thymoquinone inhibits proliferation in gastric cancer via the STAT3 pathway in vivo and in vitro. World J Gastroenterol. 2016;22(16):4149–59. https://doi.org/10.3748/wjg.v22.i16.4149.

Zuo GY, Meng FY, Hao XY, Zhang YL, Wang GC, Xu GL. Antibacterial alkaloids from Chelidonium majus Linn (Papaveraceae) against clinical isolates of methicillin-resistant Staphylococcus aureus. J Pharm Pharm Sci. 2009;11:90–4.

# Chapter 6
# Drugs and Interaction Attributes

**Caroline V. L. Moreira, Olusola J. Ayobami, John I. Ogbu, Denis B. Rosemberg, and James O. Fajemiroye**

**Abstract** Understanding the importance of drug interaction is fundamental for patient-centered care and safe use. Assessment of pharmacological profiles is an interesting strategy for new drug development and associations, treatment optimization, prevention of adverse effects, pharmacovigilance, and personalized medicine. A rational and unified system of classifying drugs and their interaction remains challenging due to confounding variables such as pathological conditions, genetic factors and microenvironments, associations of drugs, and non-drug entities (herbs, vitamins, nutrients, etc.). In this chapter, we focus on chemical classes, clinical indications, pharmacological interaction, and targets of drugs that are associated with their therapeutic or adverse outcomes. The knowledge of drugs and interaction attributes could inspire innovative pharmaceutical research and rational associations of drugs.

**Keywords** Drug classes · Structure · Targets · Attributes

## 1 Introduction

The unique property of drug- and pathology-related changes could impact their interaction attributes. As an example, treatments commonly used in pathological conditions like posttraumatic stress disorder and obsessive-compulsive disorder, panic disorder, generalized anxiety disorder (GAD), social anxiety disorder, and specific phobias involve the use of selective serotonin reuptake inhibitors (SSRIs), norepinephrine reuptake inhibitors, azapirones, mixed antidepressants,

C. V. L. Moreira · O. J. Ayobami · J. I. Ogbu · J. O. Fajemiroye (✉)
Institute of Biological Sciences, Federal University of Goias, Goiania, Goias, Brazil
e-mail: ayobami.olusola@fuoye.edu.ng; jamesfajemiroye@ufg.br

D. B. Rosemberg
Department of Biochemistry and Molecular Biology, Federal University of Santa Maria, Santa Maria, Rio Grande do Sul, Brazil

 
J. O. Fajemiroye (ed.), *Fundamentals of Drug and Non-Drug Interactions*,
https://doi.org/10.1007/978-3-031-80107-5_6

antipsychotics, alpha- and beta-adrenergic, and GABAergic medications. The actions of these drugs are in response to etiological neurochemical changes. The absorption and transport of drugs and the primary response of a certain tissue can be altered by infections (e.g., bacterial meningitis) due to changes in the blood-brain barrier permeability. Inflammation which often lowers tissue pH (acidic environment) could alter drug absorption. Local anesthetics like procaine are weak bases, and in acidic conditions, they become ionized. The ionized anesthetics may have difficulty crossing cell membranes, reducing their absorption and efficacy. Inflammation increases local blood flow and clearance of the anesthetic from the injection site, reducing its duration of action.

Therefore, providing accurate information on physicochemical properties, pharmacokinetics, and pharmacodynamics characteristics is of great relevance in deciphering interactions. The presence of distinct functional groups can influence chemical reactions, like hydroxylation, acylation, alkylation, dealkylation, and arylation, as well as help explain the nature of metabolites, toxicity liabilities, ligand interactions, and specific interactions with target proteins, thus elucidating putative mechanisms of actions (Ertl et al. 2020). The release of inactive, active, less or more potent metabolites from drugs could aid in clarifying interaction attributes. Metabolites can be toxic, such as the products of acetaminophen metabolism, which can cause liver failure, or the metabolite of meperidine, which can promote seizures. Inhibition of metabolism could result in potentially toxic concentrations of the parent compound. However, if the parent drug needs to be metabolized to the active compound (e.g., enalapril to enalaprilat; codeine to morphine) and metabolism is inhibited, a therapeutic failure exists. Enhancing drug-metabolizing enzymes could similarly induce a subtherapeutic effect by reducing drug levels and thus its efficacy while impacting the drug potency. The active metabolite of terfenadine, which is indicated for allergic rhinitis, fexofenadine, has equal potency at the histamine receptor. It is more than 50 times less active in blocking potassium channels in the heart when compared with the parent drug. In some cases, the metabolites show distinct pharmacologic actions by modulating other molecular targets not affected by the parent drug.

Drug classification on the basis of their mechanism of action is often not sufficient to establish a spectrum of drug attributes, interactions, and effects. For instance, antipsychotic drugs can elicit a reduction in hallucinations, delusions, disorganized thinking, extreme agitation and restlessness, manage mood swings, improvement in thinking, reasoning, social engagement, muscle rigidity, tremors, restlessness, involuntary movements, drowsiness and impaired alertness, fever, muscle stiffness, and confusion, hypotension, hyperlipidemia, hypertension, increased heart rate, risk of arrhythmias, dry mouth, blurred vision, constipation, urinary retention, weight gain and metabolic changes, hyperprolactinemia, leading to lactation, gynecomastia, and sexual dysfunction, agranulocytosis, and increased risk of stroke.

The action of antipsychotics on multiple targets (antagonists of the dopamine D2 receptor, antagonists or inverse agonists of the 5-HT2A receptor or other 5-HT receptor subtypes, adrenergic, muscarinic, and histamine receptors), as well as the degree of receptor occupancy, tissue specificity, or binding profiles, are sometimes accounted for when describing their effects. Aripiprazole, as a partial agonist of D2 and 5-HT1A receptors and an antagonist of the 5-HT2A receptor, was considered a dopamine-serotonin system stabilizer. Dopamine stabilizers like (-)-OSU6162 appear to act in a subpopulation of striatal D2/D3 receptors with moderate affinity. However, defining a set of mechanistic drug targets is not always straightforward (Santos et al. 2017). The binding of drugs with low, moderate, or even high affinity to multiple targets that are unrelated to the targets underlying their therapeutic effects and efficacy is often neglected (Santos et al. 2017).

Like antipsychotics, a wide range of descriptions and nomenclatures such as antihistamines, mucolytics, antiemetics, antiasthmatic; nasal decongestants, antihypertensives, calcium channel blockers (CCBs); renin-angiotensin system (RAS) inhibitors; adrenergic receptor blockers, vasodilator, cardiac glycosides, vasodilators, diuretics, antianginals, antiarrhythmics, anticoagulants and antiplatelet, antivertigo, antimalarials, antibiotics, antivirals, immunosuppressants, antidiabetics, hypoglycemics, prolactin inhibitors, anorectics, muscle relaxants, sedative, hypnotic antipyretic, analgesic, anti-inflammatory, antipsychotic or neuroleptics, antidepressants, antiepileptics, antiparkinsonians, antimigraine, and antineoplastic are commonly used to classify actions of drugs on the nervous, cardiovascular, gastrointestinal, respiratory, endocrine, and immune systems. These descriptions provide interaction attributes and insightful clues that define a wide variety of drugs (atropine, muscarine, pilocarpine, neostigmine, physostigmine, adrenaline, noradrenaline, dopamine, ephedrine, isoproterenol, morphine and other opioids, barbiturates, benzodiazepines (BZDs), insulin, quinolones, sulfonamides, macrolide, aminoglycoside, β-lactam antibiotics) and their principal targets. This chapter briefly presents overviews of representative drugs belonging to different classes with varying chemical structures, mechanisms of action, and effects (therapeutic and adverse).

## 2 Representative Drugs for CNS Disorders and Pain Control

### *2.1 Attributes of Hypnotics-Sedatives*

Hypnotic-sedative drugs are primarily indicated for conditions related to sleep and anxiety disorders. Barbiturates were once extensively used as sedative-hypnotic agents but have largely been replaced by safer alternatives like benzodiazepines and Z compounds (Peng et al. 2022). Orexin-A and Orexin-B are neurotransmitters produced by neurons in the lateral hypothalamus. These orexins are crucial for regulating the sleep cycle, promoting wakefulness, and enhancing REM and non-REM sleep. Suvorexant approved for insomnia inhibits orexin 1 and 2 receptors (Uslaner et al. 2020). Melatonin receptor agonists, although not regulated as controlled substances are approved for sleep-onset insomnia. The main types of tranquilizing and hypnotic medications prescribed for anxiety and insomnia disorders include benzodiazepines (BZDs) and non-benzodiazepines, such as diazepam and zolpidem, respectively (Geulayov et al. 2018). BZDs are widely used for their anxiolytic properties and rapid onset of action (Balon and Starcevic 2020). They are commonly prescribed for short-term relief in GAD, acute management of panic disorder, and rapid anxiolytic effects in social anxiety disorder (Geulayov et al. 2018).

The BZDs and barbiturates (Fig. 6.1) are positive allosteric modulators of GABA-A receptors (Griffin et al. 2013; Bonanno et al. 2021), which are ligand-gated chloride-selective ion channels. Upon activating these receptors, BZDs and barbiturates increase the inhibitory effect of γ-aminobutyric acid (GABA), increasing the frequency and duration of channel openings. At higher concentrations, barbiturates can directly activate GABA-A receptors (i.e. exert agonistic effect). Unlike benzodiazepines, which can act on multiple GABA-A receptor subunits, Z compounds have a higher selectivity for the α1-subunit, which is thought to mediate their hypnotic effects (Simone and Bobrin 2023). BZDs are not suitable for patients with severe respiratory or hepatic issues, sleep apnea, and myasthenia gravis, while barbiturates are not indicated for patients with porphyria, severe respiratory or liver problems, or those with a history of substance abuse. Common adverse effects of BZDs and barbiturates include drowsiness, dizziness, dependence, and abuse potential. Withdrawal symptoms can be severe for both BZDs and barbiturates medications (Pepe et al. 2021). According to Sanabria et al. (2021), BZDs should be avoided in individuals with a history of substance abuse, severe respiratory insufficiency, or sleep apnea. They can lead to drowsiness, dizziness, confusion, dependency, withdrawal symptoms, and potential overdose.

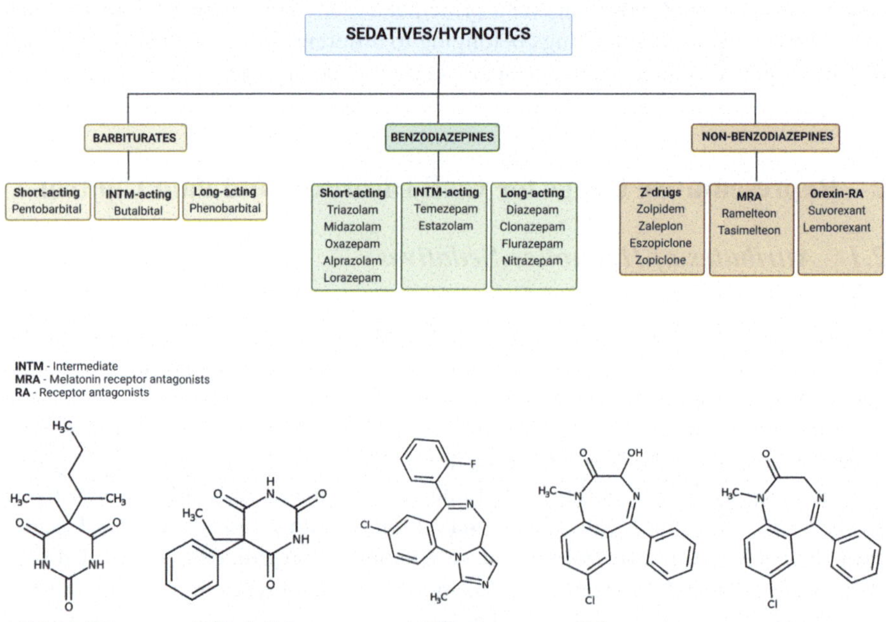


**Fig. 6.1** Some sedatives-hypnotics

## 2.2 Attributes of Antidepressant Drugs

Antidepressants including selective serotonin reuptake inhibitors (SSRIs), include medications like fluoxetine, sertraline, citalopram, escitalopram, and paroxetine (Laux 2021); serotonin and norepinephrine reuptake inhibitors (SNRIs) such as venlafaxine, duloxetine, desvenlafaxine, and levomilnacipran are all categorized based on their mechanisms of action. The class tricyclic antidepressants (TCAs) like amitriptyline, nortriptyline, imipramine, and clomipramine which modulate various neurotransmitter systems reflects chemical structure. Atypical antidepressants, which do not fit into the traditional classes, include bupropion, mirtazapine, trazodone, nefazodone, vilazodone, vortioxetine, and agomelatine. Monoamine oxidase inhibitors (MAOIs) including phenelzine, tranylcypromine, isocarboxazid, and selegiline (Fajemiroye et al. 2014, 2016; Sheffler et al. 2024) are classified based on their inhibitory role on metabolic enzymes. Selected examples of antidepressants are illustrated in Fig. 6.2a–f.

Unlike hypnotic-sedative agents, antidepressants are often prescribed for long-term management of anxiety. Anxiety-related disorders, including generalized anxiety disorder (GAD), panic disorder, and phobic disorders, have deleterious impacts on daily life. Anxiolytics, or anti-anxiety medications, are used to alleviate these symptoms. SSRIs, initially developed as antidepressants, are effective for treating GAD and other anxiety disorders (Garakani et al. 2020). While SNRIs are similarly effective for both anxiety and depression (Dell'Osso et al. 2010), SSRIs and SNRIs are first-line treatments for GAD and social anxiety disorder, providing efficacy with a lower risk of dependence. Buspirone, a non-benzodiazepine anxiolytic acting as a partial agonist at the serotonin 5-HT1A receptor, is particularly useful for GAD and has a lower risk of dependence compared to benzodiazepines (Wilson and Tripp 2024).

The regulation of serotoninergic activity by SSRIs and SNRIs contributes to anxiolytic mechanisms (Wilson and Tripp 2024). SSRIs inhibit the reuptake of serotonin (5-HT) at presynaptic terminals, increasing its availability in the synaptic cleft, which improves mood and emotional stability (Smith et al. 2023). SNRIs inhibit norepinephrine reuptake, enhancing serotonin and norepinephrine neurotransmission (Robinson et al. 2022). The TCAs block the reuptake of norepinephrine and serotonin and inhibit histamine, muscarinic, and adrenergic receptors, contributing to a broad spectrum of side effects. MAOIs inhibit the breakdown of

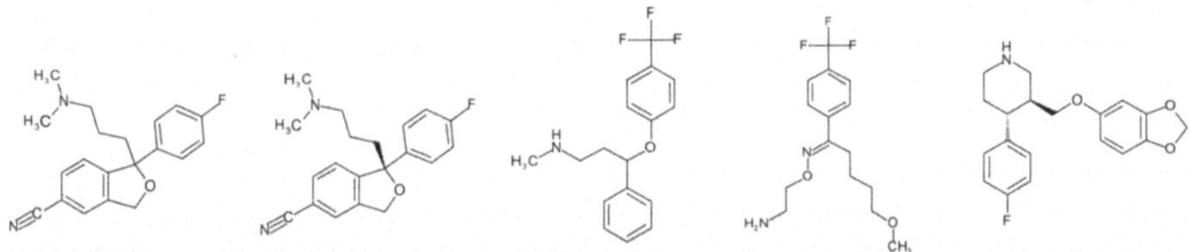

(a) Citalopram (b) Escitalopram (c) Fluoxetine (d) Fluvoxamine (e) Paroxetine

**Fig. 6.2** Some antidepressants commonly used in clinics

serotonin and noradrenaline, thereby increasing these neurotransmitters' levels (Martín-Hernández et al. 2021). Among atypical antidepressants, bupropion stands out by inhibiting the reuptake of norepinephrine and dopamine, enhancing their concentrations in the synaptic cleft. Trazodone and Nefazodone primarily act as serotonin receptor antagonists and weak serotonin reuptake inhibitors (Lin et al. 2023). Mirtazapine works uniquely by antagonizing central presynaptic α2-adrenergic receptors, increasing the release of norepinephrine and serotonin, and blocking specific serotonin receptors to mitigate anxiety and depression (Jilani et al. 2024). Ketamine, an NMDA receptor antagonist, provides rapid antidepressant effects, a significant advancement particularly relevant for treatment-resistant cases.

Antidepressants, particularly TCAs and SNRIs, and to a lesser extent SSRIs, are not recommended as monotherapy for bipolar disorder due to the risk of inducing mania or hypomania (Gomes et al. 2022). TCAs can cause sedation, dry mouth, blurred vision, constipation, urinary retention, orthostatic hypotension, exacerbate cardiac arrhythmias, and decreased seizure threshold (Kadioglu and Celik 2021). MAOIs should not be used with other medications, including SSRIs, due to the risk of serotonin syndrome and must be avoided with tyramine-rich foods to prevent hypertensive crises (Mrozek et al. 2023). Citalopram has a dose-related warning for QT prolongation not shared by escitalopram (Bardeci et al. 2023). Vortioxetine may cause gastrointestinal upset and sexual dysfunction (Diego-Adeliño et al. 2022), as nefazodone can promote liver damage (Jiang et al. 2024). Mirtazapine can cause somnolence, increase appetite, and weight gain (Davis 2020). SSRIs and SNRIs may cause nausea, headache, sexual dysfunction, increased anxiety, and serotonin syndrome.

## 2.3 Attributes of Antipsychotics

Antipsychotics are often used in the treatment of schizophrenia. They are divided into first- and second-generation medications. The patient's symptoms determine the prescription to control positive or negative symptoms. The main challenge of this treatment is overcoming the reported adverse effects to improve patient adherence and reduce disease recurrence (Kroken et al. 2009). Typical antipsychotics alleviate positive symptoms, such as hallucinations, delusions, and cognitive changes, by reducing excessive dopamine transmission. Chlorpromazine, fluphenazine, haloperidol, loxapine, perphenazine and more are examples of typical antipsychotics (Fig. 6.3a–e), as well as aripiprazole, clozapine, olanzapine, quetiapine, lumateperone, sertindole, and asenapine (atypical antipsychotics—Fig. 6.3f–j, respectively), are among the leading antipsychotic drugs:

Negative symptoms such as apathy and social withdrawal can improve with atypical antipsychotics (Meltzer and Huang 2008). However, some antipsychotics in a dose range could induce or attenuate seizure occurrence. Antipsychotic-related increase in seizure frequency is associated with the elevated blood plasma level

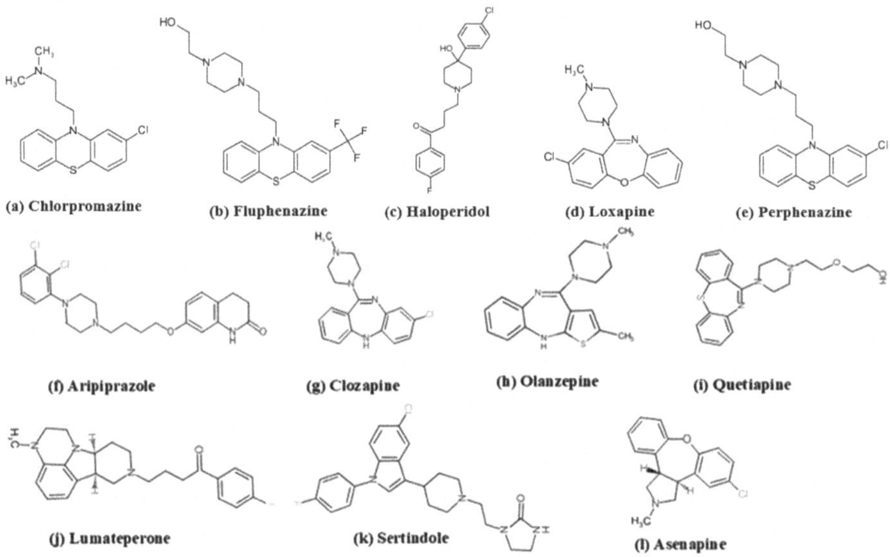


**Fig. 6.3** Some typical (**a–e**) and atypical (**f–l**) antipsychotics commonly used in clinics

effect of antipsychotics. On the other hand, it is possible to reduce the risk of seizures and detect their antiepileptic effect by titrating these drugs for a small effective dose. In general, these atypical antiepileptics are crucial in managing epilepsy, a multifactorial disease characterized by paroxysmal discharges of cerebral neurons resulting in visible and sensorial clinical events. These drugs selectively block repetitive discharges at concentrations lower than those that block normal impulse conduction. The antiepileptic mechanisms could involve the inhibition and enhancement of specific neurotransmission. As an inhibitory neurotransmitter, GABA facilitates $Cl^-$ influx, leading to hyperpolarization and suppressing epileptic discharges. Typical antiepileptic drugs could enhance GABA receptor activation and chloride ion influx, inhibit GABA transaminase and glutamate release by modulating N-methyl-D-aspartate receptors, and prolong the inactivated state of $Na^+$ channels, thereby reducing the likelihood of repetitive action potentials, among other mechanisms.

Dopamine receptors are considered as the main target of antipsychotics. The first-generation drugs have a high affinity for D2 subtype receptors, while second-generation antipsychotics antagonize D2 receptors and have varying affinities for other receptor subtypes. By antagonizing dopamine receptors, antipsychotics help regulate excessive dopamine activity, which is often associated with psychotic disorders such as schizophrenia and its clinical manifestations (Meltzer and Huang 2008). Although both classes block D2 receptors, typical antipsychotics have a higher risk of movement-related side effects compared to atypical antipsychotics. Atypical antipsychotics can also antagonize serotonin receptors, thereby reducing negative symptoms and cognitive impairment that are associated with schizophrenia.

Drug interactions are frequent with antipsychotics due to their therapeutic range. Changes in hepatic metabolism can either reduce efficacy by decreasing plasma concentration or increase toxicity. These drugs can interact or displace each other from plasma binding sites or induce metabolic enzymes. Antipsychotics can elicit dyslipidemia, impaired glucose tolerance, diabetes, weight gain, and extrapyramidal symptoms, including tremors, restlessness, and repetitive involuntary movements, especially in the elderly. Dry mouth, blurred vision, nasal congestion, constipation, urinary retention, sedation, depression, endocrine changes, and postural hypotension are among their adverse effects (Kumar et al. 2024).

## *2.4 Attributes of Analgesics*

Pain encompasses complex experiences involving signal generation, transmission, modulation, and perception of nociceptive stimuli. Pain is being managed with a wide range of analgesics including opioids, nonsteroidal anti-inflammatory drugs (NSAIDs), antiepileptics, antidepressants, and local anesthetics. Opioids are known for their potent analgesic properties, being suitable for treating severe pain, such as that experienced postoperatively or in cancer (Preuss et al. 2024). The term "opioid" encompasses a wide range of substances, including natural opiates derived from the opium poppy (e.g., morphine, codeine), semi-synthetic derivatives (e.g., oxycodone, hydromorphone), and fully synthetic compounds (e.g., fentanyl, methadone). Morphine, the prototypical opioid, serves as a benchmark for severe pain relief, along with derivatives like hydromorphone and oxymorphone. Some of these drugs are known for their high potency and rapid onset, which makes them ideal for acute pain. Methadone, a synthetic opioid with a long duration of action, is frequently used in opioid dependence therapy due to its ability to reduce withdrawal symptoms. Codeine, hydrocodone, and tramadol are used for mild to moderate pain. In addition to opioid receptor activation, tramadol also inhibits serotonin and norepinephrine reuptake. Partial agonists and mixed agonist-antagonists, such as buprenorphine, are employed in pain management and opioid dependence treatment due to their safer profile regarding respiratory depression. Structures of opioid analgesics are illustrated in Fig. 6.4.

Opioids bind to mu-opioid receptors (MOR) in the peripheral and central nervous systems, particularly in brain regions involved in pain perception, reward, and emotion including the nucleus accumbens, ventral tegmental area (VTA), and prefrontal cortex. In the VTA, activation of these receptors inhibits GABA release, which normally suppresses dopaminergic neurons. This inhibition leads to increased dopamine release in the nucleus accumbens and other reward pathways, resulting in the euphoric and pleasurable effects that are associated with opioid use. The MOR mediates pain relief and euphoria and also causes respiratory depression (Simone and Bobrin 2023). Opioids have significant contraindications and must be used

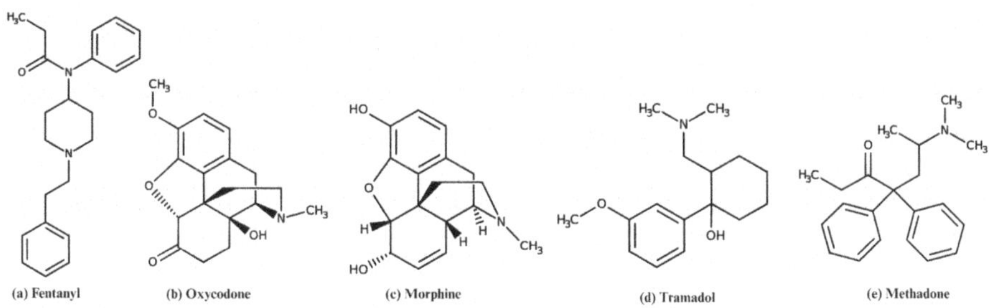


**Fig. 6.4** Representative structures of opioid analgesics

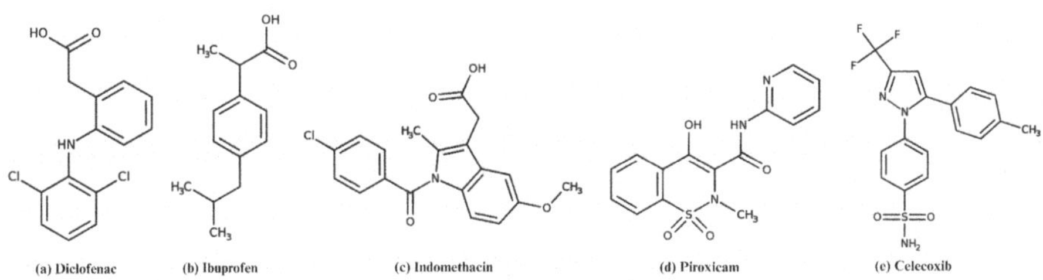


**Fig. 6.5** Representative structures of nonsteroidal anti-inflammatory drugs

cautiously. They are not suitable for patients with respiratory compromise, head injury, hypovolemia, or hypotension. They should be used with extreme caution in asthma patients. Adverse effects include respiratory depression, constipation, pruritus, allergic reactions, risk of dependency, and withdrawal symptoms.

In addition to opioids, nonsteroidal anti-inflammatory drugs (NSAIDs) (Fig. 6.5) are among the leading and widely used analgesics. Since the extraction of salicylates from willow bark in 1830 and the discovery of aspirin in 1897, NSAIDs have become widely used as over-the-counter medications, comprising about 5% of all prescriptions (Tsutsumi et al. 2004). These drugs are known for their analgesic, anti-inflammatory, and antipyretic effects. NSAIDs are used for conditions like osteoarthritis, menstrual cramps, and postoperative pain. NSAIDs have a heterogeneous structural framework but share a common feature. An acidic group attached to a planar aromatic group adapts them well to their pharmacological effects. NSAIDs are classified based on their structure (salicylates, aryl and heteroaryl acetic acid derivatives, indole/indene acetic acid derivatives, anthranilates, and enolic acid derivatives), cyclooxygenase (COX) selectivity, and plasma half-lives (Kowalski and Makowska 2015; Bindu et al. 2020).

NSAIDs function primarily by inhibiting the cyclooxygenase (COX) enzymes in the COX-prostanoid pathway, which is crucial for converting arachidonic acid into prostanoids such as thromboxane and prostaglandins (Vane 1996; Vane et al. 1998). This inhibition reduces prostaglandin synthesis, which plays a role in pain, inflammation, and fever. COX-1 is expressed constitutively, protects gastric mucosa, and maintains renal function, while COX-2 is induced during inflammatory processes

(Smyth et al. 2009). NSAIDS are not indicated in cases of gastrointestinal bleeding, ulcers, or pregnancy. Adverse effects (including gastrointestinal issues, renal effects, and CNS-related side effects) of NSAIDs can involve multiple systems (Shi and Vanhoutte 2008; Ahmadi et al. 2022). Additionally, they can lead to increased production of free radicals, contributing to tissue damage during inflammation. They are also associated with potential non-canonical effects, such as antibacterial and antineoplastic activities (Ogundeji et al. 2016; Chittepu et al. 2019).

## 3 Drugs for Cardiovascular Diseases

### *3.1 Attributes of Antihypertensives*

Hypertension is a significant risk factor for cardiovascular and kidney diseases. Genetic factors, vascular tone, and hormonal influences play roles in hypertension (Stergiou et al. 2021; Hail et al. 2024). Treatment includes the use of antihypertensives (Valenzuela et al. 2021) such as diuretics (i.e., thiazide and related agents, loop and potassium-sparing diuretics), sympatholytic agents (i.e., alpha, beta, mixed alpha, and beta-blockers, central alpha agonists), calcium channel blockers (CCBs, e.g., amlodipine and verapamil), inhibitors of the renin-angiotensin system (RAS) (i.e., angiotensin-converting enzyme inhibitors (ACEIs), angiotensin II receptor blockers (ARBs), direct renin inhibitors, and vasodilators (Reinhart et al. 2023). Generally, sympatholytic could reduce myocardial contractility, venous return, or vascular resistance by relaxing the smooth muscles in the blood vessels or indirectly interfering with the sympathetic nervous system or RAS; the diuretics often target transporters to reduce blood pressure by decreasing blood volume through increased $Na^+$ and water excretion; the CCBs inhibit the influx of $Ca^{2+}$ ions into cardiac and smooth muscle cells, resulting in the relaxation of vascular smooth muscle, decreased peripheral resistance, and reduced myocardial oxygen demand; the ACEIs lower blood pressure by inhibiting the conversion of angiotensin I to angiotensin II; the ARBs inhibit the effects of angiotensin II at its receptor sites, preventing vasoconstriction; vasodilators may reduce intracellular $Ca^{2+}$ levels by interfering with the inositol trisphosphate pathway (hydralazine) and open $K^+$ channels (minoxidil), thereby decreasing vascular resistance and blood pressure.

Some of these antihypertensives (Fig. 6.6) are known for adverse effects and contraindications. Diuretics, such as thiazides, are not indicated in patients with gout and pregnant women and may cause electrolyte imbalances and metabolic changes. Sympatholytic agents, including beta-blockers and alpha-blockers, could elicit reactive airway disease and heart failure. Beta-blockers may affect lipid profiles and cause withdrawal symptoms if abruptly discontinued, whereas alpha-blockers can cause orthostatic hypotension, particularly at the initial dose. Centrally acting sympatholytic often leads to sedation (clonidine) and depression and can induce a withdrawal syndrome if stopped suddenly. Inhibitors of RAS are generally

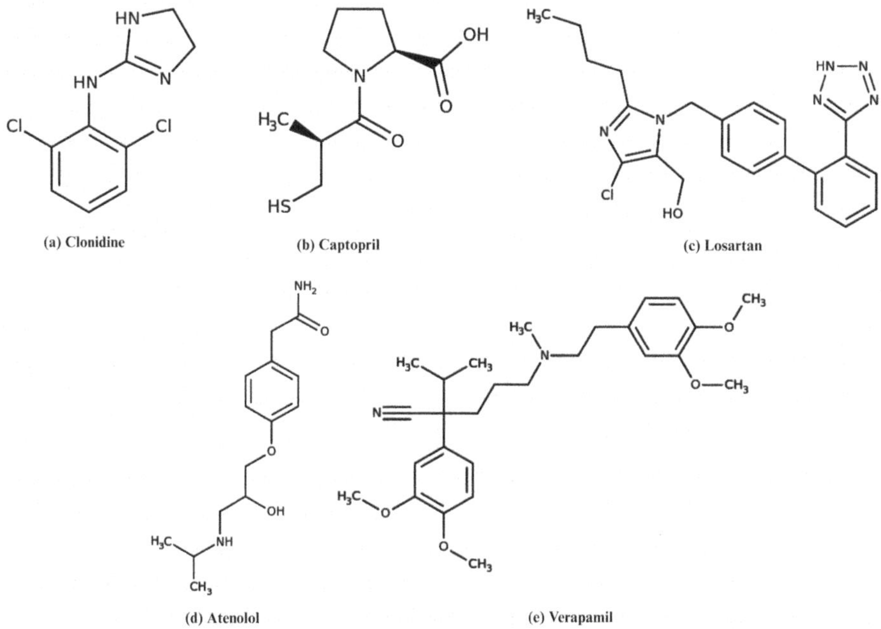


**Fig. 6.6** Representative structures of antihypertensives—clonidine (**a**), Captopril (**b**), Losartan (**c**), Atenolol (**d**), and Verapamil (**e**)

contraindicated in pregnancy, hyperkalemia conditions, etc., as vasodilators can provoke tachycardia and lupus-like syndrome, especially at high doses.

## 3.2 *Attributes of Drugs for Ischemic Heart Disease*

Globally, the most common type of heart disease is coronary artery disease, also known as ischemic heart disease. This condition is characterized by the narrowing of the coronary arteries due to the formation of plaque from deposits of fat, cholesterol, and other substances (e.g., calcium, fibrin, cellular waste products, and inflammatory cells). The reduction in blood flow to the heart can lead to angina (chest pain), heart attacks, and other serious cardiovascular complications. Coronary artery disease is a leading cause of death worldwide (Ullah et al. 2023). Platelet-inhibiting agents such as aspirin, clopidogrel, ticagrelor, vorapaxar, and rivaroxaban are commonly used. Aspirin is recommended for all patients with angina, while clopidogrel serves as an alternative for those intolerant to aspirin. Other drugs used include beta-blockers (initial treatment for unstable angina), statins, nitrates, ACEIs, ARBs, aldosterone antagonists, and others. ACEIs and ARBs improve hemodynamics and cardiac remodeling, particularly in cases of symptomatic LV dysfunction, often paired with diuretics. Beta-blockers significantly reduce morbidity and mortality by curbing sympathetic overstimulation. Mineralocorticoid receptor antagonists decrease mortality by counteracting aldosterone's effects. Hydralazine/nitrates are used where ACEIs or ARBs are unsuitable, improving hemodynamics and reducing mortality. However, low cardiac output and congestion, failure of the left ventricular (LV) due to systolic or diastolic dysfunction, accompanied by right ventricular (RV) failure marked by fluid overload, often characterize heart failure. Assessing LV function is vital for diagnosing and managing heart failure. Drugs used for treating heart failure, among others, include diuretics, beta-blockers, and mineralocorticoid receptor antagonists. Diuretics are essential for controlling fluid overload in heart failure, including conditions of preserved ejection fraction (HFpEF) and reduced ejection fraction (HFrEF). Vericiguat, a newer option, has been approved to lessen cardiovascular risks in chronic HFrEF. Sodium-glucose co-transporter 2 inhibitors have emerged as effective in reducing heart failure hospitalizations and cardiovascular deaths, particularly in HFrEF and diabetic patients. Aspirin inhibits platelet aggregation; clopidogrel blocks the P2Y12 component of ADP receptors on the platelet surface, which prevents GPIIb/IIIa receptor complex activation; ticagrelor reversibly binds to the P2Y12 receptor, preventing ADP-mediated activation of the GPIIb/IIIa receptor complex; vorapaxar inhibits thrombin-related platelet aggregation by blocking PAR-1; and rivaroxaban inhibits both free and clot-bound factor Xa, preventing thrombin generation and thrombus development. Nitrates dilate coronary arteries, increase blood flow to the heart muscle, and decrease myocardial oxygen demand by reducing preload and afterload; ranolazine inhibits the late phase of the inward sodium current in cardiac cells, reducing myocardial oxygen consumption; bivalirudin directly inhibits thrombin, preventing thrombus formation.

These medications are often used together, tailored to individual patient needs for optimal management of cardiovascular diseases. However, a series of adverse effects has been reported. For example, aspirin can cause gastrointestinal bleeding and hypersensitivity reactions; clopidogrel increases the risk of bleeding, particularly when used with aspirin; ticagrelor may cause dyspnea and bleeding. Vorapaxar is contraindicated in patients with a history of transient ischemic attack or stroke due to an increased risk of intracranial hemorrhage. Rivaroxaban increases the risk of bleeding, especially when combined with aspirin.

## *3.3 Attributes of Antiarrhythmic Drugs*

Arrhythmia is associated with abnormality in the heart's electrical activity rate, rhythm, or conduction. Dysrhythmias can be symptomatic, causing syncope, dizziness, fatigue, or palpitations, or asymptomatic. Antiarrhythmic drugs are classified into four classes (I–IV) based on their electropharmacological actions (Weirich and Wenzel 2000). Class I antiarrhythmic agents are $Na^+$ channel blockers and are further subdivided into three subclasses (Class Ia—including quinidine, procainamide, and disopyramide, which depress phase 0 depolarization, slow conduction, and prolonged repolarization; Class Ib—include lidocaine and mexiletine, which shorten repolarization without affecting conduction or refractoriness; Class Ic—include flecainide and propafenone, which depress phase depolarization more significantly and slow conduction, thereby prolonging refractoriness more than Class Ia agents); Class II agents are beta-blockers—esmolol, metoprolol, and propranolol, which decrease automaticity, prolong atrioventricular node (AV) conduction, and increase refractoriness; Class III agents, such as amiodarone, dronedarone, sotalol, dofetilide, and ibutilide, are $K^+$ channel blockers, which prolong repolarization and extend the QT interval on an electrocardiogram; Class IV agents are $Ca^{2+}$ channel blockers, including diltiazem and verapamil, which decrease automaticity and slow AV conduction, effectively treating arrhythmias by reducing the heart rate and the force of myocardial contraction (Weirich and Wenzel 2000). In addition to these classes, digoxin inhibits the Na+/K+-ATPase pump, increases intracellular calcium levels, enhances cardiac contractility, prolongs AV nodal conduction and the AV nodal refractory period, shortens the action potential, and decreases the refractoriness of the ventricular myocardium and Purkinje fibers (Obradovic et al. 2023), while adenosine, which binds to adenosine receptors, blocks AV nodal conduction, leading to a temporary pause in AV nodal conduction.

Class Ia drugs are generally contraindicated in patients with severe heart failure, marked QT prolongation, or known hypersensitivity, and they can cause serious side effects like urinary retention, QT prolongation, and blood disorders; Class Ib agents like lidocaine and mexiletine, used primarily for ventricular arrhythmias, can cause central nervous system and gastrointestinal disturbances; Class Ic drugs, including flecainide and propafenone, should not be used in patients with structural heart disease and can lead to proarrhythmic effects and other serious complications. Class II beta-blockers are not indicated in conditions like asthma and severe bradycardia and may induce bronchospasm and heart failure. Class III $K^+$ channel blockers like amiodarone can lead to thyroid dysfunction and pulmonary fibrosis due to its high iodine content and complex interactions with thyroid metabolism and lung tissue; sotalol and dofetilide can cause torsades de pointes, especially in patients with long QT syndrome. Class IV $Ca^{2+}$ channel blockers should be avoided in patients with severe heart failure. Adenosine and digoxin come with their risks, including severe bronchospastic reactions and arrhythmias.

## 4 Attributes of Drugs for Gastrointestinal-Related Diseases

Diseases of the alimentary system like peptic ulcers have prompted the need for antacids (aluminum hydroxide, calcium carbonate, magnesium salts, and sodium bicarbonate), histamine ($H_2$)-blockers, proton-pump inhibitors (PPIs), prostaglandin analogs, and sucralfate. Since their introduction, PPIs have been the mainstream choice for managing gastrointestinal disorders such as gastroesophageal reflux disease and peptic ulcer disease. PPIs, including omeprazole, are highly effective in suppressing gastric acid secretion, with omeprazole serving as a prototype for other clinically useful PPIs like esomeprazole, lansoprazole, rabeprazole, and pantoprazole (Strand et al. 2017). PPIs work by inhibiting the gastric proton pump, reducing acid secretion. They are activated in an acidic environment and react with the H+/K+-ATPase in parietal cells, converting to compounds that trap the drug and prevent acid production. This process is essential for proton-pump inhibition and increases after food ingestion (Strand et al. 2017). The $H_2$ receptor antagonists like cimetidine, ranitidine, famotidine, and nizatidine reduce gastric acid secretion in peptic ulcer disease. $H_2$-receptor antagonists inhibit acid production by competing with histamine for binding to $H_2$. They suppress 24-hour gastric acid secretion by about 70% by blocking the gastric acid-producing effect of histamine. They have rapid absorption rates and reach therapeutic levels within a few hours of dosing. Cimetidine can affect the absorption of other drugs and reduce the metabolism of certain medications. Antacids reduce gastric acidity by neutralizing gastric HCl. Some antacids, such as calcium and magnesium carbonate, demonstrate peristaltic and antiseptic activity. Besides antacids, PPIs, and $H_2$-receptor antagonists, other acid-reducing agents include misoprostol (prostaglandin analog) and sucralfate.

Metoclopramide as an effective antiemetic exhibits complex antiemetic mechanisms involving 5-$HT_4$-receptor agonism, vagal and central 5-HT antagonism, possible sensitization of muscarinic receptors on smooth muscle, and dopamine receptor antagonism. Bulk-forming agents, osmotic laxatives, stool softeners, prokinetic agents, and non-specific irritants can facilitate bowel elimination and relieve constipation. Laxatives decrease constipation by stimulating contractions, reducing fluid absorption, increasing fluid retention, facilitating bowel elimination, and emptying. Meanwhile, the therapeutic actions of these drugs are associated with several adverse effects. As an example, antacids can affect the absorption of some medications and increase urinary pH. Antacids are not recommended for patients with urinary failure, uremia, heart failure, edema, hepatic impairment, and hyperthyroidism due to potential adverse effects like constipation, bloating, diarrhea, hypotension, and systemic alkalosis. Adverse effects of PPIs include increased risk of *Clostridium difficile* infection, decreased magnesium absorption leading to hypomagnesemia, and vitamin B12 deficiency. Prolonged use of $H_2$-receptor antagonists is contraindicated and can lead to reduced sensitivity and rebound gastrinemia. Adverse effects include rashes, diarrhea, dizziness, fatigue, constipation, and muscular pain. Metoclopramide is contraindicated in hypersensitivity to the drug.

## 5 Attributes of Drugs for Respiratory-Related Diseases

Coughs can be productive with the coughs clearing of the airway or non-productive (Lee et al. 2021). Medications like pharyngeal demulcents, expectorants, and mucolytics help treat coughs. Pharyngeal demulcents soothe the throat, expectorants increase bronchial secretion, and mucolytics break down mucus for easier expulsion (Eccles 2020). Antitussives are classified into opioids and antihistamines, among others. Opioids suppress cough; nonopioids prevent addiction; antihistamines reduce coughing as adjuvant antitussives improve airflow and minimize bronchoconstriction in hyperreactivity cases. Opioid antitussives act centrally, while non-opioid antitussives like noscapine and dextromethorphan also act centrally without the risk of addiction. Antihistamines reduce coughing by blocking histamine receptors, particularly in allergic conditions. Bronchodilators like salbutamol and terbutaline reduce bronchoconstriction and improve airflow, thus relieving cough associated with airway hyperreactivity. Some of these medications should be used judiciously, as the effectiveness of many expectorants is not conclusively proven, and some combinations have been banned due to adverse effects. Opioids can cause drowsiness and constipation. Nonopioids like noscapine and dextromethorphan may cause dizziness and nausea. Antihistamines may result in sedation and dry mouth. Peripherally acting agents like prenoxdiazine can lead to gastric discomfort and rash. Bronchodilators should also be monitored to avoid potential overuse and adverse effects, including tachycardia and muscle tremors.

An inflammatory condition like bronchial asthma characterized by hyperresponsiveness of the tracheobronchial smooth muscle often leads to airway narrowing with symptoms like dyspnea, wheezing, and coughing. It can be triggered by allergens, infections, irritants, pollution, exercise, cold air exposure, and psychological factors. Medications for chronic obstructive lung disease (COPD) include bronchodilators (such as salbutamol, terbutaline, and formoterol), leukotriene antagonists (e.g., montelukast, zafirlukast), mast cell stabilizers (e.g., sodium cromoglycate and ketotifen), corticosteroids (e.g., hydrocortisone and prednisone), anti-IgE antibodies (e.g., omalizumab), anticholinergics (e.g., ipratropium bromide and tiotropium bromide), and inhibitors of phosphodiesterase 3 (e.g., theophylline) (Capra et al. 2007; Finn and Walsh 2013). The treatment of COPD is tailored to the severity of symptoms and the presence of exacerbations. High-risk COPD patients may require more intensive treatment, including smoking cessation, oxygen therapy, inhaled bronchodilators, corticosteroids, and antibiotics as needed.

Sympathomimetics, such as salbutamol, bambuterol, and salmeterol, increase cAMP formation, relax bronchial smooth muscle, and decrease inflammatory mediator release while leukotriene antagonists reduce bronchoconstriction and mucus secretion via the blockade of the cycLT1 receptor (Capra et al. 2007); mast cell stabilizers like sodium cromoglycate inhibit inflammatory cell degranulation, while corticosteroids reduce inflammation; theophylline has a narrow therapeutic range and can cause side effects like nausea, headache, and arrhythmias (Goh et al. 2023); inhaled anticholinergics could elicit dry mouth and urinary retention as anti-IgE antibody therapy with omalizumab can cause injection site reactions and potential anaphylaxis (Balbino et al. 2020).

## 6 Attributes of Drugs for Endocrine-Related Diseases

Common endocrine disorders affecting include diabetes, Addison's disease, hypothyroidism, Hashimoto's disease, hyperthyroidism, Graves' disease, and Cushing's syndrome. Diabetes mellitus is characterized by high blood glucose levels. The therapeutic options include the use of hypoglycemic agents like biguanides (e.g., metformin), sulfonylureas (e.g., tolbutamide, chlorpropamide, glibenclamide, glipizide, glimepiride, and gliclazide), thiazolidinediones (e.g., troglitazone, rosiglitazone, and pioglitazone), and α-glucosidase inhibitors (acarbose, miglitol). Metformin reduces hepatic gluconeogenesis, inhibiting respiratory chain complex 1 and suppressing ATP (Ross et al. 2016; Rena et al. 2017), as the sulfonylureas trigger hypoglycemic effects by promoting insulin release from pancreatic β cells. This is achieved by enhancing β-cell depolarization, inhibiting $K^+$ efflux, opening voltage-gated calcium channels, and reducing serum glucagon concentrations. Thiazolidinediones are selective agonists for nuclear PPARγ and activate insulin-responsive genes upon binding to the receptors. By decreasing circulating fatty acid concentrations and lipid availability in the liver and muscle, these drugs improve patient sensitivity to insulin. The α-glucosidase enzyme aids in breaking down carbohydrates into glucose and other monosaccharides, which are then absorbed into the blood. Acarbose and miglitol are both competitive inhibitors of α-glucosidase with hypoglycemic properties. They reduce postprandial plasma glucose levels in individuals with type 1 and 2 diabetes mellitus.

Graves' disease (GD), toxic multinodular goiter, and toxic adenoma are clinical indications for drugs targeting the inhibition of thyroxine synthesis (Ross et al. 2016). Several drugs, including thioamides (e.g., methimazole and propylthiouracil), iodides, radioactive iodine, and adrenoceptor-blocking agents, have been used to treat these clinical conditions and other related hyperthyroidism. Methimazole is about ten times more potent than propylthiouracil. It is absorbed orally and reaches peak concentration within 1–2 hours. Methimazole inhibits thyroid peroxidase, reducing iodine organification and preventing thyroid hormone synthesis. It may also disrupt iodine uptake by affecting iodotyrosyl residues and iodide ion oxidation (Awosika et al. 2024). The iodide-like potassium iodide is commonly employed for antithyroid actions in emergency conditions and cases of high radiation exposure (Torti and Correa 2024). Potassium iodide decreases thyroglobulin breakdown and inhibits iodine organification in the thyroid gland. Radioactive iodine targets hyperthyroidism, while adrenoceptor-blocking drugs like metoprolol, propranolol, and atenolol are effective adjuncts (Bednarczuk et al. 2021).

Corticosteroids with a mechanism of actions involving metabolic, anti-inflammatory, immunosuppressive, hormonal, and cognitive signaling processes are used to treat adrenal dysfunctions and chronic conditions like asthma, skin infections, Crohn's disease, and rheumatoid arthritis (Scherholz et al. 2019; Liu et al. 2013). Hydrocortisone and its synthetic analogs are effective when taken orally, intravenously, or intramuscularly, with minimal variations in chemical structure. Glucocorticoids through glucocorticoid nuclear receptors trigger gene transcription to regulate several activities—metabolic, electrolyte balance, cardiovascular, and skeletal muscle activities. Drugs like spironolactone and eplerenone antagonize adrenocorticoid function while mitotane inhibits cortisol synthesis. These drugs are indicated in hypercortisolism, mineralocorticoid, and aldosterone conditions. Mitotane, used in adenocarcinoma treatment, reduces tumor mass in 30% of patients. In addition, metyrapone is a drug used to treat Cushing syndrome by targeting adrenal glands and inhibiting steroid synthesis, reducing cortisol secretion and aldosterone production. It is effective for conditions related to excessive cortisol production and assesses adrenal function.

In the case of atherosclerosis, lowering cholesterol can reduce cardiovascular event rates. Statins like atorvastatin, simvastatin, rosuvastatin, and pravastatin are the cornerstone of lipid-lowering therapy. They inhibit HMG-CoA reductase, an enzyme crucial for an early step in cholesterol synthesis. The inhibition of this enzyme leads to reduced cholesterol production within the liver. When statins alone do not adequately control cholesterol, ezetimibe is often added to the regimen. Ezetimibe targets the intestinal absorption of cholesterol by specifically inhibiting the transport protein NPC1L1. This inhibition decreases the amount of cholesterol that reaches the liver from the diet, which stimulates the liver to compensate by increasing the expression of LDL receptors. Bempedoic acid is a newer class of cholesterol-lowering medication and is particularly useful in patients who experience statin intolerance (Paponja et al. 2024). Bempedoic acid inhibits ATP citrate lyase, an enzyme involved in cholesterol synthesis upstream of the HMG-CoA reductase step affected by statins. Fibrates are another important class of lipid-modifying agents, primarily effective in reducing triglycerides and raising HDL-C levels. Fibrates primarily activate peroxisome proliferator-activated receptor-alpha (PPARα). This activation leads to an increase in the expression of lipoprotein lipase and a reduction in apolipoprotein C-III. They are particularly beneficial for patients with high triglycerides and low HDL-C (Tomlinson et al. 2021). Niacin, or nicotinic acid, while effective in modifying all lipid fractions, is less commonly used due to its side effects and the less robust data supporting its cardiovascular benefits. Niacin inhibits lipolysis in adipose tissue, which decreases the transport of free fatty acids to the liver. This reduction in fatty acid flow leads to a decrease in the liver's synthesis of triglycerides and the secretion of VLDL, which in turn lowers LDL levels (Lin et al. 2022). Niacin also increases the apolipoproteins A-I and A-II levels, further enhancing lipid regulation. Bile acid sequestrants bind to bile acids within the intestine, blocking their reabsorption. This action forces the liver to convert more cholesterol into bile acids, depleting the liver's cholesterol and consequently leading to increased clearance of LDL cholesterol from the bloodstream. Proprotein convertase subtilisin/kexin type 9 (PCSK9) inhibitors disrupt the normal interaction between PCSK9 protein and LDL receptors. By blocking PCSK9's ability to bind with LDL receptors, these drugs increase the availability of LDL receptors on the surface of the liver, boosting the clearance of LDL cholesterol from the bloodstream (Pokhrel et al. 2024).

Some of these drugs are associated with adverse effects. Troglitazone has been withdrawn from use due to the severe hepatotoxicity associated with it. Metformin is not recommended for patients with cardiac failure, chronic hypoxic lung disease, or renal and hepatic diseases. Sulfonylurea is contraindicated in sulfonylurea hypersensitivity, acute liver disease, and congestive heart failure. Thiazolidinediones, inflammatory bowel disease, colonic ulceration, partial intestinal obstruction, susceptibility to intestinal blockage, or chronic digestive disorders. Common adverse effects of hypoglycemic agents include abdominal discomfort, diarrhea, metallic taste, and nausea. These drugs may cause weight gain, bone density disease, and heart failure. Prolonged consumption of carbohydrates may decrease these side effects. Methimazole is contraindicated for allergic or hypersensitive conditions, liver diseases, and pregnancy. It has low adverse effects but may cause agranulocytosis, hair loss, and dizziness. Potassium iodide is contraindicated for patients with iodine allergy, with common side effects including gastrointestinal intolerance and bitter taste. Glucocorticoids are contraindicated in patients with hypertension, heart diseases, psychoses, osteoporosis, and peptic ulcers. Adverse effects associated with chronically high doses of these drugs include hyperglycemia, hypokalemia, glycosuria, edema (due to $Na^+$ and water retention), and possible osteoporosis. Mitotane has adverse effects like diarrhea, nausea, vomiting, somnolence, and depression. Side effects of metyrapone include hirsutism, gastrointestinal disturbances, and salt/water retention (Tritos 2021; Pivonello et al. 2020). Spironolactone can cause adverse effects such as hyperkalemia, loss of libido, and gynecomastia. Eplerenone has similar actions but is more selective for aldosterone receptors and can also lead to hyperkalemia (Patibandla et al. 2024).

Statins are contraindicated in patients with severe liver disease, pregnant women, and those with a history of myopathy. They can cause muscle issues, liver enzyme elevation, and potentially diabetes and cognitive changes. Fibrates should be avoided in those with significant liver or renal dysfunction and during pregnancy and lactation. Common adverse effects include gastrointestinal disturbances, potential hepatotoxicity, and myopathy. Niacin is not recommended for individuals with liver disease, peptic ulcer, or active bleeding and should be avoided during pregnancy. It can cause flushing, gastrointestinal upset, and, in severe cases, liver toxicity. Bile acid sequestrants are not indicated in severe hypertriglyceridemia due to their potential to exacerbate the condition. They can cause gastrointestinal discomfort and may interfere with the absorption of other drugs. PCSK9 inhibitors are usually well tolerated but should be avoided in cases of hypersensitivity. Side effects include flu-like symptoms and injection site reactions. Ezetimibe should not be used by those with significant liver disease. Its side effects include diarrhea and joint pains.

## 7 Attributes of Drugs for Reproductive-Related Diseases

The female reproductive system relies on regulatory steroids like estrogen and progesterone for growth and normal functioning. Estrogens activate estrogen receptors ERα and Erβ, which trigger the activity of ligand-activated transcription factors. This results in various actions, including endometrial growth, blood vessel and skin maintenance, bone resorption reduction, and bone formation (Liang and Shang 2013; Delgado and Lopez-Ojeda 2024). Estrogens are absorbed through oral, intramuscular, and transdermal routes and bind to sex hormone-binding globulin α2 globulin. They are converted into estrone and estriol, which have a low affinity for estrogen receptors and can be excreted in bile and breast milk by nursing mothers. Progestins are synthetic compounds that mimic the action of progesterone. They are indicated in contraception and postmenopausal hormone replacement therapy (HRT) (Edwards and Can 2024). Progestin activates the progesterone receptor, which undergoes dimerization and increases DNA binding. These actions suppress GnRH and LH hormone secretion (Edwards and Can 2024). Several drugs, such as tamoxifen and related antagonists, are used in treating breast cancer through the antagonism of estrogen and progesterone receptors (Ray and Ficek 2012). Other related drugs include anastrozole, fulvestrant, raloxifene, clomiphene, and the progesterone antagonist mifepristone, which is used as an abortifacient and for managing miscarriage.

Human growth hormone (hGH) is used to treat hypopituitary dwarfism in children. Growth hormone triggers the dimerization of GH receptors, activating signaling pathways mediated by JAK tyrosine kinases and STATs (Dehkhoda et al. 2018), thereby affecting growth, body composition, and carbohydrate, protein, and lipid metabolism. It increases insulin-like growth factor-1 (IGF-1) production. Growth hormone-secreting pituitary tumors, known as somatotroph adenomas, make up approximately 30% of all functioning pituitary tumors. Somatostatin analogs are the primary treatment for alleviating excess growth hormone secretion-related symptoms and conditions (e.g., acromegaly). Drugs like octreotide, lanreotide, seglitide, and vapreotide have been designed to treat pituitary tumors and find applications in neuroendocrine tumors. Octreotide, a somatostatin analog, inhibits cell proliferation by activating the tyrosine phosphatase pathway (Buscail et al. 1994).

Gonadotropin-releasing hormone (GnRH) and its analogs stimulate the release of follicle-stimulating hormone (FSH) and luteinizing hormone (LH). These analogs include goserelin, histrelin, leuprolide, nafarelin, and triptorelin. Structural modifications of the GnRH template confer more potency and duration of action on these analogs. GnRH and its derivatives treat hypogonadism, suppress gonadotropin release and endometriosis, diagnose LH responsiveness, and protect the oocytes before starting chemotherapy (Chen et al. 2011). GnRH analogs precipitate LH surge and ovulation, making them a therapeutic option in infertile women. These drugs also suppress gonadotropin release, an action that is combined with GnRH antagonists in the treatment of prostate cancer (Nelson et al. 2023). GnRH signals through a specific G protein-coupled receptor on gonadotropes that activates Gq/11, and stimulation of the phospholipase C-inositol triphosphate-$Ca^{2+}$ pathway is associated with the effects of GnRH.

Drugs for reproductive-related diseases elicit adverse effects. Estrogens induce nausea, vomiting, abdominal cramps, and premenstrual symptoms. Its severe adverse effects include hypertension, heart attacks, and asthma (Delgado and Lopez-Ojeda 2024). Contraindications for estrogens include breast, ovarian, endometrial, and coronary diseases. Progestins are contraindicated in liver cirrhosis, pregnancy, and patients with a history of breast cancer. Patients usually report weight gain using progestins; however, the mechanisms underlying increased weight are still unclear. Somatostatin analogs are contraindicated in pregnant and lactating women. Steatorrhea, diarrhea, malabsorption, gastrointestinal cramps, and occasional nausea are adverse effects associated with these drugs. GnRH analogs with these drugs include hot flashes, gynecomastia, fatigue, weight gain, fluid retention, erectile dysfunction, decreased libido, diabetes, and osteoporosis (LiverTox 2018), which are contraindicated in patients who are hypersensitive to GnRH.

## 8 Attributes of Drugs for Other Conditions

Bone mineral homeostasis and microbial, immune, and blood-related diseases are important in describing drug attributes, interactions, and effects. Bisphonates are drugs used to treat Paget's disease of bone, osteoporosis, metastatic bone disease, multiple myeloma, and several rare bone diseases (Corral-Gudino et al. 2017). They have been classified into non-nitrogen nitroprusside, alkyl-amino bisphosphonates, and heterocyclic nitrogen bisphosphonates (Ganapathy et al. 2012). Examples include etidronate, pamidronate, alendronate, tiludronate, ibandronate, and zoledronate. Bisphosphonates inhibit bone resorption by attaching to hydroxyapatite binding sites undergoing active resorption. Bisphosphonate-impregnated bone impairs osteoclast function, inhibiting its ability to resorb bone (Drake et al. 2008). Bisphosphonates also reduce osteoclast activity by decreasing osteoclast progenitor development and recruitment, promoting osteoclast apoptosis, and incorporating non-nitrogen nitroprusside into ATP, resulting in the apoptosis of the osteoclast cells. The alkyl-amino bisphosphonates and heterocyclic nitrogen bisphosphonates inhibit bone resorption by blocking the farnesyl pyrophosphate synthase enzyme, an enzyme in the mevalonate pathway that appears to play a critical role in the survival of osteoclast cells (Ganapathy et al. 2012). Modulation of the immune system is important to alleviate immune-related diseases—infections, autoimmune diseases, allergies, asthma, and immunodeficiencies (Boraschi and Penton-Rol 2014). The immunosuppressors (e.g., glucocorticoids, immunophilin ligands, cytotoxic agents, immunosuppressive antibodies, and monoclonal antibodies) play important roles in managing the rejection of organs or tissue as well as certain diseases that arise from immune response dysregulation. As life-threatening infections could result from the suppression of the immune system, the use of antibiotics of different spectrum of actions is often recommended. Some antibiotics inhibit the biosynthesis of the bacterial cell wall (e.g., penicillin, cephalosporin, and other β-lactams) and protein (e.g., chloramphenicol, aminoglycosides, tetracyclines). Chloramphenicol interacts with the 23S rRNA of the 50S subunit, inhibiting t-RNA binding to the A site of the ribosome (Yoneyama and Katsumata 2006).

As mycobacteria are resistant to most antibiotics due to certain characteristics such as high dormancy, lipid-rich impermeable cell membranes, and intracellular protection by macrophages, special antibiotics are designed to eradicate these pathogens. Isoniazid and rifampin (broad-spectrum antimicrobial) which exert their effects by acting on cell wall proteins are used to treat tuberculosis. Polymyxin and valinomycin act on the cell membranes; fluoroquinolones affect the DNA replication process; Drugs such as streptomycin, kanamycin, amikacin, capreomycin, puromycin, linezolid, erythromycin, and chloramphenicol interfere with the translational machinery; and drugs such as pyrazinamide, para-aminosalicylic acid, and cycloserine interfere with the cell's metabolic pathways. In the case of fungal infections, polyenes, azoles, echinocandins, allylamines, and others are important antifungal agents. Polyene antifungals like Amphotericin B and nystatin are highly effective against *Candida*, *Aspergillus fumigatus*, and *Aspergillus flavus* (Kumar et al. 2018). Azoles are five-membered ring antifungal agents classified as imidazoles or triazoles. The imidazoles include the ketoconazole, miconazole, and clotrimazole. The triazoles include itraconazole, fluconazole, and voriconazole (Ping et al. 2013). Azoles inhibit the microsomal cytochrome P450 monooxygenase dependent 14-α-demethylase, leading to the fungistatic effect (Sagatova et al. 2015). Echinocandins target the 1,3-β-d-glucan synthase, disrupting the fungal cell wall and causing cell death (Mazu et al. 2016).

Herpes simplex virus (HSV), varicella-zoster virus (VSV), hepatitis, and influenza have led to the development of synthetic antiviral agents including acyclovir, valacyclovir (diacetyl ester prodrug of 6-deoxy penciclovir), and famciclovir with clinical activity against HSV-1, HSV-2, and VZV. Oral valacyclovir is metabolized by intestinal and hepatic esterase to acyclovir. Amantadine, neuraminidase inhibitors (e.g., zanamivir, oseltamivir, peramivir, and zanamivir), and cap-dependent endonuclease inhibitors (e.g., baloxavir) are among the drugs for treating influenza—a contagious disease attributed to the influenza A, B, and C virus (Świerczyńska et al. 2022). Amantadine is active against influenza A only. The guanosine analog structure of acyclovir inhibits viral DNA replication by targeting the thymidine kinase enzyme found in HSV and VZV (Abdalla et al. 2020). Penciclovir also inhibits viral DNA polymerase through phosphorylation catalyzed by virus-specified thymidine kinase in infected cells. Amantadine obstructs the A/M2 proton channel specific to the influenza A virus to interfere with viral replication, while baloxavir inhibits replication by binding to the RNA-dependent influenza virus RNA polymerase complex (Balannik et al. 2009). Besides hypersensitivity to acyclovir, patients having renal failure/impairment, immunocompromised host, potential risk of thrombotic thrombocytopenic purpura, and hemolytic uremic syndrome (Taylor and Gerriets 2024).

Generally, major classes of antiretroviral drugs include nucleoside analog reverse transcriptase (NRTIs), nucleotide analog reverse transcriptase (NERTIs), non-nucleoside analog reverse transcriptase inhibitors (NNRTIs), and fusion inhibitors. Examples of NRTIs include zidovudine, didanosine, lamivudine, and stavudine. These nucleoside analogs lack 3′-OH in their ribose moiety, which enables HIV-1 reverse transcriptase (HIV-1-RT) to differentiate them from physiologic deoxyribonucleoside triphosphate substrates (Maeda et al. 2019). NERTIs include drugs such as tetonavir, ritonavir, and atazanavir. Tetonavir, the first NERTI, is a derivative of adenosine monophosphate but lacks the ribose ring. On the other hand, NNRTIs are a broad spectrum of antiretroviral drugs. NNRTIs do not directly prevent template-primer or dNTP binding to RT but cause misalignment of the template-primer to the catalytic site, thereby preventing incoming dNTP from being incorporated. Examples of HIV protease inhibitors are amprenavir, ritonavir, indinavir, lopinavir, nelfinavir, saquinavir, atazanavir, and tipranavir. Specific precursor molecules called the Gag and Gag-Pol gene products are translated into immature or precursor polyproteins in the latter stages of the HIV growth cycle. The HIV-1 protease is an aspartyl protease required to cleave these precursor molecules to produce the final structural proteins of the mature virion core, thus completing the cycle (Maeda et al. 2019).

Another category of antiretroviral drugs is the fusion inhibitors, with enfuvirtide being the sole HIV fusion inhibitor currently on the market. It is specifically prescribed for HIV patients with signs of continuous HIV replication despite undergoing highly active antiretroviral therapy. Enfuvirtide consists of a 36-amino acid peptide analog derived from a section of the transmembrane region of gp41, which plays a role in facilitating viral particle merging with host cell membranes. Enfuvirtide must be stored at low temperature, reconstituted, and injected subcutaneously twice daily. Moreover, injection site reactions, the rapid emergence of resistant viruses, and the high cost of production have limited long-term use (Poveda et al. 2005). NRTIs have antiretroviral action through competitive inhibition and termination of polymerization (Mehellou and De Clercq 2010; Yoshida et al. 2021). They bind to HIV-1 reverse transcriptase, blocking RNA- and DNA-dependent DNA polymerase. NNRTIs do not compete with nucleoside triphosphates or require phosphorylation. HIV protease inhibitors reduce HIV-1 replication by competitively binding to the protease active site and preventing the cleavage of the Gag-Pol polyprotein. Enfuvirtide, a fusion inhibitor, blocks the interaction between the HIV gp41 protein and the host cell membrane. Most drugs are effective against both HIV-1 and HIV-2.

Contraindications of baloxavir include hypersensitivity to the drug and its ingredients. Due to the guanosine-related teratogenic effects of acyclovir, the drug is contraindicated in pregnancy. Besides nausea and vomiting, other adverse effects of acyclovir include rash, hepatitis, and neurological disturbance. Diarrhea, bronchitis, sinusitis, and headache are also associated with baloxavir. Nucleoside analogs may affect mitochondrial function, leading to lactic acidosis and liver problems. Rapidly increasing aminotransferase levels and metabolic acidosis may indicate NRTI treatment should be stopped. NRTIs have adverse effects that include bone marrow suppression, melanonychia, and lipodystrophy. NNRTIs can cause skin rash and gastrointestinal issues, including Stevens-Johnson syndrome. HIV protease inhibitors may lead to glucose intolerance and fat redistribution. Enfuvirtide is contraindicated in cases of hypersensitivity. Adverse effects of the fusion inhibitor include rash, fever, nausea, vomiting, chills, hypotension, and elevated serum liver transaminase.

Malaria is another important infectious disease, representing a significant cause of global mortality, though a slight reduction in the malaria mortality rate has been recorded (Kua et al. 2024). Antimalarials have remained effective in curbing the severity and mortality associated with the disease. Major categories of antimalarial drugs indicated in the treatment of the disease include the 4-aminoquinolines (4-AQs), the arylaminoalcohols, artesunate, and artemether. The 4-AQs demonstrate lysosomotropic activity by accumulating within the lysosomes and inhibiting malarial haem polymerase (Yoon et al. 2010). Proton-pump inhibitors may inhibit their accumulation in lysosomes. The 4-AQs are weak bases with two fused aromatic rings and are made up of commonly used drugs like chloroquine and hydroxychloroquine as well as the arylaminoalcohols (quinine, quinidine, mefloquine, and halofantrine). Chloroquine is absorbed from the gastrointestinal tract, sequestered in peripheral tissues, metabolized in the liver, and excreted via kidneys and feces (Browning 2014). Artemisinin is a sesquiterpene lactone endoperoxide whose semi-synthetic derivatives, artesunate and artemether, are effective and well-tolerated antimalarials. Artemisinin-based antimalarial drugs are activated by intraparasitic heme-iron and kill the parasite by poisoning essential malarial proteins (Ribbiso et al. 2021). These artemisinin derivatives are currently the first and second-line artemisinin-based combination therapy for treating malaria infection (Issa et al. 2023). Monotherapy involving either artesunate or artemether is not encouraged due to the high probability of malarial resistance. 4-AQs should not be used in patients with hypersensitivity to chloroquine and other drugs in this category or with retinal/vision impairment. Short-term use can cause pruritus, gastrointestinal disturbance, and a mild headache. Long-term therapy may lead to retinopathy, lichenoid skin eruption, weight loss, and ototoxicity. Quinine is contraindicated in patients with known hypersensitivity, G6PD deficiency, tinnitus, optic neuritis, prolonged QT interval, or myasthenia gravis (Hogan 2015). Artemisinin-based antimalarial drugs are not indicated for patients with allergies to active substances and intake of herbal supplements. Adverse effects may include nausea, vomiting, anorexia, dizziness, and others.

Chemotherapy, hormone therapy, radiation, and surgery are among the therapeutic options for cancer treatment (Behranvand et al. 2022). Alkylating agents, antimetabolites, DNA-binding agents, and topoisomerase inhibitors exert anticancer effects through alkylation of DNA, inhibition of metabolic pathways, activation of apoptotic mechanisms, cell cycle arrest, DNA damage, inhibition of microtubules, and topoisomerase functions (Woods and Turchi 2013; Pangs et al. 2013; Xiao et al. 2017; Behranvand et al. 2022). The spreading of tumor cells into the bloodstream and establishing metastatic colonies in secondary sites are often facilitated by angiogenesis. In addition to the tissue-specific pathological conditions in this metastatic process, some patients sometimes suffer from non-cancer-related life-threatening blood disorders like anemia, characterized by a deficiency in oxygen-carrying erythrocytes. Several hematinics used to treat anemia include iron, vitamin $B_{12,}$ and folate. Iron plays a crucial role in the body's proteins, including hemoglobin, myoglobin, CYP450s, catalase, peroxidase, and metalloflavoproteins (Turner et al. 2024). Iron is transported in the plasma bound to transferrin, a b-globulin that specifically binds two molecules of ferrous iron. It is stored primarily as ferritin in intestinal mucosal cells, macrophages, spleen, etc. Iron deficiency anemia causes pallor, fatigue, dizziness, exertional dyspnea, etc. Excessive iron deposition in the heart, liver, and other organs results in chronic iron toxicity, which is presented as hemochromatosis. Vitamin $B_{12}$ and folate are essential cofactors in the formation and maturation of red blood cells. The most common cause of deficiency is dietary or malabsorption or due to gastric surgery. Metformin and phenytoin may cause vitamin $B_{12}$ and folate deficiency, respectively (O'Leary and

Samman 2010; Azzini et al. 2021). Disruption through the enhancement or blockade of the coagulation process could provoke blood-related medical conditions. Anticoagulants include low-weight heparins such as dalteparin and enoxaparin, vitamin K antagonists like warfarin, and direct-acting oral anticoagulants such as dabigatran, rivaroxaban, apixaban, and edoxaban. They enhance antithrombin activity, reducing thrombin and factor Xa activity, thus preventing clot formation (Rodgers and Mahajerin 2023). Vitamin K antagonists inhibit the synthesis of vitamin K-dependent clotting factors (II, VII, IX, and X), reducing the body's ability to form blood clots. Dabigatran directly inhibits thrombin or factor Xa (rivaroxaban, apixaban, edoxaban), critical components in blood clotting. Fondaparinux is a synthetic molecule that functions similarly to LMWHs but is more selective for factor Xa. It binds to antithrombin and enhances its activity to inhibit factor Xa, reducing thrombin formation and thrombus development.

Adverse effects of the aforementioned drug classes like bisphosphonates include bone pain, hypocalcemia, conjunctivitis, and musculoskeletal disorders. They are not indicated in hypersensitivity to the drug and low-serum $Ca^{2+}$ levels. The immunosuppressors can increase vulnerability to infections, sterility, adverse glucocorticoid effects, carcinogenicity, teratogenicity, etc. Some of the antibiotics induce allergic reactions, diarrhea, dizziness, and nausea and are contraindicated in hypersensitivities to the drug, in children (tetracycline and quinolone), and in pregnancy and lactation, depending on critical factors. Nausea, vomiting, and diarrhea are common adverse effects of most antifungal drugs. Patients who take amphotericin B may experience fever, chills, headache, and hypotension. Other adverse effects related to amphotericin B include loss of tubular cations leading to hypokalemia, hypomagnesemia, and reversible nephrotoxicity. Common toxicities associated with azoles involve gastrointestinal upsets, rashes, hepatitis, and drug interactions related to CYP3A inhibition. The anticancer often triggers fatigue, mouth ulcers, loss of appetite, hair loss, etc. Heparins are contraindicated in patients with significant kidney disease due to their renal clearance requirements (Law et al. 2020). They can cause thrombocytopenia and osteoporosis with long-term use. Warfarin requires careful monitoring due to its narrow therapeutic window and interactions with food and other medications. DOACs are generally well tolerated but can lead to bleeding complications and are contraindicated in patients with severe renal impairment or active pathological bleeding. Fondaparinux is contraindicated in patients with creatinine clearance less than 30 mL/min due to the risk of accumulation and bleeding. It is generally associated with lower rates of heparin-induced thrombocytopenia and thrombosis than unfractionated heparin.

## 9 Conclusion

In the context of modern medicine, understanding the pharmacological attributes of highly diverse drug classes, interactions, and effects allows for optimized treatment regimens that maximize benefits while minimizing risks, ultimately improving patient health and quality of life. The complexity of pharmacological treatment outcomes reflects a wide array of medical conditions and a vast diversity of drug classes. Each drug class possesses distinct chemical structures, mechanisms of action, therapeutic uses, and adverse effect profiles. These variations necessitate careful selection and associations with pharmaceutical or non-pharmaceutical products based on the patient's individual medical needs, balancing efficacy with potential risks.

## 10 Self-Assessment Questions and Answers

### *10.1 Questions*

1. A 65-year-old patient diagnosed with neuroleptic malignant syndrome (NMS) is likely to have therapeutic benefits from
   A. Haloperidol.
   B. Aripiprazole.
   C. Olanzapine.
   D. Supportive care coupled with dantrolene or bromocriptine administration.
2. A 35-year-old patient with a medical history of generalized anxiety disorder was placed on oral lorazepam as a result of his recent complaint of chest pain, palpitations, breathlessness, and dizziness symptoms. This benzodiazepine was substituted for buspirone after 4 weeks to avoid addiction. Choose the correct alternative from the following affirmations
   A. Buspirone blocked benzodiazepine withdrawal.
   B. Buspirone often elicits anxiolytic sedation and respiratory depression.
   C. Both lorazepam and buspirone share similar interaction attributes.
   D. Lorazepam induces amnesia anterograde.
3. . A 42-year-old woman diagnosed with major depressive disorder following insomnia, lack of energy and concentration, and low appetite over the last 4 weeks was prescribed escitalopram 10 mg daily. After 3 weeks of medication without relief of depressive symptoms, she complained of the drug being ineffective to the physician. Which of the following options represents the physician's best next management step?
   A. He should explain to her that relief takes 4 to 6 weeks of escitalopram treatment and then schedule an appointment.
   B. He should substitute this drug immediately for fluoxetine.
   C. He should instruct the patient to stop taking escitalopram and schedule an appointment.
   D. Health professionals in charge should increase the dose of escitalopram.

4. A 60-year-old man with a history of coronary artery disease and heart failure is experiencing symptomatic atrial fibrillation characterized by palpitations, dizziness, and fatigue. His current medications include a beta-blocker (metoprolol) and a statin (atorvastatin). His recent echocardiogram showed a left ventricular ejection fraction of 30%, indicating significant systolic dysfunction. The patient was started on amiodarone 200 mg daily for rhythm control. After 4 weeks on amiodarone, he reports experiencing shortness of breath, dry cough, and occasional light-headedness. Which of the following represents the best next step in managing this patient?

   A. Continue amiodarone and add a diuretic for symptomatic relief.
   B. Substitute amiodarone with sotalol and schedule close follow-up appointments.
   C. Conduct a thorough evaluation, including pulmonary and thyroid function tests, and consider discontinuing amiodarone.
   D. Increase the dose of amiodarone and monitor the patient closely for further adverse effects.

5. A 58-year woman with an intact uterus complained of vaginal atrophy and vasomotor instability to a physician. She was placed on hormone replacement therapy (HRT) using small doses of estrogen with progesterone for several years. The woman faces the risk of all the following except

   A. Osteoporosis.
   B. Myocardial infarction.
   C. Stroke.
   D. Breast cancer.

## 10.2 Answers

1. *Answer:* The appropriate therapeutic approach is (D) Supportive care coupled with dantrolene or bromocriptine administration.

   Given that the patient has been diagnosed with neuroleptic malignant syndrome (NMS), a potentially life-threatening condition associated with the use of antipsychotic medications, it's essential to prioritize supportive care and the administration of medications to alleviate symptoms. In the case of NMS, continuing treatment with antipsychotic drugs such as haloperidol, aripiprazole, or olanzapine could exacerbate symptoms and worsen the condition. Therefore, Supportive care may include measures to address hyperthermia, dehydration, electrolyte imbalances, and other complications associated with NMS. Dantrolene and bromocriptine are medications commonly used to manage muscle rigidity and hyperthermia associated with NMS, and they can provide therapeutic benefits in alleviating symptoms and improving patient outcomes.
2. *Answer:* Given the options, the correct alternative is (D) Lorazepam induces amnesia anterograde. Benzodiazepines like lorazepam are known to cause

anterograde amnesia, particularly at higher doses or with prolonged use. This means they can impair the formation of new memories. However, buspirone does not typically cause significant memory impairment. Buspirone is not typically used to manage benzodiazepine withdrawal symptoms. It acts differently from benzodiazepines and does not directly target the same receptors involved in benzodiazepine dependence or withdrawal. Instead, benzodiazepine withdrawal is usually managed through a gradual tapering of the medication or sometimes with other medicines that can help alleviate withdrawal symptoms. Buspirone is an anxiolytic medication, but it does not typically cause sedation or respiratory depression. Unlike benzodiazepines, which can have sedative effects, buspirone is generally considered non-sedating and does not cause respiratory depression. However, it may take several weeks for buspirone to reach its full therapeutic effect in managing anxiety symptoms. Lorazepam and buspirone belong to different classes of medications and have different mechanisms of action. While both are used to treat anxiety disorders, they interact with different neurotransmitter systems. Lorazepam is a benzodiazepine that acts on the GABA receptors, while buspirone acts primarily as a partial agonist at serotonin 5-HT1A receptors. Therefore, they do not share similar interaction attributes.

3. *Answer*: A. He should explain to her that relief takes 4 to 6 weeks of escitalopram treatment and then schedule an appointment. Escitalopram, like other SSRIs, typically takes 4–6 weeks to demonstrate its full therapeutic effect. Since the patient has only been on the medication for 3 weeks, it is premature to deem it ineffective. Educating the patient about the expected time frame for improvement can help manage her expectations and encourage adherence to the treatment plan. Scheduling a follow-up appointment ensures that the patient's progress can be monitored while necessary adjustments can be made based on a complete evaluation of the medication's effectiveness.
4. *Answer:* C. Conduct a thorough evaluation, including pulmonary and thyroid function tests, and consider discontinuing amiodarone.

   *Explanation:* Amiodarone is associated with several serious adverse effects, including pulmonary toxicity, which can present as shortness of breath and dry cough. Thyroid dysfunction is another common issue with amiodarone. Given the patient's symptoms and the known risks of amiodarone, it is essential to evaluate for potential adverse effects through appropriate tests and consider discontinuing the drug if significant toxicity is identified. Continuing amiodarone without investigation may worsen the patient's condition. Sotalol, while an option, also carries a risk of exacerbating heart failure, particularly given the patient's low ejection fraction, so careful consideration and close monitoring are required.
5. *Answer*: The correct option is (A) because HRT with estrogen inhibits bone resorption, preventing the loss of bone or osteoporosis. However, the thrombogenic and procoagulant properties of estrogens may contribute to myocardial infarction, stroke, and blood clots. The woman is predisposed to uterine cancer because estrogen builds the endometrial lining of the uterus into a thick hyperplasia.

## References


Abdalla S, Briand C, Oualha M, Bendavid M, Béranger A, Benaboud S, Tréluyer JM, Zheng Y, Capito C, Demir Z, Foissac F, Winter S, Gana I, Boujaafar S, Bouazza N, Hirt D. Population pharmacokinetics of intravenous and oral Acyclovir and Oral Valacyclovir in pediatric population to optimize dosing regimens. Antimicrob Agents Chemother. 2020;64(12):e01426–0. https://doi.org/10.1128/AAC.01426-20. PMID: 32988829; PMCID: PMC7674054

Ahmadi M, Bekeschus S, Weltmann KD, von Woedtke T, Wende K. Non-steroidal anti-inflammatory drugs: recent advances in the use of synthetic COX-2 inhibitors. RSC Med Chem. 2022;13(5):471–96. https://doi.org/10.1039/d1md00280e. PMID: 35685617; PMCID: PMC9132194

Awosika AO, Singh G, Correa R. Methimazole. [Updated 2023 Sep 13]. In: StatPearls [Internet]. Treasure Island: StatPearls Publishing; 2024. Available from: https://www.ncbi.nlm.nih.gov/books/NBK545223/.

Azzini E, Raguzzini A, Polito A. A brief review on vitamin B12 deficiency looking at some case study reports in adults. Int J Mol Sci. 2021;22(18):9694. https://doi.org/10.3390/ijms22189694. PMID: 34575856; PMCID: PMC8471716

Balannik V, Wang J, Ohigashi Y, Jing X, Magavern E, Lamb RA, Degrado WF, Pinto LH. Design and pharmacological characterization of inhibitors of amantadine-resistant mutants of the M2 Ion Channel of Influenza A virus. Biochemistry. 2009;48:11872–82. https://doi.org/10.1021/bi9014488.

Balbino B, Herviou P, Godon O, Stackowicz J, Goff OR-L, Iannascoli B, Sterlin D, Brûlé S, Millot GA, Harris FM, Voronina VA, Nadeau KC, Macdonald LE, Murphy AJ, Bruhns P, Reber LL. The anti-IgE mAb omalizumab induces adverse reactions by engaging Fcγ receptors. J Clin Invest. 2020;130(3):1330–5. https://doi.org/10.1172/JCI129697.

Balon R, Starcevic V. Role of Benzodiazepines in anxiety disorders; 2020. p. 367–88. https://doi.org/10.1007/978-981-32-9705-0_20.

Bardeci M, Greil W, Stassen H, Willms J, Köberle U, Bridler R, Hasler G, Kasper S, Rüther E, Bleich S, Toto S, Grohmann R, Seifert J. Dear Doctor Letters regarding citalopram and escitalopram: guidelines vs real-world data. Eur Arch Psychiatry Clin Neurosci. 2023;273(1):65–74. https://doi.org/10.1007/s00406-022-01392-x.

Bednarczuk T, Brix TH, Schima W, Zettinig G, Kahaly GJ. 2021 European Thyroid Association Guidelines for the Management of Iodine-Based Contrast Media-Induced Thyroid Dysfunction. Eur Thyr J [Internet]. 2021;10(4):269–284. Available from. https://doi.org/10.1159/000517175.

Behranvand N, Nasri F, Zolfaghari Emameh R, Khani P, Hosseini A, Garssen J, Falak R. Chemotherapy: a double-edged sword in cancer treatment. Cancer Immunol Immunother. 2022;71(3):507–26. https://doi.org/10.1007/s00262-021-03013-3. Epub 2021 Aug 5. Erratum in: Cancer Immunol Immunother. 2022 Mar;71(3):527. doi: 10.1007/s00262-021-03034-y. PMID: 34355266; PMCID: PMC10992618.

Bindu S, Mazumder S, Bandyopadhyay U. Non-steroidal anti-inflammatory drugs (NSAIDs) and organ damage: a current perspective. Biochem Pharmacol. 2020;180:114147. https://doi.org/10.1016/j.bcp.2020.114147. Epub 2020 Jul 10. PMID: 32653589; PMCID: PMC7347500.

Bonanno AM, Bruno R, Filippelli A, Lombardo M. The impact of climate change on crop yield variability: evidence from a global sample of regions. ScienceDirect. 2021;210:123–32. https://doi.org/10.1016/j.foodpol.2021.102062.

Boraschi D, Penton-Rol G. Perspectives in immunopharmacology: the future of immunosuppression. Immunol Lett. 2014;161(2):211–5. https://doi.org/10.1016/j.imlet.2013.11.017. Epub 2013 Dec 12. PMID: 24333342.

Browning DJ. Pharmacology of Chloroquine and Hydroxychloroquine. In: Hydroxychloroquine and chloroquine retinopathy, vol. 4; 2014. p. 35–63. PMCID: PMC7122276. https://doi.org/10.1007/978-1-4939-0597-3_2.

Buscail L, Delesque N, Estève JP, Saint-Laurent N, Prats H, Clerc P, Robberecht P, Bell GI, Liebow C, Schally AV. Stimulation of tyrosine phosphatase and inhibition of cell prolifera-

tion by somatostatin analogs: mediation by human somatostatin receptor subtypes SSTR1 and SSTR2. Proc Natl Acad Sci USA. 1994;91:2315–9. https://doi.org/10.1073/pnas.91.6.2315.
Capra V, Thompson MD, Sala A, Cole DE, Folco G, Rovati GE. Cysteinyl-leukotrienes and their receptors in asthma and other inflammatory diseases: critical update and emerging trends. Med Res Rev. 2007;27(4):469–527. https://doi.org/10.1002/med.20071.
Chen H, Li J, Cui T. Adjuvant gonadotropin releasing hormone analogues for chemotherapy induced premature ovarian in premenopausal women. Cochrane Rev. 2011;11:CD008018.
Chittepu V, Kalhotra P, Osorio-Gallardo T, Gallardo-Velazquez T, Osorio-Revilla G. Repurposing of FDA-approved NSAIDs for DPP-4 inhibition as an alternative for diabetes mellitus treatment: computational and in vitro study. Pharmaceutics. 2019;11(5):238.
Corral-Gudino L, Tan AJ, Del Pino-Montes J, Ralston SH. Bisphosphonates for Paget's disease of bone in adults. Cochrane Database Syst Rev. 2017;12:CD004956. https://doi.org/10.1002/14651858.CD004956.pub3.
Davis MP. Mirtazapine and depressions. In: NeuroPsychopharmacotherapy. Springer International Publishing; 2020. p. 1–29. https://doi.org/10.1007/978-3-319-56015-1_102-1.
Dehkhoda F, Lee CMM, Medina J, Brooks AJ. The growth hormone receptor: mechanism of receptor activation, cell signaling, and physiological aspects. Front Endocrinol (Lausanne). 2018;9:35. https://doi.org/10.3389/fendo.2018.00035. PMID: 29487568; PMCID: PMC5816795
Delgado BJ, Lopez-Ojeda W. Estrogen. [Updated 2023 Jun 26]. In: StatPearls [Internet]. Treasure Island: StatPearls Publishing; 2024. Available from: https://www.ncbi.nlm.nih.gov/books/NBK538260/.
Dell'Osso B, Buoli M, Baldwin DS, Altamura AC. Serotonin norepinephrine reuptake inhibitors (SNRIs) in anxiety disorders: a comprehensive review of their clinical efficacy. Hum Psychopharmacol Clin Exp. 2010;25(1):17–29. https://doi.org/10.1002/hup.1074.
Diego-Adeliño J, Crespo JM, Mora F, Neyra A, Iborra P, Gutiérrez-Rojas L, Salonia SF. Vortioxetine in major depressive disorder: from mechanisms of action to clinical studies. An updated review. Expert Opin Drug Saf. 2022;21(5):673–90. https://doi.org/10.1080/14740338.2022.2019705.
Drake MT, Clarke BL, Khosla S. Bisphosphonates: mechanism of action an role in clinical practice. Mayo Clin Proc. 2008;83:1032–45.
Eccles R. What is the role of over 100 excipients in Over the Counter (OTC) cough medicines? Lung. 2020;198(5):727–34. https://doi.org/10.1007/s00408-020-00390-x.
Edwards M, Can AS. Progestins. [Updated 2024 Jan 10]. In: StatPearls [Internet]. Treasure Island: StatPearls Publishing; 2024. Available from: https://www.ncbi.nlm.nih.gov/books/NBK563211/.
Ertl P, Altmann E, McKenna JM. The most common functional groups in bioactive molecules and how their popularity has evolved over time. J Med Chem. 2020;63(15):8408–18. https://doi.org/10.1021/acs.jmedchem.0c00754.
Fajemiroye JO, Galdino PM, Florentino IF, et al. Plurality of anxiety and depression alteration mechanism by oleanolic acid. J Psychopharmacol. 2014;28(10):923–34. https://doi.org/10.1177/0269881114536789.
Fajemiroye JO, da Silva DM, de Oliveira DR, Costa EA. Treatment of anxiety and depression: medicinal plants in retrospect. Fundam Clin Pharmacol. 2016;30(3):198–215. https://doi.org/10.1111/fcp.12186. Epub 2016 Mar 4. PMID: 26851117.
Finn DF, Walsh JJ. Twenty-first century mast cell stabilizers. Br J Pharmacol. 2013;170(1):23–37. https://doi.org/10.1111/bph.12138.
Ganapathy N, Gokulnathan S, Balan N, Maheswaran T, Venkatesan. Bisphosphonates: an update. J Pharm Bioallied Sci. 2012;4(Suppl 2):S410–3. https://doi.org/10.4103/0975-7406.100309. PMID: 23066300; PMCID: PMC3467906
Garakani A, Murrough JW, Freire RC, Thom RP, Larkin K, Buono FD, Iosifescu DV. Pharmacotherapy of anxiety disorders: current and emerging treatment options. Front Psych. 2020;11 https://doi.org/10.3389/fpsyt.2020.595584.
Geulayov G, Ferrey A, Casey D, Wells C, Fuller A, Bankhead C, Gunnell D, Clements C, Kapur N, Ness J, Waters K, Hawton K. Relative toxicity of benzodiazepines and hypnotics

commonly used for self-poisoning: an epidemiological study of fatal toxicity and case fatality. J Psychopharmacol. 2018;32(6):654–62. https://doi.org/10.1177/0269881118754734. Epub 2018 Feb 14. PMID: 29442611.

Goh JW, Thaw MM, Ramim JU, Mukherjee R. Theophylline toxicity: a differential to consider in patients on long-term theophylline presenting with nonspecific symptoms. Cureus. 2023;15(11):e48480. https://doi.org/10.7759/cureus.48480.

Gomes FA, Cerqueira RO, Lee Y, Mansur RB, Kapczinski F, McIntyre RS, Yatham LN, Berk M, Milev R, Brietzke E. What not to use in bipolar disorders: a systematic review of non-recommended treatments in clinical practice guidelines. J Affect Disord. 2022;298:565–76. https://doi.org/10.1016/j.jad.2021.11.007.

Griffin CE 3rd, Kaye AM, Bueno FR, Kaye AD. Benzodiazepine pharmacology and central nervous system-mediated effects. Ochsner J. 2013;13(2):214–23. PMID: 23789008; PMCID: PMC3684331.

Hall JE, Omoto ACM, Wang Z, Mouton A, Li X, Hall ME. Pathophysiology of hypertension. In: Hypertension. Elsevier; 2024. p. 71–86. https://doi.org/10.1016/B978-0-323-88369-6.00005-0.

Hogan DB. Quinine: not a safe drug for treating nocturnal leg cramps. CMAJ. 2015;187(4):237–8. https://doi.org/10.1503/cmaj.150044. Epub 2015 Jan 26. PMID: 25623647; PMCID: PMC4347765.

Issa MS, Warsame M, Mahamat MHT, Saleh IDM, Boulotigam K, Djimrassengar H, Issa AH, Abdelkader O, Hassoumi M, Djimadoum M, Doderer-Lang C, Ndihiokubwayo JB, Rasmussen C, Menard D. Therapeutic efficacy of artesunate-amodiaquine and artemether-lumefantrine for the treatment of uncomplicated falciparum malaria in Chad: clinical and genetic surveillance. Malar J. 2023;22(1):240. https://doi.org/10.1186/s12936-023-04644-w. PMID: 37612601; PMCID: PMC10464190

Jiang A, Wei C, Zhu W, Wu F, Wu B. Adverse event profiles of drug-induced liver injury caused by antidepressant drugs: a disproportionality analysis. Ther Adv Drug Safe. 2024;15 https://doi.org/10.1177/20420986241244585.

Jilani TN, Gibbons JR, Faizy RM, Saadabadi A. Mirtazapine. In: StatPearls [Internet]. Treasure Island: StatPearls Publishing; 2024. PMID: 30085601.

Kadioglu SB, Celik T. Orthostatic hypotension and drugs: drug-induced orthostatic hypotension. In: Orthostatic hypotension in older adults. Springer International Publishing; 2021. p. 45–59. https://doi.org/10.1007/978-3-030-62493-4_6.

Kowalski ML, Makowska JS. Seven steps to the diagnosis of NSAIDs hypersensitivity: how to apply a new classification in real practice? Allergy Asthma Immunol Res. 2015;7(4):312–20.

Kroken RA, Johnsen E, Ruud T, et al. Treatment of schizophrenia with antipsychotics in Norwegian emergency wards, a cross-sectional national study. BMC Psychiatry. 2009;9:24. https://doi.org/10.1186/1471-244X-9-24.

Kua KP, Lee SWH, Chongmelaxme B. The impact of home-based management of malaria on clinical outcomes in sub-Saharan African populations: a systematic review and meta-analysis. Trop Med Health. 2024;52(1):7. https://doi.org/10.1186/s41182-023-00572-2. PMID: 38191459; PMCID: PMC10773121.

Kumar A, Zarychanski R, Pisipati A, Kumar A, Kethireddy S, Bow EJ. Fungicidal versus fungistatic therapy of invasiveCandidainfection in non-neutropenic adults: a meta-analysis. Mycology. 2018;9:116–28. https://doi.org/10.1080/21501203.2017.1421592.

Kumar P, Sheokand D, Grewal A, Saini V, Kumar A. Clinical side-effects based drug repositioning for anti-epileptic activity. J Biomol Struct Dyn. 2024;42(3):1443–54. https://doi.org/10.1080/07391102.2023.2199874. Epub 2023 Apr 12. PMID: 37042987.

Laux G. Serotonin Reuptake Inhibitors: Citalopram, Escitalopram, Fluoxetine, Fluvoxamine, Paroxetine, and Sertraline. In: NeuroPsychopharmacotherapy. Cham: Springer International Publishing; 2021. p. 1–13.

Law JP, Pickup L, Townend JN, Ferro CJ. Anticoagulant strategies for the patient with chronic kidney disease. Clin Med (Lond). 2020 Mar;20(2):151–5. https://doi.org/10.7861/clinmed.2019-0445.

Lee KK, Davenport PW, Smith JA, Irwin RS, McGarvey L, Mazzone SB, Birring SS, Abu Dabrh A, Altman KW, Barker AF, Birring SS, Blackhall F, Bolser DC, Brightling C, Chang AB, Davenport P, el Solh AA, Escalante P, Field SK, Vertigan AE. Global physiology and pathophysiology of cough. Chest. 2021;159(1):282–93. https://doi.org/10.1016/j.chest.2020.08.2086.

Liang J, Shang Y. Estrogen and cancer. Annu Rev Physiol. 2013;75:225–40. https://doi.org/10.1146/annurev-physiol-030212-183708.

Lin Y, Wang Y, Li PF. PPARα: an emerging target of metabolic syndrome, neurodegenerative and cardiovascular diseases. Front Endocrinol (Lausanne). 2022;13:1074911. https://doi.org/10.3389/fendo.2022.1074911. PMID: 36589809; PMCID: PMC9800994

Lin J, Liu W, Guan J, Cui J, Shi R, Wang L, Chen D, Liu Y. Latest updates on the serotonergic system in depression and anxiety. Front Synaptic Neurosci. 2023;15:1124112. https://doi.org/10.3389/fnsyn.2023.1124112. PMID: 37228487; PMCID: PMC10203201.

Liu D, Ahmet A, Ward L, Krishnamoorthy P, Mandelcorn ED, Leigh R, Brown JP, Cohen A, Kim H. A practical guide to the monitoring and management of the complications of systemic corticosteroid therapy. Allergy Asthma Clin Immunol. 2013;9:30.

LiverTox: clinical and research information on drug-induced liver injury [Internet]. Bethesda: National Institute of Diabetes and Digestive and Kidney Diseases; 2012. Gonadotropin Releasing Hormone (GnRH) Analogues. [Updated 2018 Mar 20]. Available from: https://www.ncbi.nlm.nih.gov/books/NBK547863/

Maeda K, Das D, Kobayakawa T, Tamamura H, Takeuchi H. Discovery and development of anti-HIV therapeutic agents: Progress towards improved HIV medication. Curr Top Med Chem. 2019;19(18):1621–49. https://doi.org/10.2174/1568026619666190712204603. PMID: 31424371; PMCID: PMC7132033

Martín-Hernández D, Ulecia-Morón C, Bris ÁG, Pereira MP, Caso JR. Monoaminergic system and antidepressants. In: The neuroscience of depression. Elsevier; 2021. p. 345–55. https://doi.org/10.1016/B978-0-12-817933-8.00093-1.

Mazu TK, Bricker BA, Flores-Rozas H, Ablordeppey SY. The mechanistic targets of antifungal agents: an overview. Mini Rev Med Chem. 2016;16:555–78. https://doi.org/10.2174/1389557516666160118112103.

Mehellou Y, De Clercq E. Twenty-six years of anti-HIV drug discovery: where do we stand and where do we go? J Med Chem. 2010;53(2):521–38. https://doi.org/10.1021/jm900492g.

Meltzer H, Huang M. In vivo actions of atypical antipsychotic drug on serotonergic and dopaminergic systems. In: Serotonin–Dopamine Interaction: Experimental Evidence and Therapeutic Relevance. Elsevier; 2008. p. 177–97.

Mrozek W, Socha J, Sidorowicz K, Skrok A, Syrytczyk A, Piątkowska-Chmiel I, Herbet M. Pathogenesis and treatment of depression: role of diet in prevention and therapy. Nutrition. 2023;115:112143. https://doi.org/10.1016/j.nut.2023.112143.

Nelson A, Lopes R, Hong H, et al. Cardiovascular effects of GnRH antagonists compared with agonists in prostate cancer: a systematic review. J Am Coll Cardiol CardioOnc. 2023;5(5):613–24.

Obradovic M, Sudar-Milovanovic E, Gluvic Z, Banjac K, Rizzo M, Isenovic ER. The Na+/K+-ATPase: a potential therapeutic target in cardiometabolic diseases. Front Endocrinol (Lausanne). 2023;14:1150171.

Ogundeji AO, Pohl CH, Sebolai OM. Repurposing of Aspirin and Ibuprofen as candidate anti-Cryptococcus drugs. Antimicrob Agents Chemother. 2016;60(8):4799–808.

O'Leary F, Samman S. Vitamin B12 in health and disease. Nutrients. 2010;2(3):299–316. https://doi.org/10.3390/nu2030299. Epub 2010 Mar 5. PMID: 22254022; PMCID: PMC3257642

Pang B, Qiao X, Janssen L, Velds A, Groothuis T, Kerkhoven R, et al. Drug-induced histone eviction from open chromatin contributes to the chemotherapeutic effects of doxorubicin. Nat Commun. 2013;4:1908. https://doi.org/10.1038/ncomms2921.

Paponja K, Pećin I, Reiner Ž, Banach M. Bempedoic acid: new evidence and recommendations on use. Curr Opin Lipidol. 2024;35(1):41–50. https://doi.org/10.1097/MOL.0000000000000911. Epub 2023 Dec 8. PMID: 38085172.

Patibandla S, Heaton J, Kyaw H. Spironolactone. [Updated 2023 Jul 4]. In: StatPearls [Internet]. Treasure Island: StatPearls Publishing; 2024. Available from: https://www.ncbi.nlm.nih.gov/books/NBK554421/.

Peng L, Morford KL, Levander XA. Benzodiazepines and related sedatives. Med Clin North Am. 2022;106(1):113–29. https://doi.org/10.1016/j.mcna.2021.08.012.

Pepe M, Lanzotti P, Mazza M. Hypnotics: course and duration of therapy, side effects, contraindications, interactions, withdrawal syndromes, and resistance to therapy. In: NeuroPsychopharmacotherapy. Springer International Publishing; 2021. p. 1–28. https://doi.org/10.1007/978-3-319-56015-1_455-1.

Ping B, Zhu Y, Gao Y, Yue C, Wu B. Second- versus first-generation azoles for antifungal prophylaxis in hematology patients: a systematic review and meta-analysis. Ann Hematol. 2013;92:831–9. https://doi.org/10.1007/s00277-013-1693-5.

Pivonello R, Fleseriu M, Newell-Price J, Bertagna X, Findling J, Shimatsu A, Gu F, Auchus R, Leelawattana R, Lee EJ, Kim JH, Lacroix A, Laplanche A, O'Connell P, Tauchmanova L, Pedroncelli AM, Biller BMK, Investigators L. Efficacy and safety of osilodrostat in patients with Cushing's disease (LINC 3): a multicentre phase III study with a double-blind, randomized withdrawal phase. Lancet Diabetes Endocrinol. 2020;8(9):748–61.

Pokhrel B, Pellegrini MV, Levine SN. PCSK9 inhibitors. In: StatPearls [Internet]. Treasure Island: StatPearls Publishing; 2024. PMID: 28846236.

Poveda E, Rodes B, Lebel-Binay S, Faudon JL, Jimenez V, Soriano V. Dynamics of enfuvirtide resistance in HIV-infected patients during and after long-term enfuvirtide salvage therapy. J Clin Virol. 2005;34:295–301. https://doi.org/10.1016/j.jcv.2005.02.004.

Preuss CV, Kalava A, King KC. Prescription of controlled substances: benefits and risks. In: StatPearls [Internet]. Treasure Island: StatPearls Publishing; 2024. PMID: 30726003.

Ray A, Ficek M. Immunomodulatory effects of anti-estrogenic drugs. Acta Pharma. 2012;62(2):141–55. https://doi.org/10.2478/v10007-012-0012-3. PMID: 22750814

Reinhart M, Puil L, Salzwedel DM, Wright JM. First-line diuretics versus other classes of antihypertensive drugs for hypertension. Cochrane Database Syst Rev. 2023;2023(7) https://doi.org/10.1002/14651858.CD008161.pub3.

Rena G, Hardie DG, Pearson ER. The mechanisms of action of metformin. Diabetologia. 2017;60(9):1577–85. https://doi.org/10.1007/s00125-017-4342-z. Epub 2017 Aug 3. PMID: 28776086; PMCID: PMC5552828.

Ribbiso KA, Heller LE, Taye A, Julian E, Willems AV, Roepe PD. Artemisinin-based drugs target the plasmodium falciparum Heme detoxification pathway. Antimicrob Agents Chemother. 2021;65(4):e02137–20. https://doi.org/10.1128/AAC.02137-20. PMID: 33495226; PMCID: PMC8097475.

Robinson C, Dalal S, Chitneni A, Patil A, Berger AA, Mahmood S, Orhurhu V, Kaye AD, Hasoon J. A look at commonly utilized Serotonin Noradrenaline Reuptake Inhibitors (SNRIs) in chronic pain. Health Psychol Res. 2022;10(3):32309. https://doi.org/10.52965/001c.32309.

Rodgers GM, Mahajerin A. Antithrombin therapy: current state and future outlook. Clin Appl Thromb Hemost. 2023;29:10760296231205279. https://doi.org/10.1177/10760296231205279. PMID: 37822179; PMCID: PMC10571690

Ross FA, MacKintosh C, Hardie DG. AMP-activated protein kinase: a cellular energy sensor that comes in 12 flavours. FEBS J. 2016;283:2987–3001.

Sagatova AA, Keniya MV, Wilson RK, Monk BC, Tyndall J. Structural insights into binding of the antifungal drug Fluconazole to Saccharomyces cerevisiae Lanosterol 14α-Demethylase. Antimicrob Agents Chemother. 2015;59:4982–9. https://doi.org/10.1128/AAC.00925-15.

Sanabria E, Cuenca RE, Esteso MÁ, Maldonado M. Benzodiazepines: their use either as essential medicines or as toxics substances. Toxics. 2021;9(2):25. https://doi.org/10.3390/toxics9020025.

Santos R, Ursu O, Gaulton A, et al. A comprehensive map of molecular drug targets. Nat Rev Drug Discov. 2017;16:19–34. https://doi.org/10.1038/nrd.2016.230.

Scherholz ML, Schlesinger N, Androulakis IP. Chronopharmacology of glucocorticoids. Adv Drug Deliv Rev. 2019;151–152:245–261. https://doi.org/10.1016/j.addr.2019.02.004. Epub 2019 Feb 21. PMID: 30797955; PMCID: PMC6703983

Sheffler ZM, Patel P, Abdijadid S. Antidepressants, 2024. https://pubmed.ncbi.nlm.nih.gov/30844209.
Shi Y, Vanhoutte PM. Oxidative stress and COX cause hyper-responsiveness in vascular smooth muscle of the femoral artery from diabetic rats. Br J Pharmacol. 2008;154(3):639–51. https://doi.org/10.1038/bjp.2008.110. Epub 2008 Apr 14. PMID: 18414395; PMCID: PMC2439510
Simone CG, Bobrin BD. Anxiolytics and sedative-hypnotics toxicity. In: StatPearls [Internet]. Treasure Island: StatPearls Publishing; 2023. PMID: 32965980.
Smith ALW, Harmer CJ, Cowen PJ, Murphy SE. The Serotonin 1A (5-HT1A) receptor as a pharmacological target in depression. CNS Drugs. 2023;37(7):571–85. https://doi.org/10.1007/s40263-023-01014-7.
Smyth EM, Grosser T, Wang M, Yu Y, FitzGerald GA. Prostanoids in health and disease. J Lipid Res. 2009;50(Suppl):S423–8.
Stergiou GS, Palatini P, Parati G, O'Brien E, Januszewicz A, Lurbe E, Persu A, Mancia G, Kreutz R. 2021 European Society of Hypertension practice guidelines for office and out-of-office blood pressure measurement. J Hypertens. 2021;39(7):1293–302. https://doi.org/10.1097/HJH.0000000000002843.
Strand DS, Kim D, Peura DA. 25 years of proton pump inhibitors: a comprehensive review. Gut Liver. 2017;11(1):27–37. https://doi.org/10.5009/gnl15502. PMID: 27840364; PMCID: PMC5221858.
Świerczyńska M, Mirowska-Guzel DM, Pindelska E. Antiviral Drugs in Influenza. Int J Environ Res Public Health. 2022;19(5):3018. https://doi.org/10.3390/ijerph19053018. PMID: 35270708; PMCID: PMC8910682.
Taylor M, Gerriets V. Acyclovir. [Updated 2023 May 7]. In: StatPearls [Internet]. Treasure Island: StatPearls Publishing; 2024. Available from: https://www.ncbi.nlm.nih.gov/books/NBK542180/.
Tomlinson B, Lin CH, Chan P, Lam CW. Personalized medicine in lipid-modifying therapy. Per Med. 2021;18(2):185–203. https://doi.org/10.2217/pme-2020-0110. Epub 2021 Feb 10. PMID: 33565320.
Torti JF, Correa R. Potassium iodide. [Updated 2023 May 1]. In: StatPearls [Internet]. Treasure Island: StatPearls Publishing; 2024. Available from: https://www.ncbi.nlm.nih.gov/books/NBK542320/.
Tritos NA. Adrenally directed medical therapies for Cushing syndrome. J Clin Endocrinol Metab. 2021;106(1):16–25. https://doi.org/10.1210/clinem/dgaa778. PMID: 33118025
Tsutsumi S, Gotoh T, Tomisato W, Mima S, Hoshino T, Hwang HJ, Takenaka H, Tsuchiya T, Mori M, Mizushima T. Endoplasmic reticulum stress response is involved in nonsteroidal anti-inflammatory drug-induced apoptosis. Cell Death Differ. 2004;11(9):1009–16.
Turner J, Parsi M, Badireddy M. Anemia. [Updated 2023 Aug 8]. In: StatPearls [Internet]. Treasure Island: StatPearls Publishing; 2024. Available from: https://www.ncbi.nlm.nih.gov/books/NBK499994/.
Ullah M, Wahab A, Khan SU, Zaman U, Rehman K, Ur Hamayun S, Naeem M, Ali H, Riaz T, Saeed S, Alsuhaibani AM, Refat MS. Stent as a Novel Technology for coronary artery disease and their clinical manifestation. Curr Probl Cardiol. 2023;48(1):101415. https://doi.org/10.1016/j.cpcardiol.2022.101415.
Uslaner JM, Herring WJ, Coleman PJ. The discovery of Suvorexant: lessons learned that can be applied to other CNS drug development efforts. ACS Pharmacol Transl Sci. 2020;3(1):161–8. https://doi.org/10.1021/acsptsci.9b00110.
Valenzuela PL, Carrera-Bastos P, Gálvez BG, Ruiz-Hurtado G, Ordovas JM, Ruilope LM, Lucia A. Lifestyle interventions for the prevention and treatment of hypertension. Nat Rev Cardiol. 2021;18(4):251–75. https://doi.org/10.1038/s41569-020-00437-9.
Vane JR. Introduction: mechanism of action of NSAIDs. Br J Rheumatol. 1996;35(Suppl 1):1–3.
Vane JR, Bakhle YS, Botting RM. Cyclooxygenases 1 and 2. Annu Rev Pharmacol Toxicol. 1998;38:97–120.
Weirich J, Wenzel W. Aktuelle Einteilung antiarrhythmisch wirksamer Substanzen [Current classification of anti-arrhythmia agents]. Z Kardiol. 2000;89 Suppl 3:62–7. German. PMID: 10810787.

Wilson TK, Tripp J. Buspirone. In: StatPearls [Internet]. Treasure Island: StatPearls Publishing; 2024. Available from: https://www.ncbi.nlm.nih.gov/books/NBK531477/ PMID: 30285372.

Woods D, Turchi JJ. Chemotherapy-induced DNA damage response. Cancer Biol Ther. 2013;14(5):379–89. https://doi.org/10.4161/cbt.23761.

Xiao M, Cai J, Cai L, Jia J, Xie L, Zhu Y, et al. Let-7e sensitizes epithelial ovarian cancer to cisplatin through repressing DNA double-strand break repair. J Ovarian Res. 2017;10(1):24. https://doi.org/10.1186/s13048-017-0321-8.

Yoneyama H, Katsumata R. Antibiotic resistance in bacteria and its future for novel antibiotic development. Biosci Biotechnol Biochem. 2006;70:1060–75.

Yoon YH, Cho KS, Hwang JJ, et al. Induction of lysosomal dilatation, arrested autophagy, and cell death by chloroquine in cultured ARPE-19 cells. Invest Ophthalmol Vis Sci. 2010;51:6030–7.

Yoshida Y, Honma M, Kimura Y, Abe H. Structure, synthesis and inhibition mechanism of nucleoside analogues as HIV-1 reverse transcriptase inhibitors (NRTIs). ChemMedChem. 2021;16(5):743–66. https://doi.org/10.1002/cmdc.202000695. Epub 2021 Feb 9. PMID: 33230979.

# Chapter 7
# Interaction from Structure-Activity Relation Perspective

**Bárbara Gonçalves Rivello, Jhon Kennedy Alves Pereira, Igor Victor Ferreira Dos Santos, Cleydson Breno Rodrigues Dos Santos, James O. Fajemiroye, and Ricardo Menegatti**

**Abstract** The well-known principles that drive the drug-drug interactions provide important clues toward our understanding of the outcome of food-drug interactions. Understanding the interplay between structure and activity is crucial in medicinal chemistry and pharmacology. By analyzing these relationships, researchers can develop more effective therapeutic agents (medications and nutrition) with interaction profiling. Hence, this chapter focuses on interactions involving drugs, cofactors, coenzymes, and compounds produced in the biochemical processes.

**Keywords** Food-drug interactions · Cofactors · Coenzymes

## 1 Introduction

Drug-drug interactions (DDIs) are often associated with enzymatic inhibition or induction. Enzymatic inhibition occurs with an interaction between a Lewis base compound and iron from cytochrome P450 (CYP), a Lewis acid. For example, the iron in heme from CYP forms strong complexes with basic compounds, which can delay drug metabolism (Hakkola et al. 2020; Zanger and Schwab 2013). As illustrated in Fig. 7.1a, it is possible to see the scheme of complex formation between the

B. G. Rivello · J. K. A. Pereira · R. Menegatti (✉)
Faculty of Pharmacy, Laboratory of Medicinal Pharmaceutical Chemistry, Federal University of Goiás, Goiania, Goias, Brazil
e-mail: jhonkennedy@discente.ufg.br

I. V. F. D. Santos · C. B. R. D. Santos
Modeling and Computational Chemistry Laboratory, Federal University of Amapá, Macapá, Amapá, Brazil
e-mail: breno@unifap.br

J. O. Fajemiroye
Institute of Biological Sciences, Federal University of Goias, Goiania, Goias, Brazil
e-mail: jamesfajemiroye@ufg.br

 
J. O. Fajemiroye (ed.), *Fundamentals of Drug and Non-Drug Interactions*,
https://doi.org/10.1007/978-3-031-80107-5_7

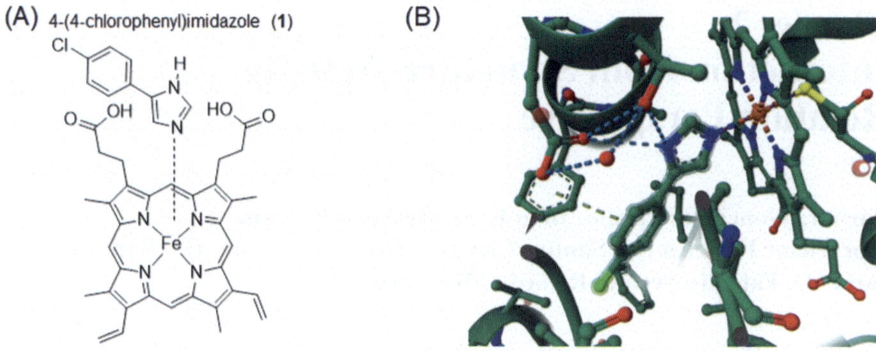


**Fig. 7.1** Scheme of CYP2B4/CPI complex formation (**a**) and the CYP2B4/CPI complex determined by X-ray crystallography (**b**). (Adapted from Scott et al. (2004))

imidazole scaffold present at the specific inhibitor 4-(4-chlorophenyl)imidazole (CPI) (**1**), which behaves as a Lewis base, and the central iron from microsomal cytochrome P450 2B4, which was determined by X-ray crystallography. The pose in Fig. 7.1b suggests a structural mechanism whereby ligands may coordinate with the catalytic site in CYP2B4. The CYP2B4/CPI complex provides some insights into the dynamics involved in the tight inhibitor binding and coordination with redox sites CYP2B4 (Scott et al. 2004). On the other hand, enzymatic induction occurs through an increase in gene expression, which elevates the CYP levels and, consequently, increases drug metabolism (Hakkola et al. 2020; Zanger and Schwab 2013). Other pharmacokinetic and/or pharmacodynamic aspects also can interfere with DDI. We can use the same approach to elucidate the essential mechanisms underlying food-drug interactions.

## 2 Cofactors

Cofactors can be defined as cationic species, which are required by some enzymes to better develop their role in living cells. They are usually ions, such as $Ca^{2+}$, $Fe^{2+}$, $Mg^{2+}$, $Mn^{2+}$, or $Zn^{2+}$, or a complex organic or metalloorganic molecule called a coenzyme like NAD(P)H/NAD(P)$^+$ and ATP/ADP (Nelson et al. 2021; Sun et al. 2023). They maintain cellular redox balance and drive metabolic (anabolic and catabolic) reactions. Calcium is important for preventing and treating osteoporosis. This mineral is obtained from dairy products and supplements, among other sources. It is known that calcium can be associated with significant drug interactions (Dawson-Hughes et al. 2002). For example, it can chelate with scaffold bidentate ligands, present in a lot of drugs, for example, tetracyclines and fluoroquinolones (Fig. 7.2). The bioavailability of ciprofloxacin (**2**) can decrease by 20–25% if taken together with calcium acetate, with a resulting decrease in the efficacy of treatment (Frost et al. 1992; Li et al. 1994; Walden et al. 2021). The chelation of tetracycline (**3**) with

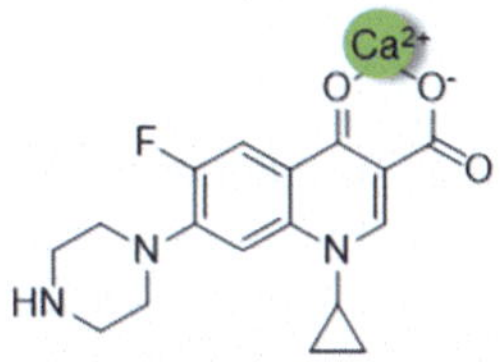

Ciprofloxacin (**2**)

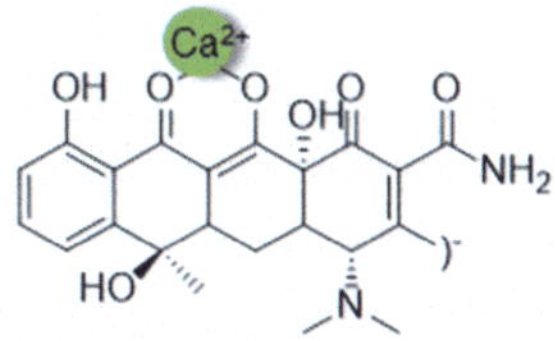

Tetracycline (**3**)

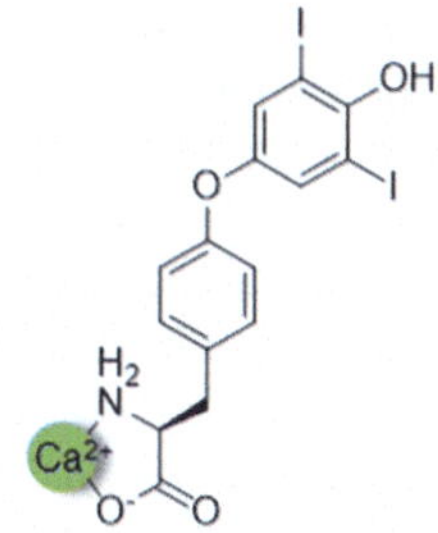

Levothyroxine (**4**)

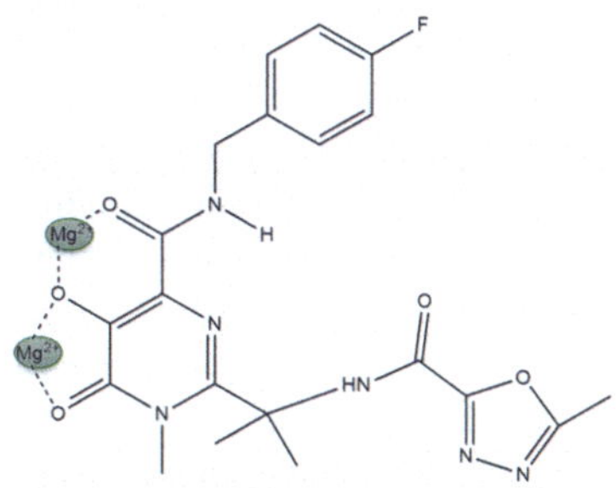

Raltegravir (**5**)

**Fig. 7.2** Formation of a complex between bidentate drugs—ciprofloxacin (**2**), tetracycline (**3**), levothyroxine (**4**), and raltegravir (**5**)—with calcium

metal ions also changes the antibiotic's pharmacological profile (Guerra et al. 2016). Coadministration of different calcium preparations can reduce levothyroxine (**4**) absorption by around 20–25% compared to that obtainable in the absence of calcium (Zamfirescu and Carlson 2011; Singh et al. 2000; Kaur et al. 2020). Drugs like phenytoin, phenobarbital, and orlistat can show similar profiles, decreasing the absorption of calcium. Calcium-affinitive and bone-seeking agents tend to accumulate in bones due to their chemical structure and properties (Sawamoto et al. 2020). The bisphosphonates (alendronate, risedronate, ibandronate) inhibit bone resorption by osteoclasts and are used to treat osteoporosis and other bone disorders. These drugs contain structures that have a high affinity for hydroxyapatite, the mineral component of bone. The strontium ranelate promotes bone formation and reduces resorption, thereby increasing bone density. Although fluoride can stimulate bone formation, it can lower bone quality if used excessively.

Integrase Strand Transfer Inhibitors (INSTIs), a class of antiretroviral drugs used to treat human immunodeficiency virus (HIV), can also be chelated by metals (Bacchi et al. 2011). This chelation can impact their effectiveness. For example, the absorption of INSTIs can be significantly reduced if taken with substances containing divalent cations such as calcium, magnesium, iron, or aluminum, particularly under fasted conditions (Song et al. 2014; Rock et al. 2020). These metals can bind to the drug, forming a chelate complex that reduces the bioavailability of the drug.

Chronic use of several drugs can interfere with the body's calcium homeostasis. Corticosteroids can decrease calcium absorption and induce osteoporosis (Jehle 2003). On the other hand, drugs such as vitamin D can increase the absorption and fixation of calcium in the bones.

Magnesium (Mg) is essential for carbohydrate metabolism, glucose transport and oxidation, intracellular insulin signaling, and the synthesis of adenosine triphosphate, proteins, and nucleic acids, among other important functions in the human body (Elin 2010). Regarding $MgCl_2$ and its association with hypermagnesemia, some foods can influence magnesium levels in the human body, contributing to the development of the disease. Food rich in phytates and dietary fibers inhibits magnesium absorption. The use of Mg supplementation to improve fitness performance is prevalent among the elderly and alcoholics compared to physically active individuals (Wang et al. 2017). Phytic acid, a derivative of phytate, chelates trace elements, inhibiting their absorption in the small intestine and reducing their bioavailability by forming soluble complexes. Besides their effects on the bioavailability of magnesium, some foods increase renal excretion of magnesium, such as diets rich in calcium and sodium, as they compete for the same reabsorption sites (Palacios et al. 2013). Studies have shown that diets with reduced sodium content led to greater magnesium retention. Regarding calcium, studies conducted in animal models have shown that increases in calcium content were able to reduce renal magnesium absorption (Seki et al. 2013; Bertinato et al. 2014; De Baaij 2023). Another relevant aspect is diets rich in animal protein. Mahalko et al. (1983) demonstrated that high amounts of this nutrient promote increased intestinal absorption and renal excretion of magnesium. In this sense, a diet poor in vegetable-based foods and rich in animal-based foods can induce acidosis, favoring increased mineral excretion (Rayssiguier et al. 2010). Magnesium, iron, and aluminum ions show a behavior related to calcium, which can form complexes with some drugs such as quinolones, tetracyclines, bisphosphonates, and levothyroxine.

Iron is an important cofactor involved in the transport of gasses, that is, oxygen and carbon dioxide, to tissues through hemoglobin and CYP-mediated oxidative metabolism. In low iron level states, people manifest signs of tiredness, shortness of breath, impaired physical performance, deficit learning, and problems with immunity. Several drugs, including proton pump inhibitors, polyphenols, antibiotics, and siderophores, affect the absorption and transport of iron. Proton pump inhibitors decrease gastric acidity; in these conditions, there is limited iron absorption and in some situations, the patients need to receive an intravenous administration of iron. As illustrated in Fig. 7.3, the absorption of iron also is compromised by polyphenols, that is baicalein (**5**), which can form complexes with iron and block iron uptake (Mladenka et al. 2011). In 2019, the antibiotic drug, cefiderocol (**6**), from Shionogi & Co, was approved by the Food and Drug Administration (FDA), which used a new drug design strategy, called siderophores (Domingues et al. 2023). Siderophores from Greek—*iron carrier*—are small organic compounds that act as ferric ion-specific chelating agents elaborated by bacteria and fungi growing under low iron stress (Neilands 1995). Cefiderocol (**6**) is a siderophore bonded to cephalosporin, which uses the iron transport system to pass through the membrane and to reach the

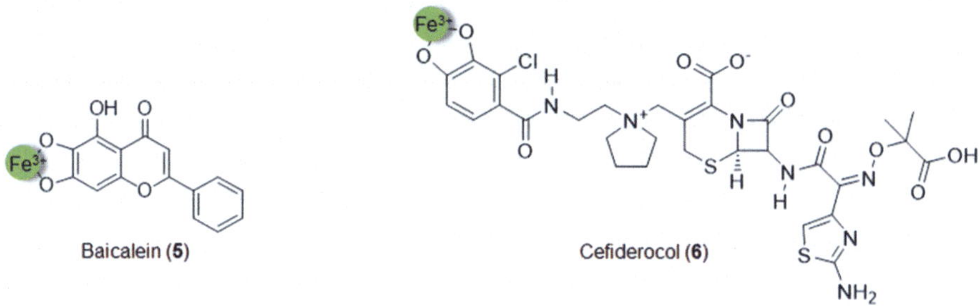


**Fig. 7.3** Formation of a complex between iron and baicalein (**5**) and cefiderocol (**6**), a new siderophore antibiotic approved by the FDA in 2019

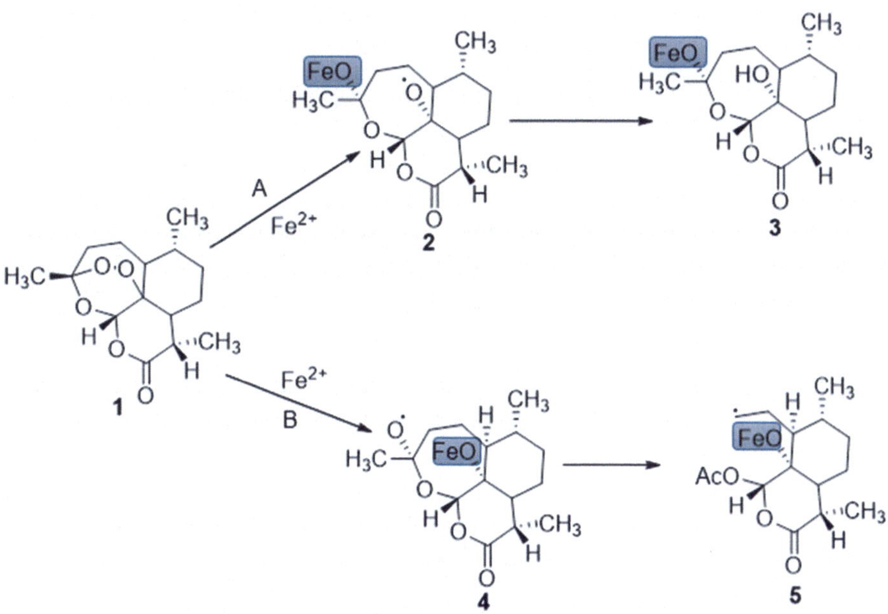


**Fig. 7.4** Iron(II)-catalyzed free radical pathways in the oxidative transformation of artemisinin (Tonmunphean et al. 2001)

periplasmic space such as a horse trojan, commercial name of "Fetroja®" (Domingues et al. 2023). The iron transport system can be used as a kind of active transport for drug delivery. Similarly, iron significantly influences the effectiveness of artemisinins, a class of antimalarial drugs. Although there is debate over whether free iron or heme is the primary activator, evidence favors heme. Artemisinin has a peroxy group that works like the Fenton reaction, where hydrogen peroxide ($H_2O_2$) is catalyzed by iron or copper to create a harmful free radical (HO·) (Li and Zhou 2010). The activation of artemisinins by ferrous iron (Fig. 7.4) enhances their cytotoxic effects, making them effective in both malaria treatment and cancer therapy (Xu et al. 2020). For instance, artemisinin derivatives can induce ferroptosis, an iron-dependent form of cell death (Ooko et al. 2015), making them potent against cancer cells with high iron uptake. This mechanism is exploited by conjugating artemisinin with transferrin to specifically target cancer cells, enhancing the drug's selectivity and potency (Emami et al. 2018).

## 3 Coenzyme

A coenzyme is an organic compound or metal-organic complex ion, which facilitates enzyme activity. They can be very tightly or even covalently bound to the enzyme protein called a prosthetic group (Nelson et al. 2021). In folk medicine, coenzymes are called vitamins. Vitamins are obtained from food, essential for the functioning of important biochemical processes in the body, especially as catalysts for chemical reactions. According to their physicochemical properties, they can be classified as fat-soluble and water-soluble vitamins. Fat-soluble vitamins include vitamins A, D, E, and K, while water-soluble vitamins include vitamin B complex and vitamin C. The fat and water-related affinities can directly interfere with the inhibition or induction of the physicochemical characteristics of medications.

Vitamin K is present in large quantities in some vegetable-based foods, oils, and animal-derived foods. In the human body, there are proteins dependent on this vitamin, such as coagulation factors and prothrombin, important for the coagulation cascade and bone metabolism. Therefore, irregular consumption of vitamin K can influence the effect of anticoagulants such as warfarin, which is used in the treatment of thrombotic disorders like atrial fibrillation and deep vein thrombosis because it is a vitamin K antagonist (Stafford 2005; De Carvalho Minighini et al. 2020; Ordi-Ros et al. 2019).

Many enzymes are responsible for the biological processes of living organisms, including cytochrome P450 enzymes, which are a superfamily of enzymes. Considering a wide variety of abundant CYP450 isoenzymes, oral anticoagulants are recommended for patients with stroke, for whom vitamin K antagonists have long been the only option on the market, such as warfarin. Since vitamin K is necessary for the synthesis of proteins that help control bleeding, new oral anticoagulants have been developed to act as non-vitamin K antagonists, such as dabigatran, rivaroxaban, apixaban, or edoxaban, introduced for patients with atrial fibrillation or venous thromboembolism (Stöllberger and Finsterer 2016; Olie et al. 2024). These new medications overcome some limitations, such as fewer food and drug interactions, intestinal absorption, and renal elimination depending on P-glycoprotein (P-gp) permeability, in addition to being substrates of hepatic cytochrome P450 3A4 (CYP3A4). Pharmacodynamic drug interactions occur when non-vitamin K antagonists are administered with drugs that affect the activity of the P-gp system or CYP3A4 as pharmacokinetic interactions involve the induction of P-gp or CYP3A4, leading to decreased serum levels of non-vitamin K antagonists, consequently, increasing the risk of embolism. Conversely, inhibition can increase serum levels of medications, leading to increased anticoagulant effects and a tendency to bleed (Hellwig and Gulseth 2013; Galgani et al. 2018; Stoellberger and Finsterer 2021).

Vitamin supplementation provides the body with the replacement of important nutrients such as vitamin D. Unlike essential vitamins, it can be produced by the body through a photosynthetic reaction when the skin is exposed to sunlight. This vitamin can be found in two forms, ergocalciferol (D2) and cholecalciferol (D3), and to become active in the body, they need to be activated in the liver and kidneys through biotransformation by adding a hydroxyl group through a photolysis reaction in the epidermis and subsequently by other hydroxylases such as CYP27B1, CYP24A1, and CYP3A4.

There are ways to obtain vitamin D, either by consuming rich foods in this nutrient or through sunlight, which synthesizes vitamin D in the skin. It is related to bone metabolism, promoting greater absorption of calcium and phosphorus in the intestine, and can regulate bone calcium stores in the presence of hypocalcemia. It is known that various medications interfere with this metabolism by activating the pregnane X receptor, which plays an important role in the detoxification of xenobiotics expressed in gastrointestinal tract cells, kidneys, and liver, such as antiepileptics, glucocorticoids, antiestrogens, or antiretroviral drugs, causing vitamin D deficiency. Some studies indicate that vitamin D affects reducing the lip activity of statins like HMG-CoA reductase and the antibacterial effect of antitubercular agents (Gröber and Kisters 2012; Silva and Furlanetto 2018).

Bone integrity disorders induced by antiepileptic drugs are particularly influenced by the type, dosage, and duration of treatment. For example, the use of carbamazepine, phenobarbital, phenytoin, and primidone increases the degradation of vitamin D by inducing microsomal enzymes in the liver mediated by the X receptor, resulting in decreased levels of 25(OH)D and 1,25(OH)2D, consequently increasing parathyroid levels and bone turnover. Medications such as phenytoin and valproic acid interfere with calcium balance by reducing intestinal absorption through renal tubular dysfunction, leading to increased calcium and phosphate losses (Junges et al. 2020; Abdullah and Mousheer 2020). Osteoporosis is one of the consequences of prolonged glucocorticoid therapy. Medications such as prednisolone can cause risks of spinal fractures depending on the dose ingested. Glucocorticoids increase osteoclast activity by increasing RANK ligand expression (nuclear factor kappa B [NFkB] receptor activator) reducing the production of osteoprotegerin and the development and differentiation of osteoblasts, as well as reducing intestinal calcium absorption and simultaneously increasing renal calcium excretion leading to decreased serum calcium levels or increased degradation of 25(OH)D and 1,25(OH)2D due to X receptor activation, for example, dexamethasone (Chalitsios et al. 2021; Kanis et al. 2004, 2018; Searing et al. 2010).

Bisphosphonates are drugs used in the treatment of osteoporosis. They tend to accumulate on the bone surface and are absorbed in regions with increased bone turnover. In patients with vitamin D deficiency, they can lead to hypomagnesemia and particularly hypocalcemia, mainly observed after administration of bisphosphonates such as zoledronic acid. In other cases, worsening of osteoporosis therapy has been observed with drug administration, for example, alendronic acid, risedronic acid, or raloxifene, altering bone density in patients with vitamin D deficiency (Gröber and Kisters 2012; Ardine et al. 2006; Hokugo et al. 2010; Hou et al. 2018). There are reports that people with human immunodeficiency virus (HIV) are at risk of osteoporosis, demonstrating that HIV-1 glycoproteins (p55-gag, gp120) impair bone calcium utilization and reduce osteoblast activity. This risk is additionally increased by antiretroviral therapy due to vitamin D metabolism disorders, particularly increasing degradation due to induction of CYP3A4 (Brown and Qaqish 2006; Brown et al. 2011; Alvarez et al. 2019; Jiménez-Sousa et al. 2018).

Medications such as aromatase inhibitors (e.g., anastrozole, letrozole, and exemestane) or modulators (e.g., tamoxifen and toremifene) are active substances used in estrogen treatment. As aromatase inhibitors block estrogen synthesis, they considerably reduce levels in bones. Estrogens promote intestinal calcium absorption and bone mineralization and also inhibit osteoclast activity. Hypovitaminosis D has been implicated in bone loss in women with breast cancer undergoing aromatase inhibitor therapy; furthermore, this deficiency is also associated with musculoskeletal symptoms and muscle strength and function (Khan et al. 2010; Plotnikoff and Quigley 2003; Park et al. 2019; Niravath et al. 2019). Finally, fat-soluble vitamin A is found in a lot of foods, mainly in yellow-orange fruits and vegetables, and it plays an important role in several biochemistry pathways. It is essential from embryogenesis to adulthood, and low levels of vitamin A can be related to some diseases, that is, celiac disease, Crohn's disease, and pancreatic disorders (Tanoury et al. 2013). Isotretinoin is a potent vitamin A derivative that is used to treat acne. However, its dermatologic use demands some care because it is teratogenic and can cause life-threatening fetal abnormalities in the early phases of pregnancy (Menezes and Almeida 2024).

The deficiency of water-soluble vitamin B6, or pyridoxine, is related to some types of anemia. The main sources of pyridoxine include meats, vegetables, grains, and fruits. These were observed in epileptic patients treated with phenytoin drug decreasing pyridoxine levels (Mintzer et al. 2012). The same was observed in folate and vitamin B12 (Kathiravan et al. 2021). Parkinson's patients can show a higher risk of having abnormal serum levels of vitamin B6, which can be associated with polyneuropathy (Modica et al. 2023). Platinum-based chemotherapeutic drugs display structural similarity to pyridoxine, interfering with its absorption, potentially leading to peripheral neuropathy and depression (Sawhney et al. 2022). Vitamin B6 supplementation could alleviate adverse effects like sensory neuropathy. Vitamin B3 or niacin is a B-complex vitamin used for the treatment of hyperlipidemia and pellagra. The main sources of vitamin B3 are animal foods, that is, meat, poultry, red fish, and vegetable's minor concentrations. The use of niacin and statins, which act as 3-hydroxy-3-methyl-glutaril-CoA reductase, can develop myopathies or rhabdomyolysis. This kind of association should be monitored because it elevates levels of multiple HDL proteins, which increases atherosclerotic risk (Ronsein et al. 2021). Vitamin B9 or folic acid aids biochemical pathways to synthesize DNA, RNA, and proteins. Folic acid supplementation is recommended during methotrexate therapy in patients with rheumatoid arthritis or psoriasis. In these situations, methotrexate inhibits dihydrofolate reductase, which is responsible for the biosynthesis of folic acid (Dhir et al. 2015). This can be related to the simultaneous binding of folic acid and methotrexate to human serum albumin and the competitive displacement of drugs (Panja et al. 2018).

## 4 Methylxanthines

Methylxanthines originate from the purine base adenine, and despite having heterocyclic nitrogen, they are not considered a true alkaloid, since they are not derived from amino acids, which is why they are called pseudo-alkaloids. Furthermore, they have an amphoteric character and can behave like acids or bases. Methylxanthines are the most widely consumed pharmacologically active compounds in the world, present in various food products and non-alcoholic drinks such as coffee, guarana, and chocolate (de Paula and Farah 2019; Silva et al. 2021). Produced in different parts of the world, with the *Coffea arabica* species being one of the most commercialized (Maurin et al. 2007), coffee is made up of the following components: caffeine, protein, polysaccharides, sucrose, monosaccharides, volatile and non-volatile acids, clonogenic lipids, trigonelline, nicotinic acid, volatile aromatic compounds, minerals, among others (Boekema et al. 1999; Nehlig 2022). Oral administration of caffeine (1,3,7-trimethylxanthine) that inhibits phosphodiesterase could relieve suffering from an asthma exacerbation with shortness of breath, hypoxia, and diffuse expiratory wheezing. The relevance of caffeine as an interferent in the pharmacokinetics of some medications is highlighted here. When the caffeine carbonyl interacts with compounds containing acidic hydrogens, dipole-dipole interactions and hydrogen bonds are formed. This complex form may be soluble depending on the nature of these organic acids. For example, caffeine can form complexes with medications that contain proton donors, such as aliphatic phenols and alcohols (Belayneh and Molla 2020).

The administration of caffeine with escitalopram, a selective serotonin reuptake inhibitor antidepressant, in an in vitro dissolution profile study, indicated that the presence of caffeine promotes a 33.15% reduction in drug release. In vivo studies demonstrated that there was a 28% reduction in drug absorption when administered with caffeine (Krishna et al. 2013). On the other hand, caffeine can increase the absorption of some medications such as acetylsalicylic acid (ASA). Administration of 60 mg of caffeine along with 650 mg of ASA increases the rate of ASA absorption as well as bioavailability (Yaghoobian et al. 2023; Yoovathaworn et al. 1986). The potentiating effect has not yet been fully elucidated; however, a possible explanation for the increased absorption is that caffeine increases the secretion of gastric juice in the stomach, thus leading to a greater release of hydrochloric acid and causing the pH of the stomach to decrease. Thus, ASA, having a pKa around 3.5, would remain in the non-ionized form, increasing the absorption of the drug in the stomach (Belayneh and Molla 2020). Caffeine has not been shown to alter the metabolism of ASA (Yoovathaworn et al. 1986), as they are metabolized by different enzymes (Yaghoobian et al. 2023). AAS is metabolized by UDP-glucuronosyltransferase

1A6 (UGT1A6), cytochrome P45 2C9 (CYP2C9), and *N*-acetyl transferase 2 (NAT2) (Palikhe et al. 2011), while caffeine is mainly metabolized by CYP1A2 (Tian et al. 2019).

Caffeine biotransformation in the liver does not occur only through CYP1A2; the isoforms CYP1A1, CYP2E1, CYP2D6-Met, and CYP3A also participate in metabolism. The main metabolites of caffeine are paraxanthine, theobromine, and theophylline, all of which are bioactive (Socala et al. 2021; Tian et al. 2019). Therefore, if caffeine and medications that are metabolized by these enzymes are administered together, there will be a reduction in the availability of these enzymes to metabolize the medications, as they are occupied by caffeine (De Paula and Farah 2019). The variability of CYP1A2 influences the rate of metabolism of substrate compounds. This interindividual variability may be due to genetic polymorphisms, race, sex, smoking, exposure to chemicals, and the environment. Moderate coffee consumption, at least three cups, and tobacco use are factors that increase enzymatic activity. In turn, oral anovulatory drugs cause CYP1A2 to be inhibited (de Paula and Farah 2019). Some drugs remained active for longer due to the occupation of the enzyme by caffeine. Among them, it is possible to mention: antidepressants (e.g., amitriptyline, clomipramine, fluvoxamine, mianserin, imipramine); antipsychotics (e.g., clozapine, olanzapine); quinolones (e.g., pefloxacin); non-steroidal anti-inflammatory drugs (e.g., phenacetin and paracetamol); and cardiovascular diseases (e.g., propranolol, propafenone, triamterene, verapamil). Therefore, increasing the concentration of these drugs can lead to an increase in the effects and adverse events (Carrillo and Benitez 2000).

## 5 Saponins

Saponins are glycosides of polycyclic steroids and triterpenes. Its chemical structure has a hydrophilic portion (sugar) and a lipophilic portion (steroids and triterpenes) and is derived from mevalonic acid. Saponins can be found in several vegetables such as licorice (*Glycyrrhiza glabra* L.), ginseng (*Panax ginseng* and *Panax quinquefolius*), marigold (*Calendula officinalis* L.), centella (*Centella asiatica* L.), and quillaia (*Quillaja saponaria* Molina). The structural diversity of saponins reflects the different pharmacological activities associated with these compounds (Mugford and Osbourn 2012). Ginseng has biological activities such as stress reduction, anti-fatigue, anti-aging, antidiabetic, immunomodulatory, and anticancer effects. In this context, the use of ginseng has spread throughout the world and has been sought after as a dietary supplement. Asian ginseng (*Panax ginseng*) and American ginseng (*Panax quinquefolius*) are the most used (Qi et al. 2011). A study using rats demonstrated that there was no influence of ginseng on the pharmacokinetics of warfarin (Zhu et al. 1999). Additionally, a prospective, randomized, double-blind, crossover study carried out to evaluate the interaction of ginseng with warfarin in humans demonstrated that although thrombin time was reduced with concomitant administration of Asian ginseng and warfarin, this reduction was not statistically significant. It is well known that Asian ginseng can inhibit the synthesis of thromboxane A2 in human thromboxane-induced platelet aggregation, so the attenuation of the effect could be because ginsenosides induce the hepatic metabolism of warfarin (Lee et al. 2010).

One of the functionalities of ginseng is to help in the treatment of hyperglycemia, which in turn is one of the adverse events of the use of HIV-1 protease inhibitors, such as indinavir. In a study seeking to evaluate the pharmacokinetic effect of American ginseng in healthy individuals using indinavir, it was revealed that the drug was able to reduce insulin sensitivity, but the reduction was not altered by the concomitant administration of American ginseng. Furthermore, American ginseng did not affect the pharmacokinetics of indinavir (Andrade et al. 2008). Finally, there is also the species called *Panax notoginseng* (PNS), also known as Chinese ginseng, which has a drug interaction with ASA. The study in question analyzed the transport of ASA and PNS in the monolayer of MDCK-MDR1 cells, in which an increase in permeability coefficients was observed when administered together. This fact indicates that PNS can increase the absorption of ASA from the gastrointestinal tract (Tian et al. 2017, 2018).

## 6 Fried Foods

Acrolein (**8**) is a highly reactive unsaturated electrophilic species and an environmental pollutant. Humans can be exposed to acrolein through food, water, cigarettes, automobile exhaust, or dermal. Acrolein (**8**) can be related to several diseases such as spinal cord injury, multiple sclerosis, Alzheimer's disease, cardiovascular disease, diabetes mellitus, and neuro-, hepato-, and nephrotoxicity. On the cellular level, acrolein (**8**) can react with DNA and protein producing adduction, oxidative stress, mitochondrial disruption, membrane damage, endoplasmic reticulum stress, and immune dysfunction (Moghe et al. 2015). Acrolein (**8**) is an organic compound that can be obtained from the oxidation of glycerol when foods are fried in fat or vegetable oil for long periods. Chemically, acrolein (**8**) reacts as a Michael acceptor, which can produce adducts (**9**) with many nucleophilic species (Michael 1887). In our body, the main sources of nucleophiles are glutathione, *N*-acetylcysteine, proteins, and DNA-RNA bases. Fried foods are very popular foods around the world and are closely related to fast foods. Fried foods are not good for human health just because of their cholesterol high levels. When fat or vegetable oil is long-term warm, it can deliver glycerol (**7**), which can be oxidized to acrolein (**8**), a scaffold that reacts as a Michael acceptor, as illustrated in Fig. 7.5. The formation of acrolein-protein adducts already was proved through mass spectrometry. The results from these experiments show that acrolein reacts rapidly with cysteine residues through Michael addition to form M+56 Da adducts (Cai et al. 2009). Acrolein (**8**) can also be generated in our biophase, that is, metabolism of cancer drugs such as cyclophosphamide (Liu et al. 2012). Acrolein (**8**) can form irreversible adductions and

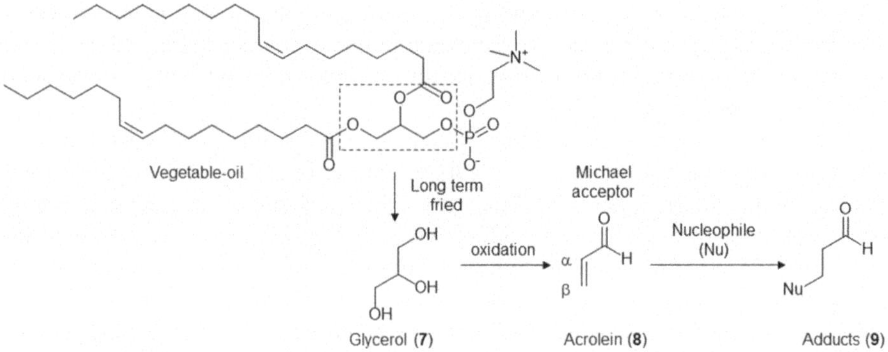


**Fig. 7.5** Scheme for the production of acrolein (**8**) from glycerol (**7**), and formation of adducts with nucleophiles

inactivation of arylamine *N*-acetyltransferases, which are xenobiotic-metabolizing enzymes (Bui et al. 2013).

## *6.1 Grilled Foods*

Benzopyrene (BP) belongs to a chemical class of polycyclic aromatic hydrocarbons, which are found in coal tar, cigarette smoke, wood smoke, and burnt foods such as coffee. Fumes from fat dripping on blistering charcoal are rich in benzopyrene, which can condense on grilled goods (Larsson et al. 1983). The cytochrome CYP1A1 and CYP1B1 isoforms metabolically activate many polycyclic aromatic hydrocarbons, such as benzo[a]pyrene (**10**), to electrophilic species, which can cause toxicity, mutagenesis, and carcinogenesis. Benzo[a]pyrene-DNA adduct (**14**) levels can be increased and found in the liver, spleen, and marrow. It happens due to a balance between the tissue-specific expression of the CYP1A1 and CYP1B1 enzymes, which are related to the sensitivity of benzo[a]pyrene toxicity and, possibly, carcinogenicity (Uno et al. 2006). As illustrated in Fig. 7.6a, through the metabolic pathway benzo[a]pyrene (**10**) is converted by CYP to benzo[a]pyrene-7,8-epoxide (**11**), which is an electrophilic species. The epoxy scaffold present at benzo[a]pyrene-7,8-epoxide (**11**) can be opened by microsomal epoxide hydrolase to benzo[a]pyrene-7,8-dihydrodiol-9,10-epoxide (**12**). The benzo[a]pyrene-7,8-dihydrodiol-9,10-epoxide (**12**) can be oxidated by CYP to benzo[a]pyrene-7,8-dihydrodiol (**13**), and the carcinogenic metabolite, which, in turn, can form 1,2,3-trihydroxy-1,2,3,4-tetrahydrobenzo [a]pyrene-guanine (**14**) adducts (Decker et al. 2009). In Fig. 7.6b, Bauer and coworkers reported a mechanism for translesion synthesis by a Y-family polymerase and the formation of BP-DNA adducts. This experiment provides insight into carcinogen-induced mutagenesis (Bauer et al. 2007). BP can increase the toxicity of cadmium caused by inhibited metallothionein (MT) synthesis in the mummichog (*Fundulus heteroclitus*) model (Van den Hurk

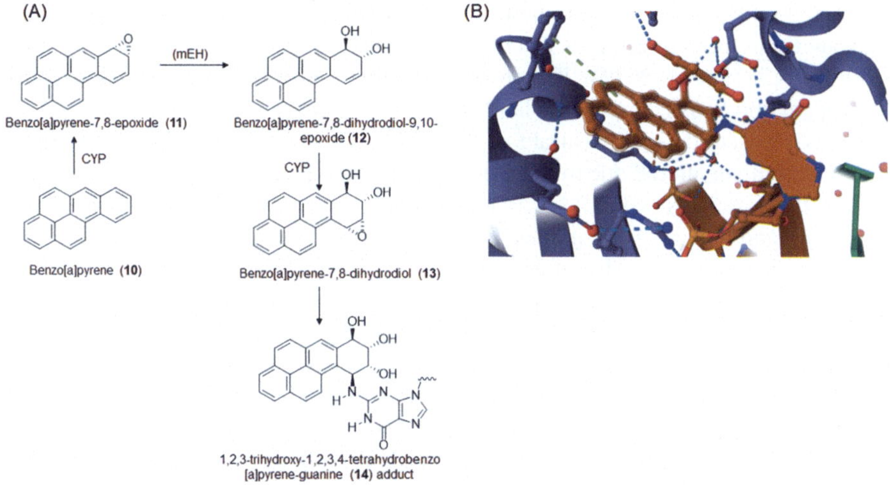


**Fig. 7.6** (**a**) Mechanism of oxidation benzo[a]pyrene (**10**). (Adapted from Decker et al. (2009)). (**b**) Formation of an adduct between benzo[a]pyrene-7,8-dihydrodiol (**13**) and DNA. (Adapted from Bauer et al. (2007))

et al. 2000). Benz[a]anthracene and BP can stimulate cell proliferation through the activation of ER and contribute to tumor promotion in estrogen-sensitive tissues (Pliskova et al. 2005).

## 7 Polyphenol

Polyphenols are antioxidant compounds characterized by the presence of the phenol group. They are commonly found in foods of plant origin such as fruits, juices, teas, coffee, red wine, legumes, cereals, chocolate, and dried legumes (Scalbert et al. 2005). Polyphenols are divided into subfamilies that have similar chemical structures, including flavonoids, flavonols, phenolic acids, phenolic alcohols, tannins, and lignans (Basheer and Kerem 2015). These organic compounds undergo biotransformation in the body and can thus influence the biotransformation and bioavailability of various drugs. Therefore, the study of this interaction becomes extremely important (Lambert et al. 2007). In this context, phloretin, a dihydrochalcone extracted from apples, is a substrate of CYP3A4 and CYP2C19. It also inhibits the catalytic activity of CYP1A1, CYP1A2, and CYP3A4. In an in vitro study, phloretin was able to inhibit the activity of UDP-glucoronosyltransferases (UGT) 1A1, 1A6, 1A7, 1A9, 2B7, and 2B15 by 95.08%, 95.34%, 96.30%, 96.92%, 95.12%, and 92.11%, respectively, at a concentration of 100 μM. UGTs are phase II metabolic enzymes, catalyzing the glucuronidation of exogenous compounds, and are responsible for the elimination of 40–70% of drugs (Chen et al. 2022).

Additionally, an in vivo prediction of the interaction of UGTs with phloretin indicated that the inhibition of UGT1A7 and UGT2B15 occurs with medium possibilities (Chen et al. 2022). Therefore, it is important to highlight drugs such as lorazepam and S-oxazepam (Rowland et al. 2013) that are substrates of UGT2B15, whereas furosemide is a substrate of UGT1A7. Therefore, such drugs may have their metabolism altered due to these enzymes (Fujiwara et al. 2018). Another polyphenol that interacts with UGTs is piceatannol, which is found in passion fruit, blueberry, and red and white grapes. Piceatannol is an analog of resveratrol and was able to inhibit several UGT isoforms in vitro. Regarding the quantitative prediction of the risks of inhibition in vivo, piceatannol was able to inhibit, with a high risk of occurrence, the isoforms UGT1A6, UGT1A7, UGT1A8, and UGT1A9, after oral administration of piceatannol (Jiang et al. 2020). Thus, the UGT1A6 substrates that may undergo changes in the pharmacokinetic profile in case of inhibition by piceatannol are paracetamol and deferiprone. Among the substrates of UGT1A9 are entacapone, indomethacin, mycophenolic acid, R-oxazepam, paracetamol, propofol, and sorafenib (Rowland et al. 2013).

Studies with UGT1A8 indicated that it is capable of metabolizing carcinogenic hydroxyestrogens in order to maintain estrogen homeostasis, thus preventing the accumulation of toxic metabolites in the body (Zhao et al. 2020). Thus, mycophenolate mofetil (MMF), a selective inhibitor of the guanine nucleotide synthetic pathways and the proliferation of T and B lymphocytes that induces apoptosis of activated lymphocytes by blocking the enzyme inosine monophosphate dehydrogenase, is a prodrug, which in organism is converted to mycophenolic acid. This medication is used in organ transplants to reduce rejection and increase patient survival. In this case, UGT1A8 converts mycophenolic acid into two metabolites: acyl glucuronide (AcMPAG) and 7-O-MPA-glucuronide (MPAG), the latter is the main metabolite of MMF (Zhou et al. 2020). Therefore, UGT inhibition should receive due attention when foods containing piceatannol are administered with medications, as it may affect the pharmacokinetics of elimination of numerous drugs, potentially causing an increase in the number of adverse events (Jiang et al. 2020). A widely studied polyphenol is resveratrol, found mostly in grape products. Resveratrol is produced due to ultraviolet radiation, plant stress, injury, and fungal infections. In general, the amount of resveratrol found in grapes is between 5 and 7 mg/kg in the grape skin. In turn, there is a smaller amount in the seed, around 1 mg/kg. White wine has approximately 0.1 mg/kg of resveratrol, while red wine has approximately 1.9 mg/L of resveratrol. Resveratrol can also be sold in the form of a food supplement and its composition can vary from 20 to 200 mg, a quantity much higher than that found in wines (Detampel et al. 2012).

The interaction of resveratrol and CYP3A4 is commonly reported, interacting with several classes of drugs increasing their bioavailability, such as benzodiazepines (e.g., midazolam, triazolam), antihistamines (e.g., terfenadine), immunosuppressants (e.g., cyclosporine and tacrolimus), antagonists calcium channel (e.g., felodipine, nicardipine, nifedipine, nitrendipine, nimodipine, and verapamil), and 3-hydroxy-3-methylglutaryl-coenzyme A reductase inhibitors (atorvastatin, lovastatin, and simvastatin) (Detampel et al. 2012). Furthermore, the interaction of

CYP3A4 in vitro proved to be efficient in several species of Amazonian fruits. Fruit juices—carambola (*Averrhoa carambola*), açaí (*Euterpe precatória*), buriti (*Mauritia flexuosa*), camu-camu (*Myritia flexuosa*), cubiu (*Solanum sessiliflorum*), cupuaçu (*Theobroma grandiflorum*), genipapo (*Genipa americana*) and tapereba (*Spondias mombin*)—were able to inhibit CYP3A4. Residual activity was observed only for carambola (20%), genipapo (54%), and cupuaçu (44%) (Costa et al. 2022). Resveratrol also inhibits the activity of the CYP2C9 enzyme, and its association with diclofenac increases the bioavailability of the latter (Bedada et al. 2015). Furthermore, the association of 10 mg/kg/day of diclofenac and 20 mg/kg/day of resveratrol for 7 days reduces the toxic effects and normalization of hepatorenal functions caused by diclofenac. Resveratrol negatively regulates miR-144 through upregulating nuclear factor erythroid 2-related factor 2 (Elbaz et al. 2022).

## 8 Alkaloid

According to the glossary from the International Union of Pure and Applied Chemistry (IUPAC), alkaloids are basic nitrogen compounds (mostly heterocyclic) from plant and animal sources (Moss et al. 1995). Alkaloids are nitrogenous compounds derived from plants and are widely used in the pharmaceutical industry as they exhibit different pharmacological activities (Kukula-Koch and Widelski 2017). Indole alkaloids are an important class of products with pharmacological activity, including vinblastine, an antineoplastic drug that inhibits the formation of microtubules in the mitotic spindle, thus resulting in an arrest of cell division in metaphase. The administration of vinblastine after a 1-week supplementation with *Capsaicin*, the active component of chili peppers, increases the plasma concentration and the area under the curve of vinblastine. A 75.5% reduction in vinblastine clearance was also observed after pre-treatment with *Capsaicin*. As vinblastine is metabolized by CYP3A, the administration of its inhibitors such as *Capsaicin* can increase plasma levels of drugs metabolized by this enzyme. Furthermore, *Capsaicin* negatively modulates the expression of P-glycoprotein (P-gp) and Multiple Resistance-Associated Protein 2 (MRP2), which alters the transport of vinblastine—a substrate of these transport proteins (Zhai et al. 2019).

## 9 Proteins

Proteins in their primary, secondary, tertiary, and quaternary structural forms could support biochemical catalysts, hormones, enzymes, building blocks, and apoptosis. The tertiary structure of protein consists of the three-dimensional shape forming a polypeptide chain that interacts with an aqueous environment while the three-dimensional arrangement of two or more independent folds of polypeptides in the quaternary structure of a protein is of clinical relevance. Hemoglobin protein with insoluble amino acids in an aqueous environment could result in the aggregation of hemoglobin, which hides newly formed hydrophobic residues. This aggregation can lead to the formation of long, inflexible polymer rods, distorting the shape of red blood cells and plasma membrane, as seen in a sickle cell crisis. Deficiency of amino acids, metabolic enzymes, and defective structure of protein can trigger a wide range of medical conditions including the breakdown of the retina leading to blindness, hearing loss, type 2 diabetes, and phenylketonuria (PKA). A PKA is a phenotypic readout of a deficiency of the phenylalanine hydroxylase, which is responsible for converting phenylalanine to tyrosine. This leads to an accumulation of phenylalanine in the body, thereby causing albinism, translucent eyes, neurological dysfunction, and a musty body odor symptom.

LaPelusa and Kaushik (2024) provide important details on the processing of proteins in the Golgi apparatus, transportation, peptide bonds resulting from the attacks between the carboxyl group of the amino acid and the amino group of the amino acid, disulfide bonds of cysteine residues or sulfhydryl group within the protein, classes of amino acids (non-essential—alanine, asparagine, aspartic acid, glutamic acid, serine; conditionally essential—arginine, cysteine, glutamine, glycine, proline, tyrosine; and essential—histidine, isoleucine, leucine, lysine, methionine, phenylalanine, threonine, tryptophan, valine), denaturation (e.g., breakdown of disulfide bonds), and loss of biological functionality among other relevance structural attributes of proteins. Eggs, dairy, meat, seafood, and Soy-based foods are important sources of essential amino acids. Higher quantities of plant-based foods may be needed to yield a similar amount of protein and amino acid profile of animal sources (Messina et al. 2023). According to LaPelusa and Kaushik (2024), increased insulin sensitivity, oxidation of fatty acids, and appetite suppression with high-protein diets can promote weight loss. Bactericidal or bacteriostatic activity can result from the binding of cephalosporins to the penicillin-binding proteins to inhibit bacterial cell wall synthesis, inhibition of bacterial protein synthesis by macrolides and lincosamides that bind to the 50s ribosomal subunit, or by aminoglycosides and tetracyclines binding to the 30s ribosomal subunit to block aminoacyl-tRNA from binding (Halawa et al. 2024).

## 10 Remarks on Structure-Activity Relations

It is important to establish relationships between biological activity (target, therapeutic and toxicity attributes) and the chemical structure (e.g., alkaloids; fat or water-soluble entities). Understanding the structural and activity relationships endpoints of various components of pharmaceutics, supplements, herbs, foods—including alkaloids, polyphenols, grilled and fried foods, saponins, methylxanthines, and vitamins among others is essential in drug development and prediction of how substances behave in biological systems. The spectrum of behavior is sometimes dependent on structure, solubility, permeability, and the degree of their absorption, distribution, metabolism, and excretion.

For instance, the two side chains of phosphate bonds in the bisphosphonate structure influence affinity or binding interactions with a bone matrix of calcium phosphate in hydroxyapatite. In addition, higher binding affinity with hydroxyapatite has been reported in alendronate with nitrogen as compared to non-nitrogen-containing ones (Nancollas et al. 2006). The alkaloids in diverse structures often exhibit different pharmacological effects and wield influence on drug bioavailability, enzyme activities, receptor binding, neurotransmitter release, uptake, or biotransformation. The polyphenols known for antioxidant properties can modulate enzyme activity and affect drug metabolism as saponins can affect membrane permeability and interact with proteins, thereby impacting absorption and metabolism. The fat-soluble coenzymes (Vitamins A, D, K) that are vital for various biochemical processes could influence enzyme functions with an overall impact on metabolic pathways.

The structural behavior that confers water solubility on vitamin B affects the microenvironment for enzymatic reactions, and subsequent metabolic outcomes and drug interactions. Iron, magnesium, and calcium, which are essential for enzyme activity, could alter drug metabolism and efficacy if deficient or excessive. From food item perspectives, varying components that present structural complexity could suffer changes in chemical structure depending on processing methods (e.g., grilled and fried foods), potentially leading to the formation of harmful byproducts capable of compromising drug effects (especially in association), functions of tissue, organs (lung, heart, kidney failures, etc.), and human health at large.

By systematically analyzing the structure, molecular interactions, and pharmacokinetics and utilizing computational methods, researchers can predict the activity spectra of various substances effectively. This approach can aid in drug development and understanding the broader implications of substances in biological systems. Predicting the spectra involves understanding various factors, including chemical structure (identifying functional groups that influence activity, e.g., hydroxyl, amine) and the use of computational models to predict conformations that may affect binding to targets, interactions with biological targets, and metabolic pathways [molecular interactions, e.g., binding affinity with docking to predict how substances interact with specific proteins or enzymes and analyze potential interactions (hydrogen bonding and hydrophobic interactions) that can enhance or reduce activity].

The prediction of biological activities was performed on some selected compounds using the PASS (Prediction of Activity Spectra for Substances) software, according to the methodological proposal established by (dos Santos et al. 2022). This powerful tool could predict more than 3500 types of biological activity, thereby allowing researchers to virtually generate bioactivity profiles of small molecules (Filimonov et al. 2014). Substrates, inhibitors, or inducers of biological activities provide additional insights into the structure and interaction with wide range of targets including UGT (1A1, 1A3,1A6, 1A7, 1A8, 1A9, 1A10, 2B15, 2B12), CYP (1A1, 1A2, 2D6, 2A11, 2C19, 2C9, 19A1, 2A4, 3A4, 1B1, 2J2, 4A11, 2C8, 2C12, 2E1), xenobiotic-transporting ATPase, 1-acylglycerol-3-phosphate O-acyltransferase, sulfotransferase, 4-Nitrophenol 2-monooxygenase, NADPH-ferrihemoprotein reductase, aldehyde dehydrogenase, bilirubin oxidase, P-benzoquinone reductase, macrophage migration inhibitory factor, 4-methoxybenzoate monooxygenase, oxidoreductase, 15-Lipoxygenase, caspase, 27-hydroxycholesterol 7alpha-monooxygenase, alcohol dehydrogenase, CDP-glycerol glycerophosphotransferase, histidine decarboxylase, xanthine oxidase, NADPH-cytochrome-c2 reductase, thyroxine 5-deiodinase, feruloyl esterase, monophenol monooxygenase, GST A, chlordecone reductase, pyruvate decarboxylase, linoleate diol synthase, ubiquinol-cytochrome-c reductase, 3,4-dihydroxyphenylacetate 2,3-dioxygenase, TNF, 4-hydroxybenzoate 3-monooxygenase,, lipid peroxidase, gluconate 2-dehydrogenase, 4-coumarate-CoA ligase, beta-carotene 15,15′-monooxygenase, 2-dehydropantoate 2-reductase, NADPH-cytochrome-c2 reductase, HIF1A, benzoylformate decarboxylase, lipid metabolism regulator, beta-adrenergic receptor kinase, G-protein-coupled receptor kinase, aspulvinone dimethylallyl transferase, protein-disulfide reductase, glutamyl endopeptidase, testosterone 17beta-dehydrogenase, phenylalanine dehydrogenase, dehydro-L-gulonate decarboxylase, acrocylindropepsin, saccharopepsin, peroxidase, GABA aminotransferase, leukotriene-B4 20-monooxygenase, acylcarnitine hydrolase, 1-alkylglycerophosphocholine O-acetyltransferase, phosphatase, macrophage colony stimulating factor, prephenate dehydrogenase, manganese peroxidase, polyporopepsin, antithrombotic, coronary trans-acenaphthene-1,2-diol dehydrogenase inhibitor, NADPH peroxidase, peroxisome proliferator-activated receptor, fatty-acyl-CoA synthase, 1,4-lactonase, sulfatase, glyoxylate reductase, beta glucuronidase,diiodophenylpyruvatereductase,steroidN-acetylglucosaminyltransferase, alkenylglycerophosphocholine hydrolase, pancreatic elastase, UDP-glucuronosyltransferase, phosphatidylcholine-retinol O-acyltransferase, diphosphomevalonate decarboxylase, ribulose-phosphate 3-epimerase, (S)-2-hydroxy-acid oxidase, 2-hydroxyquinoline 8-monooxygenase, nitrate reductase, telomerase, transcription factor, estrogen receptor, sulfite oxidase, threonine aldolase, arginine 2-monooxygenase, proteasome ATPase, lipoprotein lipase, endothelial growth factor, fibrinogen receptor, prostaglandin reductase, electron-transferring-flavoprotein dehydrogenase, choline-phosphate cytidylyltransferase, limulus clotting factor B, IgA-specific serine endopeptidase, HMG-CoA reductase, estradiol 17 beta dehydrogenase, Na+ K+ transporting ATPase, lipocortins synthesis, phospholipase, topoisomerase histone deacetylase SIRT1, peptidoglycan glycosyltransferase, vitamin-K-epoxide reductase, phosphatidylcholine-sterol O-acyltransferase, cysteine synthase, retinoic acid receptor, ferredoxin-NAD+ reductase, monoamine oxidase (MAO A and B), catechol oxidase, dopa decarboxylase, tyrosine transaminase, 5 Hydroxytryptamine uptake, etc.

The results are presented as the probability of being active and inactive (Pa and Pi, respectively). Pa indicates the likelihood that a compound belongs to the class of active compounds, meaning its structure resembles those typically found as "active" within the PASS training set. Conversely, Pi reflects the probability of a compound being inactive (Filimonov and Poroikov 2009; Lagunin et al. 2000.). Results with Pa $\geq 0.50$ were plotted and named as positive, while Pa $\leq 0.49$ was negative. The biological targets and activities of selected compounds (rosmarinic, lithospermic, salvianolic A, salvianolic B, eugenin, calycosin, wogonin, luteolin, skullcapflavone

II, rhein, emodin, aloeemodin, 9-dehydroxyeurotinone, dioscorealide B, wedelolactone, ursolic acid, lorcaserin, sitagliptin, propranolol, diclofenac, atenolol, testosterone, cypionate, isosinensetin, sciadopitysin, oroxylin A, naproxen, vitamin A, vitamin B6, vitamin B3) were antidiabetic, free radical scavenger, lipid peroxidase inhibitor, antioxidant, insulin inhibitor, insulin promoter, lipid metabolism regulator, antihypercholesterolemic, and enzyme cofactors (Fig. 7.7).

The five compounds with the best profile related to the analyzed activities were calycosin, eugenin, luteolin, rosmarinic, and wogonin. These compounds have six or seven positive predictions (Fig. 7.7). This fact indicates that they can interact with diverse molecular targets. The activities with the highest number of positive predictions were lipid peroxidase inhibitor, lipid metabolism regulator, and free radical scavenger. The wogonin compound obtained a high Pa value (0.77) for lipid peroxidase inhibition activity. In fact, there are reports in the literature that this compound is capable of suppressing lipid peroxidation induced by NADPH and consequently reducing the superoxide radical (Cho and Lee 2004; Gao et al. 1999). Rosmarinic presented a value of Pa = 0.67 for lipid metabolism regulatory activity. There are reports in the literature that this compound can exert this regulation. For

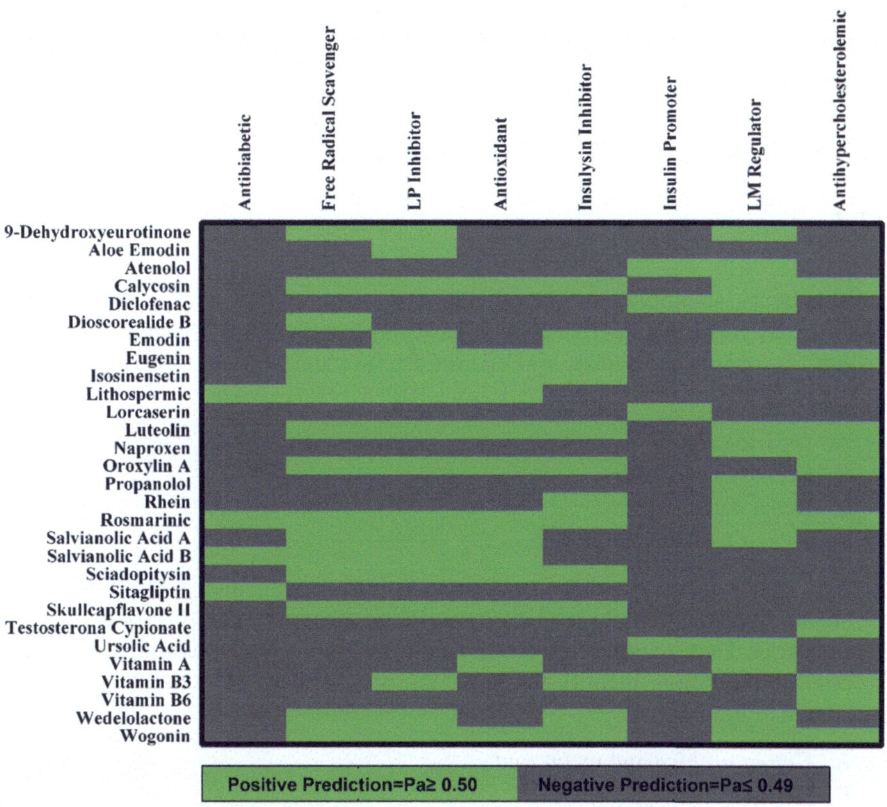


**Fig. 7.7** Predicted biological activities of compounds via PASS. LP: lipid peroxidase and LM: lipid metabolism

example, Nyandwi et al. (2021) reported that rosmarinic acid treatment significantly decreased body weight, blood glucose, and plasma levels of total cholesterol and triglycerides in mice fed a high-fat diet. Oroxylin A obtained good Pa results for antioxidant activity and free radical scavenger of 0.65 and 0.73, respectively. These data corroborate those described in the literature by (Huang et al. 2006; Kim et al. 2008).

In addition to laboratory bioassays to validate predicted therapeutic and adverse activities, the use of databases can facilitate the assessment of potential toxicity based on structural alerts associated with harmful effects. The quantitative structure-activity relationship models could facilitate the prediction of activity based on chemical structure. Understanding how these compounds, cofactors, coenzymes, and dietetic preparation methods change the interact dynamics with drug-metabolizing enzymes among other pharmacokinetic and pharmacodynamic targets is fundamental to drug design and nutritional support. The prediction of metabolic outcomes, drug interactions, and patient responses demands knowledge of food compositions and how preparations influence the chemical nature of their nutrients or components. Integrating how dietary components might alter the structure, activity, or bioavailability of other substances, and microenvironment features like pH, temperature, and biological matrices can provide clues into drug development strategies and enhance efficacy, safety, and overall therapeutic outcomes.

## 11 Self-Assessment Questions and Answers

### *11.1 Questions*

1. In patients taking platinum-based chemotherapeutic drugs with structural similarity to pyridoxine, interfering with its absorption, potentially leading to peripheral neuropathy and depression. Vitamin B6 supplements could subside which of the following adverse effects?

   A. Gastric ulcer
   B. Sensory neuropathy
   C. Renal calcifications
   D. Steven Johnson syndrome

2. A 22-year-old patient suffering from an asthma exacerbation with shortness of breath, hypoxia, and physical exam revealing tachypnea and diffuse expiratory wheezing. These symptoms were alleviated by methylxanthines with phosphodiesterase inhibition. Identify the most appropriate methylxanthine and route of administration.

   A. *Coffea arabica* and rectal route
   B. Caffeine and oral route
   C. Acetylsalicylic acid and transdermal route
   D. *Coffea arabica* and intravenous route

3. An albinism, translucent eyes, neurological dysfunction, and a musty body odor, in a 5-month-old boy diagnosed with phenylketonuria suggests a possible defect in an enzyme involved in the metabolism of which of the following molecules?
    A. Phenylalanine
    B. Histidine
    C. Vitamin B6
    D. Tryptophan
4. Which of the following compounds qualifies as a highly reactive toxic aldehyde and a Michael acceptor obtainable from the oxidation of glycerol when foods are fried in fat or vegetable oil for long periods?
    A. Phloretin
    B. Acrolein
    C. Glutathione
    D. Benzopyrene

## *11.2 Answers*

1. B. Chemotherapeutic drugs can interfere with the absorption of pyridoxine (vitamin B6), leading to deficiencies that may result in peripheral neuropathy. Supplementing with vitamin B6 can help prevent or mitigate this adverse effect.
2. B. Caffeine and oral route. Methylxanthines, such as caffeine, can help alleviate asthma symptoms through phosphodiesterase inhibition, which leads to bronchodilation. The oral route is commonly used for administering caffeine for such purposes.
3. A. Phenylalanine. Phenylketonuria results from a deficiency of the enzyme phenylalanine hydroxylase, which is responsible for converting phenylalanine to tyrosine. This leads to an accumulation of phenylalanine in the body, causing the described symptoms.
4. B. Acrolein. Acrolein is a toxic aldehyde formed during the high-temperature oxidation of fats, such as those occurring during prolonged frying. It is known for its high reactivity and potential health hazards.

# References

Abdullah AT, Mousheer ZT. Vitamin D status in epileptic children on valproic acid; a case-control study. Arch Acad Emerg Med. 2020;8(1):e13.

Alvarez N, Aguilar-Jimenez W, Rugeles MT. The potential protective role of vitamin D supplementation on HIV-1 infection. Front Immunol. 2019;10:469612.

Andrade AS, Hendrix C, Parsons TL, Caballero B, Yuan C-S, Flexner CW, Dobs AS, Brown TT. Pharmacokinetic and metabolic effects of American ginseng (*Panax quinquefolius*) in

healthy volunteers receiving the HIV protease inhibitor indinavir. BMC Complement Altern Med. 2008;8:50.
Ardine M, Generali D, Donadio M, Bonardi S, Scoletta M, Vandone AM, Mozzati M, Bertetto O, Bottini A, Dogliotti L, Berruti A. Could the long-term persistence of low serum calcium levels and high serum parathyroid hormone levels during bisphosphonate treatment predispose metastatic breast cancer patients to undergo osteonecrosis of the jaw? Ann Oncol. 2006;17(8):1336–7.
Bacchi A, Carcelli M, Compari C, Fisicaro E, Pala N, Rispoli G, Rogolino D, Sanchez TW, Sechi M, Neamati N. HIV-1 IN strand transfer chelating inhibitors: a focus on metal binding. Am Chem Soc. 2011;8(2):507–19. https://doi.org/10.1021/mp100343x.
Basheer L, Kerem Z. Interactions between CYP3A4 and dietary polyphenols. Oxid Med Cell Longev. 2015;2015:854015.
Bauer J, Ling H, Sayer JM, Xing G, Yagi H, Jerina DM. Bypass of major benzopyrene-dG adduct by Y-family DNA polymerase with unique structural gap. Proc Natl Acad Sci. 2007;104(38):14905–10.
Bedada SK, Yellu NR, Neerati P. Effect of resveratrol treatment on the pharmacokinetics of diclofenac in healthy human volunteers. Phytother Res. 2015;30(3):397–401.
Belayneh A, Molla F. The effect of coffee on pharmacokinetic properties of drugs: a review. Biomed Res Int. 2020;2020:7909703.
Bertinato J, Lavergne C, Plouffe LJ, Abou El Niaj H. Small increases in dietary calcium above normal requirements exacerbate magnesium deficiency in rats fed a low magnesium diet. Magnes Res. 2014;27(1):35–47.
Boekema JP, Samsom M, Henegouwen GPB, Smout JPM. Coffee and gastrointestinal function: facts and fiction: a review. Scand J Gastroenterol. 1999;34(230):35–9.
Brown TT, Qaqish RB. Antiretroviral therapy and the prevalence of osteopenia and osteoporosis: a meta-analytic review. AIDS. 2006;20(17):2165–74.
Brown TT, Ross AC, Storer N, Labbato D, McComsey GA. Bone turnover, osteoprotegerin/RANKL and inflammation with antiretroviral initiation: tenofovir versus non-tenofovir regimens. Antivir Ther. 2011;16(7):1063–72.
Bui LC, Manaa A, Xu X, Duval R, Busi F, Dupret JM, Rodrigues Lima F, Dairou J. Acrolein, an alpha, beta-unsaturated aldehyde, irreversibly inhibits the acetylation of aromatic amine xenobiotics by human Nat1. Drug Metab Dispos. 2013;41(7):1300–5.
Cai J, Bhatnagar A, Pierce WM Jr. Protein modification by acrolein: formation and stability of cysteine adducts. Chem Res Toxicol. 2009;22(4):708–16.
Carrillo JA, Benitez J. Clinically significant pharmacokinetic interactions between dietary caffeine and medications. Clin Pharmacokinet. 2000;39(2):127–53.
Chalitsios CV, Shaw DE, McKeever TM. Corticosteroids and bone health in people with asthma: a systematic review and meta-analysis. Respir Med. 2021;181:106374.
Chen J, Zhang H, Hu X, Xu M, Su Y, Zhang C, Yue Y, Zhang X, Wang X, Cui W, Zhao Z, Li X. Phloretin exhibits potential food-drug interactions by inhibiting human UDP-glucuronosyltransferases in vitro. Toxicol In Vitro. 2022;84:105447.
Cho J, Lee H-K. Wogonin inhibits excitotoxic and oxidative neuronal damage in primary cultured rat cortical cells. Eur J Pharmacol. 2004;485(1–3):105–10.
Costa EMA, Figueira MBA, Carrão DB, Hanbenschu MD, De Oliveira ARM, Acho LDR, Lima ES, Magalhães IRS. Phenolic and flavonoid content and in vitro inhibitory effect of some Amazonian fruit juices on CYP3A4 activity. Acta Amaz. 2022;52(2):172–8.
Dawson-Hughes B, Harris SS, Dallal GE, Lancaster DR, Zhou Q. Calcium supplement and bone medication use in a US Medicare health maintenance organization. Osteoporos Int. 2002;13(8):657–62.
De Baaij JH. Magnesium reabsorption in the kidney. Am J Physiol Renal Physiol. 2023;324(3):F227–44.
De Carvalho Minighini E, Bragança KP, Anastácio LR. Warfarin drug interaction with vitamin K and other foodstuffs. Rev Chil Nutr. 2020;47(3):470–7.

De Paula J, Farah A. Caffeine consumption through coffee: content in the beverage, metabolism, health benefits and risks. Beverages. 2019;5:37.

Decker M, Arand M, Cronin A. Mammalian epoxide hydrolases in xenobiotic metabolismo and signalling. Arch Toxicol. 2009;83:297–318.

Detampel P, Beck M, Krähenbühl S, Huwyler J. Drug interaction potential of resveratrol. Drug Metab Rev. 2012;44(3):253–65.

Dhir V, Sandhu A, Kaur J, Pinto B, Kumar P, Kaur P, Gupta N, Sood A, Sharma A, Sharma S. Comparison of two different folic acid doses with methotrexate--a randomized controlled trial (FOLVARI Study). Arthritis Res Ther. 2015;17(1):156.

Domingues S, Lima T, Saavedra MJ, da Silva GJ. An overview of cefiderocol's therapeutic potential and underlying resistance mechanisms. Life. 2023;13(7):1427.

dos Santos IVF, Borges RS, Silva GM, de Lima LR, Bastos RS, Ramos RS, Silva LB, da Silva CHTP, dos Santos CBR. Hierarchical virtual screening based on rocaglamide derivatives to discover new potential anti-skin cancer agents. Front Mol Biosci. 2022;9:836572. https://doi.org/10.3389/fmolb.2022.836572.

Elbaz EM, Ahmed KA, Abdelmonem M. Reveratrol mitigates diclofenac-induced hepatorenal toxicity in rats via modulation of miR-144/Nrf2/GSh axis. J Biochem Mol Toxicol. 2022;36(9):e23129.

Elin RJ. Assessment of magnesium status for diagnosis and therapy. Magnes Res. 2010;23(4):194–8.

Emami J, Yousefian H, Sadeghi H. Targeted nanostructured lipid carrier for brain delivery of artemisinin: design, preparation, characterization, optimization and cell toxicity. J Pharm Pharm Sci. 2018;21(1s):225s–41s. https://doi.org/10.18433/jpps30117.

Filimonov D, Poroikov V. Chapter 6. Probabilistic approaches in activity prediction. In: Chemoinformatics approaches to virtual screening. Cambridge: Royal Society of Chemistry; 2009. p. 182–216. https://doi.org/10.1039/9781847558879-00182.

Filimonov DA, Lagunin AA, Gloriozova TA, Rudik AV, Druzhilovskii DS, Pogodin PV, Poroikov VV. Prediction of the biological activity spectra of organic compounds using the pass online web resource. Chem Heterocycl Compd. 2014;50(3):444–57. https://doi.org/10.1007/s10593-014-1496-1.

Frost RW, Lettieri JT, Noe A, Shamblen EC, Lettieri JT. Effect of aluminum hydroxide and calcium carbonate antacids on ciprofloxacin bioavailability. Antimicrob Agents Chemother. 1992;36(4):830–2.

Fujiwara R, Yoda E, Tukey RH. Species differences in drug glucuronidation: humanized UDP-glucuronosyltransferase 1 mice and their application for predicting drug glucuronidation and drug-induced toxicity in humans. Drug Metab Pharmacokinet. 2018;33(1):9–16.

Galgani A, Palleria C, Iannone LF, De Sarro G, Giorgi FS, Maschio M, Russo E. Pharmacokinetic interactions of clinical interest between direct oral anticoagulants and antiepileptic drugs. Front Neurol. 2018;9:413223.

Gao Z, Huang K, Yang X, Xu H. Free radical scavenging and antioxidant activities of flavonoids extracted from the radix of Scutellaria baicalensis Georgi. Biochim Biophys Acta. 1999;1472(3):643–50.

Gröber U, Kisters K. Influence of drugs on vitamin D and calcium metabolism. Dermatoendocrinology. 2012;4(2):158–66.

Guerra W, Silva-Caldeira PP, Terenzi H, Pereira-Maia EC. Impact of metal coordination on the antibiotic and non-antibiotic activities of tetracycline-based drugs. Coord Chem Rev. 2016;327–328:188–99.

Hakkola J, Hukkanen J, Turpeinen M, Pelkonen O. Inhibition and induction of CYP enzymes in humans: an update. Arch Toxicol. 2020;94(11):3671.

Halawa EM, Fadel M, Al-Rabia MW, Behairy A, Nouh NA, Abdo M, Olga R, Fericean L, Atwa AM, El-Nablaway M, Abdeen A. Antibiotic action and resistance: updated review of mechanisms, spread, influencing factors, and alternative approaches for combating resistance. Front Pharmacol. 2024;14:1305294.

Hellwig T, Gulseth M. Pharmacokinetic and pharmacodynamic drug interactions with new oral anticoagulants: what do they mean for patients with atrial fibrillation? Ann Pharmacother. 2013;47(11):1478–87.

Hokugo A, Christensen R, Chung EM, Sung EC, Felsenfeld AL, Sayre JW, Garrett N, Adams JS, Nishimura I. Increased prevalence of bisphosphonate-related osteonecrosis of the jaw with vitamin D deficiency in rats. J Bone Miner Res. 2010;25(6):1337–49.

Hou YC, Wu CC, Liao MT, Shyu JF, Hung CF, Yen TH, Lu C, Lu KC. Role of nutritional vitamin D in osteoporosis treatment. Clin Chim Acta. 2018;484:179–91.

Huang WH, Lee AR, Yang CH. Antioxidative and anti-inflammatory activities of polyhydroxyflavonoids of Scutellaria baicalensis GEORGI. Biosci Biotechnol Biochem. 2006;70(10):2371–80. https://doi.org/10.1271/bbb.50698.

Jehle PM. Steroid-induced osteoporosis: how can it be avoided? Nephrol Dial Transplant. 2003;18(5):861–4.

Jiang L, Wang Z, Wang X, Wang S, Wang Z, Liu Y. Piceatannol exhibits potential food-drug interactions through the inhibition of human UDP-glucunosyltranferase (UGT) in vitro. Toxicol In Vitro. 2020;67:104870.

Jiménez-Sousa MÁ, Martínez I, Medrano LM, Fernández-Rodríguez A, Resino S. Vitamin D in human immunodeficiency virus infection: influence on immunity and disease. Front Immunol. 2018;9:336780.

Junges C, Machado TD, Nunes Filho PRS, Riesgo R, Mello EDD. Vitamin D deficiency in pediatric patients using antiepileptic drugs: systematic review with meta-analysis. J Pediatr. 2020;96:559–68.

Kanis JA, Johansson H, Oden A, Johnell O, De Laet C, Melton LJ III, Tenenhouse A, Reeve J, Silman AJ, Pols HAP, Eisman JA, McCloskey EV, Mellstrom D. A meta-analysis of prior corticosteroid use and fracture risk. J Bone Miner Res. 2004;19(6):893–9.

Kanis JA, Johansson H, Harvey NC, McCloskey EV. A brief history of FRAX. Arch Osteoporos. 2018;13:1–16.

Kathiravan M, Kavitha S, Shanthi R. To determine the effect of long-term antiepileptic drug on the serum folate and vitamin B12 among epileptic patients. Sci Rep. 2021;11:4393.

Kaur N, Young VG Jr, Su Y, Suryanarayanan R. Partial dehydration of levothyroxine sodium pentahydrate in a drug product environment: structural insights into stability. Mol Pharm. 2020;17(10):3915–29.

Khan QJ, O'Dea AP, Sharma P. Musculoskeletal adverse events associated with adjuvant aromatase inhibitors. J Oncol. 2010;2010:654348.

Kim DH, Kim S, Jeon SJ, Son KH, Lee S, Yoon BH, Cheong JH, Ko KH, Ryu JH. The effects of acute and repeated oroxylin A treatments on Aβ25–35-induced memory impairment in mice. Neuropharmacology. 2008;55(5):639–47. https://doi.org/10.1016/j.neuropharm.2008.05.019.

Krishna N, Gouthami B, Meyyanathan SN, Gowramma B, Elango K, Suresh B. In vitro in vivo pharmacokinetic interaction study of escitalopram oxalate when co administered with caffeine/caffeinated beverages. Open Conf Proc J. 2013;4:66–71.

Kukula-Koch WA, Widelski J. Alkaloids. In: Pharmacognosy. Boston: Academic Press; 2017. p. 163–98.

Lagunin A, Stepanchikova A, Filimonov D, Poroikov V. PASS: prediction of activity spectra for biologically active substances. Bioinformatics. 2000;16(8):747–8. https://doi.org/10.1093/bioinformatics/16.8.747.

Lambert JD, Sang S, Lu AYH, Yand CS. Metabolism of dietary polyphenols and possible interactions with drugs. Curr Drug Metab. 2007;8(5):499–507.

LaPelusa A, Kaushik R. Physiology, proteins. 2022 Nov 14. In: StatPearls [Internet]. Treasure Island: StatPearls Publishing; 2024. PMID: 32310450.

Larsson BK, Sahlberg GP, Eriksson AT, Busk LA. Polycyclic aromatic hydrocarbons in grilled food. J Agric Food Chem. 1983;31(4):867–73.

Lee YH, Lee BK, Choi YJ, Yoon IK, Chang BC, Gwak HS. Interaction between warfarin and Korean red ginseng in patients with cardiac valve replacement. Int J Cardiol. 2010;145(2):275–6.

Li J, Zhou B. Biological actions of artemisinin: insights from medicinal chemistry studies. Molecules. 2010;15(3):1378–97. https://doi.org/10.3390/molecules15031378.

Li RC, Nix DE, Schentag JJ. Interaction between ciprofloxacin and metal cations: its influence on physicochemical characteristics and antibacterial activity. Pharm Res. 1994;11(6):917–20.

Liu F, Li X, Lin T, He D, Wei G, Liu J, Li L. The cyclophosphamide metabolite, acrolein, induces cytoskeletal changes and oxidative stress in Sertoli cells. Mol Biol Rep. 2012;39(1):493–500.

Mahalko JR, Sandstead HH, Johnson LK, Milne DB. Effect of a moderate increase in dietary protein on the retention and excretion of Ca, Cu, Fe, Mg, P, and Zn by adult males. Am J Clin Nutr. 1983;37(1):8–14.

Maurin O, Davis AP, Chester M, Mvungi EF, Jaufeerally-Fakim Y, Fay MF. Towards a phylogeny for Coffea (Rubiaceae): identifying well-supported lineages based on nuclear and plastid DNA sequences. Ann Bot. 2007;100(7):1565–83.

Menezes MSS, Almeida CMM. Structural, functional, nutritional and clinical aspects of vitamin A: a review. Pharma Nutr. 2024;27:100383.

Messina M, Duncan AM, Glenn AJ, Mariotti F. Perspective: Plant-Based Meat Alternatives Can Help Facilitate and Maintain a Lower Animal to Plant Protein Intake Ratio. Adv Nutr. 2023;14(3):392–405.

Michael A. Ueber die Addition von Natriumacetessig- und Natriummalonsäureäthern zu den Aethern ungesättigter Säuren. J Prakt Chem. 1887;35:349–56.

Mintzer S, Skidmore CT, Sperling MR. B-vitamin deficiency in patients treated with antiepileptic drugs. Epilepsy Behav. 2012;24(3):341–4.

Mladenka P, Macáková K, Filipský T, Zatloukalová L, Jahodár L, Bovicelli P, Silvestri IP, Hrdina R, Saso L. In vitro analysis of iron chelating activity of flavonoids. J Inorg Biochem. 2011;105(5):693–701.

Modica JS, Déry C, Canissario R, Logigian E, Bonno D, Stanton M, Dupré N, McDermott MP, Bouchard M, Lang AE, Lizarraga KJ. A systematic review of the potential consequences of abnormal serum levels of vitamin B6 in people living with Parkinson's disease. J Neurol Sci. 2023;450:120690.

Moghe A, Ghare S, Lamoreau B, Mohammad M, Barve S, McClain C, Joshi-Barve S. Molecular mechanisms of acrolein toxicity: relevance to human disease. Toxicol Sci. 2015;143(2):242–55.

Moss GP, Smith PAS, Tavernier D. Glossary of class names of organic compounds and reactive intermediates based on structure (IUPAC Recommendations 1995). Pure Appl Chem. 1995;67(819):1307–75.

Mugford ST, Osbourn A. Saponin synthesis and function. In: Isoprenoid synthesis in plants and microorganisms. New York: Springer; 2012. p. 405–24.

Nancollas GH, Tang R, Phipps RJ, Henneman Z, Gulde S, Wu W, et al. Novel insights into actions of bisphosphonates on bone: differences in interactions with hydroxyapatite. Bone. 2006;38:617–27.

Nehlig A. Effects of coffe on the gastro-intestinal tract: a narrtive review and literature update. Nutrients. 2022;14:399.

Neilands JB. Siderophores: structure and function of microbial iron transport compounds. J Biol Chem. 1995;270(45):26723–6.

Nelson DL, Cox MM, Hoskins AA, Lehninger AL. Lehninger principles of biochemistry. 8th ed. London: W.H. Freeman and Company; 2021.

Niravath P, Hilsenbeck SG, Wang T, Jiralerspong S, Nangia J, Pavlick A, Ademuyiwa F, Frith A, Ma C, Park H, Rigden C, Suresh R, Ellis M, Kent Osborne C, Rimawi MF. Randomized controlled trial of high-dose versus standard-dose vitamin D3 for prevention of aromatase inhibitor-induced arthralgia. Breast Cancer Res Treat. 2019;177:427–35.

Nyandwi JB, Ko YS, Jin H, Yun SP, Park SW, Kim HJ. Rosmarinic acid exhibits a lipid-lowering effect by modulating the expression of reverse cholesterol transporters and lipid metabolism in high-fat diet-fed mice. Biomolecules. 2021;11(10):1470. https://doi.org/10.3390/biom11101470.

Olie RH, Winckers K, Rocca B, Ten Cate H. Oral anticoagulants beyond warfarin. Annu Rev Pharmacol Toxicol. 2024;64:551–75.

Ooko E, Saeed ME, Kadioglu O, Sarvi S, Colak M, Elmasaoudi K, Janah R, Greten HJ, Efferth T. Artemisinin derivatives induce iron-dependent cell death (ferroptosis) in tumor cells. Phytomedicine. 2015;22(11):1045–54. https://doi.org/10.1016/j.phymed.2015.08.002.

Ordi-Ros J, Sáez-Comet L, Pérez-Conesa M, Vidal X, Riera-Mestre A, Castro-Salomó A, Cuquet-Pedragosa J, Ortiz-Santamaria V, Mauri-Plana M, Solé C, Cortés-Hernández J. Rivaroxaban versus vitamin K antagonist in antiphospholipid syndrome: a randomized noninferiority trial. Ann Intern Med. 2019;171(10):685–94.

Palacios C, Wigertz K, Braun M, Martin BR, McCabe GP, McCabe L, Pratt JH, Peacock M, Weaver CM. Magnesium retention from metabolic-balance studies in female adolescents: impact of race, dietary salt, and calcium. Am J Clin Nutr. 2013;97(5):1014–9.

Palikhe NS, Kim SH, Nam YH, Ye YM, Park HS. Polymorphisms of aspirin-metabolizing enzymes CYP2C9, NAT2 and UGT1A6 in aspirin-intolerant urticaria. Allergy Asthma Immunol Res. 2011;3(4):273–6.

Panja S, Khatua DK, Halder M. Simultaneous binding of folic acid and methotrexate to human serum albumin: insights into the structural changes of protein and the location and competitive displacement of drugs. ACS Omega. 2018;3(1):246–53.

Park SH, Kim EY, Park CH. Effect of vitamin D on bone mineral density changes in patients with breast cancer receiving adjuvant aromatase inhibitor therapy. J Breast Dis. 2019;7(2):105–12.

Pliskova M, Vondracek J, Vojtesek B, Kozubik A, Machala M. Deregulation of cell proliferation by polycyclic aromatic hydrocarbons in human breast carcinoma MCF-7 cells reflects both genotoxic and nongenotoxic events. Toxicol Sci. 2005;83(2):246–56.

Plotnikoff GA, Quigley JM. Prevalence of severe hypovitaminosis D in patients with persistent, nonspecific musculoskeletal pain. Mayo Clin Proc. 2003;78(12):1463–70.

Qi LW, Wang CZ, Du GJ, Zhang ZY, Calway T, Yuan C-S. Metabolism of ginseng and its interactions with drugs. Curr Drug Metab. 2011;12(9):818–22.

Rayssiguier Y, Libako P, Nowacki W, Rock E. Magnesium deficiency and metabolic syndrome: stress and inflammation may reflect calcium activation. Magnes Res. 2010;23(2):73–80.

Rock AE, Demarais P, Vergara-Rodriguez PT, Max B. HIV-1 virologic rebound due to coadministration of divalent cations and bictegravir. Infect Dis Ther. 2020;9(3):691–6. https://doi.org/10.1007/s40121-020-00307-4.

Ronsein GE, Vaisar T, Davidson WS, Bornfeldt KE, Probstfield JL, O'Brien KD, Zhao X, Heinecke JW. Niacin increases atherogenic proteins in high-density lipoprotein of statin-treated subjects. Arterioscler Thromb Vasc Biol. 2021;41(8):2330–41.

Rowland A, Miners JO, Mackenzie PI. The UDP-glucuronosyltransferases: their role in drug metabolism and detoxification. Int J Biochem Cell Biol. 2013;45(6):1121–32.

Sawamoto K, Álvarez JV, Herreño AM, et al. Bone-specific drug delivery for osteoporosis and rare skeletal disorders. Curr Osteoporos Rep. 2020;18:515–25. https://doi.org/10.1007/s11914-020-00620-4.

Sawhney A, Singhal S, Patel R. Isolated Pyridoxine Deficiency Presenting as Peripheral Neuropathy Post-chemotherapy. Cureus. 2022;14(7):e26725.

Scalbert A, Johnson IT, Saltmarsh M. Polyphenols: antioxidants and beyond. Am J Clin Nutr. 2005;81(1):215S–7S.

Scott EE, White MA, He YA, Johnson EF, Stout CD, Halpert JR. Structure of mammalian cytochrome P450 2B4 complexed with 4-(4-chlorophenyl)imidazole at 1.9 A resolution: insight into the range of P450 conformations and coordination of redox partner binding. J Biol Chem. 2004;279(26):27294–301.

Searing DA, Zhang Y, Murphy JR, Hauk PJ, Goleva E, Leung DY. Decreased serum vitamin D levels in children with asthma are associated with increased corticosteroid use. J Allergy Clin Immunol. 2010;125(5):995–1000.

Seki N, Asano Y, Ochi H, Abe F, Uenishi K, Kudou H. Reducing effect of calcium in combination with magnesium and lactulose on body fat mass in middle-aged Japanese women. Asia Pac J Clin Nutr. 2013;22(4):557–64.

Silva MC, Furlanetto TW. Intestinal absorption of vitamin D: a systematic review. Nutr Rev. 2018;76(1):60–76.

Silva MO, Honfoga JNB, Medeiros LL, Madruga MS, Bezerra TKA. Obtaining bioactive compounds from the coffee husk (*Coffea arabica* L.) using different extraction methods. Molecules. 2021;26(1):46.

Singh N, Singh PN, Hershman JM. Effect of calcium carbonate on the absorption of levothyroxine. JAMA. 2000;283(21):2822–5.

Socala K, Szopa A, Serefko A, Poleszak E, Wlaź P. Neuroprotective effects of coffee bioactive compounds: a review. Int J Mol Sci. 2021;22:107.

Song I, Borland J, Arya N, Wynne B, Piscitelli SC. Pharmacokinetics of dolutegravir when administered with mineral supplements in healthy adult subjects. J Clin Pharmacol. 2014;55(5):490–6. https://doi.org/10.1002/jcph.439.

Stafford DW. The vitamin K cycle. J Thromb Haemost. 2005;3(8):1873–8.

Stoellberger C, Finsterer J. Update on drug interactions with non-vitamin-K-antagonist oral anticoagulants for stroke prevention in elderly patients. Expert Rev Clin Pharmacol. 2021;14(5):569–81.

Stöllberger C, Finsterer J. Interactions between non-vitamin K oral anticoagulants and antiepileptic drugs. Epilepsy Res. 2016;126:98–101.

Sun Y, Zhang T, Lu B, Li X, Jiang L. Application of cofactors in the regulation of microbial metabolism: a state of the art review. Front Microbiol. 2023;14:1145784.

Tanoury ZA, Piskunov A, Rochette-Egly C. Vitamin A and retinoid signaling: genomic and nongenomic effects. J Lipids. 2013;54(7):1761–75.

Tian Z, Pang H, Du S, Lu Y, Zhang L, Wu H, Guo S, Wang M, Zhang Q. Effect of Panax notoginseng saponins on the pharmacokinetics of aspirin in rats. J Chromatogr B. 2017;1040:136–43.

Tian Z, Pang H, Zhang Q, Du S, Lu Y, Zhang L, Bai J, Li P, Li D, Zhao M, Chen X. Effect of aspirin on the pharmacokinetics and absorption of Panax notoginseng saponins. J Chromatogr B. 2018;1074–1075:25–33.

Tian DD, Natesan S, White JR, Paine MF. Effects of common CYP1A2 genotypes and other key factors on intraindividual variation in the caffeine metabolic ratio: an exploratory analysis. Clin Transl Sci. 2019;12(1):39–46.

Tonmunphean S, Parasuk V, Kokpol S. Automated calculation of docking of artemisinin to heme. J Mol Model. 2001;7(4):26–33. https://doi.org/10.1007/s008940100013.

Uno S, Dalton TP, Dragin N, Curran CP, Derkenne S, Miller ML, Shertzer HG, Gonzalez FJ, Nebert DW. Oral benzo[a]pyrene in cyp1 knockout mouse lines: CYP1A1 important in detoxication, CYP1B1 metabolism required for immune damage independent of total-body burden and clearance rate. Mol Pharmacol. 2006;69(4):1103–12.

Van den Hurk P, Faisal M, Roberts MH Jr. Interactive effects of cadmium and benzo[a]pyrene on metallothionein induction in mummichog (*Fundulus heteroclitus*). Mar Environ Res. 2000;50(1–5):83–7.

Walden DM, Khotimchenko M, Hou H, Chakravarty K, Varshney J. Effects of magnesium, calcium, and aluminum chelation on fluoroquinolone absorption rate and bioavailability: a computational study. Pharmaceutics. 2021;13(5):594.

Wang R, Chen C, Liu W, Zhou T, Xun P, He K, Chen P. The effect of magnesium supplementation on muscle fitness: a meta-analysis and systematic review. Magnes Res. 2017;30(4):120–32.

Xu C, Zhang H, Mu L, Yang X. Artemisinins as anticancer drugs: novel therapeutic approaches, molecular mechanisms, and clinical trials. Front Pharmacol. 2020;11:529881. https://doi.org/10.3389/fphar.2020.529881.

Yaghoobian R, Sharifi M, Rezaee M, Vahidi H, Salehi N, Hosseini K. Caffeine drug interactions and its clinical implication after acute coronary syndrome: a literature review. Crit Pathw Cardiol. 2023;22(3):95–9.

Yoovathaworn KC, Sriwatanakul K, Thithapandha A. Influence of caffeine on aspirin pharmacokinetics. Eur J Drug Metab Pharmacokinet. 1986;11(1):71–6.

Zamfirescu I, Carlson HE. Absorption of levothyroxine when coadministered with various calcium formulations. Thyroid. 2011;21(5):483–6.

Zanger UM, Schwab M. Cytochrome P450 enzymes in drug metabolism: regulation of gene expression, enzyme activities, and impact of genetic variation. Pharmacol Ther. 2013;138:103.

Zhai X, Feng Y, Liu J, Li J, Zong Y, Zhan T, Gao S, Lv Y. Pharmacokinetic effects of capsaicin on vinblastine in rats mediated by CYP3A and Mrp2. Fundam Clin Pharmacol. 2019;33(4):376–84.

Zhao F, Wang X, Wang Y, Zhang J, Lai R, Zhang B, Zhou X. The function of uterine UDP-glucuronosyltransferase 1A8 (UGT1A8) and UDP-glucuronosyltransferase 2B7 (UGT2B7) is involved in endometrial cancer based on estrogen metabolism regulation. Hormones. 2020;19:403–12.

Zhou Z, Lu X, Song G, Wang Y, Chen J, Xiao C, Zhong L, Sun X, Peng Z, Wang X. In vitro study on influences of UGT1A8 gene polymorphisms on mycophenolate mofetil metabolism. Exp Clin Transplant. 2020;16(4):466–72.

Zhu M, Chan KW, Chang LSNGQ, Chang S, Li RC. Possible influences of ginseng on the pharmacokinetics and pharmacodynamics of warfarin in rats. J Pharm Pharmacol. 1999;51:175–80.

# Chapter 8
# Underlying Mechanism of Drug and Non-drug Interactions

**John I. Ogbu, Caroline V. L. Moreira, Roberto Saavedra, and James O. Fajemiroye**

**Abstract** Interactions among drugs and non-drug substances are the bane of all therapeutic and adverse endpoints in clinical practice. Alterations in drug absorption, distribution, metabolism, and excretion often characterize pharmacokinetic interactions. The drug targets (e.g., transporter, receptor, enzyme, microbiota) that contribute to the overall outcome or effects (e.g., secretions, blood clotting, blood pressure, plasma electrolyte concentration) are important components of pharmacodynamic interactions. This chapter explores the mechanism of interactions and the risk of undesirable effects. Pharmacokinetics and pharmacodynamics research could elucidate the underlying mechanisms of drug interactions, develop predictive models for identifying potential interactions, and discover new strategies for minimizing the risk of adverse effects.

**Keywords** Drugs · Foods · Targets · Pharmacokinetic · Pharmacodynamic

## 1 Introduction

Drugs and non-drug entities often interact to produce a wide range of effects. Diverse mechanisms underlying drug and non-drug interactions (DNIs) are redefining and widening the scope of pharmacological concepts. The synergistic, potentiating, agonistic, and antagonistic coverage of adverse and therapeutic outcomes demands in-depth knowledge of interactions. Understanding drug attributes and the physicochemical properties of drugs and foods among other xenobiotics is fundamental to the comprehension of DNIs. Food, nutrient or drug absorption, and metabolism could be altered by hyperglycemia-inducing drugs, lipid-altering drugs,

J. I. Ogbu · C. V. L. Moreira · J. O. Fajemiroye (✉)
Institute of Biological Sciences, Federal University of Goias, Goiania, Goias, Brazil
e-mail: jamesfajemiroye@ufg.br

R. Saavedra
Department of Basic Science, Universidad Santo Tomas, Puerto Montt, Chile



J. O. Fajemiroye (ed.), *Fundamentals of Drug and Non-Drug Interactions*,
https://doi.org/10.1007/978-3-031-80107-5_8

anabolic agents, antibiotics, mineral oils that decrease vitamin and $Ca^{2+}$ absorption, antineoplastic drugs that damage the absorptive surface, barbiturates that increase the metabolism of vitamin and decreases $Ca^{2+}$ absorption, and diuretics that induce the excretion of trace elements (Na, Mg, Cl, K, and Ca). Some food-drug interactions are mediated by several mechanisms, including the adsorption of drugs into food material, the formation of precipitates or complexes, and physical blockage at the drug's absorption site (Deng et al. 2017). For instance, digoxin adsorbs into fiber food materials, which reduces its absorption (McLachlan and Ramzan 2006). The tetracycline and quinolone antibiotics react chemically with divalent metal ions in milk (e.g., $Ca^{2+}$) and meat (e.g., $Fe^{2+}$) to form complexes or precipitates (Abbaspour et al. 2014; Uusi-Rasi et al. 2013). Oxidation and conjugation reactions have been exploited in the effect of food on drug disposition. Cruciferous vegetables enhanced the oxidative metabolism of antipyrine and phenacetin and the conjugation of acetaminophen (Pantuck et al. 1979, 1984).

Drug characteristics such as solubility and permeability are important factors that affect drug disposition (Deng et al. 2017). The aqueous or lipid solubility of drugs determines their intestinal absorption rate, especially for drugs that are associated with region-specific absorption, for example, chlorothiazide that has higher absorption in the upper intestine. Absorption and bioavailability of chlorothiazide increase due to increased solubility and permeability upon intake of food (Welling and Barbhaiya 1982). Highly lipophilic drugs have good membrane permeability, which may contribute to their potency. Some hydrophilic drugs like acetaminophen rely on their small molecular size since they are limited by membrane impermeability and slow dissolution rate. Thus, food-induced absorption of such drugs is mediated by the small size of their molecules subserved by gastric emptying (Sensoy 2021). The extraction ratio of the drug can mediate the absorptive and metabolic effect of food on drugs (Fig. 8.1). This parameter is a function of the fraction of free drugs and the intrinsic clearance rate (Birkett 2002). Drugs such as theophylline, antipyrine, and diazepam have low intrinsic clearance rates compared to the liver's blood flow. This implies that the liver removes little of such drugs (Deng et al. 2017). Thus, a postprandial increase in blood flow favors the absorption of these drugs. On the other hand, drugs with high clearance ratios, for example, propranolol and metoprolol, are poorly absorbed with low bioavailability and highly metabolized even with a propandial increase in blood flow (Dello Russo et al. 2022).

Drug solubility, permeability, and extraction ratio play a huge role in the potential impact of food, its absorption, and metabolism. Factors including the rate of gastric emptying and absorption window are enhancing factors (Fig. 8.1) (Deng et al. 2017). The presence of these factors increases the impact of food on the absorption of some drugs. These factors are mutually exclusive as their absence in the presence of food does not increase absorption of the drug like paracetamol (Moore et al. 2015). Other drugs whose absorption is enhanced by these factors include metformin, gabapentin, levodopa, ciprofloxacin, ofloxacin, and trospium chloride (Deng et al. 2017). The release of these drugs involves disintegration, dissolution, degradation, or diffusion processes (Herdiana et al. 2021). The involvement of the physiological effects of meals on modified drug release as well as the

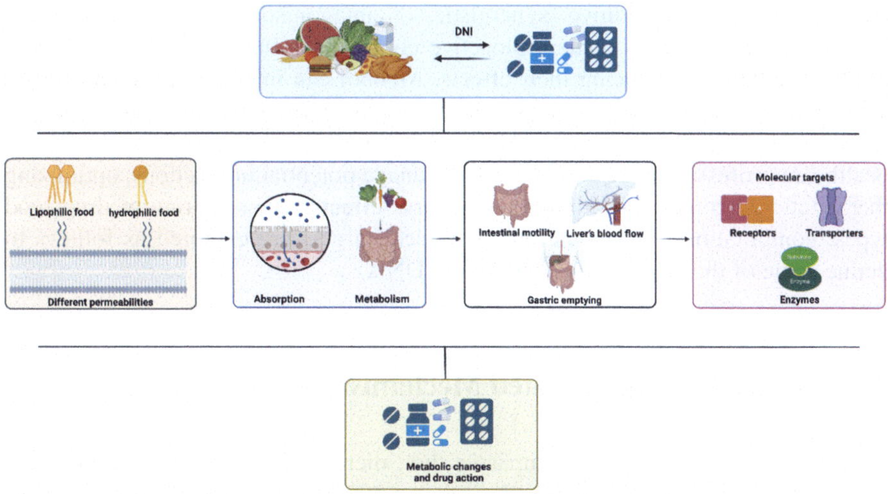


**Fig. 8.1** Some mechanisms of drug-food type interaction. DNI—drug and non-drug interaction

effect of meals on the drugs after release makes the mechanism complex. Modified-release formulations are specifically designed to control the release of drugs over a prolonged period. This provides a significant therapeutic advantage over conventional immediate-release formulations by reducing the need for frequent dosing, minimizing peaks and troughs in plasma concentrations, and overcoming the instability often associated with changes in gastric pH (Wheless and Phelps 2018). Some foods explore this mechanism to maintain the plasma concentration and bioavailability of drugs within a targeted range. For example, despite the narrow therapeutic window of theophylline, food intake increases the bioavailability and absorption of modified-release dosage forms of the drug (Varum et al. 2013).

Formulation, targets, and incompatible associations often define the mechanisms underlying DNIs. Drugs or non-drug substances can interact by altering gastric pH or competing for transport mechanisms and affecting the absorption of each other when taken orally (Fig. 8.1). Drugs, food, herbal supplements, alcohol, and environmental toxins among others can interact through physical or chemical processes, such as precipitation or complex formation, thereby altering the bioavailability or pharmacokinetics. One may displace others from plasma protein binding sites to increase the concentration of the unbound, active drug, protein, nutrients, etc. Induced or inhibited liver cytochrome P450 (CYP) enzymes could alter plasma concentrations and renal-hepatic drug-food clearance (Zarezadeh et al. 2021).

Genetic variations in drug-metabolizing enzymes, transporters, or drug targets can influence individual medication responses and may result in drug interactions. For example, polymorphisms in CYP enzymes can affect the metabolism of certain drugs, leading to variations in drug efficacy and toxicity. All these interaction mechanisms may impact bioavailability and pharmacodynamic parameters. Furthermore, drugs and non-drug substances can interact at the site of action (e.g., receptor and enzyme) to produce additive, synergistic, or antagonistic effects. For instance, herbal supplements may contain bioactive compounds that interact with medications, enhancing or reducing their effects. Alcohol can interact with medications, exacerbating sedative effects or interfering with drug metabolism. Understanding these underlying mechanisms of drug and non-drug interactions is crucial for healthcare professionals to anticipate and manage potential interactions, optimizing therapeutic outcomes and minimizing adverse effects. The spectrum of drug-food types, combinations, and numerous biological targets are explored as follows to define some of the mechanisms underlying DNIs.

## 2 Drug-Food Types Related Mechanism of Interaction

Food is a broad term that encompasses diet, dietary supplements, nutrients, and nutrient formulas ingested for health (Zohoori 2020). Unlike drugs, food was considered to possess insignificant pharmacokinetic and pharmacodynamic properties (Ernst and Magnani 2009). Despite their differences, they can interact with each other for beneficial or harmful outcomes. These interactions may be outcomes of normal ingestion of both chemicals together, as well as diet intervention in metabolic disorders such as obesity, diabetes mellitus, and metabolic syndromes. Interactions often result in clinically significant adverse reactions whose management requires an understanding of the mechanisms underlying them. Mechanistic details underlying these interactions are bidirectional—food-drug and drug-food interaction. Food-drug interactions refer to the effects of food substances on drug disposition and biological activity. On the other hand, drug-food interactions are the effects a drug exerts on food digestion, absorption, and excretion. Mechanisms mediating the reciprocity of these interactions rely on various factors including food characteristics (food type/category), molecular targets (receptors, transporters, and enzymes), physicochemical factors (including solubility, permeability, stability, physicochemical reactions), drug properties, and gastrointestinal processes (gastric emptying, enzyme secretion, intestinal motility, splanchnic and postprandial blood flow).

Mechanisms of food-drug interactions include those based on the food type/category, molecular targets, drug characteristics, physicochemical reactions, and dose formulations/regimens. The dependence of food-drug interaction mechanisms on food type or category is based on the following principles: (i) Food categories and constituents exhibit varying physicochemical reactions based on temperature, solubility, caloric content, osmolality, etc. (ii) Not all food types have the same rate and degree of absorption. (iii) Variations in nutrient content affect gastric motility, gastric emptying, splanchnic blood flow, bile secretion, intestinal motility, intestinal hyperemia, and absorption rates, which mediate drug pharmacokinetics. (iv) Postprandial responses influence drug absorption, distribution, metabolism, and excretion. Different types of food include high-fat, high-carbohydrate, high-protein, high-fiber, fluid, and special nutrients. The heterogeneity in the nutrient composition of these food types provides support for their varying capacity to alter the absorption, bioavailability, metabolism, and excretion of drugs in the body.

## 2.1 High-Fat Diet-Related Mechanism

A high-fat diet can alter lipid metabolism and increase circulating levels of fatty acids and lipoproteins. They can impair glucose metabolism, activate inflammatory pathways (saturated fats), lead to the accumulation of fat (adipose tissue, liver, and muscles), and alter the composition and function of the gut microbiota. Diets rich in polyunsaturated fatty acids can increase oxidative stress, and hormonal and cardiovascular changes, among others. Altogether, meals containing high amounts of fat generally impact the absorption, transport, and bioavailability of drugs (Cheng and Wong, 2020); the production of metabolites; nutrient absorption; and immune responses.

The role of dietary fat in drug absorption is based on the ability of the food type to alter the lipid composition of biological membranes, which are gateways for drug absorption. This mechanism works through the interplay of factors including solubility, drug characteristics (hydrophobicity or lipophilicity), and permeability. Based on drug characteristics, for instance, high-fat food enhances the absorption of hydrophobic drugs like gefitinib with a log $p$ of 4.15 (Swaisland et al. 2005; Deng et al. 2017). Fat-mediated effects on absorption may explore physiological processes such as bile secretion and intestinal lymphatic transport. Consumption of a high-fat diet can decrease the absorption of hydrophilic drugs by stimulating the secretion of bile micelles (Niederberger and Parnham 2021). Bile-mediated decrease in absorption occurs through the entrapment of drugs by the bile micelles, consequently reducing the amount of free drug available for absorption (Yeap et al. 2013). Increased absorption of some drugs such as acitretin and retinoic is also subservient to the intestinal lymphatic transport pathway, which is activated by a high-fat diet (McNamara et al. 1988; Hollander 1980). A study showed that roglitazol absorption is more gradual and sustained in the fed state than in the fasted state (Freed et al. 1999).

The presence of high fat also impacts the bioavailability and excretion of drugs, besides absorption of drugs. For example, an increase in the bioavailability of piperaquine, an antimalarial drug, after a high-fat meal has been reported (Reuter et al. 2015); a risk of theophylline toxicity has been associated with an increase in its bioavailability after high-fat meals (Bushra et al. 2011); and bioavailability of antihistamines such as esomeprazole (antiulcer drug) is decreased 15 min after taking a high-fat meal (Sostek et al. 2007). In addition to bioavailability changes, a high-fat diet could increase the excretion of HMR1031 (an integrin antagonist) via the biliary route (Shah et al. 2003). Since lipophilic drugs are highly excreted via biliary secretion, fat-induced excretion of highly lipophilic drugs will decrease bioavailability and systemic exposure (Deng et al. 2017).

Food intake and weight gain are also important consequences of interaction. Exogenous hormones have been administered in relevant hormone-related pathological conditions. One hormone, ghrelin, increases food intake and weight gain by binding to its receptor, the growth hormone secretagogue receptor. This effect is only feasible with the attachment of a fatty acid peptide as a side chain (Heppner et al. 2012). Thus, a fatty diet increases food intake and weight gain activity of ghrelin via physical interaction with the hormone (Kirchner et al. 2009). Physical mechanisms in food-hormone are not limited to food-hormone binding interaction. Major mechanisms mediating food-hormone interactions are physical interaction with hormone and receptor-mediated mechanisms. Short-chain fatty acid end products of complex indigestible carbohydrates—mainly acetate, propionate, and butyrate—bind to and activate cell-surface receptors free fatty acid receptor (FFAR2) 2 and 3 expressed on entero-endocrine L cells that produce the incretin hormone glucagon-like peptide-1 (GLP-1) and are involved in energy and glucose metabolism (Tolhurst et al. 2012; Tremaroli and Bäckhed 2012). Acetate and propionate also activate FFAR2 on adipocytes to increase the expression of the weight-reducing hormone leptin (Ryan and Seeley 2013).

## *2.2 High-Carbohydrate Diet-Related Mechanism*

Carbohydrates—monosaccharides, disaccharides, and polysaccharides—are high ATP-containing macronutrients. Their effects on drug absorption, bioavailability, metabolism, and excretion of drugs depend on their components and size. Their heterogeneity poses some difficulty in evaluating their effects on drug absorption, bioavailability, and metabolism (Niederberger and Parnham 2021). These nutrients with different molecular sizes impact functional processes such as gastric motility and pancreatic enzyme secretion. Tacrolimus, for instance, when administered with a high-carbohydrate meal, exhibited an increase in absorption rate compared to when it was administered with a high-fat meal (Bekersky et al. 2001). Consuming praziquantel with high-carbohydrate food increases bioavailability compared to high-fat meals. The reasons for this phenomenon remain unclear (Eason et al. 2022).

Carbohydrate-induced metabolic alterations of some drugs have been reported. The metabolism of antipyrine, theophylline, and propranolol increases following a high-carbohydrate diet. These alterations may be attributed to the decrease in the activity of CYP enzymes as reported by a study (Zarezadeh et al. 2021). Also, carbohydrates affect drug excretion and overall systemic exposure to drugs by altering urine pH. For example, carbohydrates increase the pH of urine making weakly basic drugs to be reabsorbed back into the circulation and favoring excretion of weakly acidic drugs (Koziolek et al. 2019). Taken together, the carbohydrate content of the diet causes pharmacokinetic alterations in the absorption, bioavailability, metabolism, and excretion of drugs. Secondary to the interaction are enzyme-dependent and independent mechanisms.

In contrast, certain drugs, including acarbose, miglitol, and orlistat, affect the metabolism of carbohydrates. Acarbose delays the digestion of carbohydrates through competitive and reversible inhibition of the pancreatic α-amylase and intestinal α-glucosidase enzymes. The pancreatic α-amylase hydrolyzes complex carbohydrates to oligosaccharides in the small intestine. Intestinal α-glucosidase enzyme breaks down oligosaccharides into disaccharides and monosaccharides. Due to acarbose's inhibition of these enzymes, the oligosaccharide drug delays the digestion of starches, oligosaccharides, and disaccharides into glucose. Acarbose also slows down glucose absorption (McIver et al. 2024). These mechanisms underlying acarbose and miglitol-carbohydrate interaction make them first-line drugs in the control of diabetes and related conditions (Standl and Schnell 2012). Orlistat, for example, improves carbohydrate metabolism both in obese otherwise healthy and obese diabetic patients but mechanisms mediating this interaction are still unclear (Bansal et al. 2024).

The underlying mechanisms of drug-carbohydrate or carbohydrate-food interactions have several health implications. A high-carbohydrate diet worsens the adverse effects of acarbose administration including diarrhea, flatulence, and abdominal pain. To reduce complications of acarbose-carbohydrate interactions, patients are advised to avoid carbohydrate-containing diets for 6 h following acarbose overdose (McIver et al. 2024). Postprandial glucose excursions, insulin-sparing effect, and consequently low risk of hypoglycemia are pharmacological fallouts of miglitol-carbohydrate interactions that need further clinical assessment (Kitano et al. 2013). Thus, carbohydrate content in diet is a basis for many drug-food and food-drug interaction mechanisms with beneficial and unhealthy implications.

## 2.3 High-Protein Diet-Related Mechanism

Protein-rich diets play a significant role in the pharmacokinetics of administered drugs. Protein-mediated mechanisms of drug pharmacokinetic alterations include an increase in the splanchnic blood flow and changes in the activities of intestinal transporters (Deng et al. 2017). This food class also interacts with several drugs through the activation or inhibition of CYP enzymes with a slight effect on plasma drug transport proteins. Protein-induced postprandial increase in splanchnic blood flow and transporter activity appears to contribute to mechanisms underlying the absorption of drugs. Protein-induced increase in splanchnic blood flow rate favors a high absorption of drugs (Lawrence and Li 2014; Niederberger and Parnham 2021). Mechanisms underlying this are still unclear. Proteins may play a huge role in facilitating the escape of highly vulnerable drugs such as propranolol and ibrutinib (a protein kinase inhibitor) from first-pass liver effects (Koziolek et al. 2019). However, further studies are needed to establish this assumption. Besides the postprandial mechanism, proteins also affect drug absorption through interaction with membrane transporters. Generally, high-protein diets could increase intestinal membrane transporters. However, protein-derived peptides might compete with drugs for transporter proteins leading to lowered absorption rates. This mechanism is relevant in the anti-Parkinson levodopa or antibiotics beta-lactam uptake, which is also dependent on transporter proteins (Koziolek et al. 2019). A decrease in phenytoin absorption following a high-protein diet follows a similar interaction mechanism (Briguglio et al. 2018).

The metabolism and bioavailability of several drugs are altered following the ingestion of a high-protein diet. The metabolism of theophylline is decreased following high-protein ingestion but mechanisms underlying the interaction are yet to be elucidated. The sensitivity of bioavailability to changes in hepatic and splanchnic blood flow (Koziolek et al. 2019) may contribute to protein-induced changes in drug metabolism. The metabolic activity of CYP on drugs is increased by a high-protein diet-induced high hepatic blood flow. One example of such an enzyme is the CYP1A2. Food containing high proteins increased the activity of the enzyme contrary to a high-carbohydrate diet. Theophylline is a major drug implicated in this form of protein-drug interaction (Zarezadeh et al. 2021).

Prolonged binding may delay the elimination of highly protein-bound drugs as their free drug concentration is low with delayed distribution, metabolism, and excretion. Dietary protein may also alter renal plasma flow, creatinine clearance, and renal tubular transport to change the fate of drugs primarily eliminated by the kidneys. These dietary proteins could alter the physicochemical properties of urine such as the pH, which affects the urinary reabsorption or elimination of drugs. These proteins tend to make the urine more acidic, thus enhancing the reabsorption of weakly acidic drugs or metabolites like phenobarbital, acetylsalicylic acid, salicylic acid (a metabolite of aspirin), and warfarin (Cao et al. 2011; Briguglio et al. 2018). This reabsorption in an acidic environment is favored by the fact that drug molecules tend to remain in their non-ionized, more lipid-soluble form, which easily crosses cell membranes. In contrast, weakly basic drugs (amphetamine, lidocaine, morphine, etc.) are poorly reabsorbed in this environment. Although a high-protein diet does not significantly alter plasma proteins, the nutritional status of proteins may alter the transport of protein-bound drugs. For example, malnutrition-induced hypoproteinemia can lead to excessive loss of plasma proteins resulting in less binding of drugs to plasma proteins (Gurevich 2013; Deng et al. 2017). These mechanisms contribute to the extended effect of drugs such as digoxin (Deng et al. 2017).

## 2.4 High-Fiber Diet-Related Mechanism

Fiber supplements including oat bran, soy fiber, and ispaghula husk decreased the bioavailability of levothyroxine in 13 hypothyroid patients, through nonspecific adsorption of the drugs (Liel et al. 1996). Fiber-enriched diets are sometimes recommended for managing obesity, cholesterolemia, diabetes, and chronic constipation (González Canga et al. 2010) and can alter drug disposition. Fibers generally delay the gastric emptying time. Some nervous system-acting, lipid-lowering, oral hypoglycemic agents, and cardiovascular-acting drugs have been implicated in fiber-drug interactions. The bioavailability of lithium and tricyclic antidepressants is affected by these interactions. In addition to a decline in the drugs' serum levels, patients became refractory to some tricyclic antidepressants such as amitriptyline, doxepin, and imipramine after ingestion of a high-fiber diet (Drossman et al. 2018). In the previous study, the bioavailability of orally administered metformin was increased after ingestion of a dietary fiber from *Plantago ovata* (Díez et al. 2017).

Several mechanisms mediate fiber-induced lowering of drug absorption. Dietary fiber may act as a physical barrier and prevent drug contact with the gastrointestinal mucosa, but it can also directly adsorb drug molecules (Deng et al. 2017). Fernández et al. (1998) reported a decrease in the absorption of ethinyloestradiol between 29% and 35% when administered simultaneously with two fiber-rich products [product A (containing 76.5% wheat bran, 12% fruit fiber, and 2% guar gum) and product B (containing *Plantago ovata* seeds—65% and husk—2.2%), respectively]. It was suggested that these fibers lowered absorption by acting as a physical barrier at the intestinal surface. The pharmacodynamic implication of the food-drug interaction is the shortening of the duration of action of the drug. Another absorption-lowering mechanism of fibers is the adsorption of the drug away as seen in digoxin (Deng et al. 2017). In addition, some fibers such as guar gum and psyllium form viscous solutions that interfere with the absorption of medications (Khalid et al. 2022).

Some oral hypoglycemic agents including glibenclamide, metformin, and digoxin have been implicated in fiber-drug interactions. Shima et al. (1983) suggested a glucomannan-induced decline in the bioavailability of glibenclamide. The antihyperglycemic effect of metformin was altered following guar gum-induced decrease in the intestinal absorption and bioavailability of the glucose-lowering drug (Gin et al. 1989). Guar gum and wheat ban fibers are also involved in food-digoxin interactions: Guar gum fiber decreased the absorption of digoxin by 16% (Huuponnen et al. 1984). In a study, wheat ban decreased area under the curve 0–24 h of the drug by about 7% but did not modify the values of peak concentration, time to reach peak concentration, or the amount detected in urine during 24 h (Johnson et al. 1987). Some interactions are considered beneficial. The beneficial effects caused by fiber-drug interactions have been explored in food-drug combination therapies. Fiber alters the pharmacodynamics of lipid-lowering drugs including gemfibrozil, colestipol (a bile acid sequestrant), and lovastatin. Fiber intake along with gemfibrozil administration elicited a synergistic lipid-lowering effect with an improved ratio of high-density lipoproteins to low-density lipoproteins (Tuomilehto et al. 1989). A 5 g of ispaghula husk has been combined with a 5 g of cholesterol, a bile acid sequestrant in a synergistic interaction that reversed moderate cholesterolemia (Turley et al. 1996). Pectin, another fiber, improved the serum lipid effect of lovastatin due to its capacity to reduce the intestinal absorption of lovastatin (Richter et al. 1991).

### 2.5 Fluid Intake and Specific Nutrient-Related Mechanism

Fluid intake is always necessary when taking medication orally. Fluid in this context includes water, beverages, milk, and any liquid consumed as food. Physicochemical factors such as solubility, dissolution processes, and pH levels are the major basis of their interaction with drugs. Fluid intake affects drug absorption and metabolism. Milk interferes with the absorption of tetracycline due to the presence of calcium that chelates or complexes with the drug. The fluid also inhibits the absorption of sotalol, a β-adrenergic receptor blocker (Pressly et al. 2023). Gastrointestinal processes also contribute to physicochemical mechanisms underlying the interaction of fluids with drugs. An example is a milk-induced decrease in the rate of absorption of theophylline via a delay in the gastric emptying time (Heimann et al. 1982). A study also confirmed the effect of specific nutrients like caffeine on drug metabolism. The study showed that the elimination of methylxanthine in caffeine-containing beverages increased the rate of metabolism of theophylline (Monks et al. 1979).

Vitamins and minerals are specific nutrients in fluid diets that play roles in the outcome of nutrition and medication. In ferrous-containing drugs, for example, ascorbic acid in citrus fruits and juices facilitates iron absorption by forming a chelate with ferric iron at an acid pH that remains soluble at the alkaline pH of the duodenum (Piskin et al. 2022). The physicochemical mechanism augments the oral bioavailability of iron-containing drugs for their therapeutic effects. Importantly, this mechanism explored by the vitamin is dependent on the type and oxidation state of the iron molecule. The oral bioavailability of heme iron is not affected as opposed to non-heme iron. Ferric iron ($Fe^{2+}$) is better absorbed compared to its redox ferrous iron ($Fe^{3+}$) counterpart (Ems et al. 2023). Other key players that may affect the interaction include gastric acidity and reducing substances. Some nutrients like phytate inhibit the absorption of iron. However, ascorbic acid possesses potent anti-chelating mechanisms against phytate-iron interaction. This indirect interaction with the trace element involves two basic processes that increase iron absorption: (i) the inhibition of iron chelation to phytates and (ii) the reduction of iron to its ferric state (Pires et al. 2023). Taken together, drug interactions with nutrients are mediated majorly by physicochemical mechanisms. In addition to pharmacokinetic change-induced interactions, specific nutrients can also initiate pharmacodynamic changes. Drugs implicated in this pharmacodynamic interaction include isoniazid (antimicrobial agent) and fluorouracil (antineoplastic drug). For instance, pyridoxine (vitamin B6) deficiency causes isoniazid-induced neuropathy via the blockage of the pyridoxine kinase (Badrinath et al. 2024). The enzyme that converts vitamin to pyridoxal phosphate blocks peripheral neuropathy induced by the antimycobacterial drug. Pyridoxal phosphate availability through vitamin B6 during isoniazid therapy can increase neurotransmitter synthesis leading to a decrease in peripheral neuropathy (Badrinath et al. 2024).

## 3 Target-Dependent Mechanisms of Interaction

Various drug targets such as transporters, enzymes, and receptors are competitive sites for food substances and drugs that can lead to an alteration in the disposition and effects of drugs. Several transporters have been characterized by their effects on food-induced dispositional changes to drugs. Among these are the peptide transporter 1 (PepT1), P-glycoprotein (P-gp), and organic anion transporters (OATPs). Being a target site for both foods and drug transport, P-gp influences food-drug interactions via competitive inhibition of drug absorption, thereby aiding their elimination through secretion into bile and urine (Fig. 8.2) and decreasing their passage into the brain. Digested products of proteinous food compete with levodopa and penicillamine for PepT1 to inhibit their absorption (Robertson et al. 1991). Food particles or components can modulate the activities of transporters. Upregulation of transporter systems in response to food intake is suggested to increase the absorption of gabapentin (Deng et al. 2017). This same effect has been reported for grapefruit juice on the P-gp transporter: The citrus drink augments the bioavailability of the angiotensin II receptor blockers such as candesartan, telmisartan, and valsartan, which is normally low (Lim et al. 2003; Romiti et al. 2004). This mechanism is similar to food effects on cephalexin and cefadroxil (β-lactam drugs) in in situ perfusion experiments (Kanan et al. 2023). OATP's roles explain a decrease in oral bioavailability of etoposide and celiprolol following interaction with grapefruit juice (Bailey 2010). Thus food-induced transporter activity impacts the absorption of drugs with potential changes in their bioavailability.

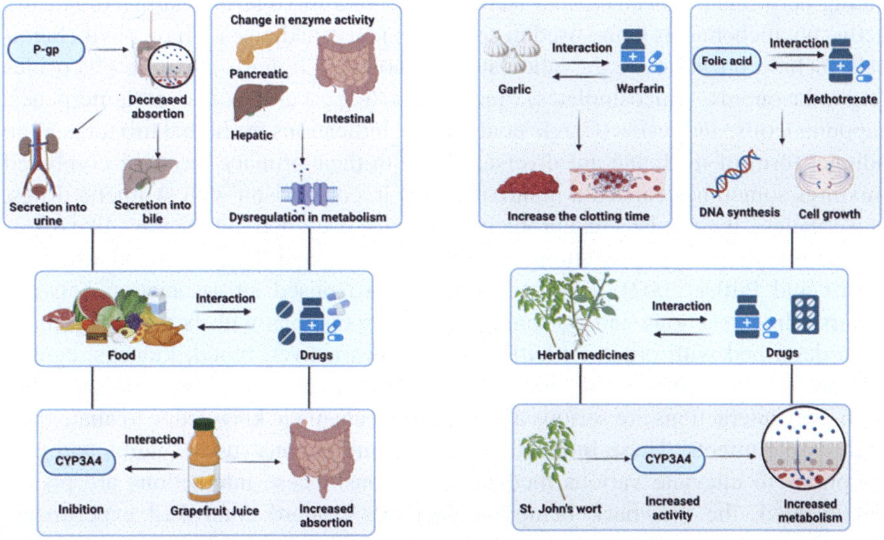


**Fig. 8.2** Some target-dependent mechanisms of interaction

Besides transporters, enzymes are also targets for food effects on drug disposition. Pancreatic, hepatic, and intestinal enzyme activities greatly affect the metabolism of different types of drugs (Fig. 8.2). Proteins increase intestinal enzyme activity resulting in increased metabolism of drugs such as theophylline (Niederberger and Parnham 2021). While some foods may not elevate enzyme activity, they competitively inhibit enzymes and mediate the pharmacological effects of drugs. This is seen in vitamin K's effect on coumarin anticoagulants. The vitamin in cruciferous vegetables antagonizes the anticoagulant effects of warfarin and other coumarin anticoagulants. Also, coumarins in some herbal teas may enhance the effects of coumarin anticoagulants (Heghes et al. 2022). Various natural products such as grape juice and herbs have recorded extensive interactions with drugs. Grapefruit juice is consumed widely as a protector against cardiovascular diseases and cancers (Kiani and Imam 2007). Since 1989 when it was first observed to alter the clearance of felodipine, a calcium channel blocker, and enhance the area under the curve of the plasma concentration-time curve (Bailey et al. 1989), the citrus drink has been known to affect the metabolism of certain drugs (Shen et al. 2020). Although it contains various constituents that vary depending on its origin, climatic conditions, and maturity, there is no indication of a specific active principle responsible for adverse grapefruit-drug interactions. The juice inhibits CYP3A4 isoenzymes most often involved in drug metabolism. Drugs that are metabolized by CYP3A when they are absorbed from the intestinal lumen are mostly affected by grapefruit juice (Fig. 8.2). The grapefruit juice-induced increase in the bioavailability of such drugs (calcium channel antagonists, statins, sedative-hypnotics, anxiolytics, psychotropics, and antihistamines) can increase the risk of overdose and dose-dependent adverse effects (Greenblatt et al. 2001).

The drug-target-dependent mechanism of interactions provides a basis for elucidating herb-related mechanisms. Herb represents a mixture of pharmacologically active phytochemicals being used in complementary medicines. These phytochemicals include phenolics (e.g., salicylates, quinolones, lignin), phenolic glycosides (e.g., flavonoids, glucosinolates), terpenoids (e.g., carotenoids, sesquiterpenes, saponins), oils, and resins (Czigle et al. 2023). Indications for herbal mixtures as an adjunct form of medicine are diverse. Herbs in their primary forms or combined mixtures sometimes carried a nontoxic status in comparison with synthetic drugs. Nevertheless, there exist some problems related to drug-herb interactions like toxicity, adulteration, contamination, misidentification, and lack of standardization (Ernst and Pittler 2002). This scenario has also raised an association between adverse drug reactions and poisoning. Most associations with herbal usage have been described with organ toxicities such as heart, liver, blood, kidneys, central nervous system, and skin and carcinogenesis (Rombolà et al. 2020). Consequently, herb-drug interactions are serious and require mechanistic knowledge to abate their undesirable effects. These interactions arise from patients' use of herbs and prescriptions to alleviate various medical conditions. These interactions are poorly documented, the drawback being the lack of standard controlled experiments involving these complex mixtures of chemicals. For instance, bleeding caused by *Gingko biloba* and warfarin interactions was not confirmed in another study conducted under controlled perspectives (Vaes and Chyka 2000; Engelsen et al. 2002). Few studies have attributed herb-drug interactions to fundamental inconsistencies in enzyme-mediated processes in which hepatic and intestinal CYPs metabolize structurally unrelated compounds (Choi and Chin 2021). The extensive oral ingestion of herbal products leads to the alteration in the absorption of drugs, interference in distribution patterns, and competition in the metabolic and excretory pathways often define the pharmacokinetic mechanisms of herbal-drug interactions (Izzo 2005).

Drugs like warfarin, digoxin, cyclosporine, tacrolimus, amitriptyline, midazolam, indinavir, and irinotecan that have a high potential to interact with herbal medicines usually have a narrow therapeutic index. Many are CYPs and P-gp substrates (Chen et al. 2011). Herbs that interact with drugs include garlic (*Allium sativum*), ginger (*Zingiber officinale*), ginkgo (*Ginkgo biloba*), ginseng (*Panax ginseng*), and St. John's wort (*Hypericum perforatum*). For example, garlic increased the clotting time and international normalized ratio of warfarin (Fig. 8.2) and induced hypoglycemia when taken with chlorpropamide. It reduced the area under the plasma concentration-time curve and the maximum concentration of saquinavir in humans. Case reports have demonstrated that ginkgo may potentiate bleeding when combined with warfarin or aspirin, increase blood pressure when combined with thiazide diuretics, and even lead to a coma when combined with trazodone. Furthermore, ginseng reduced the blood levels of warfarin and alcohol. This herb can induce mania if taken concomitantly with phenelzine, a non-selective and irreversible monoamine oxidase inhibitor. Multiple herb-drug interactions have been identified with St. John's wort, which significantly reduced the concentration-time curve and blood concentrations of warfarin, digoxin, indinavir, theophylline, cyclosporine, tacrolimus, amitriptyline, midazolam, and phenprocoumon. According to Chen et al. (2011), well-designed studies can address clinically significant herb-drug interactions.

Several drugs impact different nutrients and nutrient-deficient states. An example is the alteration of vitamin K metabolism by warfarin. Warfarin inhibits the vitamin K oxidoreductase (VKORC1), which generates vitamin K hydroquinone (KH2) required for the carboxylation and consequent activation of vitamin K–dependent proteins. VKORC1 produces KH2 in 2 reactions: reduction of vitamin K epoxide (KO) to quinone (K), and then KH2 (Rishavy et al. 2018). This interaction occurs in the intake of different foods including herbs and vegetables due to their vitamin K content. This is one of the reasons why warfarin is not administered with vitamin K. Another instance is the enzyme-mediated interaction between methotrexate and folic acid (Fig. 8.2). Due to structural similarity with folic acid, inhibition of the dihydrofolate reductase enzyme by the folate inhibitor reduces folic acid to tetrahydrofolic acid. Folic acid interferes with DNA synthesis and cell growth to elicit methotrexate toxicity (Hamed et al. 2022). This explains leucovorin (folinic acid) intervention to counteract methotrexate toxicity. Molecular modification of enzymes is also involved in enzyme-mediated mechanisms. An example is mitomycin-Zn/Cu interactions. The antineoplastic drug interferes with the Zn/Cu superoxide dismutase transcription.

Being a key player in the metabolism of some xenobiotics, the human CYP superfamily is a key regulator of herb-drug interactions (Wanwimolruk and Prachayasittikul 2014). Compounds from herbal preparations can act as substrates, inhibitors, or inducers of CYP (Zhou et al. 2003), thereby affecting the pharmacokinetics of any co-administered drugs. Induction of these enzymes (e.g., CYP3A4 by St. John's wort) by herbal products increases their activity, which translates to a rise in the rate of drug metabolism and rapid changes in the bioavailability of drugs (Madabushi et al. 2006; Fasinu et al. 2012). This mechanism explains a decrease in the plasma levels and bioavailability of substrates of CYP3A4 including warfarin, cyclosporine, tacrolimus, indinavir, and simvastatin following induction of the enzyme by St. John's wort (Fig. 8.2) (Nieminen et al. 2010; Vlachojannis et al. 2011). These herbal ingredients are metabolizable by CYP to nontoxic or poisonous metabolites. The enzyme-dependent mechanism is attributed to the overlapping substrate specificity of the liver and intestinal CYP owing to the numerous presentations of structurally unrelated compounds (Marchetti et al. 2007). CYP3A4 is expressed in the villus type of enterocytes, an important site for the absorption of orally administered drugs (Zhou et al. 2003). Apart from the CYP, St. John's wort-induced activity of p-glycoprotein intestinal enzymes to interact with other drugs such as fexofenadine (an antihistamine), phenprocoumon (an anticoagulant), tolbutamide, and mycophenolic acid (immunosuppressant) (Di et al. 2008; Hojo et al. 2011). *Ginkgo biloba* has been reported to induce CYP2C19-dependent omeprazole metabolism in healthy human subjects (Yin et al. 2004).

Inhibition of these enzymes can be competitive or non-competitive with instantaneous effects (Zhang and Wong 2005; Fasinu et al. 2012). Most herbal compounds can elicit NADPH-, time- and concentration-dependent enzyme inactivation. This mechanism is related to the biotransformation of herbal compounds into reactive metabolites by CYP. In this case, metabolites may attach to CYP irreversibly. As a result, chemical modification of the heme, the apoprotein, or both by the reactive metabolites can involve covalent binding of the modified heme to the apoprotein (Zhou et al. 2003). An example is rotenone, a naturally occurring flavonoid found in several plants that inhibits CYP by interfering with the electron transfer of the heme iron (Sanderson et al. 2004; Heo et al. 2022). Resveratrol and tryptophan also possess CYP-inhibitory properties (Wang et al. 2021). Flavonoids and coumarins in *Angelica sinensis* inhibit CYP1A2, 3A4, and P-gp to increase the bioavailability of the substrates of these enzymes (Tang et al. 2006). Transporters also contribute to the alteration of drug disposition and pharmacodynamics induced by active phytoconstituents in herbs. One example is the ATP-binding cassette (ABC) family of drug transporters, which have played significant roles in herb-anticancer drug interactions. P-glycoprotein, a member of this family, is implicated in the efflux of anticancer drugs, thus decreasing their effect. However, phytochemicals in *Rosmarinus officinalis* inhibit P-gp to decrease the efflux of tamoxifen, paclitaxel, teniposide, and other anticancer drugs. This interaction can potentiate the effect of anticancer (Ganesan et al. 2021). Curcumin in *Curcuma longa* inhibits the MRP-1, which potentiates the effects of anticancer drugs including vincristine and vinblastine (Hassanzadeh et al. 2020). Clinically, small changes in comparison to higher concentrations are considered important for drugs with a strong concentration-response relationship or narrow therapeutic index.

Certain interactions appear to be mediated by microbes. Numerous microbial species represent the gut microbiota (Scepanovic et al. 2019; Zheng et al. 2020). These populations are often harmless to their host in their homeostatic state. As recipients of exogenous substances such as food and drugs, the microbial populations participate in a non-classical disposition of orally administered drugs, leading to several drug-gut microbiota interactions (Chua et al. 2017). Like that of food, interactions between drugs and microbiota are reciprocal (gut microbiota-drug and drug-gut microbiota). For example, certain microorganisms mediate the nitroreduction of clonazepam to their corresponding 7-amino metabolite derivatives (Elmer and Remmel 1984). Microbes also cleave the isoxazole portion of the benzisoxazole ring system of risperidone, a drug used for treating schizophrenia and bipolar disorder (Dodd and Cann 2022). Additionally, *Lactobacilli*, *Enterococci*, and *Clostridia* demethylate methamphetamine in vitro (Caldwell and Hawksworth 1973). Gastrointestinal mechanisms play a crucial role in how drugs affect the body's handling of food after consumption. These processes are the pharmacological strategies explored in treating gastrointestinal-related conditions. For instance, administering cimetidine (an H2 antagonist) in patients with hypergastric secretion after small intestinal surgery can enhance nutrient reabsorption (Cortot et al. 1979) via a reduced gastric acid secretion and delayed gastric emptying rate. Additionally, some anticholinergic drugs prolonged gastric transit time and enhanced the absorption and availability of riboflavin (Levy and Jusko 1966). The intestinal microbiota can release enzymes (e.g., oxidase and hydrogenase, Table 8.1) that play roles in drug metabolism (Liang et al. 2015; Walsh et al. 2018). These enzymes can modify the structure of drugs, leading to their activation, deactivation, or reactivation (Swanson 2015). The microbiota-induced changes in drug disposition are similar to the metabolic activity of the liver on xenobiotics. These roles make the microbial population a metabolic "organ." Enzymatic activity of this metabolic "organ" with corresponding chemical reactions such as hydrogenation, dehydrogenation, and decarbonization will ultimately impact the bioavailability of the drugs and foods.

Metabolic disposition is a major pharmacokinetic mechanism in microbiota-drug interactions. Confounding factors, whether drug-related or not, may impact physicochemical properties (e.g., lipophilicity or hydrophilicity), the chemical state of the gastrointestinal tract, diseased states, density, and variation of microbial populations. Considering the huge population of the microbiota in the small intestine as compared to other parts of the body, microbiota-drug interactions may seriously impact the absorption of orally ingested lipophilic drugs. This is further influenced by the alkaline nature of the small intestine (O'Hara and Shanahan 2006). On the other hand, drug-gut microbiota interactions involve the effects of drugs on the variation, density, and activities of gastrointestinal microorganisms. Drugs may affect host metabolism or clinical efficacy by regulating gut microbiome composition or its function (Wu et al. 2017). Direct disruption of the gut microbiota, through bacteriostatic or bactericidal activities can alter the metabolic capability of the microbiota (Walsh et al. 2018). This in turn can upturn the balance between "good" and

**Table 8.1** Enzyme-dependent mechanisms of gut microbiota-drug interactions

| Interacting drug substrate | Microbiota-derived enzyme/ microbe | Hypothesized mechanisms | References |
|---|---|---|---|
| Non-steroidal anti-inflammatory drugs such as indomethacin and diclofenac | β-glucuronidase | Remove glucuronic acid moiety from hepatic phase 2 metabolites | Saitta et al. (2014) |
| Azo-containing drugs (e.g., olsalazine) and nitrofuran antibiotics (e.g., nitrofurazone and nitrofurantoin) | Azoreductase | Reduction of azo or quinone bonds | Ryan (2017) |
| Bio-activation of sulfur and selenocysteine derivatives | β-Lyase | Cleavage of C-S bond in hepatic-production cysteine S-conjugated metabolites Cysteine-conjugated metabolites | Mikov (1994) |
| Aspirin, ester-containing prodrugs | Carboxylesterase | Hydrolyze ester, thioester, amide, or carbamate-containing drugs to corresponding free acids<br>Hydrolyze esters to carboxylic acids | Imai and Ohura (2010), Laizure et al. (2013) and Kim et al. (2016) |
| 5-Aminosalicylic acid | N-acetyltransferase | Transfer of acetyl group to nitrogen or oxygen atom of primary arylamines, hydrazines, and N-hydroxylated metabolites | van Hogezand et al. (1992) and Deloménie et al. (2001) |
| Metronidazole benzodiazepines | Nitroreductase | Reduction of nitro group | Koch et al. (1979), Elmer and Remmel (1984), Takeno et al. (1990) |
| Sulfate ester hepatic metabolites | Sulfatases | Hydrolysis of sulfate esters utilizing formylglycine | Ulmer et al. (2014) and Koppel et al. (2017) |

(continued)

**Table 8.1** (continued)

| Interacting drug substrate | Microbiota-derived enzyme/ microbe | Hypothesized mechanisms | References |
|---|---|---|---|
| Levodopa, carbidopa, and entacapone | Tyrosine decarboxylases/*Enterococcus faecalis* and *Lactobacillus* | Bacteria contain tyrosine decarboxylases that can convert levodopa to dopamine | Van Kessel et al. (2019) |
| Levodopa | Molybdenum-dependent dehydroxylase/*Eggerthella lenta* | Molybdenum-dependent dehydroxylase conversion of dopamine to m-tyramine, which can cause *Eggerthella lenta* hypertensive crisis | Maini Rekdal et al. (2019) |
| Quinapril | Caecal esterase/ *Coprococcus comes* | Strong caecal esterase activities and catabolic capacity of quinapril | Yang et al. (2022) |

"bad" microbiota. For instance, the antimicrobial activity, linked to the depletion of "good" commensal gut microbiota, provides an impetus for the overgrowth of hazardous commensal bacteria. Antibiotic-microbial bacteriostatic interactions through enzyme-mediated mechanisms involve the inhibition of biosynthesis of the bacterial cell wall and protein, inhibition of nucleic acid metabolism, and interference in bacterial membrane activity. Interestingly, non-antibiotic drugs also exert some changes that affect the microbial population of the gastrointestinal tract (Table 8.2).

## 4 Final Remarks

The clinical consequence of all these drug-food types, nutrients, and target-based mechanisms of interactions varies, from being well tolerated to moderate, serious adverse, or possibly life-threatening events. Individual susceptibility to drug interactions readout is closely related to genetic variations. Understanding a patient's genetic profile can help predict their likelihood of experiencing drug interactions and guide personalized treatment decisions. Undoubtedly, timely identification of interactions involving drugs with or without food, herbs, and other associations is imperative to prevent potentially dangerous clinical outcomes.

**Table 8.2** Non-conventional antimicrobial drug-gut microbiota interactions and mechanisms

| Interacting drug | Microbes involved | Hypothesized mechanisms | References |
|---|---|---|---|
| Non-steroidal anti-inflammatory drugs such as acetylsalicylic acid, aspirin, diclofenac, and ibuprofen | *S. aureus*, *P. aeruginosa* | Anti-biofilm formation in vitro by blocking AgrA-regulated virulence, inhibiting hemolysis, hindering staphyloxanthin production, and downregulating the expression fnbA and icaA genes, which are crucial in biofilm formation in *S. aureus*; aspirin inhibits quorum-sensing in *P. aeruginosa* and salicylate reduces the production of extracellular polysaccharides required for biofilm formation | Hendricks et al. (2016), de Melo et al. (2018), and Paes Leme and da Silva (2021) |
| Bisphosphonates | *S. aureus* | Osteonecrotic microbial rinse of the jaws | Ganesan et al. (2023) |
| 5-Fluorouracil | *Escherichia coli* | Inhibits the enzyme thymidylate synthase which leads to the "thymineless death" of the bacteria | Patil et al. (2023) |
| Metformin | *Lactobacillus* | Disrupts the outer membrane of the bacteria which affects the permeability of the antibacterial agents; and reduces the inflammatory effect induced by bacteria | Silamiķele et al. (2021) |

# 5 Self-Assessment Questions and Answers

## 5.1 Questions

1. Which of these alternatives is incorrect on the drug and non-drug interactions concept?

   A. Drug-grapefruit juice interactions, which can inhibit drug metabolism enzymes and lead to increased drug concentrations; drug-dietary supplements interactions, which may potentiate or inhibit drug effects; and interactions between drugs and environmental factors such as sunlight, which can affect drug stability or metabolism.

   B. Interactions between drugs and substances other than pharmaceuticals are examples of non-drug interactions that occur only in the recipients.

   C. Pharmacodynamic interactions occur when two drugs with similar or opposing effects interact at the same receptor site, leading to enhanced or diminished effects.

   D. Examples of pharmacokinetic drug interactions include drugs that inhibit or induce cytochrome P450 enzymes, alter gastrointestinal pH, and compete for renal excretion, which could affect the metabolism, absorption, and elimination, respectively, of other drugs, food, or herbal products.

2. Identify the most likely incorrect affirmation

   A. Copper, lutein, vitamin C, vitamin E, and zinc all have been shown to reduce macular degeneration.
   B. The gut microbiota was referred to as a metabolic "organ" due to its significant enzymatic activities and chemical reactions that impact the bioavailability of drugs and foods.
   C. Factors that mediate the reciprocity of food-drug and drug-food interactions include food characteristics (food type/category, solubility, permeability, stability, physicochemical reactions), gastrointestinal processes (gastric emptying, enzyme secretion, intestinal motility, blood flow), and molecular targets.
   D. A high-fat diet always reduces the excretion of lipophilic drugs via the biliary route.

3. What are some potential outcomes of pharmacodynamic interactions?

   A. Changes in hormonal secretions.
   B. Blood clotting.
   C. Blood pressure.
   D. High volume of distribution.

4. All are correct descriptions of microbiota except.

   A. Microbiota refers to the community of microorganisms (bacteria, fungi, viruses, etc.) that inhabit various parts of the body, particularly the gut.
   B. Microbiota can metabolize drugs, herbal and dietary metabolism, or interact with the body's metabolic processes.
   C. Antibiotics can disrupt the balance of gut microbiota without alteration in drug efficacy or risk of side effects.
   D. The gut microbiota contributes significantly to nutrient metabolism, including the digestion and absorption of nutrients like carbohydrates, fats, and proteins. Microbiota-produced metabolites can also influence various physiological processes, such as immune function and neurotransmitter production.

## 5.2 *Answers*

1. The incorrect alternative is (B). This statement is inaccurate because drug and non-drug interactions are not limited to the recipients but can affect anyone exposed to the interacting substances. Therefore, the statement incorrectly implies a limitation that does not exist in the concept of drug and non-drug interactions.

2. The incorrect alternative is (D). A high-fat diet always reduces the excretion of lipophilic drugs via the biliary route.

   This statement is incorrect because a high-fat diet can increase the excretion of lipophilic drugs via the biliary route, thereby decreasing their bioavailability and systemic exposure.
3. The incorrect alternative is (D). A high volume of distribution is not a potential outcome of pharmacodynamic interactions; it is related to pharmacokinetic interactions.
4. The incorrect alternative is (C) Antibiotics can disrupt the balance of gut microbiota without alteration in drug efficacy or risk of side effects. Antibiotic use can indeed disrupt the balance of gut microbiota, leading to potential alterations in drug efficacy and increased risk of side effects.

## References

Abbaspour N, Hurrell R, Kelishadi R. Review on iron and its importance for human health. J Res Med Sci. 2014;19(2):164–74.

Badrinath M, Chen P, John S. Isoniazid toxicity. In: StatPearls [Internet]. Treasure Island (FL): StatPearls Publishing; 2024. Available from: https://www.ncbi.nlm.nih.gov/books/NBK531488/.

Bailey DG. Fruit juice inhibition of uptake transport: a new type of food-drug interaction. Br J Clin Pharmacol. 2010;70(5):645–55. https://doi.org/10.1111/j.1365-2125.2010.03722.x. PMID: 21039758; PMCID: PMC2997304.

Bailey DG, Spence JD, Edgar B, Bayleff CD, Arnold JM. Ethanol enchances the hemodynamic effects of felodipine. Clin Invest Med. 1989;12:237–62.

Bansal AB, Patel P, Al Khalili Y. Orlistat. In: StatPearls [Internet]. Treasure Island (FL): StatPearls Publishing; 2024. Available from: https://www.ncbi.nlm.nih.gov/books/NBK542202/.

Bekersky I, Dressler D, Mekki QA. Effect of low- and high-fat meals on tacrolimus absorption following 5 mg single oral doses to healthy human subjects. J Clin Pharmacol. 2001;41(2):176–82.

Birkett DJ. Pharmacokinetics made easy. Sydney: McGraw Hill Professional; 2002.

Briguglio M, Hrelia S, Malaguti M, Serpe L, Canaparo R, Dell'Osso B, Galentino R, De Michele S, Dina CZ, Porta M, Banfi G. Food bioactive compounds and their interference in drug pharmacokinetic/pharmacodynamic profiles. Pharmaceutics. 2018;10(4):277. https://doi.org/10.3390/pharmaceutics10040277. PMID: 30558213; PMCID: PMC6321138.

Bushra R, Aslam N, Khan AY. Food-drug interactions. Oman Med J. 2011;26(2):77–83. https://doi.org/10.5001/omj.2011.21. PMID: 22043389; PMCID: PMC3191675.

Caldwell J, Hawksworth GM. The demethylation of methamphetamine by intestinal microflora. J Pharm Pharmacol. 1973;25(5):422–4. https://doi.org/10.1111/j.2042-7158.1973.tb10043.x. PMID: 4146404.

Cao JJ, Johnson LK, Hunt JR. A diet high in meat protein and potential renal acid load increases fractional calcium absorption and urinary calcium excretion without affecting markers of bone resorption or formation in postmenopausal women. J Nutr. 2011;141:391–7. https://doi.org/10.3945/jn.110.129361.

Chen XW, Serag ES, Sneed KB, Liang J, Chew H, Pan SY, Zhou SF. Clinical herbal interactions with conventional drugs: from molecules to maladies. Curr Med Chem. 2011;18(31):4836–50. https://doi.org/10.2174/092986711797535317. PMID: 21919844.

Cheng L, Wong H. Food effects on oral drug absorption: Application of physiologically-based Pharmacokinetic modeling as a predictive tool. Pharmaceutics 2020;12(7):672. https://doi.org/10.3390/pharmaceutics12070672

Choi YH, Chin YW. Multifaceted factors causing conflicting outcomes in herb-drug interactions. Pharmaceutics. 2021;13(1):43. https://doi.org/10.3390/pharmaceutics13010043. Erratum in: Pharmaceutics. 2021 Mar 05;13(3): PMID: 33396770; PMCID: PMC7824553.

Chua LL, et al. Reduced microbial diversity in adult survivors of childhood acute lymphoblastic leukemia and microbial associations with increased immune activation. Microbiome. 2017;5:35.

Cortot A, Fleming CR, Malageleda JR. Improved nutrient absorption after cimetidine in short-bowel syndrome with gastric hypersecretion. N Engl J Med. 1979;300:79–80.

Czigle S, Nagy M, Mladěnka P, Tóth J, OEMONOM. Pharmacokinetic and pharmacodynamic herb-drug interactions-part I. Herbal medicines of the central nervous system. PeerJ. 2023;11:e16149. https://doi.org/10.7717/peerj.16149. PMID: 38025741; PMCID: PMC10656908.

de Melo GVP, de Oliveira BC, Magalhães Júnior AI, Thomaz-Soccol V, Soccol CR. How to select a probiotic? A review and update of methods and criteria. Biotechnol Adv. 2018;36(8):2060–76.

Dello Russo C, Bandiera T, Monici M, Surdo L, Yip VLM, Wotring V, Morbidelli L. Physiological adaptations affecting drug pharmacokinetics in space: what do we really know? A critical review of the literature. Br J Pharmacol. 2022;179(11):2538–57. https://doi.org/10.1111/bph.15822.

Deloménie C, Fouix S, Longuemaux S, Brahimi N, Bizet C, Picard B, et al. Identification and functional characterization of arylamine N-acetyltransferases in eubacteria: evidence for highly selective acetylation of 5-aminosalicylic acid. J Bacteriol. 2001;183:3417–27.

Deng J, Zhu X, Chen Z, Fan CH, Kwan HS, Wong CH, Shek KY, Zuo Z, Lam TN. A review of Food–Drug interactions on oral drug absorption. Drugs [Internet]. 2017;77(17):1833–55. Available from: https://doi.org/10.1007/s40265-017-0832-z.

Di YM, Li CG, Xue CC, Zhou SF. Clinical drugs that interact with St. John's wort and implication in drug development. Curr Pharm Des. 2008;14:1723–42. https://doi.org/10.2174/138161208784746798.

Díez R, García JJ, Diez MJ, Sierra M, Sahagun AM, Fernández N. Influence of Plantago ovata husk (dietary fiber) on the bioavailability and other pharmacokinetic parameters of metformin in diabetic rabbits. BMC Complement Altern Med. 2017;17(1):298. https://doi.org/10.1186/s12906-017-1809-x. PMID: 28592281; PMCID: PMC5463324.

Dodd D, Cann I. Tutorial: microbiome studies in drug metabolism. Clin Transl Sci. 2022;15(12):2812–37. https://doi.org/10.1111/cts.13416. Epub Sep 30. PMID: 36099474; PMCID: PMC9747132.

Drossman DA, Tack J, Ford AC, Szigethy E, Törnblom H, Van Oudenhove L. Neuromodulators for functional gastrointestinal disorders (Disorders of Gut−Brain interaction): a Rome Foundation Working Team report. Gastroenterology [Internet]. 2018;154(4):1140–1171.e1. Available from: https://doi.org/10.1053/j.gastro.2017.11.279.

Eason T, Ramirez G, Clulow AJ, Salim M, Boyd BJ. Revisiting the dissolution of praziquantel in biorelevant media and the impact of digestion of milk on drug dissolution. Pharmaceutics. 2022;14(10):2228. https://doi.org/10.3390/pharmaceutics14102228. PMID: 36297662; PMCID: PMC9609124.

Elmer GW, Remmel RP. Role of the intestinal microflora in clonazepam metabolism in the rat. Xenobiotica. 1984;14(11):829–40. https://doi.org/10.3109/00498258409151481. PMID: 6506755.

Ems T, St Lucia K, Huecker MR. Biochemistry, iron absorption. In: StatPearls [Internet]. Treasure Island (FL): StatPearls Publishing; 2023. 2024 Jan-. Available from: https://www.ncbi.nlm.nih.gov/books/NBK448204/.

Engelsen J, Nielsen JD, Winther K. Effect of coenzyme Q10 and Ginkgo biloba on warfarin dosage in stable, long-term warfarin treated outpatients. A randomised, double blind, placebo-crossover trial. Thromb Haemost. 2002;87(6):1075–6.

Ernst B, Magnani J. From carbohydrate leads to glycomimetic drugs. Nat Rev Drug Discov. 2009;8:661–77. https://doi.org/10.1038/nrd2852.

Ernst E, Pittler MH. Risks associated with herbal medicinal products. Wien Med Wochenschr. 2002;152(7–8):183–9. https://doi.org/10.1046/j.1563-258x.2002.01112.x. PMID: 12017745.

Fasinu PS, Bouic PJ, Rosenkranz B. An overview of the evidence and mechanisms of herb-drug interactions. Front Pharmacol. 2012;3:69. https://doi.org/10.3389/fphar.2012.00069. PMID: 22557968; PMCID: PMC3339338.

Fernández N, Diez MJ, Terán MT, García JJ, Calle AP, Sierra M. Influence of two commercial fibers in the pharmacokinetics of ethinylestradiol in rabbits. J Pharmacol Exp Ther. 1998;286:870–4.

Freed MI, Allen A, Jorkasky DK, DiCicco RA. Systemic exposure to rosiglitazone is unaltered by food. Eur J Clin Pharmacol. 1999;55:53–6.

Ganesan M, Kanimozhi G, Pradhapsingh B, Khan HA, Alhomida AS, Ekhzaimy A, Brindha GR, Prasad NR. Phytochemicals reverse P-glycoprotein mediated multidrug resistance via signal transduction pathways. Biomed Pharmacother. 2021;139:111632. https://doi.org/10.1016/j.biopha.2021.111632. Epub 2021 May 9. PMID: 34243600.

Ganesan K, Goyal A, Roane D. Bisphosphonate. In: StatPearls [Internet]. Treasure Island (FL): StatPearls Publishing; 2023. 2024 Jan–. PMID: 29262103.

Gin H, Orgerie MB, Aubertin J. The influence of guar gum on absorption of metformin from the gut in healthy volunteers. Horm Metab Res. 1989;21:81–3.

González Canga A, Fernández Martínez N, Sahagún Prieto AM, García Vieitez JJ, Díez Liébana MJ, Díez Láiz R, Sierra VM. Dietary fiber and its interaction with drugs. Nutr Hosp. 2010;25(4):535–9. English, Spanish. PMID: 20694287.

Greenblatt DJ, Patki KC, von Moltke LL, Shader RI. Drug interactions with grapefruit juice: an update. J Clin Psychopharmacol. 2001;21(4):357–9. https://doi.org/10.1097/00004714-200108000-00001. PMID: 11476118.

Gurevich KG. Effect of blood protein concentrations on drug-dosing regimes: practical guidance. Theor Biol Med Model. 2013;10:20. https://doi.org/10.1186/1742-4682-10-20. PMID: 23506635; PMCID: PMC3606132.

Hamed KM, Dighriri IM, Baomar AF, Alharthy BT, Alenazi FE, Alali GH, Alenazy RH, Alhumaidi NT, Alhulayfi DH, Alotaibi YB, Alhumaidan SS, Alhaddad ZA, Humadi AA, Alzahrani SA, Alobaid RH. Overview of methotrexate toxicity: a comprehensive literature review. Cureus. 2022;14(9):e29518. https://doi.org/10.7759/cureus.29518. PMID: 36312688; PMCID: PMC9595261.

Hassanzadeh K, Buccarello L, Dragotto J, Mohammadi A, Corbo M, Feligioni M. Obstacles against the marketing of curcumin as a drug. Int J Mol Sci. 2020;21(18):6619. https://doi.org/10.3390/ijms21186619. PMID: 32927725; PMCID: PMC7554750.

Heghes SC, Vostinaru O, Mogosan C, Miere D, Iuga CA, Filip L. Safety profile of nutraceuticals rich in coumarins: an update. Front Pharmacol. 2022;13:803338. https://doi.org/10.3389/fphar.2022.803338. PMID: 35140615; PMCID: PMC8818878.

Heimann G, Murgescu J, Bergt U. Influence of food intake on bioavailability of theophylline in premature infants. Eur J Clin Pharmacol. 1982;22:171–3.

Hendricks MR, Lashua LP, Fischer DK, Flitter BA, Eichinger KM, Durbin JE, Sarkar SN, Coyne CB, Empey KM, Bomberger JM. Respiratory syncytial virus infection enhances Pseudomonas aeruginosa biofilm growth through dysregulation of nutritional immunity. Proc Natl Acad Sci USA. 2016;113(6):1642–7. https://doi.org/10.1073/pnas.1516979113. Epub 2016 Jan 4. PMID: 26729873; PMCID: PMC4760822.

Heo G, Sun MH, Jiang WJ, Li XH, Lee SH, Guo J, Zhou D, Cui XS. Rotenone causes mitochondrial dysfunction and prevents maturation in porcine oocytes. PLoS One. 2022;17(11):e0277477. https://doi.org/10.1371/journal.pone.0277477. PMID: 36441709; PMCID: PMC9704683.

Heppner KM, Chaudhary N, Müller TD, Kirchner H, Habegger KM, Ottaway N, Smiley DL, Dimarchi R, Hofmann SM, Woods SC, Sivertsen B, Holst B, Pfluger PT, Perez-Tilve D, Tschöp MH. Acylation type determines ghrelin's effects on energy homeostasis in rodents.

Endocrinology. 2012;153(10):4687–95. https://doi.org/10.1210/en.2012-1194. Epub 2012 Aug 3. PMID: 22865372; PMCID: PMC3512022.

Herdiana Y, Wathoni N, Shamsuddin S, Muchtaridi M. Drug release study of the chitosan-based nanoparticles. Heliyon. 2021;8(1):e08674. https://doi.org/10.1016/j.heliyon.2021.e08674. PMID: 35028457; PMCID: PMC8741465.

Hojo Y, Echizenya M, Ohkubo T, Shimizu T. Drug interaction between St John's wort and zolpidem in healthy subjects. J Clin Pharm Ther. 2011;36:711–5. https://doi.org/10.1111/j.1365-27 10.2010.01223.x.

Hollander D. Retinol lymphatic and portal transport: influence of pH, bile, and fatty acids. Am J Phys. 1980;239(3):G210–4.

Huuponnen R, Seppälä P, Ilisalo E. Effect of guar gum, a fibre preparation, on digoxin and penicillin absorption in man. Eur J Clin Pharmacol. 1984;26:279–81.

Imai T, Ohura K. The role of intestinal carboxylesterase in the oral absorption of prodrugs. Curr Drug Metab. 2010;11:793–805.

Izzo AA. Herb–drug interactions: an overview of the clinical evidence. Fundam Clin Pharmacol. 2005;19:1–16. https://doi.org/10.1111/j.1472-8206.2004.00301.x.

Johnson BF, Rodin SM, Hoch K, Secar V. The effect of dietary fiber on the bioavailability of digoxin in capsules. J Clin Pharmacol. 1987;27:487–90.

Kanan M, Atif S, Mohammed F, Balahmar Y, Adawi Y, AlSaleem R, Farhan A, Alghoribi M, Mohammed S, Alshanbari R, Fahad M, Kallab R, Mohammed R, Alassaf D, Hazza A. A systematic review on the clinical pharmacokinetics of cephalexin in healthy and diseased populations. Antibiotics (Basel). 2023;12(9):1402. https://doi.org/10.3390/antibiotics12091402. PMID: 37760698; PMCID: PMC10526061.

Khalid W, Arshad MS, Jabeen A, Muhammad Anjum F, Qaisrani TB, Suleria HAR. Fiber-enriched botanicals: a therapeutic tool against certain metabolic ailments. Food Sci Nutr. 2022;10(10):3203–18. https://doi.org/10.1002/fsn3.2920. PMID: 36249968; PMCID: PMC9548355.

Kiani J, Imam SZ. Medicinal importance of grapefruit juice and its interaction with various drugs. Nutr J. 2007;6:33. https://doi.org/10.1186/1475-2891-6-33. PMID: 17971226; PMCID: PMC2147024.

Kim IS, Yoo DH, Jung IH, Lim S, Jeong JJ, Kim KA, et al. Reduced metabolic activity of gut microbiota by antibiotics can potentiate the antithrombotic effect of aspirin. Biochem Pharmacol. 2016;122:72–9.

Kirchner H, Gutierrez JA, Solenberg PJ, Pfluger PT, Czyzyk TA, Willency JA, Schürmann A, Joost HG, Jandacek RJ, Hale JE, Heiman ML, Tschöp MH. GOAT links dietary lipids with the endocrine control of energy balance. Nat Med. 2009;15(7):741–5. https://doi.org/10.1038/nm.1997. Epub 2009 Jun 5. Erratum in: Nat Med. 2009 Sep;15(9):1093. PMID: 19503064; PMCID: PMC2789701.

Kitano D, Chiku M, Li Y, et al. Miglitol improves postprandial endothelial dysfunction in patients with acute coronary syndrome and new-onset postprandial hyperglycemia. Cardiovasc Diabetol. 2013;12:9. https://doi.org/10.1186/1475-2840-12-92.

Koch RL, Chrystal EJT, Beaulieu BB, Goldman P. Acetamide—a metabolite of metronidazole formed by the intestinal flora. Biochem Pharmacol. 1979;28:3611–5.

Koppel N, Maini Rekdal V, Balskus EP. Chemical transformation of xenobiotics by the human gut microbiota. Science. 2017;356:eaag2770. https://doi.org/10.1126/science.aag2770.

Koziolek M, Alcaro S, Augustijns P, Basit AW, Grimm M, Hens B, Hoad CL, Jedamzik P, Madla CM, Maliepaard M, Marciani L, Maruca A, Parrott N, Pávek P, Porter CJH, Reppas C, Van Riet-Nales D, Rubbens J, Statelova M, Trevaskis NL, Valentová K, Vertzoni M, Čepo DV, Corsetti M. The mechanisms of pharmacokinetic food-drug interactions – a perspective from the UNGAP group. Eur J Pharm Sci. 2019;134:31–59. Available from: https://doi.org/10.1016/j.ejps.2019.04.003.

Laizure SC, Herring V, Hu Z, Witbrodt K, Parker RB. The role of human carboxylesterases in drug metabolism: have we overlooked their importance? Pharmacotherapy. 2013;33:210–22.

Lawrence XY, Li BV. FDA bioequivalence standards. Berlin: Springer; 2014.
Levy G, Jusko WJ. Factors affecting the absorption of riboflavin in man. J Pharm Sci. 1966;55:285–9.
Liang X, Bittinger K, Li X, Abernethy DR, Bushman FD, FitzGerald GA. Bidirectional interactions between indomethacin and the murine intestinal microbiota. elife. 2015:e08973. https://doi.org/10.7554/eLife.08973. PMID: 26701907; PMCID: PMC4755745.
Liel Y, Harman-Boehm I, Shany S. Evidence for a clinically important adverse effect of fiber-enriched diet on the bioavailability of levothyroxine in adult hypothyroid patients. J Clin Endocrinol Metab. 1996;81:857–9.
Lim GE, Li T, Buttar HS. Interactions of grapefruit juice and cardiovascular medications: a potential risk of toxicity. Exp Clin Cardiol. 2003;8(2):99–107. PMID: 19641658; PMCID: PMC2716207.
Madabushi R, Frank B, Drewelow B, Derendorf H, Butterweck V. Hyperforin in St. John's wort drug interactions. Eur J Clin Pharmacol. 2006;62:225–33. https://doi.org/10.1007/s00228-006-0096-0.
Maini Rekdal V, Bess EN, Bisanz JE, Turnbaugh PJ, Balskus EP. Discovery and inhibition of an interspecies gut bacterial pathway for levodopa metabolism. Science. 2019;364(6445):eaau6323. https://doi.org/10.1126/science.aau6323. PMID: 31196984; PMCID: PMC7745125.
Marchetti S, Mazzanti R, Beijnen JH. Concise review: clinical relevance of drug–drug and herb–drug interactions mediated by the ABC transporter ABCB1 (MDR1, P-glycoprotein). Oncologist. 2007;12:927–41. https://doi.org/10.1634/theoncologist.12-8-927.
McIver LA, Preuss CV, Tripp J. Acarbose. In: StatPearls [Internet]. Treasure Island (FL): StatPearls Publishing; 2024. Available from: https://www.ncbi.nlm.nih.gov/books/NBK493214/.
McLachlan A, Ramzan I. Meals and medicines. Aust Prescr. 2006;29(2):40.
McNamara PJ, Jewell RC, Jensen BK, Brindley CJ. Food increases the bioavailability of acitretin. J Clin Pharmacol. 1988;28(11):1051–5. 26.
Mikov M. The metabolism of drugs by the gut flora. Eur J Drug Metab Pharmacokinet. 1994;19:201–7.
Monks TJ, Caldwell J, Smith RL. Influence of methylxanthine-containing foods on theophylline metabolism and kinetics. Clin Pharmacol Ther. 1979;26:513–24.
Moore RA, Derry S, Wiffen PJ, Straube S. Effects of food on pharmacokinetics of immediate release oral formulations of aspirin, dipyrone, paracetamol and NSAIDs – a systematic review. Br J Clin Pharmacol. 2015;80(3):381–8. https://doi.org/10.1111/bcp.12628. Epub 2015 Jul 6. PMID: 25784216; PMCID: PMC4574824.
Niederberger E, Parnham MJ. The impact of diet and exercise on drug responses. Int J Mol Sci. 2021;22(14):7692. https://doi.org/10.3390/ijms22147692. PMID: 34299312; PMCID: PMC8304791.
Nieminen TH, Hagelberg NM, Saari TI, Neuvonen M, Laine K, Neuvonen PJ, Olkkola KT. St John's wort greatly reduces the concentrations of oral oxycodone. Eur J Pain. 2010;14:854–9. https://doi.org/10.1016/j.ejpain.2009.12.007.
O'Hara AM, Shanahan F. The gut flora as a forgotten organ. EMBO Rep. 2006;7(7):688–93. https://doi.org/10.1038/sj.embor.7400731. PMID: 16819463; PMCID: PMC1500832.
Paes Leme RC, da Silva RB. Antimicrobial activity of non-steroidal anti-inflammatory drugs on biofilm: current evidence and potential for drug repurposing. Front Microbiol. 2021;2:707629. https://doi.org/10.3389/fmicb.2021.707629. PMID: 34385992; PMCID: PMC8353384.
Pantuck EJ, Pantuck CB, Garland WA, et al. Stimulatory effect of Brussels sprouts and cabbage on human drug metabolism. Clin Pharmacol Ther. 1979;25:88–95.
Pantuck EJ, Pantuck CB, Anderson KE, Wattenberg LW, Conney AH, Kappas A. Effect of Brussels sprouts and cabbage on drug conjugation in humans. Clin Pharmacol Ther. 1984;35:161–9.
Patil M, Serhii K, Garzino F, Gobert Q, Giorgio S, Raimundo JM, Bolla JM, Camplo M. Synthesis and antimicrobial testing of 5-fluorouracil derivatives. Arch Pharm (Weinheim). 2023;356(7):e2300103. https://doi.org/10.1002/ardp.202300103. Epub 2023 May 18. PMID: 37199697.

Pires SMG, Reis RS, Cardoso SM, Pezzani R, Paredes-Osses E, Seilkhan A, Ydyrys A, Martorell M, Sönmez Gürer E, Setzer WN, Abdull Razis AF, Modu B, Calina D, Sharifi-Rad J. Phytates as a natural source for health promotion: a critical evaluation of clinical trials. Front Chem. 2023;11:1174109. https://doi.org/10.3389/fchem.2023.1174109. PMID: 37123871; PMCID: PMC10140425.

Piskin E, Cianciosi D, Gulec S, Tomas M, Capanoglu E. Iron absorption: factors, limitations, and improvement methods. ACS Omega. 2022;7(24):20441–56. https://doi.org/10.1021/acsomega.2c01833. PMID: 35755397; PMCID: PMC9219084.

Pressly MA, Schmidt S, Guinn D, Liu Z, Ceresa C, Samuels S, Madabushi R, Florian J, Fletcher EP. Informing a comprehensive risk assessment of infant drug exposure from human milk: application of a physiologically based pharmacokinetic lactation model for sotalol. J Clin Pharmacol. 2023;63(Suppl 1):S106–16. https://doi.org/10.1002/jcph.2242. PMID: 37317500.

Reuter SE, Evans AM, Shakib S, Lungershausen Y, Francis B, Valentini G, Bacchieri A, Ubben D, Pace S. Effect of food on the pharmacokinetics of piperaquine and dihydroartemisinin. Clin Drug Investig. 2015;35(9):559–67. https://doi.org/10.1007/s40261-015-0312-8. PMID: 26293519.

Richter WO, Jacob BG, Schwandt P. Interaction between fibre and lovastatin. Lancet. 1991;338(8768):706. https://doi.org/10.1016/0140-6736(91)91291-2. PMID: 1679514.

Rishavy MA, Hallgren KW, Wilson L, Singh S, Runge KW, Berkner KL. Warfarin alters vitamin K metabolism: a surprising mechanism of VKORC1 uncoupling necessitates an additional reductase. Blood. 2018;131(25):2826–35. https://doi.org/10.1182/blood-2017-09-804666. Epub 2018 Mar 28. PMID: 29592891; PMCID: PMC6014353.

Robertson DR, Higginson I, Macklin BS, Renwick AG, Waller DG, George CF. The influence of protein containing meals on the pharmacokinetics of levodopa in healthy volunteers. Br J Clin Pharmacol. 1991;31(4):413–7.

Rombolà L, Scuteri D, Marilisa S, Watanabe C, Morrone LA, Bagetta G, Corasaniti MT. Pharmacokinetic interactions between herbal medicines and drugs: their mechanisms and clinical relevance. Life (Basel). 2020;10(7):106. https://doi.org/10.3390/life10070106. PMID: 32635538; PMCID: PMC7400069.

Romiti N, Tramonti G, Donati A, Chieli E. Effects of grapefruit juice on the multidrug transporter P-glycoprotein in the human proximal tubular cell line HK-2. Life Sci. 2004;76(3):293–302. https://doi.org/10.1016/j.lfs.2004.06.015. PMID: 15531381.

Ryan A. Azoreductases in drug metabolism. Br J Pharmacol. 2017;174:2161–73.

Ryan KK, Seeley RJ. Physiology. Food as a hormone. Science. 2013;339(6122):918–9. https://doi.org/10.1126/science.1234062. PMID: 23430646; PMCID: PMC4240228.

Saitta KS, Zhang C, Lee KK, Fujimoto K, Redinbo MR, Boelsterli UA. Bacterial beta-glucuronidase inhibition protects mice against enteropathy induced by indomethacin, ketoprofen or diclofenac: mode of action and pharmacokinetics. Xenobiotica. 2014;44:28–35.

Sanderson JT, Hordijk J, Denison MS, Springsteel MF, Nantz MH, van den Berg M. Induction and inhibition of aromatase (CYP19) activity by natural and synthetic flavonoid compounds in H295R human adrenocortical carcinoma cell. Toxicol Sci. 2004;82:70–9. https://doi.org/10.1093/toxsci/kfh257.

Scepanovic P, et al. A comprehensive assessment of demographic, environmental, and host genetic associations with gut microbiome diversity in healthy individuals. Microbiome. 2019;7:130.

Sensoy I. A review on the food digestion in the digestive tract and the used in vitro models. Curr Res Food Sci. 2021;4:308–19. https://doi.org/10.1016/j.crfs.2021.04.004. PMID: 34027433; PMCID: PMC8134715.

Shah B, Jensen BK, Zhang J, Hunt T, Rohatagi S. Effect of food on pharmacokinetics of an inhaled drug: a case study with a mechanisms and clinical significance of food effect VLA-4 antagonist, HMR1031. J Clin Pharmacol. 2003;43(12):1341–9.

Shen X, Chen F, Wang F, Huang P, Luo W. The effect of grapefruit juice on the pharmacokinetics of tadalafil in rats. Biomed Res Int. 2020;2020:1631735. https://doi.org/10.1155/2020/1631735. PMID: 32047806; PMCID: PMC7003282.

Shima K, Tanaka A, Ikegami H, Tabata M, Sawazaki N, Kumahara Y. Effect of dietary fiber, glucomannan, on absorption of sulfonylurea in man. Horm Metab Res. 1983;15:1–3.
Silamiķele L, Silamiķelis I, Ustinova M, Kalniņa Z, Elbere I, Petrovska R, Kalniņa I, Kloviņš J. Metformin strongly affects gut microbiome composition in high-fat diet-induced type 2 diabetes mouse model of both sexes. Front Endocrinol (Lausanne). 2021;12:626359. https://doi.org/10.3389/fendo.2021.626359. PMID: 33815284; PMCID: PMC8018580.
Sostek MB, Chen Y, Andersson T. Effect of timing of dosing in relation to food intake on the pharmacokinetics of esomeprazole. Br J Clin Pharmacol. 2007;64(3):386–90. https://doi.org/10.1111/j.1365-2125.2007.02889.x.
Standl E, Schnell O. Alpha-glucosidase inhibitors 2012 – cardiovascular considerations and trial evaluation. Diab Vasc Dis Res. 2012;9:163–9.
Swaisland HC, Smith RP, Laight A, Kerr DJ, Ranson M, Wilder-Smith CH, et al. Single-dose clinical pharmacokinetic studies of gefitinib. Clin Pharmacokinet. 2005;44(11):1165–77.
Swanson HI. Drug metabolism by the host and gut microbiota: a partnership or rivalry? Drug Metab Dispos. 2015;43(10):1499–504. https://doi.org/10.1124/dmd.115.065714. Epub 2015 Aug 10. PMID: 26261284; PMCID: PMC4576677.
Takeno S, Nakagawa M, Sakai T. Teratogenic effects of nitrazepam in rats. Res Commun Chem Pathol Pharmacol. 1990;69:59–70.
Tang JC, Zhang JN, Wu YT, Li ZX. Effect of the water extract and ethanol extract from traditional Chinese medicines Angelica sinensis (Oliv.) Diels, Ligusticum chuanxiong Hort and Rheum palmatum L. on rat liver cytochrome P450 activity. Phytother Res. 2006;20:1046–51. https://doi.org/10.1002/ptr.1974.
Tolhurst G, Heffron H, Lam YS, Parker HE, Habib AM, Diakogiannaki E, Cameron J, Grosse J, Reimann F, Gribble FM. Short-chain fatty acids stimulate glucagon-like peptide-1 secretion via the G-protein-coupled receptor FFAR2. Diabetes. 2012;61(2):364–71. https://doi.org/10.2337/db11-1019. Epub 2011 Dec 21. PMID: 22190648; PMCID: PMC3266401.
Tremaroli V, Bäckhed F. 2012. Functional interactions between the gut microbiota and host metabolism. Nature. 2012;489(7415):242–9. https://doi.org/10.1038/nature11552. PMID: 22972297.
Tuomilehto J, Silvasti M, Manninen V, Uusitupa M, Aro A. Guar gum and gemfibrozil: an effective combination in the treatment of hypercholesterolaemia. Atherosclerosis. 1989;76:71–7.
Turley SD, Daggy BD, Dietschy JM. Effect of feeding psyllium and cholestiramine in combination on low density lipoprotein metabolism and fecal bile acid excretion in hamster with dietary-induced hypercholesterolemia. J Cardiovasc Pharmacol. 1996;27:71–9.
Ulmer JE, Vilén EM, Namburi RB, Benjdia A, Beneteau J, Malleron A, et al. Characterization of glycosaminoglycan (GAG) sulfatases from the human gut symbiont bacteroides thetaiotaomicron reveals the first GAG-specific bacterial endosulfatase. J Biol Chem. 2014;289:24289–303.
Uusi-Rasi K, Karkkainen MU, Lamberg-Allardt CJE. Calcium intake in health maintenance: a systematic review. Food Nutr Res. 2013;16:57. https://doi.org/10.3402/fnr.v57i0.21082.
Vaes LP, Chyka PA. Interactions of warfarin with garlic, ginger, ginkgo, or ginseng: nature of the evidence. Ann Pharmacother. 2000;34(12):1478–82. 60.
van Hogezand RA, Kennis HM, van Schaik A, Koopman JP, van Hees PA, van Tongeren JH. Bacterial acetylation of 5-aminosalicylic acid in faecal suspensions cultured under aerobic and anaerobic conditions. Eur J Clin Pharmacol. 1992;43:189–92.
van Kessel SP, Frye AK, El-Gendy AO, Castejon M, Keshavarzian A, van Dijk G, El Aidy S. Gut bacterial tyrosine decarboxylases restrict levels of levodopa in the treatment of Parkinson's disease. Nat Commun. 2019;10(1):310. https://doi.org/10.1038/s41467-019-08294-y. PMID: 30659181; PMCID: PMC6338741.
Varum FJ, Hatton GB, Basit AW. Food, physiology and drug delivery. Int J Pharm. 2013;457(2):446–60. https://doi.org/10.1016/j.ijpharm.2013.04.034. Epub 2013 Apr 21. PMID: 23612358.
Vlachojannis J, Cameron M, Chrubasik S. Drug interactions with St. John's wort products. Pharmacol Res. 2011;63:254–6. https://doi.org/10.1016/j.phrs.2010.11.011.

Walsh J, Griffin BT, Clarke G, Hyland NP. Drug-gut microbiota interactions: implications for neuropharmacology. Br J Pharmacol. 2018;175(24):4415–29. https://doi.org/10.1111/bph.14366. Epub 2018 Jun 28. PMID: 29782640; PMCID: PMC6255959.

Wang P, Hu XX, Li YH, Gao NY, Chen GQ, Chen JL. 2021. Inhibitory effect of resveratrol on the pharmacokinetics of ticagrelor in vivo and in vitro. Can J Physiol Pharmacol. 2021;99(8):821–6. https://doi.org/10.1139/cjpp-2020-0512. Epub 2021 Jan 5. PMID: 33400617.

Wanwimolruk S, Prachayasittikul V. Cytochrome P450 enzyme mediated herbal drug interactions (Part 1). EXCLI J. 2014;13:347–91. PMID: 26417265; PMCID: PMC4463967.

Welling PG, Barbhaiya RH. Influence of food and fluid volume on chlorothiazide bioavailability: comparison of plasma and urinary excretion methods. J Pharm Sci. 1982;71(1):32–5.

Wheless JW, Phelps SJ. A clinician's guide to oral extended-release drug delivery systems in epilepsy. J Pediatr Pharmacol Ther. 2018;23(4):277–92. https://doi.org/10.5863/1551-6776-23.4.277. PMID: 30181718; PMCID: PMC6117810.

Wu Y, Zhang Q, Ren Y, Ruan Z. Effect of probiotic Lactobacillus on lipid profile: a systematic review and meta-analysis of randomized, controlled trials. PLoS One. 2017;12(6):e0178868. https://doi.org/10.1371/journal.pone.0178868. PMID: 28594860; PMCID: PMC5464580.

Yang T, Mei X, Tackie-Yarboi E, Akere MT, Kyoung J, Mell B, Yeo JY, Cheng X, Zubcevic J, Richards EM, Pepine CJ, Raizada MK, Schiefer IT, Joe B. Identification of a gut commensal that compromises the blood pressure-lowering effect of ester angiotensin-converting enzyme inhibitors. Hypertension. 2022;79(8):1591–601. https://doi.org/10.1161/HYPERTENSIONAHA.121.18711. Epub 2022 May 10. PMID: 35538603; PMCID: PMC9278702.

Yeap YY, Trevaskis NL, Quach T, Tso P, Charman WN, Porter CJ. Intestinal bile secretion promotes drug absorption from lipid colloidal phases via induction of supersaturation. Mol Pharm. 2013;10(5):1874–89.

Yin OQP, Tomlinson B, Waye MMY. Pharmacogenetics and herb-drug interactions: experience with Ginkgo biloba and omeprazole. Pharmacogenetics. 2004;14:841–50. https://doi.org/10.1097/00008571-200412000-00007.

Zarezadeh M, Saedisomeolia A, Shekarabi M, Khorshidi M, Emami MR, Müller DJ. The effect of obesity, macronutrients, fasting and nutritional status on drug-metabolizing cytochrome P450s: a systematic review of current evidence on human studies. Eur J Nutr. 2021;60(6):2905–21. https://doi.org/10.1007/s00394-020-02421-y. Epub 2020 Nov 3. PMID: 33141242.

Zhang Z, Wong YN. Enzyme kinetics for clinically relevant CYP inhibition. Curr Drug Metab. 2005;6:241–57. https://doi.org/10.2174/1389200054021834.

Zheng P, et al. Gut microbial signatures can discriminate unipolar from bipolar depression. Adv Sci (Weinh). 2020;7:1902862.

Zhou S, Gao Y, Jiang W, Huang M, Xu A, Paxton JW. Interactions of herbs with cytochrome P450. Drug Metab Rev. 2003;35(1):35–98. https://doi.org/10.1081/DMR-120018248.

Zohoori FV. Chapter 1: nutrition and diet. Monogr Oral Sci. 2020;28:1–13. https://doi.org/10.1159/000455365. Epub 2019 Nov 7. PMID: 31940634.

# Chapter 9
# Clinical and Hypothetical Cases of Drug and Non-drug Interactions

**John I. Ogbu, Caroline Vitória de Lima Moreira, Adedunni W. Olusanya, and James O. Fajemiroye**

**Abstract** Drug and non-drug interaction involves co-administration of drugs with food, supplements, herbs, and alcohol, among others. These interactions occur in everyday life and are often not recognized. Although not all interactions are of clinical significance, the potential for increased morbidity and mortality from adverse outcomes and therapeutic failure is of concern. This chapter revisits drug and non-drug interactions from clinical and hypothetical perspectives to improve clinical decisions in therapeutics and drive research regarding non-drug interaction.

**Keywords** Food-drug · Interaction · Clinical outcome · Recommendation

## 1 Introduction

Drug and non-drug interactions (DNIs) could influence drug efficacy, nutritional benefits, and adverse effect profiles (Fig. 9.1). Food, supplements, herbs, alcohol, recreational drugs, and others may impact the activity of drugs when given concurrently. Drug-disease and drug-gene interactions also constitute non-drug interactions. Sometimes, these interactions may be beneficial in some instances; however, the high risk of fatalities underscores the need to review clinical experiences and hypothetical cases. Detecting DNIs involves routine inquiry during clinical evaluation about dietary habits and the use of non-pharmaceuticals. It is well known that the extent of drug and non-drug interactions varies widely among individuals, and it is difficult to predict those at risk of specific interactions.

J. I. Ogbu · C. V. de Lima Moreira · J. O. Fajemiroye (✉)
Institute of Biological Sciences, Federal University of Goias, Goiania, Goias, Brazil
e-mail: jamesfajemiroye@ufg.br

A. W. Olusanya
Department of Pharmacology, Therapeutics and Toxicology, College of Medicine, University of Lagos, Idi-Araba, Lagos, Nigeria
e-mail: dedunolusanya@unilag.edu.ng

 
J. O. Fajemiroye (ed.), *Fundamentals of Drug and Non-Drug Interactions*,
https://doi.org/10.1007/978-3-031-80107-5_9

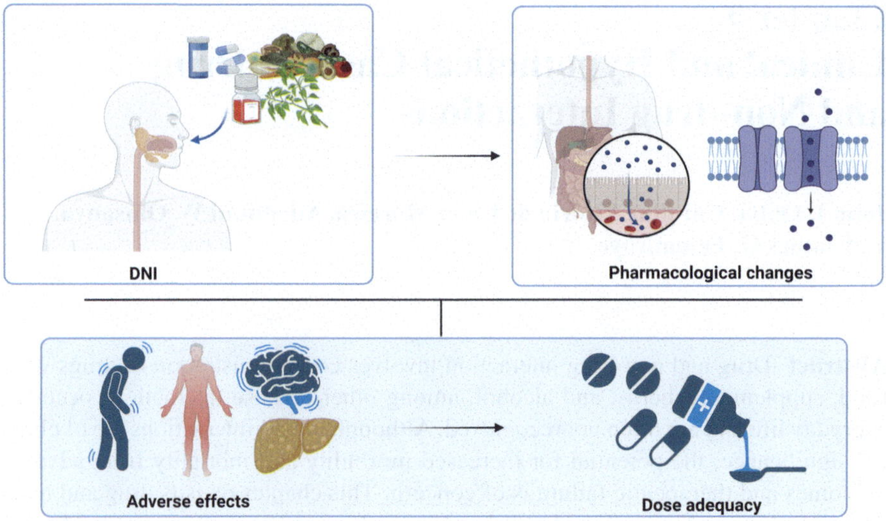


**Fig. 9.1** Clinically relevant interactions. DNI—drug-non-drug interaction

A person with diabetes often experiences significant stress. Stress can cause increased levels of cortisol and adrenaline, which can raise blood glucose levels. The person may have difficulty managing their blood sugar levels, leading to hyperglycemia. If not adequately managed, stress and hyperglycemic conditions might increase predisposition to non-drug interactions and subsequent adverse events. Sleep deprivation (a student pulls several all-nighters to study for exams, extended job shifts, etc.), which impairs cognitive function and concentration, can also alter the response to drug or nutritional interventions. In this chapter, we highlight clinically significant DNIs through drug-food, drug-supplement, drug-herb, drug–alcohol or recreational drugs, drug-disease, and drug-gene interactions.

## 2 Determining the Clinically Relevant Interactions

The assumption that a spectrum of diverse outcomes (negative, neutral, or positive) arises from non-drug interactions suggests that not all interactions are clinically relevant. Interactions are clinically significant when they affect a drug's efficacy or influence its adverse effect profile (Bushra et al. 2011; Deng et al. 2017). Such impact on drug efficacy may lead to enhanced pharmacological effects and a potential risk of increased adverse effects, possibly necessitating a reduction in drug dose. Conversely, interactions that reduce efficacy may be associated with reports of mild adverse effects and require an increase in drug dose to achieve the intended therapeutic effect. However, these scenarios are overly simplistic, as other contributing variables can influence treatment outcomes. For instance, it is well known that idiosyncratic adverse drug effects can occur from individual intrinsic factors, in which case dosage increases or reductions may not be beneficial.

In the clinical scenario, identifying DNIs can be challenging, especially in patients with multiple morbidities where the disease effect may mimic the adverse effects of drugs. These patients often take more than two drugs, further creating an exhaustive list of potential non-drug interactions that are difficult to pinpoint. Other major contributors to this challenge are the vague definition of clinically relevant interactions and the paucity of well-designed studies on DNIs. Available studies on DNIs are less emphatic on some interaction parameters and relevant factors. These investigations have encompassed single-dose studies evaluating the effect of food on drug bioavailability and studies examining drug administration at specific intervals over a short timeframe (Melander 1984). Findings from such investigations are difficult to translate into human studies due to several factors that influence a drug's pharmacokinetic and pharmacodynamic parameters. These factors include dosage duration, which is often longer in real life, dosage variation, dosing intervals, the concurrent administration of other food and non-drug substances, interindividual variations, etc. A sound pharmacological perspective and a thorough understanding of the disease process could help bridge the current gap in identifying DNIs.

To address this gap, computerized drug interaction surveillance systems have been developed to facilitate easy identification of potential interactions and clinically sound decisions in issues of drug interactions. However, their use has some limitations, including failure to detect all clinically significant drug interactions and non-identification of relevant details on risk factors for potentially adverse interaction (Van Roon et al. 2005). Thus, the interpretation of the computer-aided system needs to be done with caution, especially in patients with multimorbidity who are at a higher risk for drug and non-drug interactions.

Research indicates that bioavailability is significantly correlated with the clinical effectiveness of most drugs (Stielow et al. 2023). Potentially impacting bioavailability variables include hepatic drug-metabolizing enzymes, gastrointestinal metabolism, and drug transport. Dietary changes, herbal and supplement use, disease states, and genetic factors influencing metabolizing enzymes constitute non-drug-related interactions that may also affect bioavailability. Often, these factors interact with variables to influence pharmacokinetic parameters, leading to clinically significant non-drug interactions. Although a reduction in absorption can be clinically important if it leads to subtherapeutic serum levels (Hansten 2004), a single factor (e.g., a decrease in only the absorption rate of a drug) may not significantly impact the clinical outcomes of DNIs.

Pharmacodynamic parameters are also important determinants of clinically relevant drug and non-drug interactions. Although the effects of a drug are subservient to its pharmacokinetic state and profile, significant changes in the pharmacokinetic profile of a drug may not translate into clinical significance if the pharmacodynamics are unaffected. Such kinetic-independent outcomes may be answerable to factors such as the specificity of the drug's action site and its physicochemical properties. For example, the bioavailability of pravastatin is reduced by 31% when taken with food; however, since its lipid-lowering efficacy is unchanged, the interaction is not clinically important. This exception to the direct proportion between bioavailability and efficiency of the drug is attributed to the liver being the main target organ for the highly lipophilic statin. Thus, besides the pharmacokinetics and pharmacodynamics involved, the drug target and physical characteristics are important factors.

In addition to these pharmacological principles, drug regimen is also a determinant of clinically significant interactions (Ament et al. 2000). Drug regimen elements such as dosage, treatment schedule, and duration could determine the clinical importance of these interactions. This has not been fully explored in clinical studies since most studies are single dose under controlled conditions. This emphasizes why experimental conditions of drug-food interactions may not always be what is obtainable in clinical settings.

Drug-related adverse reactions are a major cause of morbidity, hospitalization, and death, and non-drug interaction is a major cause of these reactions. Unfortunately, this aspect is yet to be fully quantified. This emphasizes the need for high-quality pharmacotherapy and further research to mitigate unfavorable occurrences resulting from non-drug interaction (Van Roon et al. 2005). Therefore, becoming more familiar with potential drug interactions can help clinicians predict and explain a patient's drug response (Tom-Revzon and Adam 2006). Significant food effects complicate the development of new drugs, especially when clinical plans require control and monitoring of food intake to dosing. Predictions that a drug and its metabolites will impact food effects pose challenges (Lentz 2008). Clinically relevant drug-food interactions have been reported with several classes of drugs and food, and these could influence pharmacokinetic and pharmacokinetic measures. The following sections will discuss various examples of potential non-drug interactions.

## 3 Categories of Interaction

### *3.1 Drug-Food Interactions*

Several drug types, including antitubercular, anti-gout, antidepressant, and anti-inflammatory drugs, are implicated in clinical cases of drug-food interactions (Fig. 9.2). Antitubercular drugs are traditionally taken early in the morning on an empty stomach to reduce the potential for interactions that can influence absorption and, ultimately, the bioavailability of drugs and treatment efficacy. A prospective randomized cross-over pharmacokinetic study conducted in treatment-naive adults with drug-susceptible tuberculosis showed that food intake decreased the absolute bioavailability and maximum concentration of isoniazid (Saktiawati et al. 2016). An earlier study reported that exposure to isoniazid in healthy volunteers is reduced when dosed with meals (Peloquin et al. 1999). In addition, rifampicin, another antitubercular medication, has shown clinically significant interactions with food. A

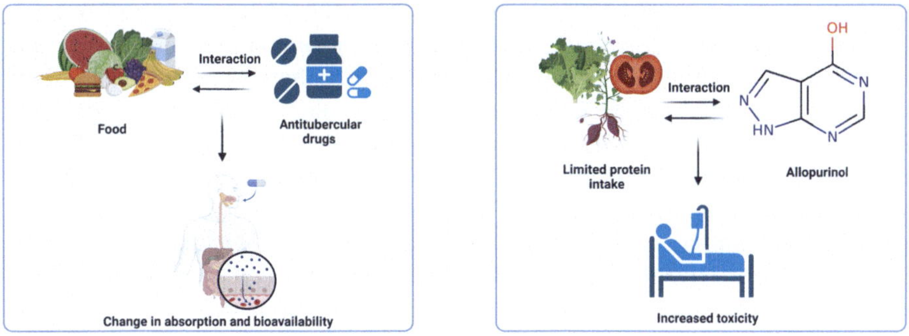


**Fig. 9.2** Drug-food interactions

study involving eight undernourished adults without tuberculosis found that rifampicin's peak plasma concentration and protein binding were notably decreased.

Adverse interactions with nutrients may occur when using anti-gout medications in cases of reduced protein levels, which can have significant clinical implications (Berlinger et al. 1985). Allopurinol, an anti-gout medication, is administered to treat gout and prevent hyperuricemia and its complications in cancer patients undergoing therapy (Pantuck et al. 1984). In a series of small cross-over studies of normally nourished men, clearances of inulin, creatinine, and oxypurinol were reduced on the low protein diet treatment (0 or 0.3 g/kg/day) versus higher protein intake (1.5–3 g/kg/day) (Berlinger et al. 1985; Park et al. 1987; Kitt et al. 1989). Authors suggested that outcomes are likely to result in an increased risk of toxicity with allopurinol use during periods of limited protein intake.

Monoamine oxidase inhibitors (MAOIs) are indicated for depression. Although older types of MAOIs are effective for treating depressive disorders, they are not widely used in clinical practice due to concerns about interactions with tyramine-containing foods such as matured cheese, red wine, ripe bananas, yogurt, shrimp paste, and salami. MAOIs interaction with these foods can trigger a condition known as the "cheese reaction," which causes a dangerous spike in blood pressure that leads to a hypertensive crisis in patients (Volz and Gleiter 1998). Previous studies have documented the presence of headaches, palpitations, subarachnoid hemorrhage, intracranial hemorrhage, cardiac arrhythmias, cardiac failure, pulmonary edema, and death in these patients (Sathyanarayana Rao and Yeragani 2009).

Another example of clinically relevant food-drug interaction is the interaction between levodopa and protein meals. Levodopa (an amino acid that crosses the blood-brain barrier via amino acid decarboxylase transporter) is used for the management of motor symptoms of Parkinsonism. Amino acids derived from protein-containing meals reduce the drug's delivery into the central nervous system (CNS) by competing with levodopa for the amino-acid transporter. To prevent this food-drug interaction, patients are advised to take their drug an hour or more before, and 2 or more hours after eating. If nausea occurs with a drug, a non-protein diet is preferred with drug administration (Ahlskog 2011).

In the clinical case of interaction between dexamethasone and grapefruit juice, patient ST [age: 55 years; gender: male; medical history: chronic obstructive pulmonary disease (COPD), recently diagnosed with asthma exacerbations; current medications: dexamethasone (steroidal anti-inflammatory), albuterol (a bronchodilator for asthma)] started drinking grapefruit juice daily, believing it to be a healthy choice and beneficial for his overall well-being. ST returns to his pulmonologist with complaints of increased side effects from dexamethasone, including severe gastrointestinal upset, mood swings, and unanticipated weight gain. Despite following the prescribed treatment plan, he has been experiencing more pronounced side effects and inadequate symptom control. Grapefruit juice inhibits certain cytochrome P450 enzymes, particularly CYP3A4, which are involved in the metabolism of medications, including corticosteroids like dexamethasone. The inhibition of CYP3A4 can lead to increased blood levels of dexamethasone by reducing its metabolism. This interaction can enhance the drug's effects and side effects, leading to an increased risk of adverse reactions such as gastrointestinal issues, mood swings, and weight gain. Dietary modification and medication review should help in reducing the elevated levels of the medication and mitigate side effects.

## *3.2 Drug-Supplement*

Cases of drug-induced nutrient depletions in hospitalized patients have been reported (Jung et al. 2022). Adverse food-drug interactions occur between drugs and substances including vitamin A, vitamin D, vitamin E, vitamin K, vitamin B1, vitamin B3, vitamin B6, vitamin B12, and folic acid (Table 9.1). Drugs such as methotrexate, phenytoin, sulfasalazine, and trimethoprim can antagonize folate utilization and inhibit its absorption or conversion to its active form resulting in folate deficiency (Khan and Jialal 2024). These reactions are more serious in chronically ill patients and patients with marginal nutritional status. The biochemical, functional, or clinical manifestation depends on the deficit degree, drug dose, and duration of use.

The potential risk of hepatotoxicity has also been documented in cancer patients, for example, the combination of vitamin A or B3 with either simvastatin or paracetamol increases the risk of hepatotoxicity (Fig. 9.3). Similarly, vitamins D and E reduce the levels of simvastatin and omeprazole via CYP3A4 enzyme induction (Lam et al. 2023). In the clinical case of interaction between chemotherapy and *Echinacea*, a patient, VC (age: 55 years; gender: female, medical history: breast cancer) undergoing chemotherapy [current medications: doxorubicin and cyclophosphamide (chemotherapy drugs), and ondansetron (antiemetic for chemotherapy-induced nausea)] resort to taking *Echinacea* (a popular herbal supplement believed to boost immune function and help with infection prevention) without consulting her oncologist. A few weeks later, she presented with symptoms of increased fatigue and frequent infections. Her blood tests show a significant drop in white blood cell count (neutropenia), which is concerning for her ongoing chemotherapy. Laboratory

**Table 9.1** Clinical evidence of drug-vitamin interactions with clinical management strategies

| Vitamin | Complicating drugs | Clinical studies/cases | Pharmacokinetic/pharmacodynamic mechanisms | Diagnosis | Clinical management strategy/recommendations | References |
|---|---|---|---|---|---|---|
| A | Orlistat, cholestyramine | A meta-analysis involving 10,631 participants from 16 trials (average body mass index (BMI): 36.3 kg/m$^2$)<br>A 4-year randomized clinical trial | Decreased absorption of vitamin A results in a reduced serum concentration and potentially a decrease in efficacy. | N/A | Daily intake of multivitamin supplements containing fat-soluble vitamins is recommended at least 2 h before or after orlistat administration | Padwal et al. (2003)<br>Torgerson et al. (2004)<br>Genentech (2015) |
| D | Corticosteroids | Cross-sectional National Health and Nutrition Examination Survey participants 2001–2006 ($n$ = 22,650) | Decreased bile salt formation and fat malabsorption<br>Direct binding and impaired absorption. | Decreased serum 25-hydroxyvitamin D<br>Increased risk for rickets<br>Serum levels of calcitriol | Limit the dose of cholestyramine<br>Vitamin D supplements<br>Allow 1–4 h between the cholestyramine dose and administering vitamin D27 | Skversky et al. (2011)<br>Gotfredsen et al. (2001)<br>Kulak et al. (2004)<br>Hoogwerf et al. (1992) |
| | Orlistat | Randomized clinical trial (30 obese subjects with a mean BMI of 47 kg/m$^2$) | | | | |
| | Cholestyramine | 268 men aged 42–68 years (cholestyramine vs. placebo for 7 ~ 10 years) | | | | |
| | Phenytoin, phenobarbital, carbamazepine | Cross-sectional (58 patients on antiepileptic therapy) | Cytochrome P450-induced increased degradation of vitamin D | Decreased serum 25-hydroxyvitamin D. Increased bone alkaline phosphatase | Vitamin D supplements in at-risk patients. | Sheth (2004)<br>Mintzer et al. (2006) |
| E | Cholestyramine | N/A | Decreased absorption of vitamin E | Serum α-tocopherol/total lipids (cholesterol + triacylglycerols + phospholipids) ratio < 0.8 mg/dl<br>Serum α -tocopherol/cholesterol ratio < 2.47 mg/dl | Limit the dose of cholestyramine<br>Vitamin E supplements<br>Allow 1–4 h between cholestyramine dose and administering vitamin E. | Horwitt et al. (1972)<br>Thurnham et al. (1986) |

(continued)

**Table 9.1** (continued)

| Vitamin | Complicating drugs | Clinical studies/cases | Pharmacokinetic/pharmacodynamic mechanisms | Diagnosis | Clinical management strategy/recommendations | References |
|---|---|---|---|---|---|---|
| K | Cholestyramine, Orlistat | Randomized clinical trial (patients aged 30–60 years with BMI ≥30 kg/m$^2$ and administered orlistat 120 mg three times daily for 4 years) | Decrease the absorption of vitamin K | Coagulopathy<br>Prolonged prothrombin time and INR | Vitamin K supplements | Bhat and Deshmukh (2003)<br>Alperin (1987) |
| | Neomycin, broad-spectrum antibiotics | Randomized clinical trial (critically ill children who received antibiotic therapy for a minimum of 14 days)<br>Nested case-control (patients aged ≥20 years with or without a hemorrhagic event after using cephalosporins) | Eradication of vitamin K-synthesizing bacteria | | | |
| B1 (Thiamine) | Loop diuretics such as furosemide, bumetanide, ethacrynic acid, and torsemide. | Prospective cohort (50 patients >18 years of age with an intensive care unit stay of at least 48 h) | Increased diuretic-induced elimination of the vitamin<br>Furosemide-induced reduction in intracellular thiamine levels by decreased cellular uptake or intestinal absorption of thiamine. | Mean serum thiamine levels in the baseline<br>Worsening congestive heart failure<br>Erythrocyte transketolase activity<br>Red blood cell thiamine concentrations<br>Neuropsychiatric syndromes including Beriberi or Wernickés encephalopathy for fluorouracil-induced thiamine deficiency | Use alternative diuretic (e.g., spironolactone)<br>Thiamine supplements | Gundogan et al. (2019)<br>DiNicolantonio et al. (2013)<br>Hanninen et al. (2006)<br>Rocha et al. (2008)<br>Isenberg-Grzeda et al. (2016) |
| | Fluorouracil | Retrospective chart review [18 patients developed Wernicke-Korsakoff Syndrome during cancer treatment (5 of whom were taking fluorouracil)] | Increased metabolism of thiamine | | | |

| | | | | | | |
|---|---|---|---|---|---|---|
| B3 (Niacin) | Isoniazid, pyrazinamide | Case report (69-year-old female with cardiac, lung, and cutaneous sarcoidosis) | Interference with the production of endogenous niacin due to the structural similarity between those drugs and vitamin B3<br>Isoniazid depletes niacin by inhibiting the intestinal absorption of the vitamin. | Pellagra, a multi-system disorder caused by niacin deficiency, has been reported in patients receiving Isoniazid and/or pyrazinamide.<br>Decreased serum nicotinic acid levels (4.5 μg/dL) after isoniazid therapy. | Niacin-enriched foods<br>Close monitoring of patients with a high risk of drug-induced niacin deficiency | Prabhu et al. (2021)<br>Tokoro et al. (2018) |
| B6 | Isoniazid | Single-arm clinical trial (20 patients with pulmonary tuberculosis administered the isoniazid among other treatments for 1 week) | Impaired niacin synthesis secondary to pyridoxine deficiency<br>Competitive inhibition of niacin Coenzymes | Plasma levels of pyridoxal phosphate<br>Prevalence of low vitamin B6 ($<5$ ng/mL)<br>Mean plasma and erythrocyte pyridoxal 5'-phosphate levels<br>Pellagra, Photosensitive dermatitis, extensive acne-like rash<br>Manic dementia and diarrhea | Discontinue isoniazid<br>Supplement with nicotinamide<br>Cautious use of isoniazid in populations with high percentage of slow isoniazid acetylators | Visser et al. (2004)<br>Mintzer et al. (2012)<br>Yasuda et al. (2022)<br>Ubbink et al. (1990)<br>Cohen et al. (1974)<br>Muratake et al. (1999)<br>Meyrick Thomas et al. (1981) |
| | Antiepileptic drugs | Cross-sectional (patients with epilepsy) | | | | |
| | Levodopa | Case report (a 75-year-old man with advanced. Parkinson's disease was administered Levodopa/carbidopa 200/20 mg/day for 45 months) | | | | |
| | Theophylline | Single-arm clinical trial (7 healthy males with 15 weeks of theophylline treatment) | | | | |

(continued)

**Table 9.1** (continued)

| Vitamin | Complicating drugs | Clinical studies/cases | Pharmacokinetic/pharmacodynamic mechanisms | Diagnosis | Clinical management strategy/recommendations | References |
|---|---|---|---|---|---|---|
| B12 | Proton pump inhibitors and histamine (H2) blockers | Meta-analysis (four case-control studies and one observational study) | Lack of gastric acid impairs the release of vitamin B12 from dietary proteins | Decreased plasma vitamin B12<br>Increased plasma methylmalonic acid | Monitor B12 status<br>Need to justify prolonged gastric acid suppression therapy<br>B12 replacement therapy | de Jager et al. (2010)<br>Aroda et al. (2016)<br>Ruscin et al. (2002)<br>Snow (1999) |
| | Metformin | Randomized clinical trial (patients with type 2 diabetes under insulin treatment) | Although still unclear, it is hypothesized that metformin interferes with calcium-dependent membrane action responsible for vitamin B12 intrinsic factor absorption in the terminal ileum. | Percentage change in vitamin B12 from baseline.<br>Incidence of vitamin B12 deficiency (≤203 pg/mL) | | |
| B9 (folic acid) | Methotrexate | Randomized clinical trial (113 patients with rheumatoid arthritis) | Decreased availability of folate secondary to inhibition of dihydrofolate reductase | Plasma homocysteine and folate levels<br>Megaloblastic.<br>Anemia | Folic acid supplementation | Van Ede et al. (2002)<br>Linnebank et al. (2011) |
| | Sulfonamides | Cross-sectional (23 outpatients with ulcerative colitis treated with sulfasalazine vs. 20 healthy controls) | | Blood concentrations of folate<br>Megaloblastic anemia | | |
| | Antiepileptic drugs such as carbamazepine, gabapentin, oxcarbazepine, phenytoin, primidone, or valproate | Prospective antiepileptic drug-treated patients ($n = 2730$) vs. AED-untreated patients with epilepsy ($n = 170$) vs. healthy controls ($n = 200$) | Inhibited absorption of pteroylpolyglutamate | Changes in mean serum folate level<br>Megaloblastic anemia | | |

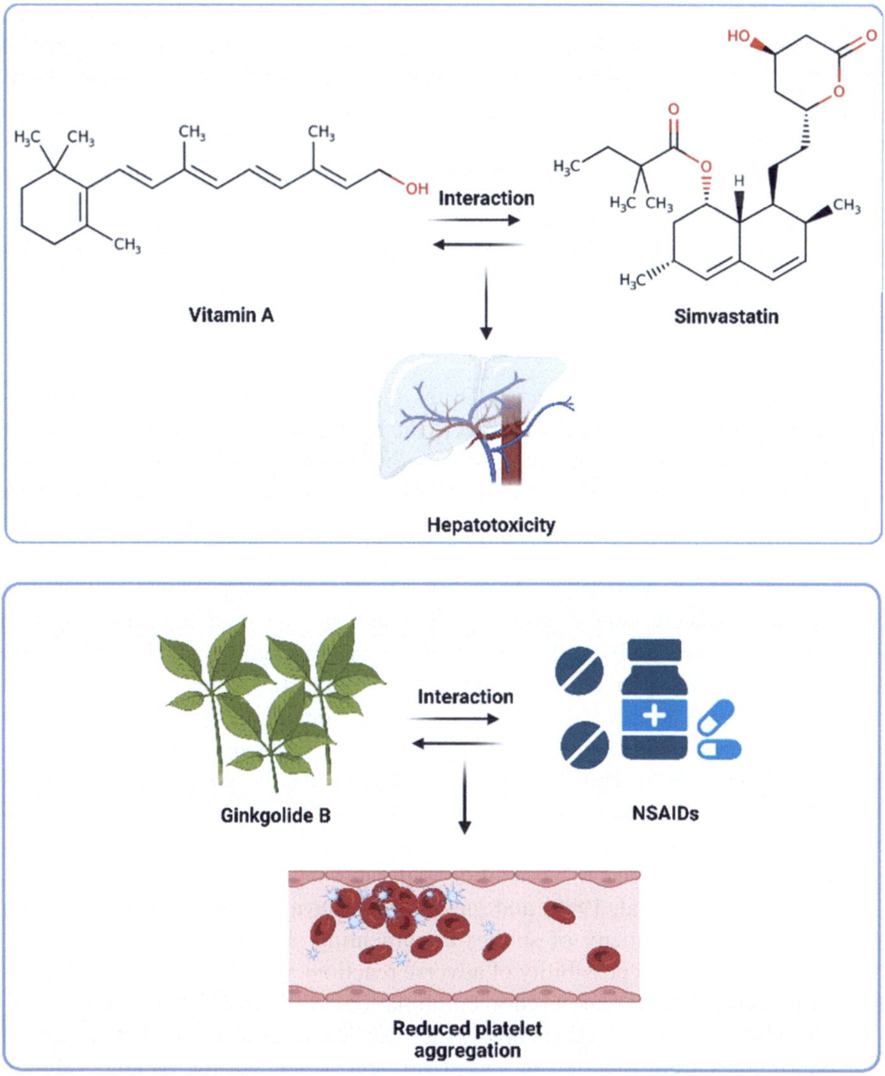


**Fig. 9.3** Supplements and herbs interaction

tests showed that white blood cell (WBC) count has decreased significantly from a baseline of 4500 cells/μL to 2500 cells/μL, indicating neutropenia. *Echinacea* is known to have immune-modulating effects. While it is often used to enhance immune function, it can also potentially interfere with the efficacy and metabolism of chemotherapy drugs. Specifically, *Echinacea* can induce the enzyme CYP3A4, which might alter the metabolism of some chemotherapy drugs. Additionally, *Echinacea* might affect the bone marrow, further exacerbating chemotherapy-induced myelosuppression (reduction in blood cell production). Immediate discontinuation of *Echinacea* with appropriate supportive care for her neutropenia, which may include growth factors to stimulate white blood cell production and antibiotics to prevent or treat infections, is essential. Additionally, optimal adjustment of the chemotherapy regimen, education about the potential interactions between herbal supplements and chemotherapy, and careful monitoring to track recovery from neutropenia and infection signs are crucial steps in managing this case.

### *3.3 Drug-Herb*

The Nonsteroidal Anti-Inflammatory Drugs (NSAIDs) interaction with herbs has been documented to be a potential threat to the health status of patients. Aspirin, for example, interferes with the clotting process by inhibiting thromboxane synthesis. Consumption of antiplatelet herbal supplements, especially by the elderly, along with aspirin has backfired clinically as the prostaglandin-inhibiting salicylate removes the gastric mucosal cytoprotective effect of prostaglandins leading to gastric bleeding (Shorr et al. 1993). In addition, ginkgo herbal products indicated in conditions associated with cognitive dysfunction, such as dementia, have adverse reactions to analgesics. Its ginkgolides (particularly ginkgolide B) have also been shown to inhibit the platelet-activating factor binding to its receptors on platelet membranes, resulting in reduced platelet aggregation. This explains the bleeding reported in patients taking ginkgo with and without concomitant administration of NSAIDs (Abebe 2002). This is confirmed in a clinical case report of a 70-year-old man who was given aspirin while taking ginkgo extracts and presented with spontaneous intraocular bleeding in his eyes (Rosenblatt and Mindel 1997). Cases of antiplatelet aggregation effects in association with other natural products such as garlic (Bordia 1978; Rose et al. 1990) and ginger (Srivastava 1984) have been reported. Although there is a paucity of studies documenting significant interactions with NSAIDs, there is a high possibility of adverse reactions with NSAIDs, and this calls for caution when NSAIDs are used in combination with these herbal drugs.

CNS-related adverse reactions involving NSAIDs, valerian (an herbal drug), and benzodiazepines were reported in a post-surgical patient; in the patient, valerian withdrawal appeared to mimic the withdrawal syndrome of benzodiazepines after the patient presented with delirium and cardiac complications following surgery (Duke 2000). Due to its mimicry of benzodiazepines, it is predicted that the herb will potentiate the sedative effects of opioids, and thus, caution is advised if these are used in combination. Another example of drug-herb interaction is the combination of HIV protease inhibitor with St. John's wort, an herbal drug known to induce CYP3A4, which metabolizes most protease inhibitors. The concomitant administration of St. John's wort with protease inhibitors could result in the induction of CYP3A4, increased metabolism, and subtherapeutic levels of the protease inhibitor which may lead to therapeutic failure.

The importance of considering potential interactions between herb and prescription medications, and the need for thorough patient education and monitoring to ensure safe and effective treatment can be illustrated in the case of John Hornwith (JH). Lately JH [age: 58 years; gender: male; medical history: hypertension, atrial fibrillation; current medications: warfarin (an anticoagulant for atrial fibrillation) and lisinopril (an angiotensin-converting enzyme inhibitor for hypertension)] reported depressive symptoms and decided to start taking St. John's Wort (an antidepressant herbal supplement) without informing his healthcare provider. Over the next few weeks, he experiences unusual symptoms such as increased bruising and bleeding from minor cuts. He also notices a change in the color of his urine and reports feeling more fatigued. Laboratory tests showed that the international normalized ratio (INR) was initially well-controlled within the therapeutic range of 2.0–3.0. After starting St. John's Wort, his INR drops to 1.5, indicating a decreased anticoagulant effect. St. John's Wort is known to induce the enzyme CYP3A4 and the P-glycoprotein transporter in the liver. This induction increases the metabolism of warfarin, reducing its blood levels and efficacy. The reduced INR suggests that St. John's Wort's anticoagulant effect is diminished with an increased risk of clot formation, bruising, and bleeding. In this case, St. John's Wort discontinuation, warfarin dose adjustment, continued monitoring of its INR levels, and education about potential interactions between herbal supplements and prescription medications are essential in managing drug-herb interactions.

## *3.4 Drug–Alcohol and Other Recreational Drugs*

Alcohol is metabolized by the CYP2E1 system, which also metabolizes other drugs like pain medication acetaminophen, the antibiotic isoniazid, and the barbiturate phenobarbital. Chronic heavy alcohol consumption induces the activity of CYP2E1 up to tenfold, leading to increased metabolism of drugs and reduced efficacy in alcoholics in the sober state (Lieber 1994). The reverse effect is, however, noticed during alcohol intoxication because the alcohol ingested competes with the medication for metabolism by CYP2E1, leading to decreased metabolic rates with the potential for significant adverse effects (Weathermon and Crabb 1999). The use of alcohol is often discouraged in people on antidiabetic drugs because of its potential to cause hypoglycemia because it suppresses hepatic gluconeogenesis.

Alcohol is known to potentiate the depressive effect of drugs acting on the CNS. A combination of alcohol with tricyclic antidepressants such as amitriptyline, doxepin, maprotiline, and trimipramine drugs result in significant sedation, causing problems with the performance of activity. A similar effect has been documented with the use of benzodiazepines, barbiturates, and different classes of antipsychotics (Simmons and Cupp 1998). Severe respiratory depression, which may be fatal, has been observed when ketamine (an antidepressant) is used in combination with alcohol. Pharmacokinetic interactions leading to reduced metabolism of amitriptyline have been described with a resultant increase in drug levels and adverse effects such as arrhythmia and reduced seizure threshold, which may be life-threatening (Weathermon and Crabb 1999). For these reasons, a physician's advice is necessary for dietary recommendations or guidelines about the simultaneous consumption of alcohol with folic acid, as excess of the former reduces absorption of the vitamin and eliminates it more quickly (Caudill 2010).

A combination of other recreational drugs with medications has also been associated with adverse consequences. One of the common presentations is hypertensive crises, which may result in myocardial infarction, intracranial bleeding, and seizures, which are seen when some classes of antidepressant drugs are combined with substances like cocaine, amphetamines, marijuana, and methylenedioxymethamphetamine. For example, a combination of any of these drugs with MAOIs will cause dangerous hypertensive crises. Interactions between selective serotonin reuptake inhibitors like citalopram and cocaine increase the risk of hypertensive crises and intracranial bleeding.

In a clinical case of benzodiazepine-alcohol-marijuana interaction involving a patient JV [age: 45 years; gender: female; medical history: generalized anxiety disorder, insomnia; current medications: diazepam 10 mg daily (a benzodiazepine for anxiety and sleep), lorazepam 1 mg as needed for severe anxiety]. JV has been using diazepam and lorazepam as prescribed. In addition to her medications, she occasionally drinks alcohol to relax and has started using marijuana to help with sleep and anxiety. JV presents to the emergency department with symptoms of severe drowsiness, confusion, and difficulty breathing. Her family reports that she has been increasingly lethargic over the past few weeks and has experienced several near-fainting episodes. Laboratory tests showed elevated blood alcohol, diazepam, and lorazepam levels, and a urine toxicology screen was positive for marijuana. Combining benzodiazepines with alcohol and marijuana significantly increases the risk of central nervous system (CNS) depression. Both alcohol and marijuana potentiate the sedative effects of benzodiazepines, leading to enhanced drowsiness, impaired motor function, and respiratory depression. This interaction can be life-threatening, as the compounded CNS depression can lead to severe respiratory issues or even coma. Immediate hospitalization for observation and supportive care, medication review with a focus on reducing dosages or switching to a less sedative medication, substance counseling to address her alcohol and marijuana use and its interaction with her medications, and education about the dangers of combining benzodiazepines with alcohol and recreational drugs are essential management measures.

## *3.5 Drug-Disease*

Macro- and micronutrient deficiencies have been reported to affect the activities of commonly used drugs like analgesics and antimicrobials. Analgesics, such as acetaminophen, salicylates, non-salicylates, and opioids, are extensively used in medical and dental practices. Both nutrient deficiency and nourished states influence the metabolism of acetaminophen. A study on children with severe protein-energy malnutrition (PEM) found that the biotransformation of acetaminophen was reduced, leading to a prolonged half-life and decreased elimination, increasing the risk of toxicity (Mehta et al. 1982). Clinical monitoring is recommended for affected patients due to potential interactions that may elevate toxicity levels. In contrast, adults with milder PEM did not show greater acetaminophen toxicity compared to those with normal nutritional status, even when administered with vitamin C (Walter-Sack and Klotz 1996). In another study, salicylate ototoxicity was enhanced with magnesium and zinc deficiencies (Gunther et al. 1988).

PEM is a common problem worldwide (Grover and Ee 2009) and is associated with an increased risk of infections; hence, antimicrobial use is common in this group of people (Trehan et al. 2013). Several clinical cases have reported antibiotic interaction with decreased levels of macro-nutrients (Table 9.2). A study reported a threefold relative increase in hepatotoxicity from isoniazid use in a cohort of 130 children with PEM compared to normally nourished controls in a longitudinal study over 3.5 years (Mehta 1990), suggesting increased adverse effects in malnourished states.

Another study of 31 children with protein-energy malnutrition reported slow acetylation of the antibiotic (Eriksson et al. 1988). Conditions associated with reduced absorption of drugs also have a significant clinical impact when drugs are administered. Diseases like heart failure, chronic liver disease, and chronic kidney diseases, which are associated with variable levels of edema of the gastrointestinal tract and ascites, will reduce drug absorption. Another example of drug-disease interaction is steroid use in patients with diabetes mellitus and other immunosuppressive diseases such as HIV infection. The prescriber should always bear in mind that steroids will worsen glycemic control and further depress immunity respectively when administered to these patients.

Pathological conditions play roles in non-drug interactions involving local anesthetics. Inflammation can reduce the effectiveness of local anesthetics through various mechanisms, including changes in tissue pH, increased blood flow, altered nerve sensitivity, and altered metabolism. Inflammation often decreases local tissue pH due to the accumulation of lactic acid and other metabolic byproducts. Local anesthetics are weak bases that are more effective in their non-ionized (neutral) form. In an acidic environment, such as inflamed tissue, the pH is lower, which reduces the proportion of the non-ionized form of the anesthetic. Consequently, less anesthetic can diffuse through nerve membranes, making the anesthetic less effective. Inflammation often causes increased local blood flow as part of the body's response to injury or infection. Increased blood flow can lead to faster absorption and clearance of the local anesthetic from the injection site. This can reduce the duration of the anesthetic effect and may necessitate higher doses or more frequent administration. Inflammation can cause changes in nerve excitability and increase the sensitivity of peripheral nerves. The heightened sensitivity of nerves due to inflammation might alter the effectiveness of local anesthetics. Even if the anesthetic is working, the increased sensitivity of the nerves may result in less effective pain relief. Inflammation can alter the concentration of proteins in the inflamed

**Table 9.2** Clinical evidence of antibiotic interactions in malnourished states

| Interacting drug | Nutrient/nutrient deficiency | Clinical studies/cases | Pharmacokinetic/pharmacodynamic mechanisms | Diagnosis/outcomes | Clinical management strategy/recommendations | References |
|---|---|---|---|---|---|---|
| Chloramphenicol | Protein-energy malnutrition (marasmic malnutrition) | Randomized clinical trial (111 children with marasmic malnutrition)<br>In 33 Ethiopian children aged 0.6–6 year, nutritional status was determined to be normal in 8, marasmus in 8, kwashiorkor in 8, and marasmic–kwashiorkor malnutrition in 9 children | Decreased absorption and clearance of chloramphenicol in malnourished children | Treatment failure was more malnourished<br>Treatment responders had a higher percentage of positive blood or lung aspirate cultures<br>Measurement of serum total protein may assist in the screening of patients who need dosage adjustment | Difficulty in separating the impact of malnutrition alone from that of concurrent infection requires special medical care<br>Drug monitoring | Mulholland et al. (1995)<br>Eriksson et al. (1983) |
| Gentamicin | Protein-energy malnutrition | 11 malnourished 3- to 10-month-old infants | Increased metabolism and elimination of gentamicin<br>Increased volume of distribution of gentamicin due to increased total body water | | Nutritional rehabilitation | Bravo et al. (1982)<br>Samotra et al. (1985)<br>Ronchera-Oms et al. (1995) |
| | | 6 malnourished children aged 4–14 years | The half-life of gentamicin was doubled | Serum levels of drug | | |
| | | 86 critically ill adult patients, those who had malnutrition (defined as low albumin and >15% weight loss) were treated with parenteral nutrition | Expansion of extracellular fluid space due to hypoalbuminemia | Increased volume of distribution with gentamicin, relative to patients without malnutrition, who received intravenous fluids | | |

| | | | | | | |
|---|---|---|---|---|---|---|
| Penicillin | Protein-energy malnutrition | Children with severe Protein-energy malnutrition | Increased half-life of penicillin reduced renal filtration | Penicillin concentrations | Nutritional rehabilitation | Mehta et al. (1980, 1982) |
| Sulfadiazine | Protein-energy malnutrition | 6 children with protein-energy malnutrition | Reduced rate of drug absorption | Peak blood levels of sulfadiazine<br>Free drugs were eliminated at normal rates, but acetylated drug elimination was reduced | Drug monitoring is required | Mehta et al. (1980, 1982) |
| | | 6 undernourished adults | Sulfadiazine absorption and renal excretion were unchanged, although its metabolism was increased, and protein binding reduced (40 vs 54% in normal controls) | | Drug monitoring is not necessary since therapeutic doses were achieved | Shastri and Krishnaswamy (1979) |
| Tetracycline | Protein-energy malnutrition | 8 malnourished adults | Tetracycline is not biotransformed and is excreted as a free drug in the urine<br>Oral drug absorption was reduced, and the elimination rate increased proposedly due to reduced protein (albumin was significantly lower in the malnourished) binding | Serum levels of drug | More frequent dosing intervals to obtain therapeutic drug concentrations | Shastri and Krishnaswamy (1976) |

tissue. Some local anesthetics bind to tissue proteins. Changes in protein binding due to inflammation can affect the distribution and duration of the anesthetic's effect. For example, increased protein binding might prolong the onset and the duration of action. Inflammatory mediators, such as histamines and prostaglandins, are released during inflammation. These mediators can affect the local tissue environment and nerve function, potentially interfering with the efficacy of local anesthetics. For instance, inflammatory mediators might influence how well anesthetics bind to their target receptors. Systemic inflammation can affect liver and renal function, which are crucial for the metabolism and excretion of many local anesthetics. Impaired metabolism or excretion can lead to altered pharmacokinetics of the anesthetic, potentially resulting in longer or shorter durations of action or increased risk of toxicity.

### *3.6 Drug-Gene*

Drugs, diseases, and gene interactions as illustrated in Fig. 9.4 also impact pharmacokinetic and pharmacodynamic mechanisms of interactions. Genetic differences are a well-known cause of variability in drug response among different populations. One of the well-known examples is the hypersensitivity reactions associated with carbamazepine, a drug used in the treatment of epilepsy, bipolar disorder, and trigeminal neuralgia. This reaction is common and typically affects the skin causing rashes, which could be fatal in severe cases. Studies have shown that the risk of hypersensitivity reaction is associated with the presence of human leukocyte antigen (*HLA)* HLA-B*1511 allele (Wang et al. 2017).

Another example is the higher risk of cardiovascular events such as in patients on clopidogrel who have genetically reduced CYP2C19 function. Clopidogrel is an antiplatelet drug that is used in the treatment and secondary prevention of some specific cardiovascular events. The enzyme CYP2C19 converts the drug to an active metabolite with antiplatelet activity. People who have lost the function of the gene coding this enzyme have diminished antiplatelet responses to clopidogrel therapy (Pereira, et al. 2024). Genetic variability affecting vitamin K epoxide reductase complex subunit 1 (VKORC1) gene, the Cytochrome P450 2C9 (CYP2C9) gene, the Cytochrome P450 4F2 (CYP4F2) gene has also been associated with the use of warfarin, an oral anticoagulant drug. Other examples of drug-gene interaction include tramadol and CYP2D6, codeine and CYP2D6, and escitalopram and (CYP2C19) gene polymorphism.

In the clinical case of drug-gene interaction between clozapine and *CYP1A2* gene polymorphism in a patient FA [age: 29 years; gender: female; medical history: schizophrenia; current medications: clozapine (an atypical antipsychotic); genetic background: FA underwent genetic testing which reveals a polymorphism in the *CYP1A2* gene. Specifically, she has the *CYP1A2* *1F/*1F genotype, which is associated with a reduced activity of the CYP1A2 enzyme]. FA has been on a stable dose of clozapine for several months and later started experiencing significant side

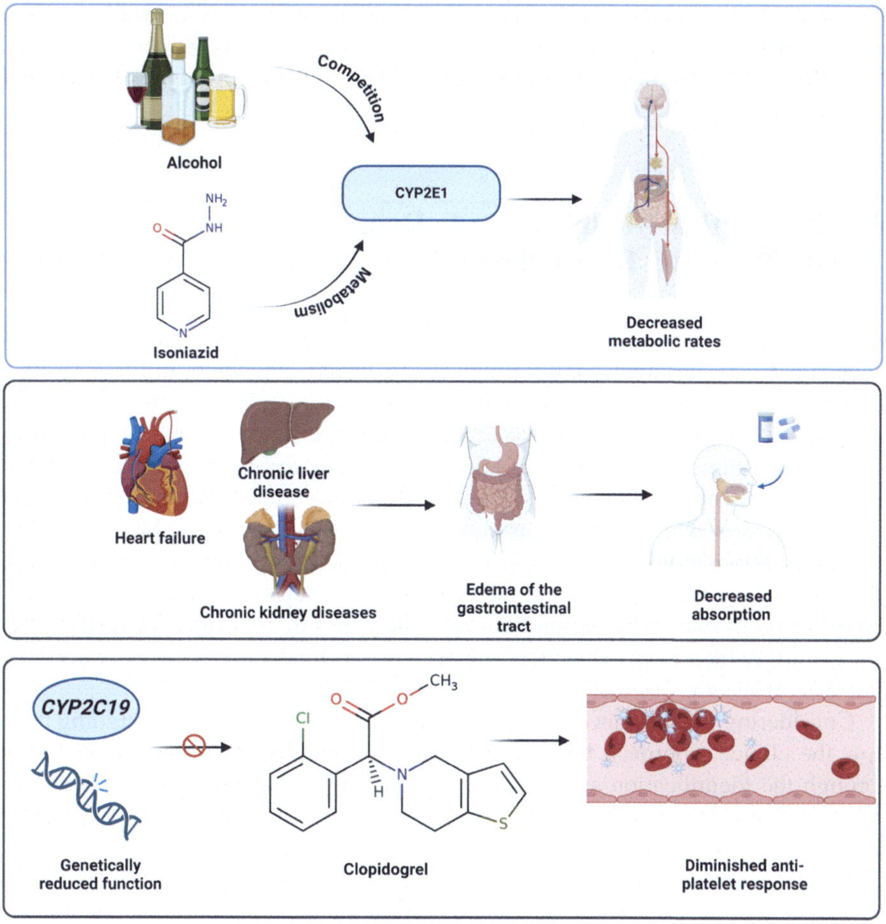


**Fig. 9.4** Drugs, diseases, and gene interactions

effects, including excessive sedation, dizziness, and an increased frequency of seizures. Despite the clozapine dose being within the recommended therapeutic range, she is experiencing these adverse effects more intensely. Laboratory tests showed elevated clozapine serum levels, indicating that the drug is not being metabolized effectively. Genetic testing confirms that FA has the CYP1A2 *1F/*1F genotype, suggesting reduced enzyme activity. The CYP1A2 enzyme is responsible for the metabolism of clozapine. The CYP1A2 *1F/*1F genotype indicates that FA has reduced enzyme activity, leading to a slower metabolism of clozapine. This results in higher plasma levels of the drug, which can cause increased sedation, dizziness, and potentially more severe side effects like seizures. The clozapine dose needs to be adjusted downward to account for the reduced metabolic rate. This helps to manage the elevated serum levels and mitigate side effects. Depending on the clinical response and ongoing management, the healthcare provider might consider an alternative antipsychotic medication with a different metabolic profile if dose adjustments are insufficient. Genetic testing can reveal how a patient metabolizes medications and facilitate personalized treatment plans to optimize efficacy and minimize adverse effects.

## 4 Hypothetical Modifications to Drug and Non-drug Interactions

Exploring hypothetical cases of drug and non-drug interactions (DNIs) can help us understand how different factors might affect health outcomes. In the case of a patient under medications of Selective Serotonin Reuptake Inhibitor (SSRI) for depression and an antibiotic such as erythromycin, there exists a chance of erythromycin inhibiting the metabolism of SSRIs, potentially increasing their levels in the blood and enhancing the risk of side effects like serotonin syndrome. This patient may experience symptoms such as agitation, confusion, and rapid heart rate. The healthcare provider might adjust the SSRI dosage or choose a different antibiotic. In the case of antihypertensive beta-blockers for hypertension and diuretic medications, a patient may suffer from diuretic-induced electrolyte imbalances (e.g., low potassium), which may reduce the effectiveness of the beta-blocker or exacerbate side effects such as dizziness.

Considering the relevance of drug-food interactions to patients undergoing therapy, the clinical approach has been to simplify the prescription details and plan through the identification and removal of food substances that may interact with prescribed drugs when taken together. Although this traditional method is clinically safe, it bears little importance in drug-nutritional status interactions. When clinicians ask patients to remove certain food items from their diet due to potential adverse reactions, they may inadvertently create a nutritional imbalance that has a negative consequence. This is due to the simplified approach of eliminating food substances in patients' diets. Therefore, there is a need for drug-food modification strategies with a comprehensive understanding of mechanisms involved in the interactions between medication and nutrient status in patients. For example, the consumption of vitamin K-rich foods (e.g., leafy green vegetables) by patients on warfarin can reduce the effectiveness of warfarin by promoting clotting. The patient might have a higher risk of clotting events. A patient taking a statin (e.g., atorvastatin) and drinking grapefruit juice regularly could have increased statin levels and a higher risk of side effects like muscle pain or liver damage. Grapefruit juice inhibits the enzyme CYP3A4, which is involved in the metabolism of many statins.

The importance of drug-food modifications, which is either to achieve desirable effects or overcome unwanted interactions, stems from the need to avoid adverse effects of nutritional deficiencies owing to simplified treatment approaches. Effects of nutrient deficiency in patients include weight loss, changes in body composition, cardiovascular diseases, hypertension, diabetes mellitus, cancer, metabolic disorders, developmental abnormalities, food intolerances, and allergies (Awuchi et al.

2020). Food supplementation and nutraceutical formulations of drug-food modifications have largely evaded drug-food interactions and drug-nutrient status interactions. This ability has helped to reverse the clinically adverse effects of nutrient deficiencies. For example, Antacids have been known to decrease the absorption of folate, however, a clinical study observed that patients taking antacids had better folate levels compared to the control group. This could be attributed to the higher consumption of folic acid through fortified breakfast cereals and vitamin supplements by those taking antacids (Bailey 1984).

Despite the great benefits of evading drug-food interactions through the nutrient-replacement approach, the non-eliminating approach presents a new level of challenge. It may not overcome interindividual variability in response to this modification strategy. This reaction is newly defined in that the demerits do not arise from either the drug or the food formula, but rather from the patient. Predicting interpatient variability in response to this approach provides a new rationale for further studies. It may encompass underlying factors including genetic polymorphisms, gene-environment interactions, the presence or absence of disease, the type and pattern of nutrient intake, and the scheduling/dosage of drug administration. This nutrient complementary approach becomes handy with detailed knowledge and comprehension of mechanisms underlying interactions between drugs and food.

## 5 Current Limitations of Clinical Studies on Food-Drug Interactions

There are insufficient data and clinical case studies on interactions between food and drugs. Human experimental studies on DNIs have not kept pace with the proliferation of new drugs that have the potential for adverse interactions with food in patients undergoing therapy. Some clinical studies discussed earlier, such as those related to antibiotics such as chloramphenicol and tetracycline, are less commonly used today. These are gradually replaced with a broad spectrum of new drugs that currently lack data on their interaction with food. Some available case studies are limited by being underpowered to detect comparators (Mehta et al. 1982; Schenker et al. 2001). This implies that conclusions drawn from these studies may still be biased, in addition to the wide variations in sample estimates that make sample results differ significantly from those of the population. In addition to these challenges, many studies lack sufficient sampling size to produce robust statistical data needed for good inferences. An influencing factor that may contribute to the low quality of some available clinical studies may include narrow eligibility criteria for participants, ethical issues related to clinical trials, and tightly controlled implementation of interventions and comparators, among others (Saldanha et al. 2022).

Finally, research into the effects of drug interactions in nutrient-deficient conditions varies from mild and short-term deficiencies to severe and long-lasting loss of fat, muscle, and organ function that can ultimately lead to death (Vanderveen 2005).

The risks involved in studying such forms of malnutrition concerning drug interactions are extremely challenging. These limitations make it difficult to predict, clinically diagnose, interpret, and recommend interactions between drugs and food. The healthcare provider could monitor electrolyte levels and the international normalized ratio to assess the effectiveness of anticoagulant medications, adjust medications, accordingly, provide dietary guidance, or suggest alternative medications or dietary adjustments.

# 6 Self-Assessment Questions and Answers

## *6.1 Questions*

1. Why are non-drug interactions often not recognized?
   A. They occur only in hospital settings.
   B. They are not listed on drug labels or prescribing information.
   C. They occur in everyday life and may not be immediately noticeable.
   D. They only happen during clinical research studies.
2. Why is it important for healthcare providers to be aware of non-drug interactions?
   A. To only prescribe medications based on clinical trials.
   B. To account for potential adverse effects and adjust treatment plans accordingly.
   C. To ignore patient-reported symptoms.
   D. To standardize all patient treatments without considering individual factors.
3. What role do underlying diseases play in non-drug interactions?
   A. They do not impact drug interactions.
   B. They can alter the effectiveness of medications and their metabolism.
   C. They only affect the absorption of oral medications.
   D. They improve the therapeutic outcomes of all drugs.
4. What can drugs like methotrexate, phenytoin, sulfasalazine, and trimethoprim cause regarding folate?
   A. Enhance folate absorption and utilization.
   B. Antagonize folate utilization and inhibit its absorption or conversion to its active form.
   C. Increase folate levels in the body.
   D. Do not affect folate levels.

5. How can healthcare providers address the challenges related to food-drug interactions?
   A. Ignore dietary factors and focus solely on medications.
   B. Monitor electrolyte levels, adjust medications, and provide dietary guidance.
   C. Avoid making any changes to patient treatments.
   D. Only use outdated studies to inform decisions.

### 6.2 Answers

1. C.
2. B.
3. B.
4. B.
5. B.

## References

Abebe W. Herbal medication: potential for adverse interactions with analgesic drugs. J Clin Pharm Ther. 2002;27(6):391–401. https://doi.org/10.1046/j.1365-2710.2002.00444.x. PMID: 12472978.

Ahlskog JE. Cheaper, simpler, and better: tips for treating seniors with Parkinson disease. Mayo Clin Proc. 2011;86(12):1211–6. https://doi.org/10.4065/mcp.2011.0443. PMID: 22134940; PMCID: PMC3228622.

Alperin JB. Coagulopathy caused by vitamin K deficiency in critically ill, hospitalized patients. JAMA. 1987;258(14):1916–9.

Ament PW, Bertolino JG, Liszewski JL. Clinically significant drug interactions. Am Fam Physician. 2000;61(6):1745–54. PMID: 10750880.

Aroda VR, Edelstein SL, Goldberg RB, Knowler WC, Marcovina SM, Orchard TJ, Diabetes Prevention Program Research Group, et al. Long-term metformin use and vitamin B12 deficiency in the diabetes prevention program outcomes study. J Clin Endocrinol Metab. 2016;101:1754–61.

Awuchi CG, Igwe VS, Amagwula IO. Nutritional diseases and nutrient toxicities: a systematic review of the diets and nutrition for prevention and treatment. Int J Adv Res. 2020;6:1–46.

Bailey A. The effects of antacid consumption on folate status. M.S. Thesis, Cornell University, Ithaca, NY; 1984.

Berlinger WG, Park GD, Spector R. The effect of dietary protein on the clearance of allopurinol and oxypurinol. N Engl J Med. 1985;313(13):771–6. https://doi.org/10.1056/NEJM198509263131302. PMID: 4033706.

Bhat RV, Deshmukh CT. A study of vitamin K status in children on prolonged antibiotic therapy. Indian Pediatr. 2003;40(1):36–40. 34.

Bordia A. Effect of garlic on human platelet aggregation in vitro. Atherosclerosis. 1978;30:355–60.

Bravo ME, Arancibia A, Jarpa S, Carpentier PM, Jahn AN. Pharmacokinetics of gentamicin in malnourished infants. Eur J Clin Pharmacol. 1982;21:499–504.

Bushra R, Aslam N, Khan AY. Food-drug interactions. Oman Med J. 2011;26(2):77–83. https://doi.org/10.5001/omj.2011.21. PMID: 22043389; PMCID: PMC3191675.

Caudill MA. Folate bioavailability: implications for establishing dietary recommendations and optimizing status. Am J Clin Nutr. 2010;91(5):1455S–60S.

Cohen LK, George W, Smith R. Isoniazid-induced acne and pellagra. Occurrence in slow inactivators of isoniazid. Arch Dermatol. 1974;109(3):377–81.

de Jager J, Kooy A, Lehert P, Wulffelé MG, van der Kolk J, Bets D, et al. Long term treatment with metformin in patients with type 2 diabetes and risk of vitamin B-12 deficiency: randomised placebo controlled trial. BMJ. 2010;340:c2181.

Deng J, Zhu X, Chen Z, Fan CH, Kwan HS, Wong CH, Shek KY, Zuo Z, Lam TN. A review of food-drug interactions on oral drug absorption. Drugs. 2017;77(17):1833–55.

DiNicolantonio JJ, Niazi AK, Lavie CJ, O'Keefe JH, Ventura HO. Thiamine supplementation for the treatment of heart failure: a review of the literature. Congest Heart Fail. 2013;19:214–22.

Duke JA. The green pharmacy herbal handbook. Emmaus: Rodale Press; 2000.

Eriksson M, Paalzow L, Bolme P, Mariam TW. Chloramphenicol pharmacokinetics in Ethiopian children of differing nutritional status. Clin Pharmacol. 1983;24:819–23.

Eriksson M, Bolme P, Habte D, Paalzow L. INH and streptomycin in Ethiopian children with tuberculosis and different nutritional status. Acta Paediatr Scand. 1988;77:890–4.

Genentech. XENICAL (orlistat) capsules for oral use [Internet]. Silver Spring (MD): U.S. Food and Drug Administration; 2015 [cited 2022 May 8]. Available from: https://www.accessdata.fda.gov/drugsatfda_docs/label/2015/020766s035lbl.pdf

Gotfredsen A, Westergren Hendel H, Andersen T. Influence of orlistat on bone turnover and body composition. Int J Obes Relat Metab Disord. 2001;25:1154–60.

Grover Z, Ee LC. Protein energy malnutrition. Pediatr Clin N Am. 2009;56(5):1055–68.

Gundogan K, Akbudak IH, Bulut K, Temel S, Sungur M, Guven M, et al. Thiamin status in adults receiving chronic diuretic therapy prior to admission to a medical intensive care unit: a pilot study. Nutr Clin Pract. 2019;34:565–71.

Gunther T, Rebentisch E, Vormann J, Konig M, Issing H. Enhanced ototoxicity of gentamicin and salicylate caused by Mg deficiency and Zn deficiency. Biol Trace Elem Res. 1988;16:43–50.

Hanninen SA, Darling PB, Sole MJ, et al. The prevalence of thiamin deficiency in hospitalized patients with congestive heart failure. J Am Coll Cardiol. 2006;47(2):354–61.

Hansten PD. Appendix II: important interactions and their mechanisms. In: Katzung BG, editor. Basic and clinical pharmacology. 9th ed. Boston: McGraw Hill; 2004. p. 1110.

Hoogwerf BJ, Hibbard DM, Hunninghake DB. Effects of long-term cholestyramine administration on vitamin D and parathormone levels in middle-aged men with hypercholesterolemia. J Lab Clin Med. 1992;119:407–11.

Horwitt MK, Harvey CC, Dahm CH Jr, et al. Relationship between tocopherol and serum lipid levels for determination of nutritional adequacy. Ann N Y Acad Sci. 1972;203:223–36.

Isenberg-Grzeda E, Alici Y, Hatzoglou V, Nelson C, Breitbart W. Nonalcoholic thiamine-related encephalopathy (Wernicke-Korsakoff syndrome) among inpatients with cancer: a series of 18 cases. Psychosomatics. 2016;57:71–81.

Jung JW, Park SY, Kim H. Drug-induced vitamin deficiency. Ann Clin Nutr Metab [Internet]. 2022;14(1):20–31.

Khan KM, Jialal I. Folic acid deficiency. [Updated 2023 Jun 26]. In: StatPearls [Internet]. Treasure Island (FL): StatPearls Publishing; 2024. Available from: https://www.ncbi.nlm.nih.gov/books/NBK535377/.

Kitt TM, Park GD, Spector R, Tsalikian E. Reduced renal clearance of oxypurinol during a 400 calorie protein-free diet. J Clin Pharmacol. 1989;29:65–71.

Kulak CA, Borba VZ, Bilezikian JP, Silvado CE, Paola L, Boguszewski CL. Bone mineral density and serum levels of 25 OH vitamin D in chronic users of antiepileptic drugs. Arq Neuropsiquiatr. 2004;62:940–8.

Lam CS, Hua R, Au-Doung PLW, Wu YK, Koon HK, Zhou KR, Loong HH, Chung VC, Cheung YT. Association between potential supplement-drug interactions and liver diseases in patients with cancer: a large prospective cohort study. Clin Nutr ESPEN. 2023;58:152–9. https://doi.org/10.1016/j.clnesp.2023.09.919.

Lentz KA. Current methods for predicting human food effect. AAPS J. 2008;10(2):282–8.

Lieber CS. Alcohol and the liver: 1994 update. Gastroenterology. 1994;106(4):1085–105. https://doi.org/10.1016/0016-5085(94)90772-2.

Linnebank M, Moskau S, Semmler A, Widman G, Stoffel-Wagner B, Weller M, et al. Antiepileptic drugs interact with folate and vitamin B12 serum levels. Ann Neurol. 2011;69:352–9.

Mehta S. Malnutrition and drugs: clinical implications. Dev Pharmacol Ther. 1990;15:159–65.

Mehta S, Nain CK, Sharma B, Mathur VS. Metabolism of sulfadiazine in children with protein-calorie malnutrition. Pharmacology. 1980;21:369–74.

Mehta S, Nain CK, Sharma B, Mathur VS. Disposition of four drugs in malnourished children. Drug Nutr Interact. 1982;1:205–11.

Melander A. Influence of food and different nutrients on drug bioavailability. World Rev Nutr Diet. 1984;43:34–44.

Meyrick Thomas RH, Rowland Payne CM, et al. Isoniazid-induced pellagra. Br Med J (Clin Res Ed). 1981;283(6286):287–8.

Mintzer S, Boppana P, Toguri J, et al. Vitamin D levels and bone turnover in epilepsy patients taking carbamazepine or oxcarbazepine. Epilepsia. 2006;47(3):510–5.

Mintzer S, Skidmore CT, Sperling MR. B-vitamin deficiency in patients treated with antiepileptic drugs. Epilepsy Behav. 2012;24:341–4.

Mulholland K, Falade AG, Corrah PT, et al. A randomized trial of chloramphenicol vs trimethoprimsulfamethoxazole for the treatment of malnourished children with community-acquired pneumonia. Pediatr Infect Dis J. 1995;14:959–65.

Muratake T, Watanabe H, Hayashi S. Isoniazid-induced pellagra and the N-acetyltransferase gene genotype. Am J Psychiatry. 1999;156(4):660.

Padwal RS, Rucker D, Li SK, Curioni C, Lau DCW. Long-term pharmacotherapy for obesity and overweight. Cochrane Database Syst Rev. 2003;4:CD004094.

Pantuck EJ, Pantuck CB, Weissman C, Askanazi J, Conney AH. Effects of parenteral nutritional regimen on oxidative drug metabolism. Anesthesiology. 1984;60:534–6.

Park GD, Berlinger W, Spector R, Kitt TM, Tsalikian E. Sustained reductions in oxipurinol renal clearance during a restricted diet. Clin Pharmacol Ther. 1987;41:616–21.

Peloquin CA, Namdar R, Dodge AA, et al. Pharmacokinetics of isoniazid under fasting conditions, with food, and with antacids. Int J Tuberc Lung Dis. 1999;3:703–10.

Pereira NL, Cresci S, Angiolillo DJ, Batchelor W, Capers Q 4th, Cavallari LH, Leifer D, Luzum JA, Roden DM, Stellos K, Turrise SL, Tuteja S, American Heart Association Professional/Public Education and Publications Committee of the Council on Genomic and Precision Medicine, Council on Arteriosclerosis, Thrombosis and Vascular Biology, Council on Cardiovascular and Stroke Nursing, Council on Clinical Cardiology, Council on Peripheral Vascular Disease, Stroke Council. CYP2C19 Genetic testing for Oral P2Y12 inhibitor therapy: a scientific statement from the American Heart Association. Circulation. 2024. https://doi.org/10.1161/CIR.0000000000001257. Advance online publication.

Prabhu D, Dawe RS, Mponda K. Pellagra a review exploring causes and mechanisms, including isoniazid-induced pellagra. Photodermatol Photoimmunol Photomed. 2021;37:99–104.

Rocha RM, Silva GV, de Albuguergue DC, et al. Influence of spironolactone therapy on thiamine blood levels in patients with heart failure. Arq Bras Cardiol. 2008;90(5):324–8.

Ronchera-Oms CL, Tormo C, Ordovas JP, Abad J, Jimenez NV. Expanded gentamicin volume of distribution in critically ill adult patients receiving total parenteral nutrition. J Clin Pharm Ther. 1995;20:253–8.

Rose KD, Croissant PD, Parliament CF, Levin MP. Spontaneous spinal epidural hematoma with associated platelet dysfunction from excessive garlic ingestion: a case report. Neurosurgery. 1990;26:880–2.

Rosenblatt M, Mindel J. Spontaneous hyperema associated with ingestion of Ginkgo biloba extract. N Engl J Med. 1997;336:1108–15.

Ruscin JM, Page RL 2nd, Valuck RJ. Vitamin B(12) deficiency associated with histamine(2)-receptor antagonists and a proton-pump inhibitor. Ann Pharmacother. 2002;36(5):812–6.

Saktiawati AM, Sturkenboom MG, Stienstra Y, Subronto YW, Sumardi, Kosterink JG, van der Werf TS, Alffenaar JW. Impact of food on the pharmacokinetics of first-line anti-TB drugs in treatment-naive TB patients: a randomized cross-over trial. J Antimicrob Chemother. 2016;71(3):703–10. https://doi.org/10.1093/jac/dkv394. Epub 2015 Dec 11. PMID: 26661397.

Saldanha IJ, Skelly AC, Ley KV, et al. Inclusion of nonrandomized studies of interventions in systematic reviews of intervention effectiveness: an update [Internet]. Rockville (MD): Agency for Healthcare Research and Quality (US); 2022. Strengths and Limitations of RCTs. Available from: https://www.ncbi.nlm.nih.gov/books/NBK584466/

Samotra K, Gupte S, Raina RK. Pharmacokinetics of gentamicin in protein-energy malnutrition. Eur J Clin Pharmacol. 1985;29:255–6.

Sathyanarayana Rao TS, Yeragani VK. Hypertensive crisis and cheese. Indian J Psychiatry. 2009;51(1):65–6.

Schenker S, Speeg KV, Perez A, Finch J. The effects of food restriction in man on hepatic metabolism of acetaminophen. Clin Nutr. 2001;20:145–50.

Shastri RA, Krishnaswamy K. Undernutrition and tetracycline half-life. Clin Chim Acta. 1976;66:157–64.

Shastri RA, Krishnaswamy K. Metabolism of sulphadiazine in malnutrition. Br J Clin Pharmacol. 1979;7:69–73.

Sheth RD. Metabolic concerns associated with antiepileptic medications. Neurology. 2004;63(10 Suppl 4):S24–9.

Shorr RI, Ray WA, Daughtery JR, Griffen MR. Concurrent use of nonsteriodal antinflammatory drugs and oral anticoagulants places elderly persons at high risk for hemorrhagic peptic ulcer disease. Arch Intern Med. 1993;153:1665–70.

Simmons MM, Cupp MJ. Use and abuse of flunitrazepam. Ann Pharmacother. 1998;32(1):117–9. https://doi.org/10.1345/aph.17027.

Skversky AL, Kumar J, Abramowitz MK, Kaskel FJ, Melamed ML. Association of glucocorticoid use and low 25-hydroxyvitamin D levels: results from the National Health and Nutrition Examination Survey (NHANES): 2001-2006. J Clin Endocrinol Metab. 2011;96:3838–45.

Snow CF. Laboratory diagnosis of vitamin B12 and folate deficiency: a guide for the primary care physician. [see comment]. Arch Intern Med. 1999;159(12):1289–98.

Srivastava KC. Aqueous extract of onion, garlic and ginger inhibit platelet aggregation and alter arachidonic acid metablosm. Biomed Biochim Acta. 1984;43:335–46.

Stielow M, Witczyńska A, Kubryń N, Fijałkowski Ł, Nowaczyk J, Nowaczyk A. The bioavailability of drugs-the current state of knowledge. Molecules. 2023;28(24):8038.

Thurnham DI, Davies JA, Crump BJ, et al. The use of different lipids to express serum tocopherol: lipid ratios for the measurement of vitamin E status. Ann Clin Biochem. 1986;23(Pt 5):514–20.

Tokoro S, Namiki T, Miura K, Yokozeki H. A case of isoniazid-induced pellagra aggravated by secondary anorexia presenting with an amino acid and vitamin-deficient profile. Eur J Dermatol. 2018;28:832–3.

Tom-Revzon C, Adam HM. Drug interactions. Pediatr Rev. 2006;27(8):315–7.

Torgerson JS, Hauptman J, Boldrin MN, Sjöström L. XENical in the prevention of diabetes in obese subjects (XENDOS) study. A randomized study of orlistat as an adjunct to lifestyle changes for the prevention of type 2 diabetes in obese patients. Clin Diabetol. 2004;5:95–104.

Trehan I, Goldbach HS, LaGrone LN, Meuli GJ, Wang RJ, Maleta KM, Manary MJ. Antibiotics as part of the management of severe acute malnutrition. N Engl J Med. 2013;368(5):425–35.

Ubbink JB, Delport R, Bissbort S, Vermaak WJ, Becker PJ. Relationship between vitamin B-6 status and elevated pyridoxal kinase levels induced by theophylline therapy in humans. J Nutr. 1990;120:1352–9.

Van Ede AE, Laan RF, Blom HJ, Boers GH, Haagsma CJ, Thomas CM, et al. Homocysteine and folate status in methotrexate-treated patients with rheumatoid arthritis. Rheumatology (Oxford). 2002;41:658–65.

Van Roon EN, Flikweert S, le Comte M, Langendijk PN, Kwee-Zuiderwijk WJ, Smits P, Brouwers JR. Clinical relevance of drug-drug interactions: a structured assessment procedure. Drug Saf. 2005;28(12):1131–9.

Vanderveen JE. Handbook of drug-nutrient interactions. Am J Clin Nutr. 2005;81(1):200. Available from: https://doi.org/10.1093/ajcn/81.1.200.

Visser ME, Texeira-Swiegelaar C, Maartens G. The short-term effects of anti-tuberculosis therapy on plasma pyridoxine levels in patients with pulmonary tuberculosis. Int J Tuberc Lung Dis. 2004;8:260–2.

Volz HP, Gleiter CH. Monoamine oxidase inhibitors. A perspective on their use in the elderly. Drugs Aging. 1998;13(5):341–55.

Walter-Sack I, Klotz U. Influence of diet and nutritional status on drug metabolism. Clin Pharmacokinet. 1996;31:47–64.

Wang Q, Sun S, Xie M, Zhao K, Li X, Zhao Z. Association between the HLA-B alleles and carbamazepine-induced SJS/TEN: a meta-analysis. Epilepsy Res. 2017;135:19–28.

Weathermon R, Crabb DW. Alcohol and medication interactions. Alcohol Res Health. 1999;23(1):40–54.

Yasuda H, Hatano T, Honda T, Tsutsui M, Hattori N, Ando M, et al. Vitamin B6 deficiency anemia attributed to levodopa/carbidopa intestinal gel therapy for Parkinson's disease: a diagnostic pitfall for myelodysplastic syndrome with ring sideroblasts. Intern Med. 2022;61(24):3719–22.

# Chapter 10
# Omics Approaches to Drug and Drug-Non-Drug Interactions

**Angela Adamski da Silva Reis and Rodrigo da Silva Santos**

**Abstract** Interindividual variability in drug response is driven by several extrinsic and intrinsic factors promoted by changes in the activity or availability of drug-metabolizing enzymes (DMEs), receptors, channels, and other proteins involved in drug pharmacokinetics and pharmacodynamics, respectively. Genomics promotes a thorough understanding of various molecular mechanisms and pathways in the human system, supported by pharmacogenomics (PGx) and personalized medicine. However, with the technological advances of the genomic era and the ability to decipher the molecular makeup of cells at increasing precision and resolution, it has become clear that the "one gene, one protein, one function" paradigm does not fully explain the complex functional phenotypes of organisms. However, combining high-throughput techniques, especially next-generation sequence (NGS), with analytical tools will allow complex understanding of the interactions between drugs and non-drug substances for an effective treatment. NGS studies have integrated a systems biology approach and combined sequencing data with other types of information, for instance, protein family information, pathway, or protein-protein interaction (PPI) networks, in a computation integrative analysis. Additionally, advancements in artificial intelligence (AI) and machine learning (ML) methods have been greatly aided in analyzing, learning, and explaining pharmaceutical-related big data in the drug discovery process.

**Keywords** Molecular interactions · Personalized medicine · Biological system · Multi-omics

A. A. da Silva Reis (✉) · R. da Silva Santos
Department of Biochemistry and Molecular Biology, Neurogenetics Research Center – NeuroGene, Institute of Biological Sciences II, Federal University of Goiás, Goiânia, Goiás, Brazil
e-mail: angela@ufg.br; rdssantos@ufg.br



J. O. Fajemiroye (ed.), *Fundamentals of Drug and Non-Drug Interactions*,
https://doi.org/10.1007/978-3-031-80107-5_10

# 1 Introduction

A fundamental goal of molecular life science research is to understand the complex relationship between genotype and phenotype (Bludau and Aebersold 2020). On the other hand, precision medicine (PM) highlights the importance of coupling established clinical indexes with molecular profiling to craft diagnostic, prognostic, and therapeutic strategies specific to the needs of each group of patients (Seyhan and Carini 2019). Implementation of multi-omics data introduces new informatics and interpretation challenges. Specifically, novel analytical and statistical methods are needed for combining disparate data sets, as well as standardized quality control metrics (Bludau and Aebersold 2020). Correct interpretation of the data is necessary for the best use of the PM, combining omics and clinical data to determine the best course of action to be taken for each specific patient group, including an effective treatment (Seyhan and Carini 2019).

Additionally, preclinical models are integral for the study testing novel therapies and validating biomarkers of treatment response. These models include both in vitro and in vivo systems. The ideal model is easily manipulable, cost-effective, and most importantly, recapitulates the biology of human disease (Liu et al. 2021). Moreover, a drug gets approved after a lengthy regulatory process. One way to address this problem is to focus on a selected group of patients; thus, Phase III clinical studies can be conducted with a small group of patients rather than thousands and thousands of patients typically needed for Phase III studies (Fountzilas et al. 2022).

This approach should potentially guarantee a more rapid and expeditious way to perform drug development of next-generation pharmacotherapy. A narrower focus on a specific patient's group at the stage of the regulatory approval process should facilitate streamlining the regulatory approval, resulting in greater clinical and economic success (Seyhan and Carini 2019). New tools now allow the characterization of large biological systems in great detail and with unprecedented resolution, rather than focusing research efforts on individual chemicals, metabolic pathways, or cells of interest.

The goal of omics technology is to study changes in DNA, RNA, proteins, and other biological molecules of various types in different species and individuals. Continued developments in sequencing technology and the incorporation of high-throughput Omics data provided invaluable data to untangle the complexity of biological systems at several dimensions. In the following paragraphs, multi-omics approaches and drug-non-drug interactions were described.

# 2 Single Omics Approaches

Genomic data plays a crucial role in drug development. Additionally, for diseases with unknown targets and most small-molecule drugs, we still lack a comprehensive understanding of their mechanisms and interconnections. In the past few decades, the development of high-throughput sequencing technology has led to the improvement of gene expression profile databases, enabling us to measure and analyze large-scale gene expression data. Thus, exploring the connections between diseases, genes, and drugs is currently a hotspot in scientific research (Fang et al. 2023).

Interindividual variability in drug response is driven by several extrinsic and intrinsic factors, with genetic variations being increasingly recognized among these factors that lead to changes in the activity or availability of drug-metabolizing enzymes (DMEs), receptors, channels, and other proteins involved in drug pharmacokinetics and pharmacodynamics, respectively (Kabbani et al. 2023). A vast number of susceptibility loci have been discovered as a result of genome-wide association studies (GWAS), which have highlighted the significance of genetic variations in the pathogenesis of chronic diseases (Gholami et al. 2022).

Pharmacogenomics (PGx), a study of interindividual variations in DNA sequence related to drug efficacy and toxicity, provides routine personalized medicine (PM) by considering a patient's genetic background (Kabbani et al. 2023; Kim et al. 2022). Additionally, PGx allows patients to be treated based on their genetic makeup (Kabbani et al. 2023). Interestingly, the declining costs of next-generation sequencing technologies are accelerated by the development of genomic profiles, such as single-nucleotide variants (SNVs) and copy-number variants (CNVs), from a large population (Fang et al. 2023).

Efforts to apply to personalized medicine relied upon studies providing robust evidence for the benefit of PGx-guided therapeutic strategies. Moreover, the population variome resources allow the construction of genotyping panels to analyze drug metabolism or gene-drug/Drug-drug-gene interaction pairing functionality (Kim et al. 2022). Thus, PGx has become an effective tool to fulfill the promise of personalized medicine and is crucial for understanding human variation and genotyping arrays to define therapeutic strategies. Despite all emerging evidence and efforts enabling PGx, its clinical implementation has been suboptimal worldwide, especially in developing countries (Kabbani et al. 2023).

The Clinical Pharmacogenetics Implementation Consortium (CPIC) provides terminology standards for clinical pharmacogenetic test results. This database provides clinical evidence of the pharmacogenetic and corresponding drug interactions. PGx has also been involved in silico approaches in cancer genomics, for example. On the other hand, EUROPGx, a consortium of European research initiatives, has developed 1931 PGx biomarkers among the European population. Even today, variome data for diagnosis and treatment of disease cohorts are steadily accumulating (Kim et al. 2022).

Pharmacogene Variation Consortium (PharmVar) initially analyzed cytochrome P450 (CYP) enzymes that function as drug metabolizers and later expanded to other pharmacogenes (Pratt et al. 2023). East Asian (EA) populations are genetically different, having originated from the admixture of different ancestral populations. This suggests that the EA population differs in genetic epidemiology and the diversity of risk alleles for drugs. This necessitates the characterization of the EA-PGx variome (Kim et al. 2022).

In addition, it has been shown that adverse drug reactions (ADRs) and hospitalization resulting from drug toxicity can be better controlled by applying PGx. More specifically, PGx-guided treatment of antidepressants and antipsychotic medications revealed the cost-effectiveness and cost-saving of PGx testing, respectively. What sets PharmacoGx apart is its ability to compare query results from multiple datasets in a unified database. This functionality makes it a powerful tool to assist researchers in pharmacogenomics research, such as screening sensitizers, for example, in cancer radiotherapy and studying in vitro drug sensitivity, as well as developing new methods and functionalities to better understand drug-induced effects (Fang et al. 2023).

A plethora of drugs and toxic substances are metabolized by CYP enzymes. These are coded by highly variable genes abundant with SNVs and small insertions/deletions (indels) that affect the functionality of the enzymes, increasing or decreasing their activity (Skadrić and Stojković 2019). CYP gene genotyping, followed by haplotype inference, provides substrate-specific metabolic phenotype prediction. However, a high number of alleles in the CYP450 superfamily and interethnic variability in frequency distribution require precise gene panel customization (Pratt et al. 2021).

Drug metabolism is the process of altering molecules chemically after entering the body (Backman et al. 2016). In general, the metabolism of drugs decreases their therapeutic effects. Most of the drugs' lipophilic centers are converted to hydrophilic centers during drug biotransformation, which can increase their water solubility, to allow elimination in urine or bile. This is important progress for drug metabolism because the lipophilic nature of drugs can keep them staying for longer in the body, which may in turn lead to toxicity. On the other hand, CYPs are also capable of affecting drug responses by influencing drug action, safety, bioavailability, and drug resistance through metabolism, in both metabolic organs and local sites of action (Zhao et al. 2021).

In addition to the basic metabolic effects for elimination, CYPs are also capable of affecting drug responses by influencing drug action, safety, bioavailability, and drug resistance through metabolism, in both metabolic organs and local sites of action. Structures of CYPs have recently provided new insights into both understanding the mechanisms of drug metabolism and exploiting CYPs as drug targets. (Zhao et al. 2021). Genetic polymorphisms and epigenetic changes in *CYP* genes and environmental factors may be responsible for interethnic and interindividual variations in the therapeutic efficacy of drugs (Zhou et al. 2008).

CYPs, as the most diverse catalysts known in biochemistry, contribute to interindividual variations in drug responses, resulting from genetic and epigenetic variants, as well as environmental factors, such as gender, age, nutrition, disease states, and pathophysiological factors (Zhou et al. 2008). In particular, CYPs can be inhibited or induced by concomitant drugs and circulating metabolites, which can influence treatment outcomes through drug-drug interaction (DDI), drug-gene interaction (DGI), and drug-drug-gene interaction (DDGI) (Zhao et al. 2021).

Polymorphisms in the CYP family may have had the most impact on the fate of pharmaceutical drugs (Zhao et al. 2021). *CYP2D6*, *CYP2C19*, and *CYP2C9* gene polymorphisms and gene duplications account for the most frequent variations in phase I metabolism of drugs since nearly 80% of drugs in use today are metabolized by these enzymes. Approximately 5% of Europeans and 1% of Asians lack CYP2D6 activity, and these individuals are known as poor metabolizers. CYP2C9 is another clinically significant drug-metabolizing enzyme that demonstrates genetic variants. Studies into *CYP2C9* polymorphism have highlighted the importance of the *CYP2C9*2* and *CYP2C9*3* alleles (Zhou et al. 2008).

Thus, the goals of the PGx Working Group are to define the key attributes of pharmacogenetic alleles recommended for clinical testing and to determine a minimal set of variants that should be included in clinical PGx genotyping assays. The document series revealed recommendations on a minimal panel of variant alleles (Tier 1) and an extended panel of variant alleles (Tier 2) that will aid clinical laboratories in designing assays for PGx testing. The current document is focused on CYP2D6 and is intended to guide clinical laboratorians and assay manufacturers who develop, validate, and/or offer clinical CYP2D6 genotyping assays, and therefore to promote the standardization of PGx testing across clinical laboratories (Pratt et al. 2021).

This document should be implemented together with other relevant clinical guidelines, including those published by the Clinical Pharmacogenetics Implementation Consortium (CPIC), the Dutch Pharmacogenetics Working Group of the Royal Dutch Pharmacists Association (DPWG), the Canadian PGx Network for Drug Safety, and the American College of Medical Genetics and Genomics, which have mostly been focused on the interpretation of PGx test results and therapeutic recommendations on specific drug-gene pairs (Pratt et al. 2021).

*CYP2D6* is a major gene member of this superfamily, as it carries out the metabolism of 25% of drugs currently available in the market. Contrary to the concept of specificity of enzyme action, these can metabolize substrates of different chemistry. Substrates of CYP2D6 are usually lipophilic bases with a planar hydrophobic aromatic ring and a nitrogen atom that can be protonated at physiological pH and have a negative molecular electrostatic potential above the planar part of the molecule (Pratt et al. 2021).

Structurally, the nitrogen atom is considered essential for electrostatic interactions with the carboxylate group of amino acids in the active site of CYP2D6 (Pratt et al. 2021). The prototypical substrates of CYP2D6 are debrisoquine and sparteine, which are widely used to determine the phenotype of CYP2D6-mediated metabolism. Six substrate recognition sites (SRS) are present in P450, which display distinct differences in their substrate-binding residues (Gopisankar 2017).

CYP2D6 has nine exons and is located on chromosome 22q13.2 next to two highly homologous pseudogenes, CYP2D7 and CYP2D8. It is highly polymorphic, with over 130-star allele haplotypes defined by the PharmVar. Most of the CYP2D6 star alleles are defined by single-nucleotide variants or small insertions/deletions (Gopisankar 2017). CYP2D6 can also harbor copy-number variants (CNVs; deletions and duplications) and gene conversions with the nearby CYP2D7 pseudogene, leading to CYP2D6-2D7 and CYP2D7-2D6 hybrid alleles (Pratt et al. 2023).

CYP2D6 is mostly non-inducible and is subjected to inhibition by many drugs and compounds that lead to drug-drug interactions (Cavallari et al. 2019). The drug-drug interaction can be minor, mild, or fatal. For example, CYP2D6-mediated venlafaxine-propafenone interaction may cause hallucinations and psychomotor agitation. Table 10.1 summarizes the main inhibitors of the CYP2D6 enzyme (Gopisankar 2017). Importantly, the CYP2D6 metabolizer phenotype significantly affects drug safety and effectiveness, prompting the CYP2D6 gene to be broadly included in the US Food and Drug Administration (FDA) (Pratt et al. 2023).

A study in over 100,000 patient samples across all regions of the United States showed that structural variants, including CNVs, accounted for 7% of all CYP2D6 variants and could affect CYP2D6 metabolizer status. Therefore, the variability in assay design, with or without the detection of CNVs and structural variants, can result in discrepancies in star allele calls and haplotype assignment, which directly affect phenotype translation, interpretation, and ultimately a patient's care. However, there are specific challenges unique to CYP2D6 testing given the complexity of the gene and its relevance to multiple medications (Cavallari et al. 2019). Thus, genomics could contribute to the variation of CYP phenotypes among humans and can partly explain the interindividual differences in drug response. Identification of CYP variants is a promising future for disease treatment (Bludau and Aebersold 2020).

Individuals' different allelic variants of *the CYP2D6* gene determine the metabolism status, such as "extensive metabolizers," "intermediate metabolizers," "poor metabolizers," and "ultra-rapid metabolizers." The poor metabolizer phenotype has two mutant alleles while those with two wild alleles are labeled as extensive metabolizers. Poor metabolizers are unable to break down the drug into an excretable form and, as a result, have a higher concentration of the medication in their plasma. This results in greater and more frequent adverse drug reactions and toxicity. The range in SNV leads to up to a tenfold difference in the required dose of medicine necessary to achieve plasma concentrations in poor metabolizers. On the other hand, extensive and ultra-rapid metabolizers break down medications at a very fast rate, causing low plasma levels of the medication, and thus higher dosage of medication administration is necessary for efficacy (Gummadi and Guddati 2021).

**Table 10.1** The main inhibitors of the CYP2D6 enzyme (Gopisankar 2017)

| Inhibition of the CY2D enzyme and compounds that lead to drug-drug interactions |
|---|
| Quinidine inhibitor-mediated metabolic conversion of dextromethorphan, thereby increasing systemic bioavailability and fewer ADRs when used for the pseudobulbar effect. |
| Lansoprazole, a potent inhibitor, reduces the conversion of dextromethorphan to dextrorphan. |
| Antipsychotics like olanzapine, chlorpromazine, fluphenazine, haloperidol, thioridazine, risperidone, clozapine, and trifluperidol. |
| Terfenadine, antifungal agents like terbinafine. |
| Irreversible inhibitors: cimetidine, paroxetine, pimozide, methamphetamine, metoclopramide, and desethylamiodarone. |
| Steroids: progesterone, testosterone. |
| *Panax ginseng* and *Ginkgo biloba* inhibit CYP2D6-mediated debrisoquine metabolism. |

Variation in *CYP2C19* expression might give rise to different plasma concentrations of the active metabolites, potentially affecting both drugs' efficacy and tolerability. The interindividual variability in CYP2C19-mediated metabolism may affect drug concentrations/drug elimination, which might potentially impact both drug efficacy and safety. Various antidepressant agents are metabolized by the CYP2C19 enzyme, including citalopram and escitalopram. Notably, the low metabolizers present low adherence, frequent changes in treatment regimens, and more side effects than the rapid metabolizers. In this sense, CYP2C19 genotyping in escitalopram and citalopram therapy personalization supports the potential clinical utility (Mahajna et al. 2023).

N-acetyltransferase (NAT) is an enzyme involved in the addition of various chemical groups to facilitate the inactivation of a xenobiotic and biotransformation. Therefore, NAT is responsible for catalyzing the addition of acetyl groups to xenobiotics toward their excretion. NAT is controlled by the *NAT*-1 or the *NAT-2* gene. The *NAT-2* gene presents two SNVs, *NAT2-A* and *NAT2-B* gene. These SNVs produce two groups of patients—one variant produces fast acetylators while the other variant produces slow acetylators. Thus, the rate of acetylation affects the amount of drug concentration that remains in the plasma and can produce many different clinical outcomes. Slow acetylators present more side effects because they remain in the plasma longer, while fast acetylators may have less of a clinical response because they are excreted more quickly. In highlight, slow acetylators include dapsone and procainamide, and individuals with this variation present severe symptoms, for example, red cell toxicity.

Apart from their essential role as detoxification enzymes, glutathione S-transferase (GST) participates in several other important functions such as cell signaling, post-translational modifications, and chemotherapeutic drug resistance (Mazari et al. 2023). Variants on the GST alter the metabolism of chemotherapeutic drugs and modify the effectiveness of therapy (Gummadi and Guddati 2021). Genetic variants emerged as viable therapeutic targets because specific GST isoenzymes have been shown to be overexpressed in numerous tumors. Efforts were made to develop specific and potent inhibitors of GSTs to enhance the cytotoxic effects of existing chemotherapeutic agents, thus diminishing tumor growth. Advances in molecular studies have been prioritized to design isotype-specific inhibitors since different GST isoforms have unique glutathione binding sites (Mazari et al. 2023).

The exponential proliferation of the drug market demands feasible and effective methods to determine drug toxicity. Despite animal-based models, researchers are increasingly using in vitro cellular systems to explore different drug doses and measure specific parameters such as cytotoxicity, mitochondrial damage, oxidative stress, and apoptosis. In this sense, multi-omics approaches could suggest potential drugs without having to first identify the modes of action of a compound. Thus, omics technologies, when compared with in vitro and multiplexed assays, can effectively provide in-depth information about biological perturbations to understand the biological relevance of drug-induced hepatotoxicity and yet provide effective strategies for drug discovery at the early stage. Furthermore, the promotion of non-animal testing also encourages a strategic shift in risk assessment that promotes omics to use in regulatory decision-making (Nguyen et al. 2022).

On the other hand, some molecules can affect the metabolic pathways, potentially contributing to alterations in therapeutic response (McQuade and Campbell 2021). However, most ADRs refer to DDI, and drug-nutrient-gene interactions (DNGIs) may pose a considerable risk that is often disregarded in clinical practice. DNGIs can modulate pharmacokinetic (e.g., nutrient affects drug absorption, distribution, metabolism, and excretion) and pharmacodynamic (e.g., nutrient-driven modulation of drug action at a receptor or signaling level) parameters, while the scientific evidence available implicates largely the former (Stouras et al. 2022).

Micronutrients are indispensable for a variety of vital functions. In most cases, deficiencies are treatable with supplementation of the elements in lack. The most frequent deficiency is that of iron, followed by vitamin A, folate, iodine, and zinc deficiency, and in minors, vitamin and folate deficiencies dominate across all age groups. Notably, micronutrient deficiencies (MDs) are not a problem in developing countries exclusively. Drug-nutrient interactions (DNIs) can also lead to micronutrient reduction or depletion by various pathways. Some factors that may enhance the type and intensity of DNIs include patient-related variables such as age, sex, comorbidities, nutritional status, and drug- and nutrient-associated factors such as route of administration, nutrient status, and pharmacological/toxicological profile of the drug. A clinically significant drug-nutrient interaction is one related to an impaired physiologic process (quantifiable alteration of the kinetic and/or dynamic profile of a drug or a nutrient), which may result in malnutrition, therapy failure, adverse events, or even a life-threatening situation (Karadima et al. 2016).

Genetic variants in the above enzymes or dietary deficiency of B-vitamin cofactors and their substrates could be responsible for elevations in serum total homocysteine (tHcy) levels. Elevated serum tHcy levels, referred to as hyperhomocysteinemia (HHcy), are associated with a wide range of health problems, including cardiovascular disease, deep vein thrombosis or pulmonary embolism, neurocognitive disorders, pregnancy complications, birth defects, and osteoporotic fractures (Mazokopakis et al. 2023; Raghubeer and Matsha 2021). The *methylenetetrahydrofolate reductase* (*MTHFR*) gene encodes for the enzyme, which catalyzes the reduction of 5,10-methylenetetrahydrofolate to 5-methyltetrahydrofolate participant in one-carbon metabolism, the primary form of folate in the blood and a required element (along with cobalamin/vitamin B12) in the conversion of homocysteine (Hcy) to methionine (Stouras et al. 2022; Raghubeer and Matsha 2021).

Sufficient dietary and supplemental riboflavin (RF), a member of the B-vitamin family, intake appears to have a protective effect on various medical conditions. RF also participates in one-carbon metabolism in its active co-factor forms, flavin adenine dinucleotide (FAD) and flavin mononucleotide (FMN). Pyridoxine-phosphate oxidase requires FMN for the formation of the active form of vitamin B6, pyridoxal 5′-phosphate, from pyridoxine phosphate. MTHFR enzyme requires FAD as a co-factor and converts 5,10-methyleneTHF to 5-methylTHF, which is subsequently converted to tetrahydrofolate (THF) in a reaction catalyzed by methionine synthase, completing the cycle. The latter conversion also requires vitamin B12 (i.e., methylcobalamin) as a co-factor and simultaneously enables the remethylation of homocysteine to methionine and subsequently S-adenosylmethionine (SAM). This universal methyl donor is essential for a range of methylation processes, including DNA methylation (McMahon et al. 2016). Thus, apart from folate, three other B vitamins play essential roles in one-carbon metabolism, as they are required for the activity of the various enzymes within the folate cycle (McMahon et al. 2016).

The common *C677T* variant in the gene encoding the folate-metabolizing enzyme methylenetetrahydrofolate reductase (MTHFR) is the most well-known genetic factor influencing folate status (Stouras et al. 2022). The prevalence of the TT genotype varies across ethnic groups and regions. In Europe, Asia, Central America, and South America, the prevalence of the TT genotype ranges from 10% to 32%, whereas different African populations have a prevalence of only 0–3% (Wilcken et al. 2003). Binia et al. (2014) reported that the prevalence rates of the TT genotype were 25% and 57% in the Mexican Mestizo and American Indian populations, respectively (Hiraoka and Kagawa 2017).

The cobalamin, often known as vitamin B12, is a water-soluble vitamin that contains cobalt and functions as a co-factor for enzymes that are important for metabolism. There is a drug-non-drug interaction (DNI) between metformin and B12 (McMahon et al. 2016). Metformin is one of the first-line options and the most widely prescribed oral diabetic pharmacological mono- or combination therapy (Owen et al. 2021). It is recommended for most T2DM patients for its effectiveness, low risk of hypoglycemia, neutrality toward body weight, cardioprotective effect, and affordability (Pratama et al. 2022).

Metformin is not easily metabolized after entering the human body; 80–100% of metformin will be discharged from the renal tubules in the form of the prototype, thus the drugs that affect the renal function will affect the metabolism of metformin. Studies have shown that the elimination of metformin is mediated by the transporters MATE1, MATE2K, and OCT2 in the kidney. Cimetidine, a drug for the treatment of peptic ulcers, can lead to metformin accumulation by competing with metformin for OCT2 or MATE1, which can result in a significant increase in the concentration of lactate in the blood (Luo et al. 2021).

Therefore, metformin blocks the absorption of vitamin B12 through a mechanism that has not been established but could be because of interference with the calcium-dependent binding of the intrinsic factor vitamin B12 complex to the Cuban receptor in the terminal ileum (Mazokopakis et al. 2023). Additionally, the subsequent deficiency of vitamin B12 may cause or accelerate distal symmetrical and autonomic neuropathy in diabetic patients (Pecikoza et al. 2021). Individuals with the TT genotype for the *MTHFR* gene (677C/T) have significantly higher tHcy levels and lower folate levels in serum than those with the CT and CC genotypes. However, administration of folic acid has been shown to eliminate these differences. Moreover, data has suggested that interventions based on genotype (TT genotype) may be effective for supplementation indication (Hiraoka and Kagawa 2017).

Therefore, periodic monitoring of vitamin B12 is recommended in all patients who utilize metformin, particularly if metformin has been used for over 5 years, at which stage hepatic stores of vitamin B12 would probably be depleted (Aroda et al.

2016). Factors that accelerate the loss of hepatic vitamin B12 stores are proton pump inhibitors, bariatric surgery, being elderly, and having an increased turnover of red blood cells. If serum vitamin B12 levels are borderline, measurement of methylmalonic acid and homocysteine levels can detect vitamin B12 deficiency at its earliest stage (McMahon et al. 2016).

Therapies for neurodiabetic patients in treatment with metformin include prophylactic calcium and vitamin B12 supplementation by intramuscular or oral therapy. In addition, it is necessary for a regular assessment of vitamin B12 levels to be monitored. Supplementation of vitamin B12 promotes the level of adequate cobalamin, providing improved symptoms of neuropathy; however, in some patients, the symptoms stabilize or do not improve (Pecikoza et al. 2021). In addition, a positive association between the TT mutant genotype of the *MTHFR* C677T variant requires the indication of B12 supplementation intending to benefit diabetic patients undergoing treatment with metformin and prevent diabetic neuropathy (Raghubeer and Matsha 2021). Figure 10.1 describes the mechanisms involved in *MTHFR* C677T for the TT mutant genotype and vitamin B12 depleted in association with metformin treatment.

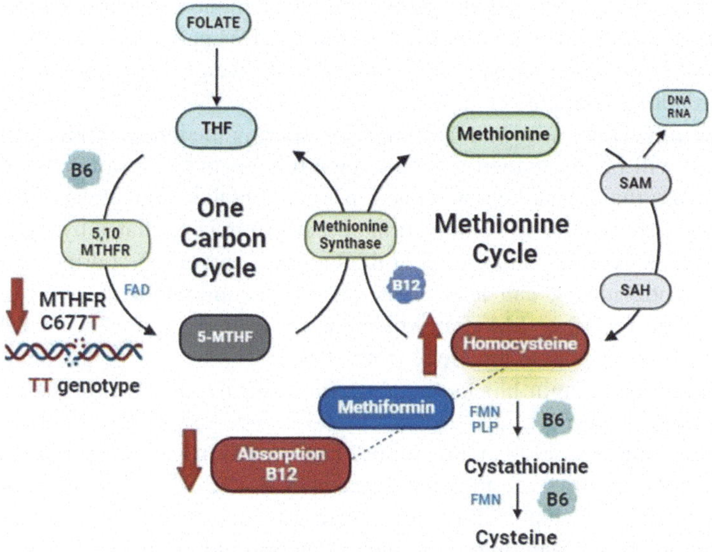


**Fig. 10.1** One-carbon pathway will circulate normally with the methionine cycle. The activity of the MTHFR enzyme is reduced in individuals carrying the SNV *MTHFR* (C677T), and this can lead to a reduction in 5-MTHF and the accumulation of homocysteine, causing hyperhomocysteinemia. Homozygotes TT genotypes had significantly higher homocysteine concentrations than heterozygotes (CT) and homozygotes for *MTHFR* C677T. Abbreviations: MTHFR, methylenetetrahydrofolate reductase; PLP, plasma pyridoxal phosphate; MTHFR, methylenetetrahydrofolate reductase; FAD, flavin adenine dinucleotide; FMN, flavin mononucleotide; SAM, S-adenosylmethionine; SAH, S-adenosyl-homocysteine hydrolase. (Figure created with BioRender.com)

Why do people with the same disease show different responses to the treatment with the same drug? Many available medical treatments are designed for a wide range of patients with the same condition that, after prescribing medication, the following states may be observed: (i) the prescribed drug may reduce the symptoms of the disease and improve the treatment process, (ii) the expected response may fail to occur, and (iii) the prescribed medication can cause side effects in patients (Khodadadian et al. 2020).

Interindividual variations in response to drug treatment often present a challenge in the treatment process. It is estimated that the genome of a typical individual consists of 14 million SNVs. Every SNV presents a persistent source of variation in a patient's response to drugs. Single-base variations in the genes coding for metabolizing enzymes, drug transporters, as well as target sites can affect how a drug responds in the patient's body. Thus, SNVs present an ongoing challenge and opportunities for the treatment of patients; genotyping of well-known variants should be comprehensively studied and should be widely used for treatment plans (Gummadi and Guddati 2021).

Advances in genetic screening technologies, such as genome-wide association studies (GWASs) and whole-genome sequencing studies, have facilitated the rapid identification of genetic variants associated with increased risk for many complex human diseases. Although these variants and their related proteins could become targets for drug discovery, most of them have small effect sizes, and their functions remain largely unknown (Astarita et al. 2023).

Since the completion of the human genome project and the development of next-generation sequencing (NGS) techniques, the structure and functions of the genome, as well as mechanisms regulating genes' expression, have been extensively explored (Supplitt et al. 2021). Combining high-throughput techniques, especially NGS, with appropriate analytical tools has allowed gaining a deeper systematic understanding of various biological levels, most importantly genomics, transcriptomics, and epigenetics (Park et al. 2021). Thus, the power of high-scale "genomics" analyses coupled with the ample portfolio of targeted drugs under clinical development and already approved offers more effective individualized therapy. It is rewarding to observe that sophisticated genomic studies reinforce the already-known clinical strategy to use drug combinations to treat disease specifics (Supplitt et al. 2021).

Furthermore, more sophisticated analysis tools based on computational modeling are introduced to decipher underlying molecular mechanisms for drug interaction to provide precision therapeutic strategies. The increasing size and complexity of the data required the adaptation of bioinformatics processing pipelines for higher efficiency and sophisticated data mining methodologies, particularly for large-scale NGS datasets (Supplitt et al. 2021). Nowadays, several NGS studies integrate a systems biology approach and combine sequencing data with other types of information, for instance, protein family information, pathway, or protein-protein interaction (PPI) networks with a computation integrative analysis (Park et al. 2021). Figure 10.2 summarizes omics approaches and outcomes associated with DDI and DNI.

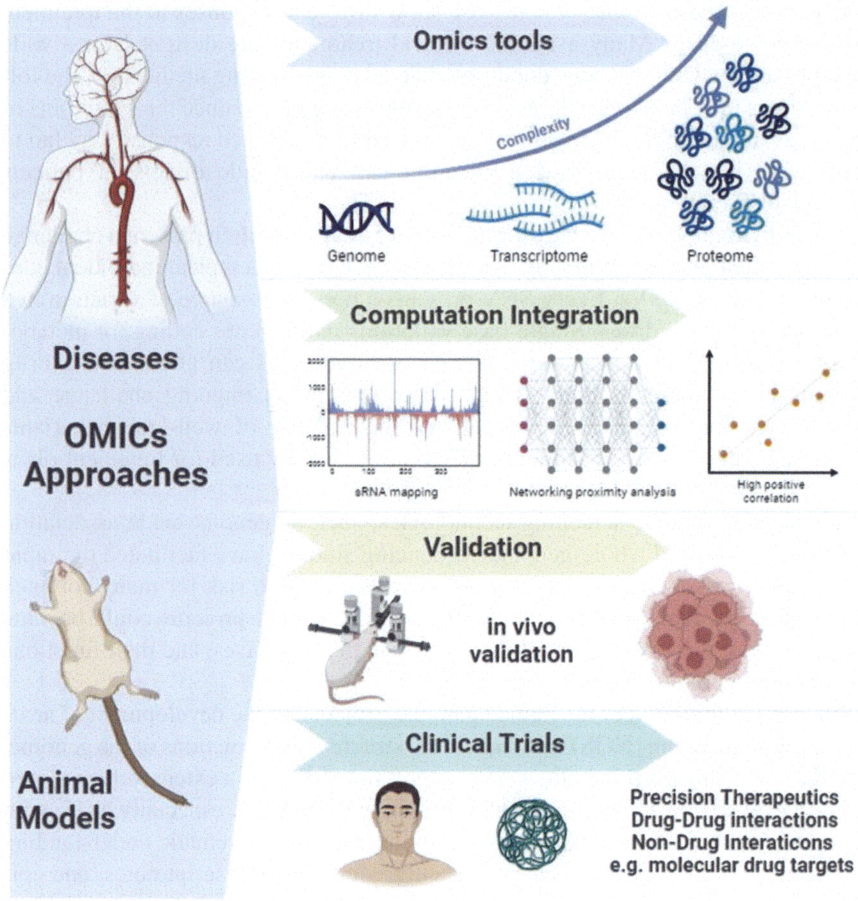


**Fig. 10.2** Sophisticated analysis tools based on multi-omics and computational modeling are introduced to decipher underlying molecular mechanisms for drug interaction to provide precision therapeutic strategies. (Figure created with BioRender.com)

Experimentally validated knowledge in systems biology may enhance analysis models and guide them to uncover novel findings. Such integrated analyses have been useful to extract essential information from high-dimensional NGS data (Park et al. 2021). To deal with the increasing size and complexity, the application of machine learning (ML), is an area of artificial intelligence (AI) that uses statistical techniques such as complex algorithms to make predictions of compound toxicity based on big data (Chiarella et al. 2023), and specifically deep learning methodologies, has become state-of-the-art in NGS data analysis (Ngiam and Khor 2019). They can be classified into three main groups: affinity enrichment-based methods, restriction enzyme-based methods, and bisulfite conversion-based methods. The affinity enrichment-based methods group uses either antibodies (methylated DNA immunoprecipitation, MeDIP-Seq) or proteins (methylated-CpG domain-binding proteins, MBD-Seq) to compile methylated genomic regions for sequencing (Nguyen et al. 2022).

## 3 Transcriptomics and Proteomics Approaches

The relationship between the transcriptome and the genome suggests that the information of an organism is stored in the DNA (genome) and expressed by RNA (transcriptome) (Saviano et al. 2020). Transcriptomics is a continuum that complements genomics and fills the gap between genomics and proteomics in the precision medicine era (Supplitt et al. 2021). One of the main reasons for the study of the transcriptome is that most human genes undergo a process called alternative splicing (AS). AS is a particular type of splicing in which two or more rearrangements (removal of introns and reconnection of exons) can occur in one mRNA in separate tissues or different stages of cell (tissue) growth (Lowe et al. 2017).

AS contributes to diversifying and regulating cellular responses to environmental conditions and developmental cues by differentially producing multiple mRNA and protein isoforms from a single gene (Jeon et al. 2022). Thus, transcriptome analysis enables the study of gene expression in human tissues and is a valuable tool for identifying prognostic markers or signatures and facilitating the discovery of new therapeutic targets. The main focus of transcriptomics is to discover how transcripts of a cell, tissue, or living organism are influenced by disease or environmental factors, such as drugs (Saviano et al. 2020).

Non-coding RNA is another very important aspect of transcriptomics. These functional elements play an important role in the occurrence of a variety of diseases and their response to treatment (Khodadadian et al. 2020). The transcriptomics includes the post-transcriptional era; hence, the alterations that cannot be detected at the genomics level can be revealed at the transcriptomics level. Then, if these changes have pathogenic importance, they should be identified to design more reliable treatments (Saviano et al. 2020).

Drug-induced transcriptome (DIT) data is the response of cells to drug-related properties; it can well characterize the macromolecular drugs. Thus, using DIT data is a promising approach to building an accurate macromolecular drug-related DDI prediction model (Luo et al. 2021). Thus, transcriptomic analysis assists in the identification of molecular targets characterizing genes that can be direct or even indirect targets of drug and non-drug molecules. The elucidation of these targets contributes to the characterization of the molecular processes of therapeutic responses (Cicchese et al. 2020).

The first approach to studying human transcriptomes describes the quantification of mRNA with Northern blotting. This time-consuming method uses gene-specific DNA probes, which are hybridized to the RNA and are capable of analyzing a maximum of a few gene transcripts at the same time (Supplitt et al. 2021). Subsequently, newer methods based on complementary probe hybridization were introduced to transcriptome analysis. Sequencing technologies vary from low to high throughput and allow whole-transcriptome analysis (Kankanige et al. 2022).

Different tools have been developed to measure gene transcription and expression in specific cellular states (Table 10.2). Sequencing technologies vary from low to high throughput and allow whole-transcriptome analysis. RNA-seq coupled with bioinformatics approaches has been widely used in different studies to reveal the exact process of cellular transcription (Horta et al. 2023; Lowe et al. 2017). Additionally, RNA-seq data have been used for modeling gene interactions through coexpression networks, a powerful method for making functional inferences about genes involved in a certain process (Lowe et al. 2017).

Most studies focus on one protein, ignoring the multiple splice-variants or protein isoforms, which might contribute to unexpected therapeutic activity or adverse side effects. Identifying and evaluating the right target are the most important factors in the early drug discovery phase. In recent years, the cost of identifying a new compound and converting it into an FDA-approved drug has increased enormously, with an estimated cost to drug-makers of $2.6 billion, which includes the cost of the compounds that failed to make it to the market. Despite great advances in multi-omics technologies and medicinal chemistry assays, together with screening and secondary assays, they are generating enormous amounts of data and knowledge. While the ever-increasing datasets should have aided in silico experimentation to expedite the overall search for new drugs, the rate of new drugs entering the market is falling, and many drugs approved by regulators are being withdrawn due to toxicity and safety concerns (Ji et al. 2020).

Drug similarity studies rely on the assumption that drugs with similar pharmacological properties are similar in their mechanism of action, share similar side effects, and are indicated for the treatment of similar diseases (Ji et al. 2020). Moreover, in silico analysis has been derived for a variety of applications, including drug-target identification, side effect prediction, drug-drug interaction prediction, and drug repositioning (Azad et al. 2020). With the current availability of several drug-related data sources, there is a need for a comprehensive, contemporary knowledge base integrating diverse information from a wide array of evidence sources to derive multimodal drug-drug similarities (Ji et al. 2020).

**Table 10.2** Different methodologies to measure gene transcription and gene expression (Khodadadian et al. 2020)

| One or few genes whole | Whole transcriptome |
|---|---|
| Tissue in situ hybridization<br>Cellular in situ hybridization<br>Northern blot hybridization<br>RNA dot-blot hybridization<br>Ribonuclease protection assay<br>Reverse transcription polymerase chain reaction (RT-PCR) 5′ and 3′ RACE<br>Quantitative (Real-Time) reverse transcriptase-PCR (qRT-PCR)<br>cDNA synthesis and Sanger sequencing | Expressed sequence tag (EST) sequencing<br>Serial analysis of gene expression (SAGE)<br>Microarrays<br>cDNA synthesis and RNA sequencing (NGS technology) |

Many new drug compounds have been successfully developed using computational methods. In silico approaches have been attractive due to their potential to accelerate drug discovery in terms of time, labor, and costs. Almost all the experimental and/or computational studies assume the "one gene-one protein" paradigm, ignoring the true dynamic complexity of the proteome, which includes alternative protein isoforms generated from the same gene by mechanisms known as alternative transcription and AS (Sabe et al. 2021).

The use of computer-aided drug discovery (CADD) techniques in preliminary studies by leading pharmaceutical companies and research groups has helped to expedite the drug discovery and development process, minimizing the costs and failures in the final stage (Gurung et al. 2021). Thus, CADD techniques are attracting increasing attention as they can help mitigate the scale, time, and cost issues faced by conventional experimental approaches (Sabe et al. 2021). The application of rational drug design as an integral part of CADD provides useful insights into the understanding of the binding affinity and molecular interaction between the target protein and ligand. Additionally, lead identification in pharmaceutical research has been facilitated by the availability of supercomputing facilities, parallel processing, and advanced programs, algorithms, and tools (Sabe et al. 2021; Macalino et al. 2015).

The advent of whole-transcriptome expression profiling techniques that catalog protein-coding genes expressed within cells and tissues has enabled the development of individual algorithms for particular tasks. For example, (i) gene ontology algorithms that predict gene/protein subsets involved in related cell processes; (ii) algorithms that predict intracellular protein interaction pathways; and (iii) algorithms that correlate druggable protein targets with known drugs and/or drug candidates (Kankanige et al. 2022).

Improved discovery of candidate druggable proteins within specific cell, tissue, and disease contexts will aid the development of new treatments. Additionally, protein interaction pathways and networks are critically required for a vast range of biological processes. Current efforts to combine algorithms into pipelines to enable the identification of druggable targets and associated known drugs have been used in gene expression datasets. In doing so, new opportunities are identified for the development of powerful algorithm pipelines, suitable for wide use by no bioinformaticians, that can predict protein interaction networks, druggable proteins, and related drugs from user gene expression datasets. Predicting protein interaction networks from gene expression data can provide valuable insights into normal and disease biology (Kankanige et al. 2022).

Proteins present a modular organization made up of several domains. Apart from the domains playing catalytic functions, many others are crucial to recruiting interactors. The latter domains can be defined as "protein interaction domains" (PIDs) and are responsible for pivotal outcomes in signal transduction and a certain array of normal physiological and disease-related pathways. Targeting such PIDs with small molecules and peptides able to modulate their interaction networks may represent a valuable route to discover novel therapeutics (Vincenzi et al. 2021).

Proteomics approaches are powerful tools for analyzing DDI and non-drug interaction (NDI). Several compounds can be investigated at the molecular level, showing the interaction with specific proteins in the human organism (Lu et al. 2020; Mulder et al. 2018). Studies of protein-protein interaction (PPI) can be applied to identify drug-target proteins as well as possible interactions with other proteins, helping to understand the mechanisms of action of drugs and to predict possible collateral effects. Moreover, proteomics analysis can be applied to observe how non-drug substances interact with drugs and affect gene expression regulation and metabolism (Feng et al. 2017; Lu et al. 2020; Xie et al. 2023).

Proteomics investigates the functionality of all expressed proteins (Richard and Horgan 2011). Based on protein sequencing, the most used tool is mass spectrometry (MS), which can determine the amino acid sequence and protein conformation through mass-charge quantification (Alfaro et al. 2021; Wilhelm et al. 2014). High-resolution MS methods include matrix-assisted laser desorption ionization-time of flight (MALDI-TOF), with a fast-scanning speed and high throughput but present a low resolving power. Low-resolution methods include the ion trap, which measures the nominal mass (De Cesare et al. 2020).

The data-generated interpretation by proteomic approaches also requires software and computational analyses. Between some platforms, MaxQuant and Scaffold quantify and identify proteins from mass spectrometry data (Searle 2010; Tyanova et al. 2016). Some platforms also allow in silico molecular modeling of proteins and predicting PPI, such as Rosetta, AlphaFold (Drake et al. 2022), and STRING (Szklarczyk et al. 2019).

Furthermore, recent advancements in artificial intelligence and machine learning methods have greatly aided in analyzing, learning, and explaining pharmaceutical-related big data in the drug discovery process. Different methods employed in the identification of new inhibitors from chemical databases include pharmacophore modeling, quantitative structure-activity relationship (QSAR), molecular docking, quantum mechanics, and statistical learning methods. Thus, CADD can be broadly divided into structure-based and ligand-based drug design approaches; both have been widely used in the drug discovery process in the identification of suitable lead molecules. While the structure-based drug design relies on the three-dimensional structure of the target receptor and its active sites to understand the molecular interaction between the receptor and ligand, the ligand drug design depends on the knowledge of ligands interacting with the given target receptor (Gurung et al. 2021).

Additionally, in silico analysis of the databases obtained from transcriptomics and metabolomics allows us to predict drugs that might prevent or treat a pathology. Comprehensive genetic and metabolic studies can help elucidate the nature of the differences in response between individuals. While transcriptomics determines the functional response and helps predict its master regulators, metabolomics provides a downstream and closer to phenotype description of the biological processes (Malachowska et al. 2023).

# 4 Metabolomics and Interactomics Approaches

Metabolites have a crucial role in cellular energy production by providing substrates for glycolysis/fermentation and oxidative phosphorylation. Additionally, they act as building blocks for large biomolecules and cell membranes, as well as signaling messengers. A biological sample can contain thousands of metabolites, covering a diverse range of biochemical classes, such as amino acids, biogenic amines, organic acids, nucleotides, and lipids (Wishart et al. 2018). Alterations in metabolite levels are intricately tied to diverse diseases, revealing insights into disease etiology, status, therapeutic response, and biological process changes (Astarita et al. 2023).

The systematic study of metabolites found within biological systems is termed "metabolomics" (Astarita et al. 2023). Metabolomics is associated with the identification and quantification of metabolites of endogenous origin (e.g., small peptides, lipids, and nucleic acid precursors and degradation products to chemical intermediates in biosynthesis and catabolism) generally considered the downstream elements of any toxicological mechanism (Chiarella et al. 2023).

In other words, whereas metabolomics aims to identify and quantify the composition of metabolites, lipidomics is a subset of metabolomics that focuses specifically on lipid metabolites. The profile of metabolites, or the "metabolome," is determined by two factors: the biological activity of cells, which reflects the genomic variation in biological systems (e.g., individuals), and the impact of environmental factors, such as drugs, diet, lifestyle, and the microbiome, on those biological systems (Astarita et al. 2023).

Metabolomics studies involving knocking down or knocking in a specific gene or variant in cellular and animal models provide physiological and pathological functions of genes or variants of interest. These resulting metabolite profiles promote a better understanding of the physiological and pathological functions of the gene or variant and can also serve as biomarkers for target and pathway engagement (Wishart et al. 2018). Most diseases result in metabolic changes; in many cases, these changes play a causative role in disease progression. Additionally, metabolomics can point to potential new sites for therapeutic intervention based on the identification of pathological metabolic changes. In this perspective, drug discovery has been essential for the survival and growth of common eukaryotic cell types; most enzymes and metabolites have been identified, for example, via knockout screens in yeast (Rabinowitz et al. 2011).

The integration of genetics, transcriptomics, proteomics, metabolomics, or even clinical data in a systems approach may present a comprehensive metabolic network for cardiometabolic disease, including type 2 diabetes (T2DM). A significant fraction of T2DM patients often present advanced complications that can be difficult to manage and costly to treat (Ortiz-Martínez et al. 2022). Metabolomic studies have identified biomarkers for the screening, diagnosis, and prediction of diabetes and its complications. On the other hand, some metabolites could also be used for monitoring the therapeutic effects of treatment. Thus, efforts toward preventing or improving the clinical management of T2DM and its complications using metabolomics tools show a powerful approach that can deepen the molecular understanding of this disease (Jin and Ma 2021).

Drug metabolite identification is a critical component of drug discovery and development and provides a variety of inputs, including in vitro metabolite profiling, in vitro/in vivo correlation, cross-species comparison, characterization of major circulating metabolites, rationalizing drug-drug interactions, and identifying pharmacologically active or toxic metabolites and the mechanisms by which they are formed (Mortishire-Smith et al. 2011). Metabolomics approaches can also evaluate DDIs and NDIs at the complete set of metabolites in a biological system, which can arise from external responses, such as to drugs or substances present in other compounds (Dai and Shen 2022).

Metabolites are the classification of low molecular weight (<1500 kDa) biomolecules in a biological system that participate in intracellular endogenous metabolism in functions ranging from energy synthesis to intermediates with modulatory functions in proteins (epiproteomics) (Dai and Shen 2022). Due to the vast diversity of possible drug metabolites, the demand for increasing throughput in drug discovery, and the challenge of characteristically low concentrations of drug metabolites amidst a high abundance of endogenous matrix background (Mortishire-Smith et al. 2011).

In addition, it can aid in comprehending how non-drug compounds can affect the human systems, evaluating nutrient metabolization, and how NDIs can modulate the efficacy of drugs or propose drug adverse events. These analyses can promote an understanding of the metabolic pathways involved in the interactions and identify metabolites that are influenced by the drugs, collaborating with the identification of possible adverse effects and synergistic or antagonistic actions (Gonzalez-Covarrubias et al. 2022; Wishart 2016).

Metabolomic analyses are based on spectroscopy and mass spectrometry (MS) tools for the evaluation of metabolites important for cellular homeostasis. The evaluation comprises all targeted (specific) or non-targeted metabolites except molecular biology dogma molecules (genome to proteome) and micronutrients (Dai and Shen 2022). Spectrometry can be applied to metabolomics, being the most workable tool due to its capacity to detect a wide range of metabolites (Johnson et al. 2016). MS methods are usually coupled to liquid or gas chromatography (Carriot et al. 2021; Taya et al. 2021). Nevertheless, both techniques promote gene identification and a better understanding mechanism for therapeutic efficacy (Meng et al. 2021).

Compared with only a few decades ago, we can now measure metabolites in biological samples hundreds of times faster and with greater specificity. The remarkable advances in the sensitivity and specificity of MS and nuclear magnetic resonance (NMR) spectroscopy have led to the development of sophisticated analytical tools that can precisely measure an ever-increasing number of metabolites in biological samples. This presents an unprecedented opportunity to describe molecular phenotypes accurately and comprehensively, thereby enhancing the potential for discovering novel targets for therapeutic interventions and valuable biomarkers (Astarita et al. 2023).

There is a strong need for spatial metabolomics (SM) approaches, which will furthermore help enable a spatial framework for personalized/precision medicine initiatives (Wheeler et al. 2024). SM allows for the visualization of the spatial composition of metabolites within different regions and subregions of biological tissue samples. SM has been used to determine the spatial pharmacodynamic response to drug treatments and to investigate drug absorption, distribution, and metabolism in specific organs or regions of animal tissues. For example, a recent study examined the spatial localization of metabolite-based biomarkers of pathway engagement in response to enzyme replacement therapy (ERT) targeting the lysosomal enzyme iduronate 2-sulfatase (IDS) (Astarita et al. 2023).

This approach has many applications in drug discovery, including determining the spatial distribution and composition of endogenous metabolites in response to drug treatment and investigating the absorption, distribution, and metabolism of drugs in tissue subregions. On the other hand, SM analysis has often been used to investigate drug distribution. Drug efficacy and drug safety are thus dependent on drug tissue distribution. In addition, for example, some chemical constituents of eye drops are toxic to parts of the eye, like benzalkonium chloride. Some constituents of eye drops do not reach the correct part of the eye for treatment, such as atropine, which must reach the sclera (Wheeler et al. 2024).

Studies with animal models are used to comprehend the spatial distribution profiles of drugs. Notably, Seneviratne et al. 2020 described a particular HIV treatment that involves the combinations of emtricitabine (FTC), tenofovir (TFV), efavirenz (EFV), or rilpivirine (RPV). After treatment using either the TFV-FTC-EFV combination or the TFV-FTC-RPV combination in mice, matrix-assisted laser desorption/ionization-mass spectrometry imaging (MALDI-MSI) methodology enabled the visualization of distribution patterns of CYP-dependent monooxygenated metabolites of EFV and RPV, EFV-OH and RPV-OH. Additionally, hydroxylated EFV, which encompasses the CYP-dependent monooxygenated metabolites of EFV, was detected in the liver, brain, spleen, and heart tissue sections. Additionally, the tissue distribution profiles of RPV and hydroxylated RPV following in vivo dosing of TFV-FTC-RPV were different from EFV/hydroxylated EFV despite RPV belonging to the same drug class as EFV. Thus, it was observed that the spatial distribution profiles of the study drugs agree with their safety profiles in humans.

Analysis and prediction of drug-target interactions (DTIs) play an important role in understanding drug mechanisms, as well as drug repositioning and design. Machine learning (ML)-based methods for DTIs prediction can mitigate the shortcomings of time-consuming and labor-intensive experimental approaches while providing new ideas and insights for drug design. These computational methods in DTIs prediction have gained more attention because carrying out in vitro and in vivo experiments on a large scale is costly and time-consuming (Abbasi et al. 2021). Interestingly, metabolite analyses can also evaluate NDIs, showing high potential in nutrition. Several nutrients and bioactive ingredients can interact with molecules in a biological system. The process of analyzing micronutrients includes sample preparation to data processing and assists in the identification of chemical compounds relevant to the human body (Hu et al. 2022). Figure 10.3 describes an interactome analysis model to provide a new drug design.

In addition, nutrients and drugs affect the activity of cytochrome P450 enzymes involved in drug metabolism, also contributing to important interaction. For almost 60 years, vitamin K antagonists (VKAs) were the mainstay of anticoagulation therapy, but in recent years, direct oral anticoagulants (DOACs) have become the anticoagulant treatment of choice. However, nutrients, dietary supplements, or herbs may contain substances that, when administered concomitantly with direct DOACs, can potentially affect the plasma concentration of the drugs. DOACs act as direct factor Xa inhibitors (rivaroxaban, apixaban, edoxaban, and betrixaban) or direct thrombin inhibitors (dabigatran). DOACs effect is more predictable and stable (i.e., less dependent on interactions with food, herbal supplements, and other drugs) compared to warfarin and acenocoumarol (Grześk et al. 2021).

Drugs can influence food intake, nutrient digestion, absorption, distribution, metabolism to active forms, function, catabolism, and excretion. Additionally, the long-term use of prescription and over-the-counter drugs can induce subclinical and clinically relevant micronutrient deficiencies, which may develop gradually over months or even years (Mohn et al. 2018). In addition to flavonoids and other polyphenols, isoflavones are secondary metabolites of plants (Pandey and Rizvi 2009), which can modulate drug response and resistance and protein expression and interaction (Do Valle et al. 2021). In this sense, metabolomics can be applied to understand pathophysiological mechanisms, in drug discovery, target, and monitor therapeutic response (Wishart 2016).

In high light, potential DNIs and known risk factors are described, such as acid-suppressing drugs that promote the decrease of these micronutrients: vitamin B12, vitamin C, iron, calcium, magnesium, zinc, and β-carotene. Another drug category, non-steroids and anti-inflammatories, influence the decrease of vitamin C and iron. The interaction between omeprazole treatment and vitamin C was more pronounced

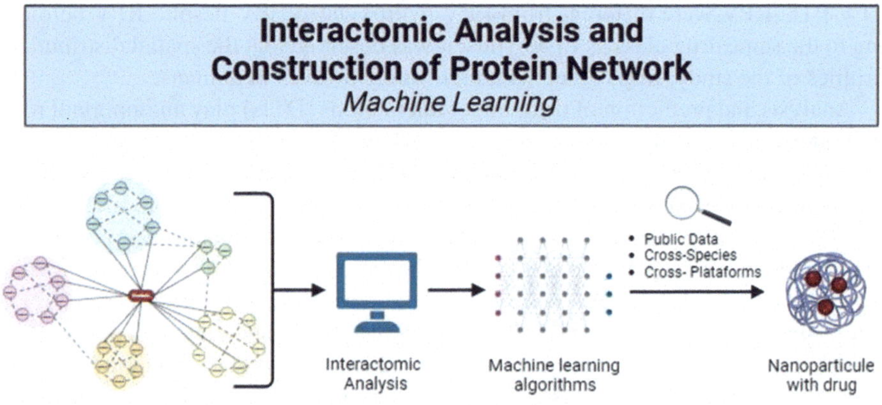


**Fig. 10.3** Interactome analysis model using machine learning to provide new drug design. (Figure created with BioRender.com)

in patients with H. pylori infection, as these subjects also experienced significant reductions in total vitamin C concentrations in their gastric juice. This finding was later confirmed in another short-term omeprazole intervention study of a similar design (Mohn et al. 2018).

Recent developments in molecular biology are allowing researchers to gain a more comprehensive understanding of the biological and chemical processes involved in DDI and DTI, which have been a major driver in the development of several public databases (Chen et al. 2021). Given the large number of medications currently available, the number of studies examining potential drug-nutrient interactions (DNIs) is quite limited (Mohn et al. 2018). Dependence on the use of pharmacologically active substances—including prescription medications, over-the-counter pharmaceuticals, and dietary supplements—to restore or preserve health or to manage medical disorders has increased in the past several decades. Yet, clinicians rarely advise their patients about what to eat and what not to eat while taking these prescribed medications. There has been a major emphasis on the application of current knowledge about drug-drug interactions; however, except for a few notable drug-nutrient interactions (e.g., with grapefruit juice, alcohol, and caffeine), few healthcare practitioners are aware of or advise their patients on how to minimize the risks associated with drug-nutrient interactions (Boullata and Armenti 2004).

A fundamental goal of molecular life science research is to understand the complex relationship between genotype and phenotype. With the technological advances of the genomic era and the ability to decipher the molecular makeup of cells at increasing precision and resolution, it has become clear that the "one gene, one protein, one function" paradigm does not fully explain the complex functional phenotypes of organisms. Thus, the underlying assumptions of systems biology are that the functional diversity of cells results from multiple layers beyond the genome and that, to gain a comprehensive understanding of complex biological processes, molecules and their relationships must be studied as part of integrated systems. The systems-based approach is dependent on the detection and precise quantification of the molecular diversity of the cell at the levels of the transcriptome, proteome, and interactome (Bludau and Aebersold 2020).

Cellular functions and responses are coordinated by proteins working in concert through networks of PPIs, often involving higher-order complexes. Understanding the architecture of this interactome from a dynamic, topological, and quantitative perspective is key to discerning biological processes and their involvement in disease, such as drug development (Salas et al. 2020). Several proteins perform biochemical functions as part of multimolecular assemblies. These assemblies form a large network of molecular interactions, including interactions between molecules of the same type, e.g., PPIs, or between molecules of different types, e.g., protein-DNA, protein-RNA, protein-lipid, or protein-metabolite interactions. The vast array of molecular interactions in the cell is referred to as its interactome (Bludau and Aebersold 2020).

An interactome consists of the evaluation of all interactions between biomolecules in an organism. However, the protein-protein interaction is the most extensively studied interaction within cells once it shows a more functional connection between them (Cho et al. 2012; Dai and Shen 2022). PPI is characterized by an MS-based method coupled with chromatography, enabling faster identification of multiple interacting proteins without the need for antibodies. This approach offers high throughput and allows for the confirmation of molecules through immunofluorescence (Bader and Hogue 2002).

The methods to define the protein interactome of oligonucleotide (ON)-based drugs are more limited. An in vitro assay is often applied in which an ON drug is immobilized on beads and incubated with cell lysate, which is followed by the elution of the protein binders with a competing ON drug. This assay was successfully used to identify a large part of the currently approximately 80 known antisense oligonucleotide (ASO) binders and forms the basis of today's knowledge on ON drug-protein interactions and how they modulate the pharmacological properties of ON drugs. Nevertheless, the assay is limited to in vitro application and, thus, cannot detect interactome changes induced by stress or interactomes specific to subcellular compartments (Hanswillemenke et al. 2024).

Multi-omics has the advantage of managing several data with high throughput, useful for the description and interpretation of any biological systems. In the near future, the application of bioinformatics approaches could be a fundamental tool to support novel genomic technologies in toxicogenomics studies (Chiarella et al. 2023). Toxicogenomics, the application of genomics to toxicology, was described as "a new era" for toxicology. Genetic toxicity testing is an essential part of drug safety assessment since DNA damage can lead to genetic changes, including mutations and chromosome damage (David 2020).

In this sense, toxicogenomics can be considered as a subdiscipline of toxicology that assesses toxicity using a combination of genomics, advanced computational technologies, and other high-throughput techniques (Chiarella et al. 2023). Toxicogenomics also combines the study of genetics with the toxicological aspect of drugs/chemicals and plays an important role in connecting personalized medicine and toxicology. Moreover, this field is concerned with the study of the mechanisms of toxicity, the identification of susceptible populations, and the development of strategies for risk assessment and risk management (Singh et al. 2023).

Regarding genetic toxicology, toxicogenomics, through the use of proteomics and metabolomics, has highlighted the value of these novel methodologies (David 2020). Metabolomics and lipidomics are more recent and less developed, although technical improvements made in the last few years, including pathway-driven bioinformatics and expanding curated databases for small biomolecules, are steadily convincing toxicologists of their potentialities. On the other hand, toxicoproteomics can be applied to global protein expression analysis technologies for toxicological and clinical research (Chiarella et al. 2023).

The computational technologies represent a multidisciplinary tool that integrates and incorporates information from various sources such as databases to develop mathematical-statistical and computer-based models (i.e., ML algorithms) (Chiarella et al. 2023). ML is a branch of artificial intelligence (AI) in which computers can learn from data and make predictions. This can be fast, cost-effective, and highly accurate, and ML is being used to predict toxicity. NGS generates large data sets that are applicable to ML and could be used to classify compounds and predict toxicity (David 2020). AI tools can make predictions on adverse outcomes using extensive volumes of data with high accuracy and efficiency (Chiarella et al. 2023).

In the technologies context, the Comparative Toxicogenomics Database (CTD; http://ctdbase.org/) is a robust, toxicogenomics publicly available database that aims to advance understanding of how environmental exposures affect human health. It provides manually curated information about chemical-gene/protein interactions and chemical-disease and gene-disease relationships. These data are integrated with functional and pathway data to aid in the development of hypotheses about the mechanisms underlying environmentally influenced diseases. Moreover, CTD harmonizes cross-species heterogeneous data for chemical exposures and their biological repercussions by manually curating and interrelating chemical, gene, phenotype, anatomy, disease, taxa, and exposure content from the published literature (Davis et al. 2023).

In the last update 2023, CTD reports a 20% increase in overall CTD content for 17,100 chemicals, 54,300 genes, 6100 phenotypes, 7270 diseases, and 202,000 exposure statements. This database has the goal of exploring ways to provide additional functionality to the CTD Tetramers tool, such as allowing users to start the query with a chemical of interest (in addition to the existing phenotype- and/or disease-based functionality) to generate tetramers for that compound (Davis et al. 2023). Omics tools represent the future, in association with emerging and innovative technologies to reduce costs and manage huge data sets, to understand DDI, DNI, DGI, DTI, and DNI concerning individual characteristics, making up PM.

## 5 Concluding Remarks

The key elements required for a full understanding of precision medicine. (i) Genomics, to get a deeper understanding of the disease by discovering specific genetic changes (in the genome of the patients) that have an association with the disease. (ii) Transcriptomics, a technique that obtains information on the abundance of multiple mRNA transcripts within a biological sample simultaneously. (iii) Proteomics, completing and confirming the information obtained through genomics by studying all of the proteins. (iv) Metabolomics investigates the cellular metabolites from biological processes to obtain a physiological snapshot of the cell. (v) Interactome consists of the evaluation of all interactions between biomolecules in an organism. Several molecules work together to maintain cellular functions, and the connection between them involves interactions between DNA-RNA, RNA-RNA, DNA-protein, RNA-protein, protein-protein, and protein-metabolite.

# 6 Self-Assessment Questions and Answers

## 6.1 Questions

1. Why do people with the same disease show tissue responses to the treatment with the same drug?
2. What are applications (PGx) for interindividual variations in DNA sequence related to drug response?
3. What is the applicability of transcriptomics and proteomics for drug-drug interactions?
4. How can in silico analysis accelerate drug discovery?
5. Most diseases promote metabolic alterations. How to explain this in drug interaction and metabolite?
6. What is the relation between interactome and drugs?

## 6.2 Answers

1. Answer: Pharmacogenetics is perhaps the first precision medicine application and is defined as a study of genetic differences and their effect on drug metabolism pathways, which ultimately lead to different responses to the same drug in different individuals according to their genetic characteristics.
2. Answer: PGx tools assess the interindividual variations in DNA sequence related to drug efficacy and toxicity and provide routine precision medicine.
3. Answer: The unified analyses of both these data types have been rewarding for several biological objectives like improving genome annotation, predicting RNA-protein quantities, deciphering gene regulations, and discovering drug targets.
4. Answer: Data-generated interpretation by proteomic approaches requires software and computational analyses. Moreover, in silico approaches have been of considerable interest in terms of time, labor, and costs. Many new drug compounds have been successfully developed using computational methods.
5. Answer: Most diseases induce metabolic alterations, significantly impacting drug interactions and metabolite profiles. These alterations can change the activity of metabolic enzymes, such as cytochrome P450, which are essential for drug metabolism. For instance, liver diseases may impair the liver's ability to metabolize drugs, leading to increased drug levels in the bloodstream and potential toxicity. Additionally, gastrointestinal disorders can affect drug absorption, as altered gut flora or changes in pH may influence the solubility and bioavailability of medications, resulting in variations in therapeutic effects.
6. Answer: An interactome is associated with molecular interactions, including interactions between molecules of the same type or between molecules of different types. This interaction can influence the drug effect or drug response, respectively.

## References

Abbasi K, Razzaghi P, Poso A, Ghanbari-Ara S, Masoudi-Nejad A. Deep learning in drug target interaction prediction: current and future perspectives. Curr Med Chem. 2021;28(11):2100–13. https://doi.org/10.2174/0929867327666200907141016.

Alfaro JA, Bohländer P, Dai M, Filius M, Howard CJ, van Kooten XF, et al. The emerging landscape of single-molecule protein sequencing technologies. Nat Methods. 2021;18(6):604–17. https://doi.org/10.1038/s41592-021-01143-1.

Aroda VR, Edelstein SL, Goldberg RB, Knowler WC, Marcovina SM, Orchard TJ, et al. Long-term metformin use and vitamin B12 deficiency in the diabetes prevention program outcomes study. J Clin Endocrinol Metab. 2016;101(4):1754–61. https://doi.org/10.1210/jc.2015-3754.

Astarita G, Kelly RS, Lasky-Su J. Metabolomics and lipidomics strategies in modern drug discovery and development. Drug Discov Today. 2023;28(10):103751. https://doi.org/10.1016/j.drudis.2023.103751.

Azad AKM, Dinarvand M, Nematollahi A, Swift J, Lutze-Mann L, Vafaee F. A comprehensive integrated drug similarity resource for in-silico drug repositioning and beyond. Brief Bioinform. 2020;22(3):bbaa126. https://doi.org/10.1093/bib/bbaa126.

Backman JT, Filppula AM, Niemi M, Neuvonen PJ. Role of cytochrome P450 2C8 in drug metabolism and interactions. Koulu M, editor. Pharmacol Rev. 2016;68(1):168–241. https://doi.org/10.1124/pr.115.011411.

Bader GD, Hogue CW. Analyzing yeast protein-protein interaction data obtained from different sources. Nat Biotechnol. 2002;20:991–7. https://doi.org/10.1038/nbt1002-991.

Binia A, Contreras AV, Canizales-Quinteros S, Alonzo VA, Tejero ME, Silva-Zolezzi I. Geographical and ethnic distribution of single nucleotide polymorphisms within genes of the folate/homocysteine pathway metabolism. Genes Nutr. 2014;9(5):421. https://doi.org/10.1007/s12263-014-0421-7.

Bludau I, Aebersold R. Proteomic and interactomic insights into the molecular basis of cell functional diversity. Nat Rev Mol Cell Biol. 2020;21(6):327–40. https://doi.org/10.1038/s41580-020-0231-2.

Boullata JI, Armenti VT. Approaching drug–nutrient interactions. In: Handbook of drug-nutrient interactions (2004); ISBN 978-1-60327-363-3. e-ISBN 978-1-60327-362-6. https://doi.org/10.1007/978-1-60327-362-6

Carriot N, Paix B, Greff S, Viguier B, Briand JF, Culioli G. Integration of LC/MS-based molecular networking and classical phytochemical approach allows in-depth annotation of the metabolome of non-model organisms – the case study of the brown seaweed Taonia atomaria. Talanta. 2021;225:121925. https://doi.org/10.1016/j.talanta.2020.121925.

Cavallari LH, Van SL, Prows CA, Bishop JR, Limdi NA, Pratt VM, et al. Multi-site investigation of strategies for the clinical implementation of CYP2D6 genotyping to guide drug prescribing. Genet Med. 2019;21(10):2255–63. https://doi.org/10.1038/s41436-019-0484-3.

Chen C, Shi H, Jiang Z, Salhi A, Chen R, Cui X, et al. DNN-DTIs: improved drug-target interactions prediction using XGBoost feature selection and deep neural network. Comput Biol Med. 2021;136:104676. https://doi.org/10.1016/j.compbiomed.2021.104676.

Chiarella P, Capone P, Sisto R. Contribution of Genetic Polymorphisms in Human Health. Int J Environ Res Public Health. 2023;20(2):912. https://doi.org/10.3390/ijerph20020912.

Cho DY, Kim YA, Przytycka TM. Chapter 5: Network biology approach to complex diseases. PLoS Comput Biol. 2012;8(12):e1002820. https://doi.org/10.1371/journal.pcbi.1002820.

Cicchese JM, Sambarey A, Kirschner D, Linderman JJ, Chandrasekaran S. A multi-scale pipeline linking drug transcriptomics with pharmacokinetics predicts in vivo interactions of tuberculosis drugs. Sci Rep. 2020;11(1):5643. https://doi.org/10.1038/s41598-021-84827-0.

Dai X, Shen L. Advances and trends in omics technology development. Front Med. 2022;9:911861. https://doi.org/10.3389/fmed.2022.911861.

David R. The promise of toxicogenomics for genetic toxicology: past, present and future. Mutagenesis. 2020;35(2):153–9. https://doi.org/10.1093/mutage/geaa007.

Davis AP, Wiegers TC, Johnson RJ, Sciaky D, Wiegers J, Mattingly CJ. Comparative Toxicogenomics Database (CTD): update 2023. Nucleic Acids Res. 2023;51(D1):D1257–62. https://doi.org/10.1093/nar/gkac833.

De Cesare V, Moran J, Traynor R, Knebel A, Ritorto MS, Trost M, et al. High-throughput matrix-assisted laser desorption/ionization time-of-flight (MALDI-TOF) mass spectrometry-based deubiquitylating enzyme assay for drug discovery. Nat Protoc. 2020;15(12):4034–57. https://doi.org/10.1038/s41596-020-00405-0.

do Valle IF, Roweth HG, Malloy MW, Moco S, Barron D, Battinelli E, et al. Network medicine framework shows that proximity of polyphenol targets and disease proteins predicts therapeutic effects of polyphenols. Nat Food. 2021;2(3):143–55. https://doi.org/10.1038/s43016-021-00243-7.

Drake ZC, Seffernick JT, Lindert S. Protein complex prediction using Rosetta, AlphaFold, and mass spectrometry covalent labeling. Nat Commun. 2022;13:7846. https://doi.org/10.1038/s41467-022-35593-8.

Fang J, Chen Q, Wu G. Genomics-based tools for drug discovery and development: from network maps to efficacy prediction. J Holist Integr Pharm. 2023;4(3):199–209. https://doi.org/10.1016/j.jhip.2023.11.001.

Feng Y, Wang Q, Wang T. Drug target protein-protein interaction networks: a systematic perspective. Biomed Res Int. 2017;2017:1289259. https://doi.org/10.1155/2017/1289259.

Fountzilas E, Tsimberidou AM, Vo HH, Kurzrock R. Clinical trial design in the era of precision medicine. Genome Med. 2022;14(1):101. https://doi.org/10.1186/s13073-022-01102-1.

Gholami N, Haghparast A, Alipourfard I, Nazari M. Prostate cancer in omics era. Cancer Cell Int. 2022;22(1):274. https://doi.org/10.1186/s12935-022-02691-y.

Gonzalez-Covarrubias V, Martínez-Martínez E, Del Bosque-Plata L. The potential of metabolomics in biomedical applications. Metabolites. 2022;12(2):194. https://doi.org/10.3390/metabo12020194.

Gopisankar MG. CYP2D6 pharmacogenomics. Egypt J Med Hum Genet. 2017;18(4):309–13. https://doi.org/10.1016/j.ejmhg.2017.03.001.

Grześk G, Rogowicz D, Wołowiec Ł, Ratajczak A, Gilewski W, Chudzińska M, et al. The clinical significance of drug–food interactions of direct oral anticoagulants. Int J Mol Sci. 2021;22(16):8531. https://doi.org/10.3390/ijms22168531.

Gummadi AC, Guddati AK. Genetic polymorphisms in pharmaceuticals and chemotherapy. World J Oncol. 2021;12(5):149–54. https://doi.org/10.14740/wjon1405

Gurung AB, Ali MA, Lee J, Farah MA, Al-Anazi KM. An updated review of computer-aided drug design and its application to COVID-19. Biomed Res Int. 2021;2021:8853056. https://doi.org/10.1155/2021/8853056.

Hanswillemenke A, Hofacker DT, Sorgenfrei M, Fruhner C, Franz-Wachtel M, Schwarzer D, et al. Profiling the interactome of oligonucleotide drugs by proximity biotinylation. Nat Chem Biol. 2024;20(5):555–65. https://doi.org/10.1038/s41589-023-01530-z.

Hiraoka M, Kagawa Y. Genetic polymorphisms and folate status. Congenit Anom. 2017;57(5):142–9. https://doi.org/10.1111/cga.12232.

Horta MAC, Pimenta RJG, Almeida DA, Rosolen RR, Aono AH, Filho JF, et al. Transcriptomic analysis of genes: expression and regulation. In: Ajmal AM, Lee J, editors. Transcriptome profiling. 1st ed. London: Academic Press; 2023. p. 1–41.

Hu S, Liu C, Liu X. Innovative application of metabolomics on bioactive ingredients of foods. Foods. 2022;11(19):2974. https://doi.org/10.3390/foods11192974.

Jeon J, Kim KT, Choi J, Cheong K, Ko J, Choi G, et al. Alternative splicing diversifies the transcriptome and proteome of the rice blast fungus during host infection. RNA Biol. 2022;19(1):373–85. https://doi.org/10.1080/15476286.2022.2043040.

Ji Y, Mishra RK, Davuluri RV. In silico analysis of alternative splicing on drug-target gene interactions. Sci Rep. 2020;10(1):134. https://doi.org/10.1038/s41598-019-56894-x.

Jin Q, Ma RCW. Metabolomics in diabetes and diabetic complications: insights from epidemiological studies. Cells. 2021;10(11):2832. https://doi.org/10.3390/cells10112832.

Johnson CH, Ivanisevic J, Siuzdak G. Metabolomics: beyond biomarkers and towards mechanisms. Nat Rev Mol Cell Biol. 2016;17:451–9. https://doi.org/10.1038/nrm.2016.25.

Kabbani D, Akika R, Wahid A, Daly AK, Cascorbi I, Zgheib NK. Pharmacogenomics in practice: a review and implementation guide. Front Pharmacol. 2023;14:189976. https://doi.org/10.3389/fphar.2023.1189976.

Kankanige D, Liyanage L, O'Connor MD. Application of transcriptomics for predicting protein interaction networks, drug targets and drug candidates. Front Med Technol. 2022;4:693148. https://doi.org/10.3389/fmedt.2022.693148.

Karadima V, Kraniotou C, Bellos G, Tsangaris GT. Drug-micronutrient interactions: food for thought and thought for action. EPMA J. 2016;7(1):10. https://doi.org/10.1186/s13167-016-0059-1.

Khodadadian A, Darzi S, Haghi-Daredeh S, Sadat Eshaghi F, Babakhanzadeh E, Mirabutalebi SH, et al. Genomics and transcriptomics: the powerful technologies in precision medicine. Int J Gen Med. 2020;13:627–40. https://doi.org/10.2147/IJGM.S249970.

Kim J, Choi JP, Kim MS, Bhak J. PharmaKoVariome database for supporting genetic testing. Database. 2022;2022:baac092. https://doi.org/10.1093/database/baac092.

Liu L, Tang Y, Baxter GD, Yin H, Tumilty S. Complementary and alternative medicine – practice, attitudes, and knowledge among healthcare professionals in New Zealand: an integrative review. BMC Complement Med Ther. 2021;21(1):63. https://doi.org/10.1186/s12906-021-03235-z.

Lowe R, Shirley N, Bleackley M, Dolan S, Shafee T. Transcriptomics technologies. PLOS Comput Biol. 2017;13(5):e1005457. https://doi.org/10.1371/journal.pcbi.1005457.

Lu H, Zhou Q, He J, Jiang Z, Peng C, Tong R, et al. Recent advances in the development of protein-protein interactions modulators: mechanisms and clinical trials. Signal Transduct Target Ther. 2020;5(1):213. https://doi.org/10.1038/s41392-020-00315-3.

Luo Q, Mo S, Xue Y, Zhang X, Gu Y, Wu L, et al. Novel deep learning-based transcriptome data analysis for drug-drug interaction prediction with an application in diabetes. BMC Bioinformatics. 2021;22(1):318. https://doi.org/10.1186/s12859-021-04241-1.

Macalino SJY, Gosu S, Hong S, Choi S. Role of computer-aided drug design in modern drug discovery. Arch Pharm Res. 2015;38(9):1686–701. https://doi.org/10.1007/s12272-015-0640-5.

Mahajna M, Abu Fanne R, Berkovitch M, Tannous E, Vinker S, Green I, Effect of CYP2C19 pharmacogenetic testing on predicting citalopram and escitalopram tolerability and efficacy: a retrospective, longitudinal cohort study. Biomedicines. 2023;11(12):3245. https://doi.org/10.3390/biomedicines11123245.

Malachowska B, Yang WL, Qualman A, Muro I, Boe DM, Lampe JN, et al. Transcriptomics, metabolomics, and in-silico drug predictions for liver damage in young and aged burn victims. Commun Biol. 2023;6(1):597. https://doi.org/10.1038/s42003-023-04964-2.

Mazari AMA, Zhang L, Ye ZW, Zhang J, Tew KD, Townsend DM. The multifaceted role of glutathione S-transferases in health and disease. Biomolecules. 2023;13(4):688. https://doi.org/10.3390/biom13040688.

Mazokopakis EE, Papadomanolaki MG, Papadakis JA. The effects of folinic acid and l-methylfolate supplementation on serum total homocysteine levels in healthy adults. Clin Nutr ESPEN. 2023;58:14–20. https://doi.org/10.1016/j.clnesp.2023.09.002.

McMahon A, McNulty H, Hughes C, Strain J, Ward M. Novel approaches to investigate one-carbon metabolism and related B-vitamins in blood pressure. Nutrients. 2016;8(11):720. https://doi.org/10.3390/nu8110720.

McQuade BM, Campbell A. Drug prescribing: drug-drug interactions. FP Essent. 2021;508:25–32.

Meng X, Xu J, Zhang M, Du R, Zhao W, Zeng Q, et al. Third-generation sequencing and metabolome analysis reveal candidate genes and metabolites with altered levels in albino jackfruit seedlings. BMC Genomics. 2021;22:543. https://doi.org/10.1186/s12864-021-07873-y.

Mohn E, Kern H, Saltzman E, Mitmesser S, McKay D. Evidence of drug–nutrient interactions with chronic use of commonly prescribed medications: an update. Pharmaceutics. 2018;10(1):36. https://doi.org/10.3390/pharmaceutics10010036.

Mortishire-Smith RJ, Zhang H, Bateman KP. High-resolution mass spectrometry and drug metabolite identification. In: Lee MS, Zhu M, editors. Mass spectrometry in drug metabolism and disposition: basic principles and applications. 1st ed. New Work: Wiley; 2011. p. 407–48.
Mulder C, Leijten N, Lemeer S. Proteomic tools to study drug function. Curr Opin Syst Biol. 2018;10:9–18. https://doi.org/10.1016/j.coisb.2018.05.002.
Ngiam KY, Khor W. Big data and machine learning algorithms for health-care delivery. Lancet Oncol. 2019;20:e262–73. https://doi.org/10.1016/S1470-2045(19)30149-4.
Nguyen N, Jennen D, Kleinjans J. Omics technologies to understand drug toxicity mechanisms. Drug Discovery Today. 2022;27(11):103348. ISSN 1359-6446. https://doi.org/10.1016/j.drudis.2022.103348
Ortiz-Martínez M, González-González M, Martagón AJ, Hlavinka V, Willson RC, Rito-Palomares M. Recent Developments in Biomarkers for Diagnosis and Screening of Type 2 Diabetes Mellitus. Curr Diab Rep. 2022;22(3):95–115. https://doi.org/10.1007/s11892-022-01453-4.
Owen MD, Baker BC, Scott EM, Forbes K. Interaction between metformin, folate and vitamin B12 and the potential impact on fetal growth and long-term metabolic health in diabetic pregnancies. Int J Mol Sci. 2021;22(11):5759. https://doi.org/10.3390/ijms22115759.
Pandey KB, Rizvi SI. Plant polyphenols as dietary antioxidants in human health and disease. Oxidative Med Cell Longev. 2009;2(5):270–8. https://doi.org/10.4161/oxim.2.5.9498.
Park Y, Heider D, Hauschild AC. Integrative analysis of next-generation sequencing for next-generation cancer research toward artificial intelligence. Cancers. 2021;13(13):3148. https://doi.org/10.3390/cancers13133148.
Pecikoza U, Tomić M, Nastić K, Micov A, Stepanović-Petrović R. Synergism between metformin and analgesics/vitamin B12 in a model of painful diabetic neuropathy. Biomed Pharmacother. 2021;153:113441. https://doi.org/10.1016/j.biopha.2022.113441.
Pratama S, Lauren BC, Wisnu W. The efficacy of vitamin B12 supplementation for treating vitamin B12 deficiency and peripheral neuropathy in metformin-treated type 2 diabetes mellitus patients: a systematic review. Diabetes Metab Syndr. 2022;16(10):102634. https://doi.org/10.1016/j.dsx.2022.102634.
Pratt VM, Cavallari LH, Del Tredici AL, Gaedigk A, Hachad H, Ji Y, et al. Recommendations for clinical CYP2D6 genotyping allele selection: a joint consensus recommendation of the association for molecular pathology, college of american pathologists, Dutch pharmacogenetics working group of the royal dutch pharmacists association, and the european society for pharmacogenomics and personalized therapy. J Mol Diagn. 2021;23(9):1047–64. https://doi.org/10.1016/j.jmoldx.2021.05.013.
Pratt VM, Cavallari LH, Fulmer ML, Gaedigk A, Hachad H, Ji Y, et al. CYP3A4 and CYP3A5 genotyping recommendations: a joint consensus recommendation of the association for molecular pathology, clinical pharmacogenetics implementation consortium, college of american pathologists, Dutch pharmacogenetics working group of the royal dutch pharmacists association, european society for pharmacogenomics and personalized therapy, and pharmacogenomics knowledgebase. J Mol Diagn. 2023;25(9):619–29. https://doi.org/10.1016/j.jmoldx.2023.06.008.
Rabinowitz JD, Purdy JG, Vastag L, Shenk T, Koyuncu E. Metabolomics in drug target discovery. Cold Spring Harb Symp Quant Biol. 2011;76:235–46. https://doi.org/10.1101/sqb.2011.76.010694.
Raghubeer S, Matsha TE. Methylenetetrahydrofolate (MTHFR), the one-carbon cycle, and cardiovascular risks. Nutrients. 2021;13(12):4562. https://doi.org/10.3390/nu13124562.
Richard P, Horgan LCK. "Omic" technologies: genomics, transcriptomics, proteomics and metabolomics. Obstet Gynaecol. 2011;13:189–95. https://doi.org/10.1576/toag.13.3.189.27672.
Sabe VT, Ntombela T, Jhamba LA, Maguire GEM, Govender T, Naicker T, et al. Current trends in computer aided drug design and a highlight of drugs discovered via computational techniques: A review. Eur J Med Chem. 2021;224:113705. https://doi.org/10.1016/j.ejmech.2021.113705.

Salas D, Stacey RG, Akinlaja M, Foster LJ. Next-generation interactomics: considerations for the use of co-elution to measure protein interaction networks. Mol Cell Proteomics. 2020;19(1):1–10. https://doi.org/10.1074/mcp.R119.001803.

Saviano A, Henderson NC, Baumert TF. Single-cell genomics and spatial transcriptomics: discovery of novel cell states and cellular interactions in liver physiology and disease biology. J Hepatol. 2020;73(5):1219–30. https://doi.org/10.1016/j.jhep.2020.06.004.

Searle BC. Scaffold: a bioinformatic tool for validating MS/MS-based proteomic studies. Proteomics. 2010;10(6):1265–9.

Seneviratne HK, Hamlin AN, Heck CJS, Bumpus NN. Spatial distribution profiles of emtricitabine, tenofovir, efavirenz, and rilpivirine in murine tissues following in vivo dosing correlate with their safety profiles in humans. ACS Pharmacol Transl Sci. 2020;3(4):655–65. https://doi.org/10.1021/acsptsci.0c00015.

Seyhan AA, Carini C. Are innovation and new technologies in precision medicine paving a new era in patients centric care? J Transl Med. 2019;17(1):114. https://doi.org/10.1186/s12967-019-1864-9.

Singh AV, Chandrasekar V, Paudel N, Laux P, Luch A, Gemmati D, et al. Integrative toxicogenomics: Advancing precision medicine and toxicology through artificial intelligence and OMICs technology. Biomed Pharmacother. 2023;163:114784. https://doi.org/10.1016/j.biopha.2023.114784.

Skadrić I, Stojković O. Defining screening panel of functional variants of CYP1A1, CYP2C9, CYP2C19, CYP2D6, and CYP3A4 genes in Serbian population. Int J Legal Med. 2019;134(2):433–9. https://doi.org/10.1007/s00414-019-02234-7.

Stouras I, Papaioannou T, Tsioufis K, Eliopoulos A, Sanoudou D. The challenge and importance of integrating drug–nutrient–genome interactions in personalized cardiovascular healthcare. J Pers Med. 2022;12(4):513. https://doi.org/10.3390/jpm12040513.

Supplitt S, Karpinski P, Sasiadek M, Laczmanska I. Current achievements and applications of transcriptomics in personalized cancer medicine. Int J Mol Sci. 2021;22(3):1422. https://doi.org/10.3390/ijms22031422.

Szklarczyk D, Gable AL, Lyon D, Junge A, Wyder S, Huerta-Cepas J, et al. STRING v11: protein-protein association networks with increased coverage, supporting functional discovery in genome-wide experimental datasets. Nucleic Acids Res. 2019;47(D1):D607–13. https://doi.org/10.1093/nar/gky1131.

Taya N, Katakami N, Omori K, Arakawa S, Hosoe S, Watanabe H, et al. Evaluation of change in metabolome caused by comprehensive diabetes treatment: a prospective observational study of diabetes inpatients with gas chromatography/mass spectrometry-based non-target metabolomic analysis. J Diabetes Investig. 2021;12:2232–41. https://doi.org/10.1111/jdi.13600.

Tyanova S, Temu T, Cox J. The MaxQuant computational platform for mass spectrometry-based shotgun proteomics. Nat Protoc. 2016;11:2301–19. https://doi.org/10.1038/nprot.2016.136.

Vincenzi M, Mercurio FA, Leone M. Protein interaction domains: structural features and drug discovery applications (Part 2). Curr Med Chem. 2021;28(5):854–92. https://doi.org/10.2174/0929867327666200114114142.

Wheeler K, Gosmanov C, Jimenez Sandoval M, Yang Z, McCall LI. Frontiers in mass spectrometry-based spatial metabolomics: current applications and challenges in the context of biomedical research. TrAC Trends Analyt Chem. 2024;175:117713. https://doi.org/10.1016/j.trac.2024.117713.

Wilhelm M, Schlegl J, Hahne H, Gholami AM, Lieberenz M, Savitski MM, et al. Mass-spectrometry-based draft of the human proteome. Nature. 2014;509:582–7. https://doi.org/10.1038/nature13319.

Wilcken B, Bamforth F, Li Z, Zhu H, Ritvanen A, Renlund M, et al. Geographical and ethnic variation of the 677C>T allele of 5,10 methylenetetrahydrofolate reductase (MTHFR): findings from over 7000 newborns from 16 areas world wide. J Med Genet. 2003;40(8):619–25. https://doi.org/10.1136/jmg.40.8.619.

Wishart DS. Emerging applications of metabolomics in drug discovery and precision medicine. Nat Rev Drug Discov. 2016;15(7):473–84. https://doi.org/10.1038/nrd.2016.32.

Wishart DS, Feunang YD, Marcu A, Guo AC, Liang K, Vázquez-Fresno R, et al. HMDB 4.0: the human metabolome database for 2018. Nucleic Acids Res. 2018;46:D608–17. https://doi.org/10.1093/nar/gkx1089.

Xie X, Yu T, Li X, Zhang N, Foster LJ, Peng C, et al. Recent advances in targeting the "undruggable" proteins: from drug discovery to clinical trials. Signal Transduct Target Ther. 2023;8(1):335. https://doi.org/10.1038/s41392-023-01589-z.

Zhao M, Ma J, Li M, Zhang Y, Jiang B, Zhao X, et al. Cytochrome P450 enzymes and drug metabolism in humans. Int J Mol Sci. 2021;22(23):12808. https://doi.org/10.3390/ijms222312808.

Zhou SF, Ming DY, Chan E, Du YM, Chow V, Xue C, et al. Clinical pharmacogenetics and potential application in personalized medicine. Curr Drug Metab. 2008;9(8):738–84. https://doi.org/10.2174/138920008786049302.

# Chapter 11
# Photosensitizers Interactions: From Basic Concepts to Clinical Applications

**Anselmo F. R. Rodriguez, Anderson L. Ramos, and Christiane Pavani**

**Abstract** Photodynamic therapy (PDT) is a therapeutic modality for cancer and various non-malignant diseases, including infections, which uses an interaction of light with photosensitizers (PS) and photoactivation using a specific wavelength of light in the presence of molecules. The PS agent is first administered to the patient, either orally or topically, and is then absorbed by the abnormal cells. Photosensitizers are non-toxic dyes that are activated by the absorption of visible light to initially form the excited singlet state, followed by the transition to the long-lived excited triplet state generating cytotoxic reactive oxygen species (ROS), thus inducing apoptosis, necrosis, and autophagy of the cells. PDT has been approved as a non-invasive option for the treatment of anti-tumors and other diseases, such as lung, esophageal, and bladder cancer.

**Keywords** Natural photosensitizers · Photodynamic therapy · Photochemistry · Photophysics

## 1 Introduction

The origin of photodynamic therapy (PDT) can be traced back to the pioneering experiments conducted in the early nineteenth century by Oscar Raab and Tappenier in Munich (Kessel 2019). In a series of fortuitous experiments, it was observed that microorganisms such as paramecia, incubated with specific dyes, succumbed to

A. F. R. Rodriguez (✉) · A. L. Ramos
Laboratory of Nanobiotechnology, Federal University of Acre, Rio Branco, AC, Brazil
e-mail: anselmo.rodriguez@ufac.br

C. Pavani
Post Graduation Program in Biophotonics-Medicine, Universidade Nove de Julho – UNINOVE, São Paulo, Brazil



J. O. Fajemiroye (ed.), *Fundamentals of Drug and Non-Drug Interactions*,
https://doi.org/10.1007/978-3-031-80107-5_11

sunlight exposure while remaining unharmed in darkness (Sperandio et al. 2013). Specifically, a 1:20,000 acridine solution would kill paramecia exposed to direct sunlight in 6 min. In diffused light, it took 1 h to kill the paramecia, while there was no lethal action in the dark (Pucelik and Dąbrowski 2021). The light was later shown to be involved in the excitation of the PS with a specific wavelength, triggering reactions that were later better elucidated (Algorri 2021).

The PDT combines three components: light, PS, and oxygen, which trigger photophysical and photochemical mechanisms generating a cascade of oxidative reactions that trigger a biological response (Moan and Peng 2003; Monge-Fuentes et al. 2017). At the first stage, the PS absorbs light at a specific wavelength, reaching an electronically excited singlet state, following the transition to a triplet state, which interacts with oxygen by two main mechanisms: type I, by electron transfer, producing reactive oxygen species such as hydroxyl radicals, superoxide anion radical, hydrogen peroxide, and others; and type II, by energy transfer, generating singlet oxygen (Grandi et al. 2022; Westermayr and Marquetand 2021). These oxidative molecules damage important macromolecules from the cell structure and function, such as lipids, proteins, and nucleic acids, causing the death of microorganisms and cancer cells (Dolmans et al. 2003).

Visible light is the main part of the electromagnetic spectrum used for PS activation (Austin et al. 2021). Especially, around 600–800 nm, the "therapeutic window," penetrates deeper into the soft tissue and contains suitable energy to activate the PS (Alvarez and Sevilla 2024).

A wide range of molecules may be used as PS: porphyrins, chlorins, phthalocyanines, phenothiazinium dyes, xantenes, metal complexes, and natural products (Lima and Reis 2023). The ideal PS for PDT possess a constellation of indispensable characteristics, each essential for the safety and efficacy of the treatment (Gallardo-Villagrán et al. 2019).

Foremost, the PS must exhibit non-toxicity in the absence of light exposure (Muniyandi et al. 2020). This mandates that its application should not elicit harmful effects, and its metabolic processes should refrain from generating toxic by-products (Gurcan Gunaydin et al. 2021). Moreover, the PS must be devoid of carcinogenic and mutagenic properties to ensure patient safety, thereby mitigating the risk of long-term adverse effects (Agostinis et al. 2011).

Additionally, an optimal PS should demonstrate exquisite target selectivity, preferentially accumulating in the target tissue while sparing adjacent healthy tissues (Celli et al. 2010). This targeted localization not only enhances therapeutic efficacy but also minimizes collateral damage, thereby optimizing patient outcomes (Polat and Kang 2021). Furthermore, swift elimination from the body is imperative to mitigate the risk of patient photosensitivity, underscoring the importance of efficient clearance pathways (Gunaydin et al. 2021).

Practical considerations also come into play, necessitating that the synthesis and purification of the PS be straightforward (Sarbadhikary et al. 2021). Ease of production facilitates scalability and widespread accessibility, ensuring that cost considerations do not impede the utilization of PDT as a therapeutic modality (Hernández-Bule et al. 2024). Thus, the ideal PS embodies a delicate balance of safety, efficacy, and practicality. Considering the first PS discovered and the recent developments, PSs may be classified into three generations:

## 1.1 First-Generation PS

The first generation of PS emerged from hematoporphyrin in 50 Schwarz (Daniell and Hill 1991) treated hematoporphyrin with acetic acid and 5% sulfuric acid at room temperature for 30 min, obtaining a purple-colored mixture that was later called hematoporphyrin derivatives (HPD) (Fig. 11.1).

Subsequently, Lipson and collaborators used this preparation injected into an animal model in four types of tumors at the time: rhabdomyosarcoma, mammary adenocarcinoma, Walker 256 carcinoma, and Flexner-Jobling Shawarz carcinoma, s/d apud (Lipson et al. 1961). During the surgical experiment, the tumor cells stood out due to the fluorescence emitted against the white light, which optimized the visualization of the tissue during the procedure (Moan et al. 1982). This first generation of PS was studied for a long time in several studies and clinical experiments (Kato 1996).

During these experiments, it was found that hematoporphyrin had some disadvantages such as low selectivity with around 0.1–3% of the total amount of PS being found in the tumor tissue (Wöhrle et al. 1998); prolonged skin sensitivity, which made it necessary to avoid sunlight for weeks (Baas et al. 1995); the low absorption in the red region of the spectrum above 600 nm; and little penetration into the treated tissues, around 5 mm (Spikes 1990). There were still some uncertainties related to the predominant structures in these HPDs, whether ethers or esters, and whether the side chains were formed mainly by vinyl groups or hydroxyethyl groups (Dougherty et al. 1998). Subsequently, many of these limitations were improved through HPLC

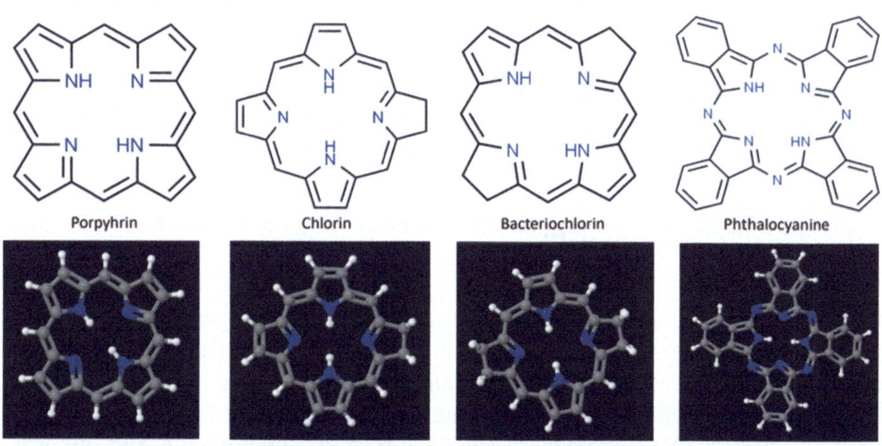


**Fig. 11.1** Basic structure of cyclic tetrapyrrolic of porphyrins and analogs (Souris et al. 2023)

purification and gel permeation chromatographic techniques (Ismail et al. 1998). Some monomer fractions were removed to give rise to the highly active commercial form that is currently known as Photofrin® (Kennedy et al. 1990).

## *1.2 Second-Generation PS*

At the beginning of the 1980s, research related to the next generation of photosensitizers was already beginning, and hundreds of candidate substances were already participating in clinical tests, as they presented characteristics of interest; but the number of substances officially approved for the clinical treatment of cancer by TDF is always small (Kwiatkowski et al. 2018). The modification of porphyrins, hematoporphyrins, and protoporphyrins already available was the most appropriate path for the development of new PS that could be prepared by partial syntheses, which generated economic and environmental advantages when compared to total and complex syntheses (Nyman and Hynninen 2004). This second generation of PS was already being sought with new criteria in mind, such as being unique substances, easy to interpret and reproduce, yet with great selectivity for the tumor tissue, and quickly excreted from the body (Gomer 1991). Longer light absorption in the red range of 675–800 nm is preferable, as there is deeper penetration into tissues, around 2–3 cm; resulting in a generation of quantum singlet oxygen ($\varphi\Delta$) greater than 0.3, characteristic of this class of PS (Wöhrle et al. 1998). Phthalocyanines, naphthalocyanines, chlorins, verdins, and different other types of hematoporphyrin derivatives already appeared in studies in several works as "second-generation" sensitizers for PDT (Spikes 1990).

These compounds, in general, have good photophysical properties and may contain metallic ions in their constitution, which may or may not be transitional, such as zinc, aluminum, magnesium, ruthenium, and osmium (Derosa and Crutchley 2002). Although this generation of PS has significantly improved compared to the first generation, a very important characteristic was not yet found in second-generation PS: the capacity for selective accumulation in a given tumor tissue (Riesz and Krishna 1988). The binding of PS in specific tissues was only possible with the development of carriers in the third generation of PS (Mfouo-Tynga et al. 2021).

## *1.3 Third-Generation PS*

In addition to the characteristics found in first- and second-generation PS, third-generation photosensitizers have properties of targeting and binding to specific types of tumor tissues (Spikes 1990; Mfouo-Tynga et al. 2021). This new capacity of PS occurs through the covalent binding of various types of biological modifiers such as amino acids, monoclonal antibodies (mABs), lipoproteins, and oligosaccharides, or they can be formulated in drug delivery systems (Drug Delivery) as

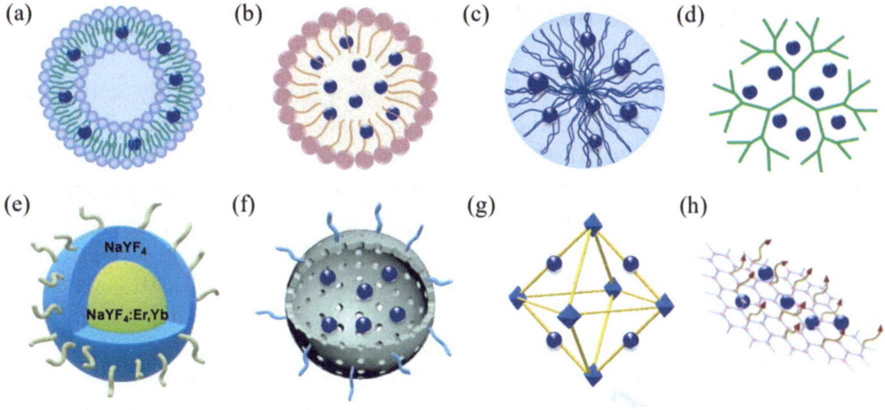


**Fig. 11.2** Schematic representation of PS delivery systems, including (**a**) lipossomes, (**b**) nanomicelles, (**c**) polymeric NPs, (**d**) dendrimers, (**e**) upconversion NPs, (**f**) mosoporus silica, (**g**) metal-organic frameworks, and (**h**) graphene-based nanomaterials (Zhang et al. 2018)

solid lipid nanoparticles—NLS, liposomes, polymeric nanoparticles, cyclodextrins, emulsions, etc. (Fig. 11.2) (Seo et al. 2023)—with the aim of improving the photophysical and pharmacokinetic properties of PS (Yankovsky et al. 2016).

Tumor cells have different antigens on their surface than normal cells; this allows the exploration of the use of monoclonal antibodies in the efficient delivery of photosensitizing biomolecules (Mfouo-Tynga et al. 2021). PS conjugated to mABs selectively bind to tumor cells, facilitating photosensitization while normal tissue remains intact (O'Connor et al. 2009). PS as pyro-pheophorbides were functionalized by Zhang et al. (2003) with different saccharide sequences. These conjugates aim to bind to glycoproteins differentially expressed on the surface of various tumor cells (Guo et al. 2023). Many of these classes of substances, such as porphyrins, are hydrophobic and tend to aggregate in aqueous systems, which ends up reducing the efficiency of PS action (Castano et al. 2004).

Recently, biodegradable and biocompatible nanomaterials combined with PS have received considerable attention, as they can overcome critical limitations presented by these molecules (Lim et al. 2012). Taking these limitations as an example, Lima et al. (2013) encapsulated Hipericin in NLS and tested it on HEp-2 and B16-F10 tumor lines. Pramual et al. (2017) incorporated 5,10,15,20-tetrakis(4-hydroxyphenyl)-21H,23H-porphin (pTHPP) into a polymer-lipid-PEG hybrid for use in PDT against cell lines of thyroid carcinoma.

## 1.4 Photosensitizers Based on Plant Extracts

Plant species considered toxic are often a source of pharmaceutically active compounds that are some of our medicines and are isolated from plants (Kubrak et al. 2022; Rodrigues et al. 2015). Due to the presence of chromophores, they easily absorb light and fluoresce blue-yellow under long waves of ultraviolet light (Zacharioudaki et al. 2022). These compounds can be produced by the plant as a protective mechanism against large doses of sunlight, which is why they are often used in sunscreens and cosmetics for this purpose (Cavinato et al. 2017). Brazil is very rich in plant biodiversity and yet there is minimal effort on the part of local scientists to exploit its potential for developing medicines based on biodiversity (Valli et al. 2018). Some studies have explored the use of photosensitizers derived from plant extracts, such as chlorophyll and other photosensitive compounds present in plants (Aziz et al. 2023). These extracts may offer a natural and potentially safer source of photosensitizers for use in PDT, avoiding synthetic and potentially toxic substances (de Oliveira et al. 2023). However, more research is needed to evaluate the efficacy and safety of these plant-based photosensitizers (Cogno et al. 2020). In recent decades, some photosensitizers based on plant extracts and natural components have been developed with some screening of plant extracts that possess the ideal characteristics of photosensitizers that can be used for antimicrobial PDT in relation to light absorption properties (Villacorta et al. 2017; Hwang et al. 2022; Dimmer et al. 2019). Photosensitizers of plant origin are products of natural origin that have been used to treat the most diverse ailments since ancient times in traditional medicine (Braga 2021). These products can be of mineral, animal, or plant origin and are generally associated with strong popular knowledge (Koehn and Carter 2005). Even with the advance of biotechnology and other techniques for obtaining new molecules, natural products continue to play a highly significant role in the discovery of new drugs and the development of new processes (Newman and Cragg 2007). During the nineteenth century, knowledge about the chemical constituents present in plants became more evident, and at the beginning of the twentieth century, the photodynamic activities of these compounds were better clarified (Senapathy et al. 2020). Chlorophyll a (Chl a), the predominant green pigment in all organisms that carry out oxygenic photosynthesis, is composed of a porphyrinated chlorine skeleton ring and is responsible for these photophysical properties (Ebermann et al. 1996). Many species of plants make up the biodiversity of the Brazilian Amazon, and this wide variety of plant organisms can contribute significantly to research and the discovery of new molecules with potential applications in PDT (Elisabetsky and Shanley 1994).

## 1.5 Basic Photophysics

Photophysics is a branch of physics that focuses on the intricate interactions between light and matter (Weight et al. 2023). It explores the fundamental processes that occur when light interacts with atoms, molecules, and solid-state materials (Ueda et al. 2019). This field encompasses a wide range of phenomena, including absorption, excitation, emission, light scattering, fluorescence, phosphorescence, energy transfer, photoionization, and photofragmentation (Ji et al. 2024). One of the fundamental aspects of photophysics is the study of transitions between excited states and a molecule's ground state (Quina and Silva 2021). These transitions play a fundamental role in determining the optical and electronic properties of materials (Gul et al. 2023). Understanding these transitions is crucial for unraveling the behavior of molecules under light excitation and designing materials with specific photophysical properties (Gong et al. 2007). This area of research has significant implications for various disciplines, including chemistry, materials science, biophysics, and engineering (Deniz et al. 2008). Basic photophysics is found within photobiology, which takes place in the processes when sunlight interacts, passes through the Earth's atmosphere is filtered by it, and interacts with the matter (atoms and molecules) present on Earth (Derosa Crutchley 2002). The spectrum of solar radiation or electromagnetic radiation that reaches the Earth ranges from 100 nm to $10^6$ nm (1 nm = $10^{-9}$ m) and can be divided into ultraviolet (UV) (100 nm to 400 nm), visible (400 nm to 700 nm), and infrared (IR) (700 nm to 106 nm) (Nunomura and Sousa 2023). UV radiation has both harmful and beneficial effects on living beings and is also responsible for the photochemical reaction that leads to the production of the protective ozone layer in the atmosphere (Vanhaelewyn et al. 2020). The visible part of the spectrum is the light that human eyes can detect on the μm scale (Sliney 2016). An important part of the electromagnetic radiation that reaches the Earth is infrared radiation (IR) (Horton et al. 2023). Photophysics is explained by the Jablonski diagram, in which atoms or molecules undergo excitation and emission (Guerrero-Rubio et al. 2020). Photoexcitation (straight up arrow) to an excited state is followed by rapid non-radiative steps (wavy arrows) in which energy is dissipated in vibrations through intra- and intermolecular processes until the lower level of the excited states is reached (no change in spin multiplicity) (Marin-Beloqui et al. 2023). Fluorescence generally occurs from this state in polyatomic molecules (although exceptions are known); this principle is known as Kasha's rule (Kasha 1950). Changes in spin multiplicity, known as intersystem crossing, can access even lower excited states, leading to phosphorescence (Penfold et al. 2018).

## 2 Photoexcitation

Photoexcitation is a fundamental process in photophysics in which a molecule or atom undergoes an energy transition from its ground state to a higher energy state when it absorbs a photon (Yong et al. 2020). This process follows the principles of Planck's quantum theory, which describes the quantization of energy associated with electromagnetic radiation (Srivastava 2020). When a photon with energy corresponding to the energy difference between the ground state and an excited state is absorbed by a quantum system, the system gains energy and is promoted to the excited state (Westermayr and Marquetand 2021). This energy transition occurs due to the interaction between the photon and the quantum system, which can be a molecule, an atom, or a solid-state material (Khan et al. 2019). The incident photons have different energies, depending on the wavelength of the incident light (Fig. 11.1) (Wang et al. 2014). Only when the photon's energy matches or exceeds the energy difference between the ground state and an excited state, can the electron be excited to a higher energy level (Demmig-Adams et al. 2020). At the atomic or molecular level, photoexcitation occurs when the photon energy is too low to cause photoionization, the removal of one or more electrons from an atom or molecule after the absorption of light (Xin et al. 2022) (Fig. 11.3).

The energy of the absorbed photon must be greater than or equal to the ionization energy, which is the energy required to remove an electron from the neutral particle (Rioux and DeKock 1998). The ionization energy depends on the specific atom or molecule and is usually in the ultraviolet or X-ray range of the electromagnetic spectrum (Evans 1955). In the case of molecules, the photoionization cross-section can be determined by the Franck-Condon principle, when a photon beam with angular frequency $\omega$ and energy $\Delta E$ hits a molecule, the photon flux is $N\Delta E$ $\mathrm{cm}^{-2}\mathrm{s}^{-1}$. Then, the photoionization rate is given as Eq. (11.1) (Baig 2022):

$$R = \sigma(\omega) N \Delta E \tag{11.1}$$

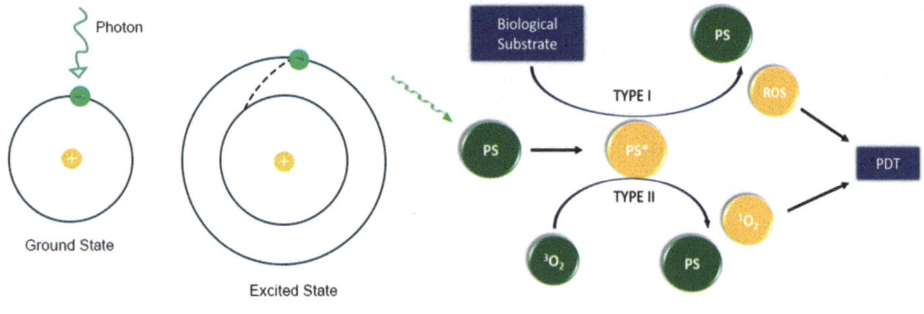


**Fig. 11.3** Schematic of the single-pulse pump-probe scheme: Photoexcitation occurs when the photon energy is too low to cause photoionization, the removal of one or more electrons from an atom or molecule after light absorption (Li et al. 2021)

# 3 Photofragmentation

Photofragmentation refers to a process in which a molecule is broken down into smaller fragments or particles through the absorption of light (Xin et al. 2022). When a molecule absorbs photons with sufficient energy, it can undergo various photochemical reactions, leading to the breaking of chemical bonds and the formation of smaller fragments or particles (Chatani et al. 2014).

# 4 Basic Photochemistry

The study of the interactions of light with atoms and molecules is a branch of chemistry concerned with the study of chemical reactions initiated or influenced by light (Mai and Gonzales 2020). These reactions involve the absorption of photons by molecules, leading to electronic or structural changes in the molecules and, consequently, the formation of new chemical species (Kumpulainen et al. 2017). It deals with the absorption, excitation, and emission of photons by atoms, atomic ions, molecules, molecular ions, etc. This process includes the absorption and subsequent emission of a photon by an atom (Kennedy et al. 2021).

## *4.1 Fundamentals of Photochemistry*

### 4.1.1 Light Absorption

The photochemical reaction begins when photons (individual units of light) are absorbed by molecules, and the photon energies must be able to move the electrons from a fundamental state to an excited state in the molecule (Roy et al. 2024).

### 4.1.2 Electronic States

Molecules have different electronic energy states, including the ground state (lowest energy) and the excited state (highest energy). When a molecule absorbs a photon, it moves from the ground state to the excited state (Blázquez-Castro 2017).

### 4.1.3 Jablonski Diagram

The Jablonski diagram is a graphical representation that illustrates the energy levels as well as the transitions involved in photochemistry, ground state, excited singlet states (S1, S2, etc.), and excited triplet states (T1, T2, etc.) in typical cases (Ariese et al. 2017) (Fig. 11.4).

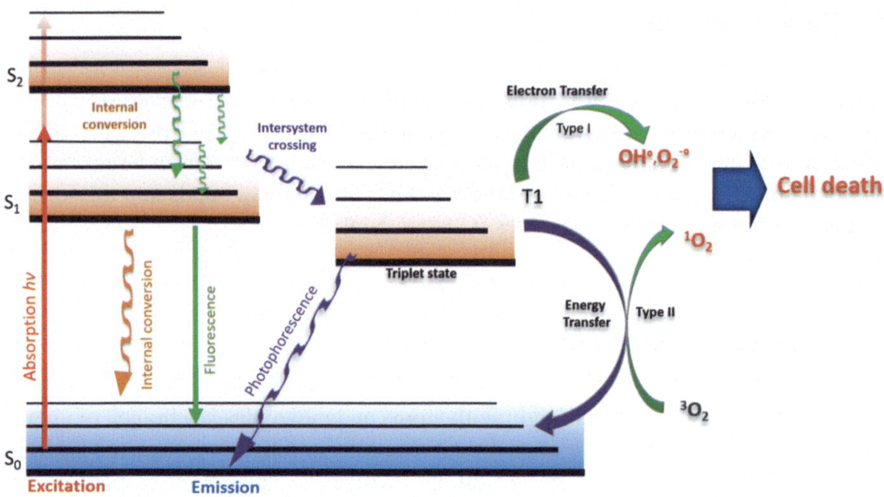


**Fig. 11.4** Jablonski energy level diagram for PDT. (Adapted Zhao et al. 2021)

### 4.1.4 Singlet and Triplet States

Singlet states have paired electrons in their molecular orbitals, so they are usually more reactive in photochemical reactions. Triplet states have unpaired electrons, so they tend to be longer-lived, but are usually less reactive than singlet states (Brash and Goncalves 2023).

## *4.2 Photophysical Processes*

Molecules in the excited state after absorption can undergo various photophysical processes, such as internal conversion, intersystem crossing, and fluorescence emission. The rapid relaxation of an excited molecule to a lower energy excited state or the ground state through non-radiative processes is called an internal conversion. Intersystem crossing involves the transition from a singlet to a triplet or vice versa. Fluorescence is simply the emission of light when a molecule descends from a high singlet level to its ground state (Yun-Tse et al. 2016).

## *4.3 Photochemical Reactions*

Chemical transformations of molecules in the excited state can take place during photochemical reactions that are unlikely to occur under thermal (non-photolytic) conditions. These reactions often involve the breaking and making of chemical bonds. Photochemical reactions include photodissociation, photoisomerization, and photoreduction (Ravelli et al. 2015).

## *4.4 Quantum Yield*

This is a measure of efficiency that quantifies the number of molecules that undergo a desired photochemical transformation divided by the number of photons absorbed per quantum yield (Rubin and Braslavsky 2010).

### 4.4.1 Factors Affecting Photochemical Reactions

Factors such as the wavelength and intensity of the incident light, the concentration of reactants, and the presence of sensitizers or catalysts determine the efficiency and outcome of photochemical reactions (Buglioni et al. 2022).

# 5 Quantum Confinement

This refers to the phenomenon where the movement of electrons in a material is restricted in all three dimensions, leading to quantized energy levels (Uprety and Abrahamse 2022). This effect is commonly observed in nanoparticles, where the particle size is of the order of the electron's de Broglie wavelength (Karagianni et al. 2023). In the context of PDT, quantum confinement can be used to increase the efficiency of photosensitizers, which are compounds that, when exposed to specific wavelengths of light, produce reactive oxygen species (ROS) that can selectively kill cancer cells. By incorporating photosensitizers into nanoparticles with quantum confinement (Fig. 11.5), the efficiency of the photosensitization process can be increased due to the greater absorption of light and confinement of the excited electrons (Chen et al. 2020).

Lately, the emergence of biocompatible nanoparticles has revolutionized the prospects for photodynamic therapy (PDT) in clinical trials (Hamblin et al. 2015). Consequently, much research is now being focused on the development of non-toxic and biocompatible nanoparticle-based photosensitizers for effective cancer

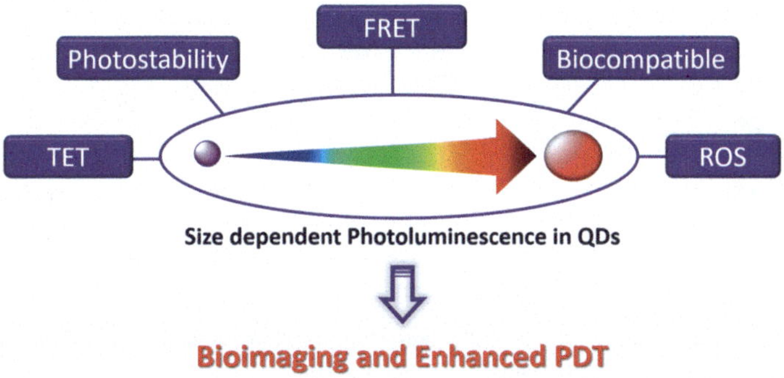


**Fig. 11.5** Size-dependent photoluminescence in quantum dots (Uprety and Abrahamse 2022)

treatments using PDT, in this respect, semiconductor and carbon quantum dots have shown encouraging results (Zeng et al. 2023). Quantum dots are artificial semiconductor and carbon nanocrystals with distinct chemical and physical properties, their optical properties can be adjusted by varying their size, which generally ranges from 1 to 10 nm (Uprety and Abrahamse 2022). They have many advantages over conventional photosensitizers, mainly their emission properties can be manipulated in the near-infrared region, as opposed to the visible region by the former (Wang et al. 2023a, b). Consequently, low-intensity light can be used to penetrate deeper tissues due to low scattering in the near IR region (Dang et al. 2019). Recently, successful reports on cancer imaging and PDT using carbon-based (carbon, graphene) and metallic quantum dots increasing treatment accuracy and reducing side effects (Cd- and carbon-based) are promising (Ahirwar et al. 2020). We conclude that quantum dots have great potential for application in PDT for the treatment of cancer (Tripathi et al. 2015). Roch Doi and collaborators investigated the production capacity of medicines based on natural products. Due to the many pharmacologically active secondary metabolites produced by plants, natural extracts are mainly used by humans to help fight cancer by inhibiting cell division or even causing cell apoptosis (Atanasov et al. 2021). Plant extracts can be excellent PSs that are used in PDT and are characterized as a promising technique when combining light and antimicrobial phototherapeutics (Fig. 11.6) (Zang et al. 2023). This process (Fig. 11.6) involves the elimination of cancer cells by apoptosis or necrosis caused by a chemical reaction produced by the combination of light, oxygen, and PSs (Girotti 2001), which accumulates in the target tissue. Therefore, they produce oxidation products that can cause cell death (Maharjan and Bhattarai 2022) (Fig. 11.7).

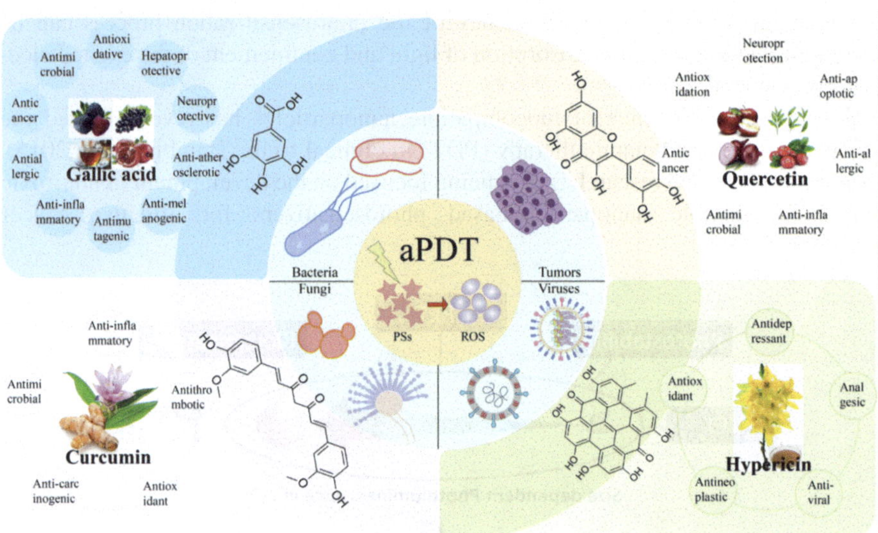


**Fig. 11.6** The representative polyphenolic natural products as photosensitizers for antimicrobial photodynamic therapy. (Adapted from Zang et al. 2023)

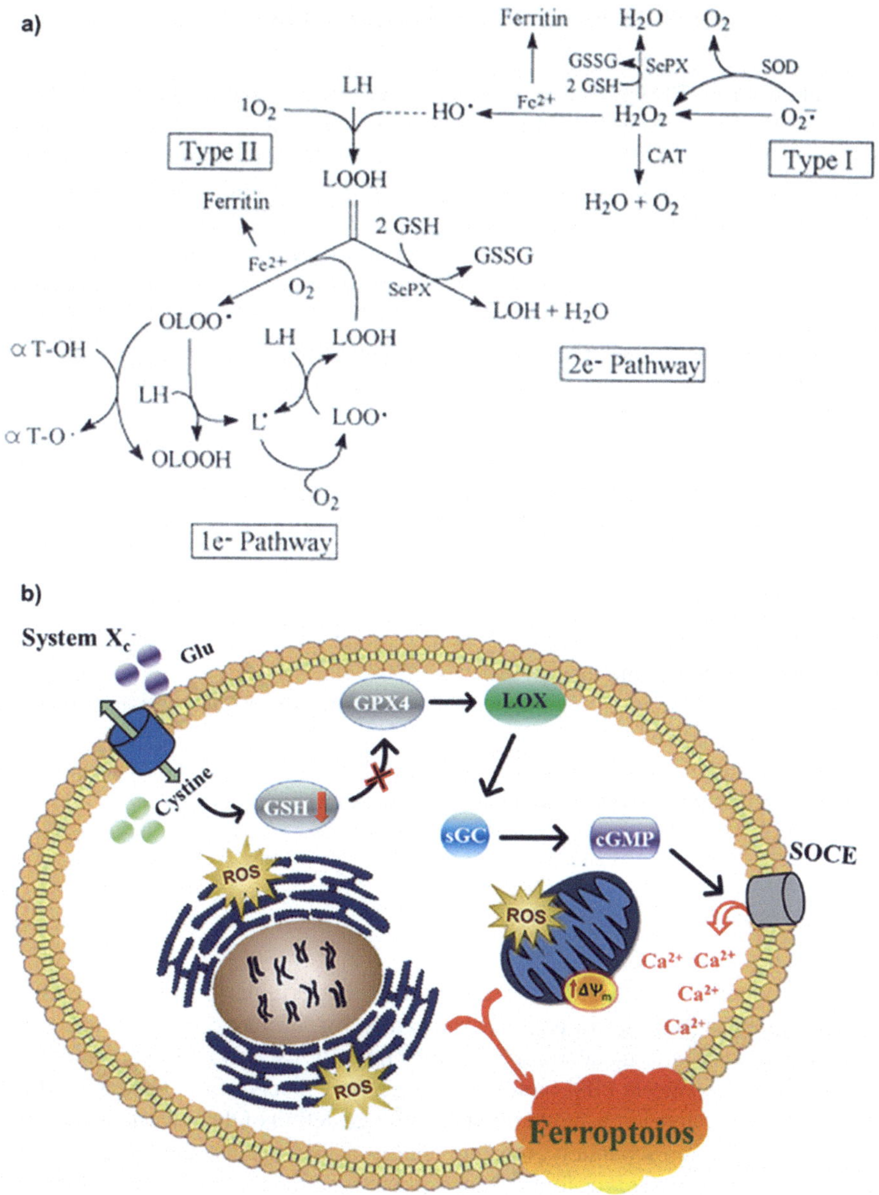


**Fig. 11.7** (**a**) Diagram showing possible routes of LOOH formation and turnover in photodynamic activated cells. (**b**) Mechanism of iron-dependent, oxidative death. (**a**) Reproduced with permission (Girotti 2001) Copyright 2001, Elsevier Science B.V

# 6 Laser Applications for Photodynamic Therapy

The light used for PDT can come from a laser or other sources (Mang 2004). Laser light can be directed through fiber optic cables (thin fibers that transmit light) to deliver light to areas inside the body (Guo et al. 2022). For example, a fiber optic cable can be inserted through an endoscope (a thin, lighted tube used to view tissue inside the body) into the lungs or esophagus to treat cancer in these organs (Lee et al. 2010). Other light sources include light-emitting diodes (LEDs), which can be used for superficial tumors such as skin cancer. Upon exposure to laser irradiation, the photosensitizer can absorb photons and then be excited from the singlet ground state S° to the excited singlet state S1. The excited triplet state T1 of PS is formed by the intersystem crossing (ISC) process (Castano et al. 2005; Robertson et al. 2009; Goldman et al. 1963).

# 7 Self-Assessment Questions and Answers

## *7.1 Questions*

1. What is photodynamic therapy?
2. What are the benefits of photodynamic therapy in cancer treatment and what are the current challenges?
3. What photodynamic therapy treatments are currently approved for cancer?
4. What are the potential evolutions of PDT in cancer treatment?

## *7.2 Answers*

1. Photodynamic therapy (PDT) is a treatment that originated in the early 1980s and has been growing in popularity recently for various conditions like acne, psoriasis, or acute macular degeneration. Initially used in cancer for the treatment of skin cancers (and rapidly extended for cancers of the prostate and the respiratory system), it is based on the combined action of three harmless components when taken separately: oxygen, photosensitive agents, and light.
2. Despite being studied for four decades, photodynamic therapy is still considered a promising cancer treatment, notably for its potential to improve patients' quality of life. This is where its role in anti-tumor strategy is growing, due to its positive effects on healing and quality of life of those treated.
3. The first photodynamic therapy approved for cancer was Porfimer Sodium, in 1993 for bladder cancer. Since then, it has been extended to other cancers like esophageal cancer and non-small cell lung cancer and is still studied for continued expansion. A second generation emerged, like Aminolevulinic acid for pre-

cancerous skin diseases, and many chlorin-based PS like Temoporfin or Talaporfin for various types of cancers (exclusively for skin or luminal organs) have been approved.

4. The main challenge is to extend this treatment option to all cancers. Since light does not pass through tissue, this involves bringing the light source as close as possible to the tumor to be treated. Different strategies have been elaborated and seem promising.

## References

Agostinis P, Berg K, Cengel KA, Foster TH, Girotti AW, Gollnick SO, Hahn SM, Hamblin MR, Juzeniene A, Kessel D, Korbelik M, Moan J, Mroz P, Nowis D, Piette J, Wilson BC, Golab J. CA Cancer J Clin. 2011;61(4):250–81.

Ahirwar S, Mallick S, Bahadur D. Photodynamic therapy using graphene quantum dot derivatives. J Solid State Chem. 2020;282:121107.

Algorri JF, Ochoa M, Roldán-Varona P, Rodríguez-Cobo L, López-Higuera JM. Light technology for efficient and effective photodynamic therapy: a Critical Review. Cancers. 2021;13:3484.

Alvarez N, Sevilla A. Current advances in photodynamic therapy (PDT) and the future potential of PDT-combinatorial cancer therapies. Int J Mol Sci. 2024;25:1023.

Ariese F, Roy K, Kumar VR, Sudeeksha HC, Kayal S, Umapathy S. Time-resolved spectroscopy: instrumentation and applications. In: Encyclopedia of analytical chemistry, 2006–2017. Wiley; 2017.

Atanasov AG, Zotchev SB, Dirsch VM, et al. Natural products in drug discovery: advances and opportunities. Nat Rev Drug Discov. 2021;20:200–16.

Austin E, Geisler AN, Nguyen J, Kohli I, Hamzavi I, Lim HW, Jagdeo J. Visible light part I. Properties and cutaneous effects of visible light. J Am Acad Dermatol. 2021;84(5):1219–31. https://doi.org/10.1016/j.jaad.2021.02.048.

Aziz B, Aziz I, Khurshid A, Raoufi E, Esfahani FN, Zahra Jalilian MR, Mozafari ET, Ikram M. An overview of potential natural photosensitizers in cancer photodynamic therapy. Biomedicines. 2023;11:224.

Baas P, Mansom I, Tinteren H, Stewart FA, Zandwijk N. Effect of N- acetylcysteine on photofrin-induced skin photosensitivity in patients. Lasers Surg Med. 1995;16(4):359–67.

Baig MA. Measurement of photoionization cross-section for the excited states of atoms: a review. Atoms. 2022;10:39.

Blázquez-Castro A. Direct 1O2 optical excitation: a tool for redox biology. Redox Biol. 2017;13:39–59.

Braga FC. Brazilian traditional medicine: historical basis, features and potentialities for pharmaceutical development. J Tradit Chinese Med Sci. 2021;8(1):44–50.

Brash DE, Goncalves LCP. Chemiexcitation: mammalian photochemistry in the dark. Photochem Photobiol. 2023;99:251–76.

Buglioni L, Raymenants F, Slattery A, Zondag SDA, Noël T. Technological innovations in photochemistry for organic synthesis: flow chemistry, high-throughput experimentation, scale-up, and photoelectrochemistry. Chem Rev. 2022;122:2752–906.

Castano AP, Demidova TN, Hamblin MR. Mechanisms in photodynamic therapy: part one-photosensitizers, photochemistry and cellular localization. Photodiagn Photodyn Ther. 2004;1(4):279–93.

Castano AP, Demidova TN, Hamblin MR. Mechanisms in photodynamic therapy: part two-cellular signaling, cell metabolism and modes of cell death. Photodiagn Photodyn Ther. 2005;2:1–23.

Cavinato M, Waltenberger B, Baraldo G, Grade CVC. Plant extracts and natural compounds used against UVB-induced photoaging. Biogerontology. 2017;18:499–516.

Celli JP, Spring BQ, Rizvi I, Evans CL, Samkoe KS, Verma S, Pogue BW, Hasan T. Chem Rev. 2010;110(5):2795–838.

Chatani S, Kloxin CJ, Bowman CN. Polym Chem. 2014;5:2187–201.

Chen J, Wentian W, Zhang F, Zhang J, Liu H, Zheng J, Guo S, Zhang J. Graphene quantum dots in photodynamic therapy. Nanoscale Adv. 2020;2:4961–7.

Cogno IS, Gilardia P, Cominib L, Núñez-Montoya SC, Cabrerac JL, Rivarola VA. Natural photosensitizers in photodynamic therapy: in vitro activity against monolayers and spheroids of human colorectal adenocarcinoma SW480 cells. Photodiagn Photodyn Ther. 2020;31:101852.

Dang X, Bardhan NM, Qi J, et al. Deep-tissue optical imaging of near cellular-sized features. Sci Rep. 2019;9:3873.

Daniell MD, Hill SJ. A history of photodynamic therapy. Aust N Z J Surg. 1991;61(5):340–8.

de Oliveira AB, Ferrisse TM, França GG, de Annunzio SR, Kopp W, Fontana CR, Brighenti FL. Potential use of Brazilian green propolis extracts as new photosensitizers for antimicrobial photodynamic therapy against cariogenic microorganisms. Pathogens. 2023;12:155.

Demmig-Adams B, Stewart JJ, López-Pozo M, Adams WW. Zeaxanthin, a molecule for photoprotection in many different environments. Molecules. 2020;25:5825.

Deniz AA, Mukhopadhyay S, Lemke EA. Single-molecule biophysics: at the interface of biology, physics and chemistry. J R Soc Interface. 2008;5:15–45.

Derosa MC, Crutchley RJ. Photosensitized singlet oxygen and its applications. Coord Chem Rev. 2002;233–234:351–71.

Dimmer J, Cabral FV, Sabino CP, Silva CR, Núñez-Montoya SC, Cabrera JL, Ribeiro MS. Natural anthraquinones as novel photosentizers for antiparasitic photodynamic inactivation. Phytomedicine. 2019;61:152894.

Dolmans DEJGJ, Fukumura D, Jain RK. Nat Rev. 2003;3:381.

Dougherty TJ, et al. Photodynamic therapy. J Natl Cancer Inst. 1998;90(12):889–905.

Ebermann R, Alth G, Kreitner M, Kubin A. Natural products derived from plants as potential drugs for the photodynamic destruction of tumor cells. J Photochem Photobiol B Biol. 1996;36(2):95–7.

Elisabetsky E, Shanley P. Ethnopharmacology in the Brazilian Amazon. Pharmacol Ther. 1994;64(2):201–14.

Evans R. The atomic nucleus. McGraw-Hill; 1955.

Gallardo-Villagrán M, Leger DY, Liagre B, Therrien B. Photosensitizers used in the photodynamic therapy of rheumatoid arthritis. Int J Mol Sci. 2019;20:3339. https://doi.org/10.3390/ijms20133339.

Girotti A. J Photochem Photobiol B. 2001;63:103.

Goldman L, et al. Pathology of the effect of the laser beam on the skin. Nature. 1963;197:912–4.

Gomer CJ. Preclinical examination of first and second generation photosensitizers used in photodynamic therapy. Photochem Photobiol. 1991;54(6):1093–107.

Gong Y, Guo X, Wang S, Su H, Xia A, Qingguo HE, Bai F. Photophysical properties of photoactive molecules with conjugated push–pull structures. J Phys Chem A. 2007;111:5806.

Grandi V, Corsi A, Pimpinelli N, Bacci S. Cellular mechanisms in acute and chronic wounds after PDT therapy: an update. Biomedicines. 2022;10(7):1624.

Guerrero-Rubio MA, García-Carmona JEF, Gandia-Herrero F. Light emission in betalains: from fluorescent flowers to biotechnological applications. Trends Plant Sci. 2020;25(2)

Gul B, Khan MS, Ahmad B, Rahaman M, Lolika PO, Wafae G, Ahmad H. A first-principles study of the electronic, optical, and transport properties of novel transitionmetal dichalcogenide. Mater Adv. 2023;4:4204–15.

Gunaydin G, Gedik ME, Ayan S. Photodynamic therapy – current limitations and novel approaches. 2021;9:691697.

Guo Y, Jia W, Yang J, Zhan X. Cancer glycomics offers potential biomarkers and therapeutic targets in the framework of 3P medicine. Front Endocrinol. 2022;13:970489.

Guo Y, Zheng J, Wang Z, Chen G, Hou K, Meifang Z. Biocompatible optical fiber for photomedical application. Giant. 2023;16:100195.

Hamblin MR, Chiang LY, Lakshmanan S, Huang Y-Y, Garcia-Diaz M, Karimi M, de Souza Rastelli AN, Chandran R. Nanotechnology for photodynamic therapy: a perspective from the Laboratory of Dr. Michael R. Hamblin in the Wellman Center for Photomedicine at Massachusetts General Hospital and Harvard Medical School. Nanotechnol Rev. 2015;4(4):359–72.

Hernández-Bule ML, Naharro-Rodríguez J, Bacci S, Fernández-Guarino M. Unlocking the power of light on the skin: a comprehensive review on Photobiomodulation. Int J Mol Sci. 2024;25:4483.

Horton L, Brady J, Kincaid CM, Torres AE, Lim HW. The effects of infrared radiation on the human skin. Photodermatol Photoimmunol Photomed. 2023;39:549–55.

Hwang MH, Seo JW, Park BJ, Han KJ, Lee JG, Kim NY, Kim MJ, Seong ES. Evaluation of growth characteristics and biological activities of 'Dachul', a hybrid medicinal plant of Atractylodes macrocephala × Atractylodes japonica, under different artificial light sources. Plan Theory. 2022;11(15):2035.

Ismail MS, Weitzel H, Berlien H-P. Applied photosensitizers classification for clinical photodynamic therapy. Berlin/Heidelberg: Springer; 1998. p. 300–4.

Ji X, Wang N, Wang J, Wang T, Huang X, Hao H. Non-destructive real-time monitoring and investigation of the self-assembly process using fluorescent probes. Chem Sci. 2024;15:3800–30.

Karagianni A, Tsierkezos NG, Prato M, Terronesf M, Kordatos KV. Application of carbon-based quantum dots in photodynamic therapy. Carbon. 2023;203:273–310.

Kasha. Characterization of electronic transitions in complex molecules. Faraday Soc Discuss. 1950;9:14–9.

Kato H. History of photodynamic therapy past, present and future. Gan To Kagaku Ryoho. 1996;23(1):8–15.

Kennedy JC, Pottier RH, Pross DC. Photodynamic therapy with endogenous protoporphyrin IX: basic principles and present clinical experience. J Photochem Photobiol B. 1990;6(1–2):143–8.

Kennedy ET, Costello JT, Mosnier J-P. Interaction of ionizing photons with atomic and molecular ions. Atoms. 2021;9:111.

Kessel D. Photodynamic therapy: a brief history. J Clin Med. 2019;8:1581.

Khan N, Kalair E, Abas N, Kalair AR, Kalair A. Energy transition from molecules to atoms and photons. Eng Sci Technol Int J. 2019;22:185–214.

Koehn FE, Carter GT. The evolving role of natural products in drug discovery. Nat Rev Drug Discov. 2005;4(3):206–20.

Kubrak TP, Kołodziej P, Sawicki J, Mazur A, Koziorowska K, Aebisher D. Some natural photosensitizers and their medicinal properties for use in photodynamic therapy. Molecules. 2022;27:1192.

Kumpulainen T, Lang B, Rosspeintner A, Vauthey E. Ultrafast elementary photochemical processes of organic molecules in liquid solution. Chem Rev. 2017;117(16):10826–939.

Kwiatkowski S, et al. Photodynamic therapy – mechanisms, photosensitizers and combinations. Biomed Pharmacother. 2018;106:1098–107.

Lee CM, Engelbrecht CJ, Soper TD, Helmchen F, Seibel EJ. Scanning fiber endoscopy with highly flexible, 1 mm catheterscopes for wide-field, full-color imaging. J Biophotonics. 2010;3:385–407.

Li S, Driver T, Alexander O, Cooper B, Garratt D, Marinelli A, Cryan JP, Marangos JP. Time-resolved pump–probe spectroscopywith spectral domain ghost imaging. Faraday Discuss. 2021;228:488.

Lim CK, Heo J, Shin S, Jeong K, Seoh Y, Jang WD, Park SY, Kim S, Kwon IK. Nanophotosensitizers toward advanced photodynamic therapy of cancer. Cancer Lett. 2012;334(2):176–87.

Lima E, Reis LV. Photodynamic therapy: from the basics to the current progress. Molecules. 2023;28:5092.

Lima A, Dal Pizzol C, Filippin MF, Creczynski-Pasa T, Andrade G, Ribeiro A, Perussi JR. Hypericin encapsulated in solid lipid nanoparticles: phototoxicity and photodynamic efficiency. J Photochem Photobiol B. 2013;125:146–54.
Lipson RL, Baldes EJ, Olsen AM. The use of a derivative of Hematoporphyrin in tumor detection. JNCI J Natl Cancer Inst. 1961;26(1):1–11.
Maharjan PS, Bhattarai HK. Singlet oxygen, photodynamic therapy, and mechanisms of cancer cell death. Hindawi J Oncol. 2022;20:7211485.
Mai S, Gonzales L. Molecular photochemistry: recent developments in theory. Angew Chem Int Ed. 2020;59:16832–46.
Mang TS. Lasers and light sources for PDT: past, present and future. Photodiagn Photodyn Ther. 2004;1:43–8.
Marin-Beloqui JM, Congrave DG, Toolan DTW, Montanaro S, Guo J, Wright IA, Clarke TM, Bronstein H, Dimitrov SD. Generating long-lived triplet excited states in narrow band gap conjugated polymers. J Am Chem Soc. 2023;145:3507–14.
Mfouo-Tynga IS, Dias LD, Inada NM, Kurachi C. Features of third generation photosensitizers used in anticancer photodynamic therapy: review. Photodiagn Photodyn Therapy. 2021;34(102091)
Moan J, Peng Q. An outline of the hundred-year history of PDT. Anticancer Res. 2003;23:3591–600.
Moan J, Christensen T, Sommer S. The main photosensitizing components of hematoporphyrin derivative. Cancer Lett. 1982;15(2):161–6.
Monge-Fuentes V, Muehlmann LA, Longo JPF, Silva JR, Fascineli ML, de Souza P, Faria F, Degterev IA, Rodriguez A, Carneiro FP, Lucci CM, Escobar P, Amorim RFB, Azevedo RB. J Photochem Photobiol B Biol. 2017;166:301–10.
Muniyandi K, George B, Parimelazhagan T, Abrahamse H. Role of photoactive phytocompounds in Photodynamic therapy of cancer. Molecules. 2020;25:4102.
Newman DJ, Cragg GM. Natural products as sources of new drugs over the last 25 years. J Nat Prod. 2007;70(3):461–77.
Nunomura CY, Sousa SJ d FE. Ultraviolet radiation and the human eye. Arq Bras Oftalmol. 2023;86(6):e2021–0354.
Nyman ES, Hynninen PH. Research advances in the use of tetrapyrrolic photosensitizers for photodynamic therapy. J Photochem Photobiol B Biol. 2004;73(1–2):1–28.
O'Connor A, Gallagher W, Byrne AT. Porphyrin and nonporphyrin photosensitizers in oncology: preclinical and clinical advances in photodynamic therapy. Photochem Photobiol. 2009;85:1053–74.
Penfold TJ, Gindensperger E, Daniel C, Marian CM. Spin-vibronic mechanism for intersystem crossing. Chem Rev. 2018;118:6975–7025.
Polat E, Kang K. Natural photosensitizers in antimicrobial photodynamic therapy. Biomedicines. 2021;9:584.
Pramual S, Lirdprapamongkol K, Svasti J, Bergkvist M, Jouanhureaux V, Arnoux P, Frochot C, Barberi-HEYOB M, Niamsiri N. Polymer-lipid-PEG hybrid nanoparticles as photosensitizer carrier for photodynamic therapy. J Photochem Photobiol B Biol. 2017;173:12–22.
Pucelik B, Dąbrowski JM. Photodynamic inactivation (PDI) as a promising alternative to current pharmaceuticals for the treatment of resistant microorganisms. Adv Inorg Chem. 2021;79
Quina F, de Medeiros Silva GT. The photophysics of photosensitization: a brief overview. J Photochem Photobiol. 2021;7:100042.
Ravelli D, Protti S, Albini A. Energy and molecules from photochemical/photocatalytic reactions. An overview. Molecules. 2015;20(1):1527–42.
Riesz CP, Krishna M. Phthalocyanines and their sulfonated derivatives as photosensitizers in photodynamic therapy. In: Proceedings Volume 0847. New Directions in Photodynamic Therapy; 1988.
Rioux F, DeKock RL. The crucial role of kinetic energy in interpreting ionization energies. J Chem Educ. 1998;75:537.

Robertson CA, Evans DH, Abrahamse H. Photodynamic therapy (PDT): a short review on cellular mechanisms and cancer research applications for PDT. J Photochem Photobiol B Biol. 2009;96:1–8.

Rodrigues MC, Muehlmann LA, Longo JPF, Silva RC, Graebner IB, Degterev IA, Lucci CM, Azevedo RB, Garcia MP. Photodynamic therapy based on Arrabidaea chica (Crajiru) extract nanoemulsion: in vitro activity against monolayers and spheroids of human mammary adenocarcinoma MCF-7 cells. J Nanomed Nanotechnol. 2015;6:3–6.

Roy P, Sardjan AS, Browne WR, Feringa BL, Meech SR. Excited state dynamics in unidirectional photochemical molecular motors. J Am Chem Soc. 2024;146(18):12255.

Rubin MB, Braslavsky SE. Quantum yield: the term and the symbol. A historical search. Photochem Photobiol Sci. 2010;9:670–4.

Sarbadhikary P, George BP, Abrahamse H. Recent advances in photosensitizers as multifunctional theranostic agents for imaging-guided photodynamic therapy of cancer. Theranostics. 2021;11(18)

Senapathy GJ, George BP, Abrahamse H. Exploring the role of phytochemicals as potent natural photosensitizers in photodynamic therapy. Anticancer Agents Med Chem. 2020;20(15):1831–44.

Seo Y, Lim H, Park H, Lee T. Recent progress of lipid nanoparticles-based lipophilic drug delivery: focus on surface modifications. Pharmaceutics. 2023;15:772.

Sliney D. What is light? The visible spectrum and beyond. Eye. 2016;30:222–9.

Souris JS, Leoni L, Zhang HJ, Pan A, Tanios E, Tsai H-M, Balyasnikova IV, Bissonnette M, Chen C-T. X-ray activated nanoplatforms for deep tissue photodynamic therapy. Nanomaterials. 2023;13:673.

Sperandio FF, Huang Y-Y, Hamblin MR. Antimicrobial photodynamic therapy to kill Gram-negative bacteria. Recent Pat Antiinfect Drug Discov. 2013;8(2):108–20.

Spikes JD. New trends in photobiology: chlorins as photosensitizers in biology and medicine. J Photochem Photobiol B Biol. 1990;6(3):259–74.

Srivastava SK. Quantum and quantization (Quantum theory of everything). Chiang Mai J Sci. 2020;47(6)

Tripathi SK, Kaur G, Khurana RK, Kapoor S, Singh B. Quantum dots and their potential role in cancer theranostics. Crit Rev Ther Drug Carrier Syst. 2015;32(6):461–502.

Ueda K, Sokell E, Schippers S, Aumayr F, Sadeghpour H, Burgdörfer J, Lemell C, Tong X-M, Pfeifer T, Calegari F. Roadmap on photonic, electronic and atomic collision physics: I. Light–matter interaction. J Phys B At Mol Opt Phys. 2019;52:171001.

Uprety B, Abrahamse H. Semiconductor quantum dot for photodynamic therapy: recent advances. Front Chem. 2022;11(10):946574.

Valli M, Russo HM, Bolzani VS. The potential contribution of the natural products from Brazilian biodiversity to bioeconomy. An Acad Bras Cienc. 2018;90(1 Suppl. 1):763.

Vanhaelewyn L, van der Straeten D, de Coninck B, Vandenbussche F. Ultraviolet radiation from a plant perspective: the plant microorganism context. Front Plant Sci. 2020;11:597642.

Villacorta RB, Roque KFJ, Tapang GA, Jacinto SD. Plant extracts as natural photosensitizers in photodynamic therapy: in vitro activity against human mammary adenocarcinoma MCF-7 cells. Asian Pac J Trop Biomed. 2017;7(4):358–66.

Wang J, Ming T, Jin Z, et al. Photon energy upconversion through thermal radiation with the power efficiency reaching 16%. Nat Commun. 2014;5:5669.

Wang S, Zhang C, Fang F, Fan Y, Yang J, Zhang J. Beyond traditional light: NIR-II light-activated photosensitizers for cancer therapy. J Mater Chem B. 2023a;11:8315–26.

Wang X, Wang L, Reza F, Lu Z, Xian J, Gu H, Xiang W. Polyphenolic natural products as photosensitizers for antimicrobial photodynamic therapy: recent advances and future prospects. Front Immunol. 2023b;14:1275859.

Weight BM, Li X, Zhang Y. Theory and modeling of light-matter interactions in chemistry: current and future. Phys Chem Chem Phys. 2023;25:31554–77.

Westermayr J, Marquetand P. Machine learning for electronically excited states of molecules. Chem Rev. 2021;121:9873–926.

Wöhrle D, et al. Photodynamic therapy of cancer: second and third generations of photosensitizers. Russ Chem Bull. 1998;47(5):807–16.

Xin P, Qiu T, Ma H, Yuan H, Liu H. Photoionization of electrons in degenerate energy level of hydrogen atom induced by strong laser pulses. Photo-Dermatology. 2022;9:256.

Yankovsky I, Bastien E, Yakavets I, Khludeyev I, Lassalle H-P, Gräfe S, Bezdetnaya L, Zorin V. Inclusion complexation with β-cyclodextrin derivatives alters photodynamic activity and biodistribution of meta-tetra(hydroxyphenyl)chlorin. Eur J Pham Sci. 2016;91:172–82.

Yong H, Zotev N, Ruddock JM, et al. Observation of the molecular response to light upon photoexcitation. Nat Commun. 2020;11:2157.

Yun-Tse W, Shanmugam C, Tseng W-B, Hiseh M-M, Tseng W-L. A gold nanocluster-based fluorescent probe for simultaneous pH and temperature sensing and its application to cellular imaging and logic gates. Nanoscale. 2016;8:11210–6.

Zeng L, Gowda BHJ, Ahmed MG, et al. Advancements in nanoparticle-based treatment approaches for skin cancer therapy. Mol Cancer. 2023;22:10.

Zacharioudaki D-E, Fitilis I, Kotti M. Review of fluorescence spectroscopy in environmental quality applications. Molecules. 2022;27:4801.

Zhang M, et al. Pyropheophorbide 2-Deoxyglucosamide: a new photosensitizer targeting glucose transporters. Bioconjug Chem. 2003;14(4):709–14.

Zhang Y, Wang B, Zhao R, et al. Multifunctional nanoparticles as photosensitizer delivery carriers for enhanced photodynamic cancer therapy. Mater Sci Eng C. 2018; https://doi.org/10.1016/j.msec.2020.111099.

Zhao Y-Y, Chen J-Y, Hu J-Q, Zhang L, Lin A-L, Wang R, Zheng BY, Ke MR, Li X, Huang JD. The substituted zinc(II) phthalocyanines using "sulfur bridge" as the linkages. Synthesis, red-shifted spectroscopic properties and structure-inherent targeted photodynamic activities. Dyes Pigments. 2021;189:Article 109270.

# Chapter 12
# New Investigative Prospects for Drug–Non-drug Interactions

**Edeildo Ferreira da Silva-Júnior and Thiago Mendonça de Aquino**

**Abstract** The "Drug–Non-drug Interactions Concept," including drug-disease or drug-food interactions which has been widely studied over the past decades in which the patient has been the central focus of studies, is becoming an increasingly current subject. In this context, it is known that some diseases can affect organs associated with drug metabolism, reducing the success rate of treatments. A similar effect is observed during patient aging, where age should be considered as an additional factor in the patient's pathology. In other situations, components of patients' diets can interfere with drug pharmacokinetics, affecting their pharmacodynamic profile and resulting in treatment failures. In this chapter, an overview is provided focusing on this issue, explaining the interactions and the drug classes most affected due to patient diet components, including antivirals, anti-inflammatories, antidiabetics, and anticancers among others. Additionally, how these interferences affect CYP-related enzymes, organic anion transporters (OATs), intestinal P-glycoprotein, and human serum albumin (HSA) are discussed. Phenolic and flavonoid compounds are widely studied, as they are phytochemicals easily found in various food preparations. Furthermore, their interactions with physiological macromolecules are presented, making clear the chemical nature of these interactions. Finally, a perspective on the theme is presented at the end of this chapter, as well as some questions are provided to aid in understanding this relevant topic.

**Keywords** Drug-food interaction · Drug-disease interaction · In silico approaches · CYP-interference

---

E. F. da Silva-Júnior (✉)
Group of Research on Biological and Molecular Chemistry, Institute of Chemistry and Biotechnology, Federal University of Alagoas, Alagoas, Maceió, Brazil
e-mail: edeildo.junior@iqb.ufal.br

T. M. de Aquino
Group of Research on Therapeutic Strategies, Institute of Chemistry and Biotechnology, Federal University of Alagoas, Alagoas, Maceió, Brazil
e-mail: thiago.aquino@iqb.ufal.br

 
J. O. Fajemiroye (ed.), *Fundamentals of Drug and Non-Drug Interactions*,
https://doi.org/10.1007/978-3-031-80107-5_12

# 1 Introduction

In the complex realm of modern pharmacotherapy, where a myriad of medications intersect with diverse physiological states, lies a domain both intricate and consequential: drug-disease interactions. While the term "drug-drug interaction" typically evokes images of one pharmaceutical entity perturbing another within the intricate landscape of human physiology, the equally significant realm of drug-disease interactions often remains in the shadows, yet with profound implications. Defined as the exacerbation by drugs of preexisting diseases, syndromes, or conditions, drug-disease interactions represent a critical facet of clinical pharmacology, especially accentuated among older adults burdened with multiple chronic pathologies and polypharmacy. In this captivating exploration, we embark on a journey through the nexus of medications and physiological states, unraveling the intricate dance between pharmacokinetics, pharmacodynamics, and the complexities of aging. Delving deeper into this domain, we uncover multifaceted interactions weaving through the fabric of geriatric medicine, where the delicate balance of therapeutic efficacy and adverse outcomes hangs in the balance. From the modulation of β-blockers in depression to the nuanced interplay of opioid analgesics in dementia, manifestations of drug-disease interactions paint a vivid tapestry of clinical challenges and opportunities for optimization.

In the intricate web where culinary delights intertwine with pharmacological efficacy, a captivating domain emerges—the realm of herb/food-drug interactions. This intersection, shrouded in biochemical intricacies, serves as a gateway to understanding the profound implications that dietary choices hold for medication outcomes and therapeutic effectiveness. Delving beyond mere recognition of what to consume alongside medications, exploration of herb/food-drug interactions unveils a labyrinth of biochemical nuances governing absorption, metabolism, and efficacy. At the heart of this exploration lies a fundamental understanding of how food modulates the absorption, distribution, metabolism, and excretion (ADME) of medications, shaping their therapeutic impact. These interactions, ranging from subtle modifications to profound alterations, can either potentiate the benefits of medications or precipitate adverse reactions, echoing the delicate balance of nature's symphony.

Welcome to the captivating journey within the pages of this book chapter, where the convergence of pharmaceutical science, clinical medicine, and geriatric care unveils the intricacies of drug-disease interactions, herb-drug interactions, and food-drug interactions. Drawing from a rich tapestry of research, from foundational studies elucidating drug-nutrient interactions to contemporary investigations probing the influence of dietary habits on substance abuse, we embark on an odyssey through this multifaceted terrain. In this exploration, we are compelled to adopt a nuanced approach, weaving pharmacological insights with dietary considerations to decipher the molecular intricacies underlying these interactions. As we chart the course ahead, we stand poised to unlock the full potential of understanding drug-disease, herb-drug, and food-drug interactions, heralding a new era of precision therapeutics and patient-centered care. Join us as we navigate this captivating landscape, where each revelation brings us closer to personalized therapeutic interventions and pharmacological enlightenment, ultimately enhancing patient outcomes and transforming the modern medical landscape.

## 2 Drug-Disease Interactions

Conventionally, the term drug-drug interaction refers to the effects of a drug on the bioavailability of another one, interfering with its pharmacokinetics—absorption, distribution, metabolization, or excretion (ADME), or affecting its pharmacodynamics in terms of pharmacological potency (Benet et al. 2019). In parallel, drug-disease interactions have equal importance, in which patients present specific physiopathological conditions that affect both pharmacokinetics and pharmacodynamics parameters (Canio 2022; Nwose and Yee 2016). In this context, drug-disease interactions can be defined as exacerbations by drugs of preexisting diseases, syndromes, or conditions, which are more common in older adults (6 to 30%) since they frequently have multiple chronic pathologies and use several different drugs (Chin et al. 1999; Giron et al. 2001; Gosney and Tallis 1984; Hanlon et al. 2002). Additionally, drug-disease interactions cause more deleterious clinical impacts on older adults since their physiological systems are more sensitive (Fried and Walston 2003).

Overall, different classes of drugs can be affected by aging factors. Although aging has not been considered a disease, it is still a complicating factor for drug pharmacokinetics or pharmacodynamics (Craig Brater 1986; Goldberg et al. 1996). Regarding this, Lindblad et al. (2006) developed an analysis of a 2-round modified Delphi survey and cross-sectional study involving 1388 patients (≥65 years old) from 11 veterans affairs medical centers within the following criteria: dependence in ≥1 activity of daily living, previous falls, stroke within 3 months, malnutrition, difficulty ambulating, depression, dementia, prolonged bed rest, unplanned admission in the past 3 months, and incontinence (Lindblad et al. 2006). Thus, the authors verified that 15.3% of the frail older veteran patients presented more than one drug-disease interaction. In general, older patients frequently exhibited drug-disease interactions within 28 pathological conditions, in which the actions of β-blockers (e.g., propranolol and atenolol), α-blockers (e.g., reserpine), digoxin, benzodiazepines, and barbiturates can interfere in depression; opioid analgesics and tricyclic antidepressants can interfere in dementia; atypical antipsychotics, corticosteroids, and thiazide diuretics can interfere in diabetes; whereas CNS stimulants (e.g., methylphenidate), non-aspirin NSAIDs, and pseudoephedrine can interfere in hypertension. For the entire list of diseases and their drug interferences, see the ref. (Lindblad et al. 2006). Finally, they verified that the two most common drug-disease interactions in the patients in their study involved first-generation calcium channel blockers in patients having congestive heart failure, whereas aspirin in patients with peptic ulcer disease.

## 2.1 Drug-Inflammation Interaction

Graig Brater wrote a complete review paper focused on non-steroidal anti-inflammatory drugs (NSAIDs) and their associations with drug-drug and drug-disease interactions (Craig Brater 1986). Concerning their drug-disease interactions, Brater provides an interesting overview regarding aging aspects, including bioaccumulation, and diseases affecting the liver and kidneys. In this sense, their albumin binding and elimination have been in-depth discussed for several NSAIDs, such as ibuprofen, naproxen, oxaprozin, ketoprofen, piroxicam, etodolac, benoxaprofen, and azapropazone (Craig Brater 1986).

It is known that inflammatory conditions have reduced the cardiac response to β-adrenergic antagonists in rats, such as sotalol (Kulmatycki and Jamali 2001) and propranolol (Guirguis and Jamali 2003), whereas verapamil has been negatively affected in humans (Mayo et al. 2000). The reduced responses of verapamil and propranolol could be associated with the elevation in their concentrations due to the inflammatory process (Guirguis and Jamali 2003; Mayo et al. 2000). Based on this, Clements and Jamali (2007), considering the anti-inflammatory properties of pravastatin, hypothesized that it could reduce both potency and reverse the reduced clearance of propranolol in the inflammation animal model. Besides, they evaluated the effects of pravastatin on propranolol pharmacokinetics in the pre-adjuvant arthritis animal model (Clements and Jamali 2007). For this purpose, they used two groups of adult male Sprague-Dawley rats, one for pharmacokinetics and another for pharmacodynamics experiments. The authors verified that propranolol (on day 8) causes meaningful prolongation of the PR interval and presents reduced potency in inflamed rats when compared with the healthy condition. Still, pravastatin restored the low response of propranolol in inflamed rats. Additionally, it did not interfere with the pharmacokinetics of propranolol in healthy animals. Similarly, it did not affect the unbound propranolol concentration in inflamed rats (Clements and Jamali 2007). Finally, the authors affirmed that pravastatin reversed the β-adrenergic reduced response during the inflammatory process and increased the potency of propranolol to prolong PR interval in inflamed rats. Since pravastatin does not affect propranolol pharmacokinetics, it could interfere with the pharmacodynamics of it (Clements and Jamali 2007).

Hepatocellular damage induced by steatosis, lipotoxicity, or inflammation can modify the expression of Cytochrome P450 (CYP450) enzymes, nuclear factors, and transporters, as well as cause alterations in the pharmacokinetics of xenobiotics, including drugs (Fisher et al. 2009a, b). In this sense, it has been evidenced that the drug bioavailability can be affected in nonalcoholic fatty liver disease (NAFLD) since it affects the protein and mRNA expressions (Canet et al. 2014; Fisher et al. 2009a). This pathologic condition can lead to varying changes in liver organic anion transporting polypeptide (OATP), which interferes with the OATP-mediated drug transport from the bloodstream into hepatocytes (Fisher et al. 2009a). However, this physiopathological mechanism is poorly understood in terms of OATP expression in NAFLD. Then, Li et al. (2021) decided to explore the activity of Ginkgolides A-C and Ginkgo flavonols Quercetin, Kaempferol, and Isorhamnetin in male

Sprague-Dawley rats (Li et al. 2021), since these compounds have been associated with OATP/1B3, 1A2, and 2B1 mediated uptake (Mandery et al. 2010; Ude et al. 2013; Wu et al. 2012). The authors verified that ginkgo flavanols and Ginkgolides A-C in NAFLD rats' plasma are found to increase in a dose-dependent manner when 3.6–32.4 mg/kg of extract of *Ginkgo biloba* L. is orally administrated. Ginkgolides A-C demonstrated a shorter half-life (2–3 h) in comparison with the flavonols (~20 h). Additionally, the plasma concentration of pitavastatin was found to be 50% reduced in NAFLD rats due to the increased OATP/1B2 expression in the liver. Moreover, isorhamnetin and kaempferol were capable of inhibiting OATP1B1-mediated uptake of $^{3}$H-Estrone sulfate ($^{3}$H-ES), exhibiting 50% inhibitory concentration ($IC_{50}$) values of 46.12 ± 5.25 and 3.28 ± 1.08 μM, respectively. Lastly, the authors concluded that both extracts of *G. biloba* and NAFLD can interfere with the OATP/1B2 transporter in combination or not (Li et al. 2021).

## 2.2 *Drug-Anticoagulation Disturb Interaction*

Bleeding complications during chronic anticoagulant treatment by using Warfarin occur in 7.6–16.5 individuals *per* 100 patients years (Cannegieter et al. 1995; Hirsh et al. 1998; Palareti et al. 1996). High sensitivity to oral anticoagulant agents can increase the risk of bleeding, and this enhanced physiological aspect can be associated with genetic factors, hepatic dysfunction, chronic renal failure, and old age (Landefeld and Beyth 1993; Mcmahan et al. 1998). Warfarin is the first-choice drug to prevent bleeding in patients, and its (*S*)-enantiomer is responsible for four-fold more anticoagulant activity than its (*R*)-enantiomer since Warfarin is a racemate composition (Choonara et al. 1986). It has been evidenced that the in vitro CYP2C9 intrinsic clearance presents >20-fold variation toward Warfarin (Takahashi et al. 1998; Yamazaki and Shimada 1997), although it has two wild-type allelic variants due to the genetic polymorphism in the gene encoding CYP2C9, resulting in CYPC9*1 ($Arg^{144} \rightarrow Cys^{144}$ variation) and CYP9C9*2 ($Ile^{359} \rightarrow Leu^{359}$ variation) (Stubbins et al. 1996). Additionally, it has been demonstrated that the CYP2C9*3 allele is associated with the in vivo clearance of CYP2C9 substrates, including Warfarin (Scordo 2002). Thus, Muszkat et al. (2007) decided to investigate how some additional conditions could affect Warfarin metabolism, such as congestive heart failure, chronic hepatic dysfunction, use of antibiotics, and renal failure (Muszkat et al. 2007). Moreover, they hypothesized that CYP2C9 genetic polymorphism could affect Warfarin dosage in patients. Their research involved 119 non-dialysis patients (≥65 years old), having comorbid conditions (congestive heart failure, CrCl <50 mL/min, and hepatic dysfunction), and treatment with warfarin for at least 3 months before admission. The authors verified that the warfarin dosage required for CYP2C9*1/*3 patients is 33% lower than that for CYP2C9*1/*1 ones. Additionally, it was verified that Warfarin dosage requirements are affected by chronic conditions, such as renal failure, use of antibiotics, and congestive heart failure (Muszkat et al. 2007).

# 3 In Silico Approaches for Predicting Drug-Disease Interactions

## *3.1 Machine-Learning and Deep-Learning Approaches*

There is an estimative of 0.8–4.5 billion USD for developing a new FDA-approved drug (Wouters et al. 2020) since the traditional De novo drug discovery/development requires up to 17 years with less than a 10% success rate due to high failure rate in the preclinical stages (Ashburn and Thor 2004). Several in silico methods have been developed to aid the drug discovery process or even when repurposing drugs, such as structural bioinformatics, molecular dynamics, and molecular dynamics simulations (Cavasotto and Di Filippo 2021; Dos Santos Nascimento et al. 2021, 2022). However, *machine-* and *deep-learning* models have demonstrated significant accuracy. In this context, drug repurposing represents a coming-on-strong alternative to De novo approaches, in which approximately 30% of new FDA-approved agents and/or vaccines came from drug repurposing studies (Chong and Sullivan 2007).

A machine-learning approach refers to an approach where the researchers use computer software to make the machine more "intelligent," in which it could learn about biological and biomedical data to discover new drug-disease interactions for repurposing drugs (Papanikolaou et al. 2016) or drug indications (Wong et al. 2021). In this context, Papanikolaou et al. (2016) used text mining on the DrugBank database, aiming to identify drug-drug interactions (Papanikolaou et al. 2016). Segler et al. (2018) applied a recurrent neural network model to repurpose drugs (Segler et al. 2018). Additionally, Lim et al. (2016) conducted a large-scale study to propose off-target drug interactions, repurposing drugs for different applications (Lim et al. 2016). Moreover, network analysis can be used to predict drug-disease interactions, as reported by Hu and Agarwal (2009), in which repurposed drugs and discovered drug side effects (Hu and Agarwal 2009). Notwithstanding these findings, Jarada et al. (2021) applied SNF-CVAE, integrating similarity selection, similarity measures, similarity network fusion (SNF), and finally collective variational autoencoder (CVAE), to perform a non-linear analysis for predicting drug-disease interactions (Jarada et al. 2021). Then, they identified 3179 interactions involving 719 diseases and 781 drugs, using the stratified five-fold cross-validation for their methods, in which the training and test groups corresponded to 80 and 20% of the entire dataset, respectively. Their SNF-CVAE model presented a precision of 0.902 and an AUC-ROC of 0.958. As a result, the authors ranked the 40 top drug-disease interactions predicted by their machine-learning model, with references supporting their findings. For example, Cefaclor (sinusitis), Cefazolin and Ceftriaxone (*Haemophilus* infections), Clomipramine (bulimia), Meloxicam and Piroxicam (juvenile rheumatoid arthritis), Pantoprazole (stomach ulcer), and Quinacrine (*vivax malaria*), in which all their findings were supported by DrugBank, PubMed, or ClinicalTrials.gov. Besides, the authors ranked the 10 top predicted drug candidates against Alzheimer's disease and juvenile idiopathic arthritis, being Bupropion (pred-score of 0.84) and Piroxicam (pred-score of 0.99), respectively (Jarada et al. 2021).

Deep-learning approach is well-known to be a subclassification of artificial intelligence (AI) that utilizes multilayers of abstractions to "*learn*" the representations of the input data, which has been successfully applied to solve biomedical problems (Lecun et al. 2015; Naveen Sundar et al. 2022; Wen et al. 2017), introducing calculated similarities between the nodes (Yu et al. 2021).

By using these approaches, it has been determined that drugs with similar chemical structures tend to target similar proteins or diseases (Patel and Gorman 2009; Schuffenhauer et al. 2003). In this context, information on the targets, pathways, enzymes, proteins, and other drug-related data can be used as input data to build a binary classification for repurposing drugs, as it was performed by Napolitano et al. (2013), which used a machine-learning-based strategy to predict the correlation between drugs and diseases by integrating multilayers of drug-related information (Napolitano et al. 2013). Using a model based on logistic regression, Kim et al. (2019) developed a predictive model including three types of disease-disease similarities and four types of drug-drug similarities (Kim et al. 2019). Thus, Ghasemi and Lakizadeh (2023) developed a deep-learning method named LHGCE to predict drug-disease associations, in which it was capable of absorbing the structural data from nodes in a heterogeneous network from distinct biological origins (Ghasemi and Lakizadeh 2023). This heterogeneous network was built by crossing information concerning drug-disease, drug-target, drug-pathway, drug-enzyme, and drug-structure. Then, a dataset containing 18,416 drug-disease associations, 598 diseases, and 269 drugs. The drug-disease data were obtained from the comparative toxicogenomic database (CTD), whereas the drug-related targets, enzymes, substructures, and pathways were extracted from DrugBank. The LHGCE method developed by the authors was constituted of two levels, being the (*i*) initial extraction of optimal feature vectors and, then, (*ii*) the prediction of drug-disease interactions (Ghasemi and Lakizadeh 2023). Still, the k-fold cross-validation method was used to determine the accuracy of their model. In general, the authors ranked the top-20 drug-disease associations predicted by the proposed method with a probability of correlations higher than 0.9 (or >90%), in which each drug was associated with one disease, for example, naloxone and diazepam (seizures), metoprolol (left ventricular dysfunction), diazepam (hallucinations), risperidone (tremor), amitriptyline (delirium), and propranolol (bradycardia). For the entire list, the reader should access the ref. (Ghasemi and Lakizadeh 2023). Furthermore, their deep learning was able to predict morphine-disease association with five different indications, being seizures, hyperalgesia, hyperkinesis, hallucinations, and dizziness.

## 3.2 Deep-Learning Models on Drug-SARS-CoV-2 Interaction

Das et al. (2022) described a study reporting an innovative deep-learning model for predicting drug-SARS-CoV-2 interactions, aiming to understand the chemical mechanism of how SARS-CoV-2 interacts with drugs and discover novel promising treatments (Das et al. 2022). Within this purpose, the authors used Favipiravir, Hydroxychloroquine, Oseltamivir, Remdesivir, Tocilizumab, Azithromycin, and other drugs in their study using a geometric deep-learning model, which is a well-known strategy to predict the virus-drug relationship (Townshend et al. 2021). Still, they built a message-passing neural network-based (MPNN) architecture to initially discover the genomic basis of the virus-drug interactions. A dataset containing 3318 drugs was built from PubChem and MoleculeNet databases, in which 1659 of them were positive samples for SARS-CoV-2, whereas 1659 of them were negative ones. In total, 80% of this dataset was used as a training group, 10% as a five-fold cross-validation group, and 10% as a testing group, resulting in an accuracy of 97.31% of the geometric deep-learning model, expressed in terms of area under the curve. As a result, UDP galactosamine, trimethyllysine, and *N*-(4-fluorophenyl)-2-pyrrolidin-1-ylacetamide, which are positive samples active against SARS-CoV-2, were correctly identified for their in silico model, which was capable of predicting virus-drug interactions based on their bond structures to be formed according to their valence electron, atomic, and hydrogen numbers (Das et al. 2022).

## 3.3 Deep-Learning Models on Host-Microbe-Drug Interaction

The gastrointestinal human microbiota protects against invading pathogens by producing antimicrobials, preserving the gut barrier, depleting nutrients, and guiding metabolic development (Roja and Chellapandi 2024; Thursby and Juge 2017). Imbalances in gut microbiota have been associated with food-borne infections and gastrointestinal disorders since the gut microbiota is related to the immune system and regulation of human physiology (Jandhyala 2015). In this context, it is known that antimicrobial agents can cause imbalances in electrolytes and probiotics in microbial communities, leading to antibiotic resistance and increasing the diversity of microorganisms (Amalaradjou and Bhunia 2012; Beceiro et al. 2013), as well as resulting in microbial infections in humans (Pérez-Cobas et al. 2014). Thus, it becomes important to understand the host-microbe-drug interactions to effectively provide molecular contributions of the gut microbiota and its interactions with the human gastrointestinal system (Chellapandi and Prisilla 2018; Valdes et al. 2018). Then, Roja et al. (2024) decided to explore the interplay between gut microbes and human immune responses, as well as repurposing antibiotics based on their food-borne-drug interactions (Roja et al. 2024). They built a system medicine framework containing 2654 edges and 1609 nodes, comprising 1370 interacting human genes.

By modeling their deep-learning approach, they found 39 interactions involving human target genes for food-borne bacteria, nine targets for probiotic bacteria, and 11 targets for gut methanogens. Also, the authors reported the direct relationship between linoleic acid and the abundance of food-borne bacteria. Ultimately, by interfering with gut microbiota, it was identified that nabiximol, fusidic acid, ampicillin, chloroquine, oxacillin, cefdinir, and phenoxymethylpenicillin could be repurposed to treat food-borne infections (Roja et al. 2024).

## *3.4 Data-Mining Models on Gene-Drug-Disease Interaction*

All recent advances in precision medicine and pharmacogenomics improved the availability of accurate data focused on gene-drug-disease interactions (Koscielny et al. 2017; Lee et al. 2014; Mooney 2015; Relling and Evans 2015; Samwald et al. 2012; THORN et al. 2010). Since the complex scale of task for collecting and organizing all these manually, along with the effort- and time-consuming, the modern data-mining approaches have emerged to algorithmically identify and extract gene interactions from literature for scaling these to big data, making this process more efficient, accurate, and with reduced bias (Davis et al. 2013; Mallory et al. 2016; Tari et al. 2010; Vazquez et al. 2011). The main advantages of the mining process are (*i*) the data volume does not overload the user; (*ii*) interactions and evidence of each one of them are made easily available; (*iii*) the data quality threshold can be controlled; and (*iv*) exploratory and longitudinal studies can be developed with a well-known foundation (Davis et al. 2013; Mallory et al. 2016; Tari et al. 2010; Vazquez et al. 2011). For these facts, Wong et al. (2021) created and developed GeneDive, a free web-based software (Model-View-Controller—MVC, architecture) used for retrieval, filtering, discovery of topologies, and visualization tool, by processing over 3.2 million gene-drug-disease interactions and providing the supporting evidence (Wong et al. 2021). Their repository comprises gene-disease, gene-drug, and gene-gene interactions of clinical relevance. The authors presented several case studies in which the accuracy of GeneDive was demonstrated. For example, consider a case of a co-infected patient with HIV and Hepatitis C genotype I, who received a prescription of ribavirin and peginterferon alfa for treating Hepatitis C infection, while atazanavir, raltegravir, or abacavir to treat HIV infection. By initially analyzing peginterferon alfa and ribavirin, GeneDive was capable of identifying a possible drug-drug interaction since both of them interfere with the gene *IFNL3* (also named IL28B). In contrast, no drug-drug interactions were found for abacavir, atazanavir, and raltegravir since these interact with *HLA-B*, *UGT1A1*, and *ABCB1*, respectively. Finally, GeneDive was demonstrated to be a powerful tool for identifying drug-drug and drug-disease interactions based on genes (Wong et al. 2021).

# 4 Food-Drug Interactions

In the realm where culinary delights meet pharmacological efficacy lies a fascinating intersection known as food-drug interactions. This intricate interplay between what we consume and the medications we ingest holds profound implications for health outcomes and therapeutic efficacy. Understanding food-drug interactions is not merely about recognizing what to eat with certain medications but delving into the biochemical nuances that govern absorption, metabolism, and efficacy (Boullata and Hudson 2012). This background delves into the concepts, importance, and clinical relevance of food-drug interactions, shedding light on their multifaceted nature. At their core, food-drug interactions encompass the alterations in the absorption, distribution, metabolism, and excretion (ADME) of medications induced by concurrent food consumption. These interactions can either enhance or diminish the therapeutic effects of drugs or trigger adverse reactions (Fig. 12.1). Understanding the various mechanisms underlying food-drug interactions is crucial, ranging from changes in gastrointestinal pH affecting drug solubility to food components modulating drug metabolism enzymes (BUSHRA et al. 2011; Ötles and Senturk 2014). In this context, Anadón et al. (2016) produced a nice book chapter focused on drug-nutrient interactions, regarding different types of interactions and their chemical components associated with them (Anadón et al. 2016). In parallel, it has been demonstrated that drugs of abuse (e.g., methamphetamine, heroin, and alcohol) are affected by food interactions in rats, as described by Townsend et al. 2021.

Although some food-drug interactions could provide benefits to patients, these lead mostly to drug-induced adverse reactions (Bushra et al. 2011; Fasinu et al. 2012). The significance of comprehending food-drug interactions extends far beyond mere clinical considerations. These interactions have the potential to significantly impact medication adherence, efficacy, and safety. Failure to recognize and manage food-drug interactions can lead to suboptimal treatment outcomes, compromised patient safety, and increased healthcare costs. Moreover, with the rising prevalence of polypharmacy and the diverse dietary habits across populations, the relevance of food-drug interactions in clinical practice is more pronounced than ever (Schmidt and Dalhoff 2002). Navigating food-drug interactions in clinical practice necessitates a nuanced approach that integrates pharmacological knowledge with dietary considerations. Healthcare professionals must educate patients

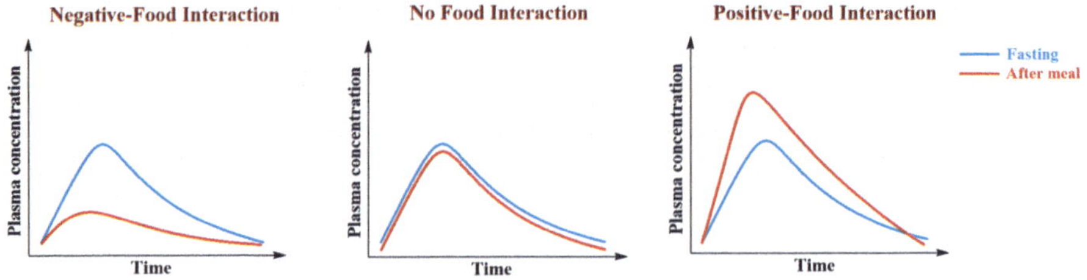


**Fig. 12.1** Comparison of different food effects on the pharmacokinetics of drugs

about potential food-drug interactions associated with their medications and provide tailored dietary recommendations to optimize therapeutic outcomes. Moreover, pharmacists play a pivotal role in screening for potential food-drug interactions and collaborating with other healthcare providers to mitigate risks and optimize medication regimens (Boullata and Hudson 2012; Schmidt and Dalhoff 2002).

Food-drug interactions are crucial considerations in pharmacotherapy, impacting the efficacy and safety of orally administered pharmacotherapy. Understanding the specific mechanisms underlying these interactions is essential for optimizing drug dosing regimens and patient outcomes. Thus, Koziolek et al. (2019) wrote a complete review paper to describe detailed drug-food interactions at a molecular level, focusing on pharmacokinetic aspects involving drugs and their interactions with common food components such as milk, grapefruit juice, alcohol, and functional foods (Koziolek et al. 2019). They identified a set of drugs known to exhibit significant food interactions and common food components that are likely to influence drug absorption and metabolism, mainly associated with CYP3A and CYP2C9 enzymes. Computational tools were used to simulate the interactions between selected drugs and food components at a molecular level, performing docking studies to predict their binding affinities, target sites, and amino acid residues involved in the drug-food interactions. Then, in vitro experiments were performed to validate their in silico predictions by measuring drug concentrations in the presence of different food components to assess changes in absorption and metabolism. Posteriorly, they quantified the impact of food components on drug bioavailability by determining concentration values under various conditions. Calculate parameters such as plasma-concentration-time curve (AUC), concentration in the plasma ($C_{max}$), and time at which this concentration is observed ($t_{max}$) to evaluate the extent of drug-food interactions. It is known that food intake can lead to changes in the physiological performance of the human gastrointestinal tract, interfering with the pharmacokinetic aspects of drugs by altering their ADMET profile (Koziolek et al. 2019). In this context, the authors stated that altered volumes of fluids into the small intestine and stomach could alter the bioavailability of drugs, affecting their $C_{max}$ and $t_{max}$ parameters. Furthermore, the authors suggest discussing how understanding these interactions can guide personalized medicine approaches, recommending future research directions, such as exploring novel computational methods for predicting drug-food interactions and investigating the clinical relevance of identified interactions in diverse patient populations.

## 4.1 CYP-Food Interactions

A study performed by Baltes et al. (2001) decided to investigate the inhibitory effects of lemon juice on CYP3A4 (Fig. 12.2) activity—CYP450 monooxygenase isoform 3A (CYP3A) is known to have an essential role in the metabolism of many drugs (Zhou et al. 2007), a crucial enzyme involved in drug metabolism (Baltes 2001). The study compared the inhibitory effects of lemon juice with classical

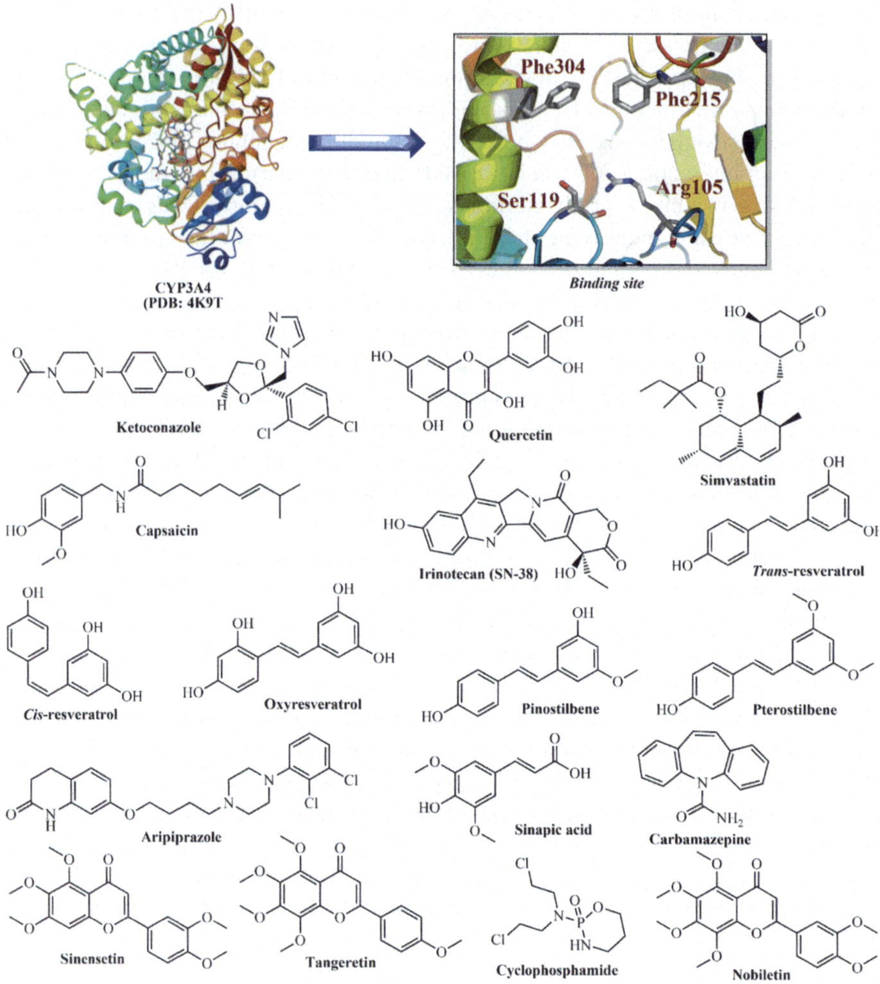


**Fig. 12.2** 3D-Structure of CYP3A4 with its binding site, and CYP-related drugs that have their pharmacokinetic profiles affected by food intake. The binding site of CYP3A4 (PDB: 4K9T) was reported by Sevrioukova and Poulos (2013)

inhibitors like ketoconazole (0.34 μM), quercetin (24 μM) (Fig. 12.2), and the well-known inhibitor grapefruit juice. Understanding these interactions is essential for predicting potential adverse effects and optimizing drug therapy. The researchers used a human lymphoblastoid cell line (h3A4/OR) expressing CYP3A4 activity. Additionally, the tested inhibitors including ketoconazole, quercetin, grapefruit juice, lemon juice, tomato juice, cherry juice, and orange juice were investigated, and high-performance liquid chromatography (HPLC) was employed for their metabolites analysis. Still, metabolite formation with co-addition of inhibitors was quantified as a percentage ratio relative to the control without inhibitors. As a result, lemon juice was identified as a potent inhibitor of CYP3A4 activity, comparable to grapefruit juice. Their study revealed significant inhibition percentages for lemon juice, indicating its potential to interfere with drug metabolism. However, the authors could not provide insights into the binding interactions between the inhibitors and key amino acid residues in the CYP3A4 active site, elucidating the mechanism of inhibition. In conclusion, the study underscored the importance of considering lemon juice as a potential inhibitor of CYP3A4 activity in drug therapy, emphasizing the need for further investigations into drug-food interactions (Baltes 2001). Furthermore, their in vitro cell model using living cells provided a robust tool for studying CYP3A4 inhibition, offering a cost-effective and efficient approach for assessing drug interactions. Although in vivo studies are warranted to validate the clinical relevance of the identified drug-food interactions and to explore the specific components in lemon juice responsible for CYP3A4 inhibition.

Simvastatin (Fig. 12.2), which is a prodrug targeting hydroxy-3-methylglutaryl CoA (HMG-CoA) reductase, prescribed for the treatment of hyperlipidemia, hypercholesterolemia, and to reduce the risk of cardiovascular events (Neuvonen 2010; Wang et al. 2011). Its active form is simvastatin acid, which is found in the liver, intestinal mucosa, and plasma (Zhang et al. 2010a, b), majority metabolized by CYP3A, whereas a minor proportion is also metabolized by CYP2C8 (Neuvonen et al. 2006). It has been demonstrated that the in vivo pharmacokinetics of the CYP3A substrate is altered by oral administration of capsaicin (Fig. 12.2) in rats (Hirotani et al. 2007), as well as it can inhibit the CYP3A in vitro activity (Takanohashi et al. 2010). Then, considering that to understand how dietary components, such as capsaicin, could influence drug metabolism is crucial for optimizing treatment outcomes. In this context, Zhai et al. (2013) evaluated the effects of capsaicin on the pharmacokinetics of simvastatin in rats, shedding light on potential interactions between dietary constituents and medications (Zhai et al. 2013). With this purpose, the rats were divided into groups receiving different doses of capsaicin (3, 8, or 25 mg/kg) for seven consecutive days. On the seventh day, rats were orally administered simvastatin at a dose of 80 mg/kg. Subsequently, pharmacokinetic parameters, including AUC, $C_{max}$, clearance, and half-life, were calculated using non-compartmental analysis. The authors observed that the AUC of simvastatin was significantly decreased in rats pre-treated with capsaicin, with reductions of 67.06, 73.21, and 77.49% observed for 3, 8, and 25 mg/kg doses, respectively. Moreover, capsaicin pretreatment led to a decrease in $C_{max}$ of simvastatin and its active metabolite compared to the control group. Posteriorly, statistical analysis revealed significant differences in pharmacokinetic parameters between capsaicin-treated and control groups, indicating a pronounced effect of capsaicin on simvastatin metabolism (Zhai et al. 2013). Finally, their findings suggest that capsaicin can alter the pharmacokinetics of simvastatin in rats, potentially impacting the efficacy of cholesterol-lowering therapy. However, further investigations are warranted to elucidate the underlying mechanisms of this interaction and assess its clinical relevance in human populations.

Posteriorly, Wang et al. (2018) studied the effects of capsaicin on the pharmacokinetics of irinotecan (SN-38, Fig. 12.2) and its main metabolites in rats (Wang et al. 2018). Then, the effects of capsaicin on the pharmacokinetics of irinotecan and its metabolites, shedding light on the underlying mechanisms and clinical implications, were investigated by them. The authors utilized then human liver microsomes to investigate the inhibitory effects of capsaicin on CYP3A4, administering 3 mg/kg capsaicin for 7 days, and assessing the systemic exposure and toxicity of irinotecan and its metabolites. Subsequently, AUC, CL, and $T_{1/2}$ pharmacokinetic parameters were determined, including $IC_{50}$ values. As a result, they verified that in the treatment with 3 mg/kg of capsaicin, the $AUC_{0\rightarrow\infty}$ of SN-38 significantly increased by 1.48-fold in rats, indicating enhanced systemic exposure. Regarding its mechanisms, it was suggested that capsaicin exhibits dual inhibition of CYP3A and P-glycoprotein, leading to altered metabolism and transport of irinotecan and its metabolites. Additionally, the authors investigated the capsaicin effects on the activity of carboxylesterase 2 (Ces2) and uridine diphosphate glucuronosyl transferase 1A1 (Ugt1A1) enzymes in both short- and long-term treatment. Finally, they found that capsaicin $IC_{50}$ values for Ces2 and Ugt1A1 enzymes are 60.74 and 408.2 µM, respectively, revealing that capsaicin is a poor Ces2 inhibitor and almost inactive toward the Ugt1A1 enzyme (Wang et al. 2018).

*Trans*-resveratrol (Fig. 12.2) is a well-known phytochemical in various dietary sources like red wine, grapes, blueberries, peanuts, and cranberry juice. It has been reported to interact with CYP450 enzymes, particularly CYP3A4, which are crucial for drug metabolism (Basheer et al. 2017; Bavaresco et al. 2016). A study focused on investigating the potential risk of drug-food interactions associated with *trans*-resveratrol and other natural stilbenes commonly found in food sources was developed by Hyrsova et al. (2019). This study aimed to assess the impact of natural stilbenes on CYP450 enzyme activities, specifically CYP3A4 and CYP2B6, and their potential to induce or inhibit these enzymes in human hepatocytes (Hyrsova et al. 2019). Five primary human hepatocyte cultures were utilized in the research, isolated from human liver resected from adult multiorgan donors and commercial cultures of long-term human hepatocytes. The hepatocytes were treated with different stilbenes to assess their effects on CYP3A4 and CYP2B6 mRNA induction and enzyme activities. Total RNA was isolated from the hepatocytes, and the levels of CYP3A4, CYP2B6, and GAPDH mRNAs were determined using quantitative reverse transcriptase polymerase chain reaction (qRT-PCR). Additionally, Western blotting experiments were conducted to analyze the expression of CYP3A4 and CYP2B6 proteins in response to the tested stilbenes. The inhibitory activities of stilbenoids on luciferase activities were assessed using HepG2 cells transfected with Firefly and Renilla luciferase expression vectors. Thus, the authors verified that *trans*-resveratrol was found to inhibit CYP3A4 activity significantly, with a $K_i$ value lower than 10 µM. This inhibition was observed for both testosterone 6$\beta$-hydroxylation and Midazolam 1′-hydroxylation CYP3A4/5 activities in micromolar concentrations ($IC_{50}$: 5.5 µM) via non-competitive binding mode. Other natural stilbenes like *cis*-resveratrol, oxyresveratrol, pinostilbene, and pterostilbene (Fig. 12.2) also inhibited CYP3A4/5 enzymatic activities, but only *trans*-resveratrol showed significant inhibition in micromolar concentrations by a non-competitive mechanism. Still, the authors suggest a potential risk of food-drug interactions with

CYP3A4/5 substrates due to the inhibitory effects of *trans*-resveratrol on these enzymes. Some natural stilbenes have been identified as an agonist of pregnane X receptor (PXR) (Dolezelova et al. 2017; Dring et al. 2010; Kluth et al. 2007; Smutny and Pavek 2014; Yu et al. 2011), in which *trans*-resveratrol being a more potent PXR ligand than the *cis*-isomer (Yu et al. 2011). The authors demonstrated that polymethoxylated stilbenes were identified as PXR ligands, while natural stilbenes like *trans*- and *cis*-Resveratrol did not significantly induce CYP3A4 or CYP2B6 expression in primary human hepatocytes via PXR interaction. While natural sources like red wine contain relatively low concentrations of *trans*-resveratrol (1.9 mg/L), dietary supplements or medications can provide doses of up to 2–5 g/day, leading to micromolar concentrations of *trans*-resveratrol in plasma. Finally, they speculated that portal vein blood concentrations may exceed 10 μM after the intake of certain food supplements or natural sources, indicating a potential for higher exposure of hepatocytes or enterocytes to *trans*-resveratrol (Hyrsova et al. 2019). These findings highlighted the differential effects of *trans*-resveratrol compared to other natural stilbenes on CYP3A4 activity and suggested a need for further investigation into the implications of these interactions on food-drug interactions.

Aripiprazole (Fig. 12.2) is a second-generation antipsychotic used in the treatment of various psychiatric disorders. Sinapic acid (Fig. 12.2), a bioavailable phytoconstituent with antioxidant properties, has been shown to inhibit CYP3A4 enzymes (Kirschbaum et al. 2008; Shapiro et al. 2003; Swainston Harrison and Perry 2004). Given the potential for sinapic acid to interact with hepatic enzymes involved in Aripiprazole metabolism, Raish et al. (2019a, b) investigated the specific effects of sinapic acid on Aripiprazole pharmacokinetics in rats and its modulation of hepatic CYP450 expression. Moreover, they evaluated the impact of sinapic acid on the absorption and metabolism of Aripiprazole in rats, the modulation of hepatic CYP3A2 and CYP2D6 expression by sinapic acid, and the potential food-drug interaction between sinapic acid and Aripiprazole and its implications for clinical practice (Raish et al. 2019a). Initially, Wistar rats were categorized into groups for single and multiple-dose studies, and UPLC-MS/MS techniques were employed to measure Aripiprazole plasma concentrations and calculate pharmacokinetic parameters. Besides, hepatic protein expression of CYP3A2, CYP2D6, and β-Actin was assessed using specific antibodies and immunoblot analysis. Finally, data were expressed as mean ± SD, and significance was determined using appropriate statistical tests. The authors demonstrated that the co-administration of sinapic acid with Aripiprazole led to increased absorption and metabolism of Aripiprazole, as indicated by changes in $C_{max}$, AUC, volume of distribution (Vd), and total clearance (CL). Sinapic acid treatment influenced the expression of CYP3A2 and CYP2D6 in rat liver tissues, suggesting a potential modulation of drug metabolism pathways. Their findings underscore the importance of considering food-drug interactions in medication management, particularly when co-administering Aripiprazole with compounds like sinapic acid (Raish et al. 2019a).

Posteriorly, Raish et al. (2019a, b) developed a new study involving sinapic acid, although focusing on its effects on hepatic CYP450 3A2, 2C11, and intestinal

P-glycoprotein on the pharmacokinetics of oral carbamazepine (Fig. 12.2) in rats (Raish et al. 2019b). Carbamazepine, a drug commonly used in epilepsy treatment, has a narrow therapeutic window and is susceptible to interactions with dietary supplements and herbal medicines (Ketter et al. 1991). It is capable of inducing different drug-metabolizing enzymes, such as CYP450 2B and 3A in the liver (Roujeau et al. 1995), in which CYP3A is responsible for converting carbamazepine into its active 10,11-epoxide, 3-hydroxyl-carbamazepine metabolite (Sullivan and Mcelhatton 1977), whereas it is also metabolized by intestinal P-glycoprotein (Hudson et al. 2000). Is it known that the prolonged use of carbamazepine can induce liver injuries, which is commonly verified by increased levels of alanine aminotransferase (ALT), aspartate aminotransferase (AST), and γ-glutamyltransferase (γGT) (Raish et al. 2017). Additionally, the authors aimed to elucidate the potential mechanisms of food-drug interactions between sinapic acid and carbamazepine. Then, 24 male Wistar rats were divided into four groups for the study, regarding the oral administration of carbamazepine (80 mg/kg) and sinapic acid (20 mg/kg), in which the plasma concentration-time plots were generated. Several pharmacokinetic parameters of carbamazepine were determined with and without sinapic acid pretreatment. Protein expression analysis of hepatic CYP450 and intestinal P-glycoprotein was conducted using Western blotting. Thus, the authors observed that the sinapic acid pretreatment reduced intestinal MDR1 protein expression compared to normal rats. Moreover, carbamazepine administration induced a significant increase in intestinal MDR1 protein expression. Besides, sinapic acid pretreatment inhibited the carbamazepine-induced MDR1 protein expression, indicating a potential interaction between sinapic acid and carbamazepine at the protein level. The study provides valuable insights into the complex interactions between sinapic acid and carbamazepine, emphasizing the importance of monitoring drug concentrations to mitigate potential side effects. Further research is warranted to elucidate the molecular mechanisms underlying food-drug interactions and improve the safety and efficacy of epilepsy treatment regimens (Raish et al. 2019b).

It is known that food-drug interactions can significantly impact the pharmacokinetics and pharmacodynamics of medications, leading to altered drug efficacy and safety profiles (Rodríguez-Fragoso et al. 2011). In particular, the interaction between citrus fruits, such as Clementines, and certain drugs has gained attention due to their potential effects on key drug-metabolizing enzymes (Theile et al. 2017). In this context, Weiss et al. (2020) investigated specific compounds in Clementines that may induce food-drug interactions, focusing on their effects on metabolic enzymes and potential implications for drug therapy (Weiss et al. 2020). The authors aimed to identify the main compounds in clementines responsible for food-drug interactions, evaluate their impact on key drug-metabolizing enzymes, and also investigate the potential mechanisms underlying the observed interactions through in vitro and in silico analyses. In vitro assays were carried out to assess the effects of the identified compounds on key drug-metabolizing enzymes, such as CYP1A1, CYP1A2, CYP3A4, and ABCB1. Additionally, molecular docking studies were performed to predict the binding interactions between the identified compounds and the active sites of the target enzymes. Moreover, they used LS180 cells to investigate the induction or inhibition of metabolic enzymes by the clementine compounds and to measure the expression levels of AhR and PXR-regulated genes in response to the clementine compounds. Then, nobiletin, sinensetin, and tangeretin (Fig. 12.2) were identified as the main perpetrators in clementines causing food-drug interactions, in which these compounds exhibited varying degrees of inhibition on CYP1A1, CYP1A2, CYP3A4, and ABCB1, with nobiletin showing the most potent inhibitory effects. Molecular docking simulations revealed specific amino acid residues within the active sites of the enzymes that interact with the clementine compounds, providing insights into their binding mechanisms, in which nobiletin was found to interact with residues in the active site of CYP1A1, forming hydrogen bonds with residues such as $Ser^{119}$ and $Thr^{124}$. In the binding pocket of CYP1A2, nobiletin exhibited hydrophobic interactions with residues like $Phe^{226}$ and $Phe^{260}$, contributing to its inhibitory effects on the enzyme. Still, it was revealed that nobiletin formed π-π stacking interactions with aromatic residues like $Phe^{304}$ and $Phe^{108}$ in the active site of CYP3A4, influencing its inhibitory activity. Besides, LS180 cells exposed to the compounds showed altered expression of AhR and PXR-regulated genes, indicating their potential to modulate drug metabolism pathways. The study elucidated the molecular basis of food-drug interactions induced by clementine compounds, emphasizing the need for caution when consuming citrus fruits alongside medications. Understanding the mechanisms of enzyme inhibition by specific compounds in clementines can guide personalized medicine approaches to mitigate potential drug interactions (Weiss et al. 2020).

Gougis et al. (2022) investigated specific drug-food and drug-herb interactions in 95 cancer patients by applying a survey, focusing on their molecular targets, concentration values, and potential implications for clinical practice (Gougis et al. 2022). They identified specific drug-food interactions based on known pharmacokinetic and pharmacodynamic properties of the drugs (paclitaxel, vinorelbine, letrozole, cyclophosphamide, lenvatinib, dexamethasone, lapatinib, ibrutinib, and vinblastine) and the potential impact of food components (green tea, pomelo, soy, grapefruit, coffee, bitter orange, cranberry, and licorice) on the metabolism. The authors identified that 53% of patients consume CYP-interacting food. Also, it was observed that four patients simultaneously used cyclophosphamide (Fig. 12.2) and coffee (Raaska et al. 2004). It is known that coffee inhibits CYP1A2, which is one of the enzymes associated with cyclophosphamide activation since it is a prodrug associated with also other CYP, such as CYP3A and CYP2B6 (Gougis et al. 2021). In contrast, letrozole seems not been affected by coffee. In summary, all foods able to inhibit CYP enzymes should be supervised by the medical assistance staff. The authors concluded that self-assessment of herbs, food used as medicine, and dietary supplements in cancer patients is feasible and should be considered in the daily routine. However, they recommended more studies to have an enlightened overview (Gougis et al. 2022).

## 4.2 UDP-Glucuronosyltransferase-Food Interaction

UDP-glucuronosyltransferases (UGTs) play a crucial role in the detoxification and elimination of various endogenous and exogenous compounds, including drugs (Zhang et al. 2015). Piceatannol (Fig. 12.3), a polyphenol compound found in passion fruit and blueberry, has shown various biological activities, including antitumor and antioxidant effects (Matsui et al. 2010; Viñas et al. 2011). Previous studies have indicated that piceatannol may inhibit CYPs activities. However, the impact of piceatannol on UGT enzymes, which are essential for the clearance of numerous compounds, including drugs, remains unclear (Bastianetto et al. 2010; Piotrowska et al. 2012). Regarding this fact, Jiang et al. (2020a, b) aimed to assess the inhibitory effects of piceatannol on human UGT enzymes and evaluate the potential risk of food-drug interactions through UGT inhibition (Jiang et al. 2020b). Within this purpose, various UGT supersomes (UGT1A1, UGT1A3, UGT1A4, UGT1A6, UGT1A7, UGT1A8, UGT1A9, and UGT1A10) were employed. Moreover, the inhibitory effects of piceatannol on UGTs were evaluated using high-performance liquid chromatography (HPLC) to measure the formation rates of 4-MUG and imipramine *N*-glucuronide. Subsequently, the inhibition mechanisms of piceatannol on different UGT isoforms were also determined. Thus, it was observed that piceatannol exhibited broad-spectrum inhibition against human UGT enzymes. Surprisingly, piceatannol displayed a non-competitive binding mode toward UGT1A6 and UGT1A7, while on UGT1A8 and UGT1A9, it exhibited competitive and mixed inhibition mechanisms, respectively. Still, quantitative analysis predicted potential food-drug interactions when piceatannol was co-administered with drugs

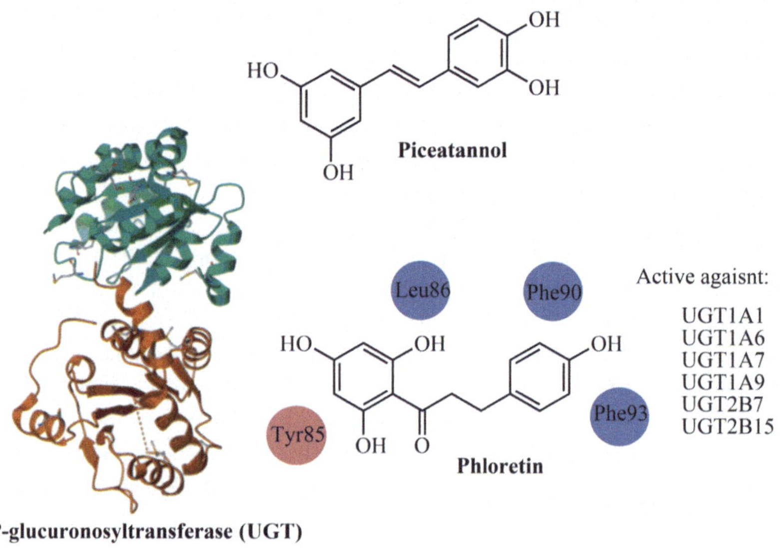


**Fig. 12.3** Inhibitors of human UDP-glucuronosyltransferase (UGT), in which phloretin presents hydrophobic contacts (blue) and hydrogen-bonding interactions (red)

metabolized by UGT1A6, UGT1A7, UGT1A8, and UGT1A9. Finally, the authors highlighted the importance of considering piceatannol's inhibitory effects on UGT enzymes in the context of food-drug interactions to avoid potential adverse effects (Jiang et al. 2020b).

Pterostilbene (see Fig. 12.2), a natural stilbene found in small berries and food supplements, has been reported to inhibit UGT1A9 enzymes, raising concerns about potential interactions with drugs metabolized by this enzyme (Hyrsova et al. 2019). Then, Jiang et al. (2020a, b) performed a study to investigate the inhibitory effects of pterostilbene on UGT1A9 enzymes and evaluate the risk of food-drug interactions due to UGT inhibition (Jiang et al. 2020a). The study assessed the inhibitory effects of pterostilbene on various UGT enzymes, including UGT1A1, UGT1A3, UGT1A6, UGT1A8, UGT1A9, and UGT2B15. Moreover, $IC_{50}$ values ranging from 0.92 to 28.66 μM were determined, with pterostilbene showing the strongest inhibition on UGT1A9. Further kinetic studies revealed that pterostilbene exerted potent non-competitive inhibition of 4-MU glucuronidation by UGT1A9, with a Ki value of 0.52 ± 0.04 μM. Still, they predicted that oral doses of pterostilbene above 50 mg/day, in combination with drugs metabolized by UGT1A9, could lead to a more than 50% increase in the AUC of the co-administered drug. Finally, their results underscored the importance of considering the inhibitory effects of pterostilbene on UGT enzymes in pharmacotherapy to mitigate potential adverse effects and ensure therapeutic efficacy (Jiang et al. 2020a).

Phloretin (Fig. 12.3), a bioactive polyphenol found in apples, has garnered attention for its potential to influence drug metabolism through the inhibition of human UDP-glucuronosyltransferases (UGTs) (Mariadoss et al. 2019). In this context, Chen et al. (2022) performed a study aiming to elucidate if procyanidin B2, catechins, epicatechins, chlorogenic acid, caffeic acid, *p*-coumaric acid, phloretin, and phlorizin had any effects on human UGTs (Chen et al. 2022). The authors employed in vitro models to evaluate the inhibitory effects of phloretin on different UGT isoforms. Enzyme kinetics studies were conducted to determine the inhibitory potency of phloretin toward specific UGT enzymes involved in drug metabolism. Concentration-response experiments were carried out to elucidate the binding kinetics and potential amino acid residues involved in their interaction. The comprehensive analysis revealed that phloretin exhibited significant inhibitory effects on multiple UGT isoforms, including UGT1A1, 1A6, 1A9, 2B7, 2B15, and 1A7. Regarding the inhibition of UGT caused by phloretin, it was found to be performed by a non-competitive binding mode toward UGT1A1, 1A6, 1A9, 2B7, and 2B15, while uncompetitive for UGT1A7. Besides, it was determined $K_i$ values ranged from 2.46 to 17.74 μM, indicating varying affinities of phloretin toward different UGT isoforms. In silico analyses revealed that phloretin predominantly interacts with UGT isoforms via hydrophobic interactions with key residues such as $Leu^{86}$, $Phe^{90}$, and $Phe^{93}$. Additionally, hydrogen-bonding interactions with the $Tyr^{85}$ residue in UGT1A1 contribute to the binding affinity of phloretin to the UGT enzymes. These specific amino acid residues and their interactions play a crucial role in determining the binding mode and inhibitory effects of phloretin on UGT isoforms, providing valuable insights into the molecular mechanisms underlying these interactions. The study also highlighted the potential risk of food-drug interactions when phloretin-rich foods are co-administered with drugs primarily metabolized by UGT1A7 and/or UGT2B15. In conclusion, the authors highlighted the importance of considering dietary polyphenols, such as phloretin, in the context of drug therapy to minimize the risk of food-drug interactions and associated adverse effects. Still, the structure-dependent inhibition pattern of phloretin toward UGTs provides valuable insights into the molecular mechanisms underlying these interactions. Understanding the impact of phloretin on drug metabolism can inform personalized medicine strategies to mitigate potential adverse effects and optimize therapeutic outcomes (Chen et al. 2022).

## 4.3 Organic Anion Transporters (OAT)-Food Interaction

Within the renal proximal tubule, specific transport mechanisms are responsible for facilitating the secretion of a range of drugs, encompassing antibiotics, antivirals, antitumor agents, and diuretics. These transporters also play a crucial role in controlling the concentrations of these drugs in the bloodstream. Among these transporters, human organic anion transporters hOAT1 (SLC22A6, Fig. 12.4) and hOAT3 (SLC22A8) are prominently involved. Situated on the basolateral membrane of proximal epithelial cells, they are primarily responsible for facilitating the uptake of

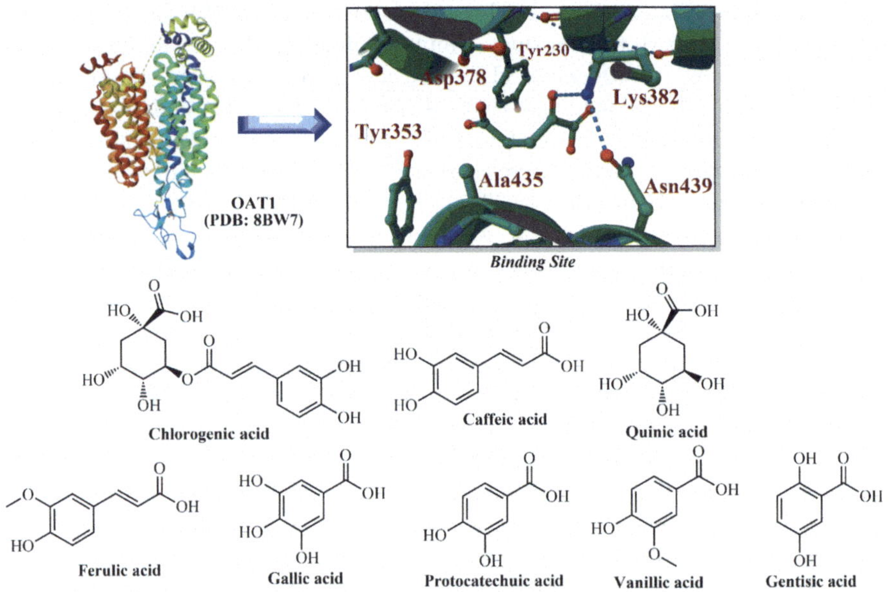


**Fig. 12.4** 3D-Struture of Organic Anion Transporter 1 (OAT1) with its binding site and inhibitors of hOAT1, hOAT3, and hOAT4. OAT1 structure (PDB: 8BW7) bound to alpha-ketoglutaric acid was reported by Parker et al. (2023)

these drugs from the bloodstream into the tubules. Consequently, the concurrent administration of a drug that inhibits hOAT1 and hOAT3 alongside their substrate drugs could precipitate a drug-drug interaction (El-Sheikh et al. 2008). While numerous components found in food and herbs have demonstrated interactions with drug transporters in the small intestine and metabolic enzymes in both the small intestine and liver, there remains a dearth of information regarding food-drug interactions involving hOAT1 and hOAT3 (Hong et al. 2007; Whitley et al. 2005). In 2011, considering that only Ellagic acid and flavonoids have been documented to interact with these transporters, Uwai et al. (2011) decided to investigate the inhibitory effects of caffeic, quinic, and chlorogenic acids (Fig. 12.4) on hOAT1 and hOAT3, key transporters involved in renal drug uptake, by performing uptake experiments using *Xenopus laevis* oocytes and then to evaluate the clinical significance of these effects (Uwai et al. 2011). Initially, these oocytes were injected with hOAT1 and hOAT3 cRNA to stimulate the uptake of typical substrates (p-aminohippurate and estrone sulfate, respectively) and subsequently exposed to varying concentrations of chlorogenic, caffeic, and quinic. Then, the authors observed that caffeic acid exhibited the strongest inhibitory effect on hOAT1 and hOAT3 compared to chlorogenic and quinic acids within a concentration-dependent inhibition, in which caffeic acid significantly reduced transporter activity of hOAT1 and hOAT3 with $IC_{50}$ values of 16.60 ± 3.7 and 5.4 ± 1.3 μM, respectively. In contrast, quinic acid demonstrated no influence on hOAT1 or hOAT3 activity. Moreover, caffeic acid significantly reduced the transport of antifolates by hOAT1 and hOAT3, in a concentration-dependent inhibition manner. Interestingly, it was seen that the presence of caffeic acid also led to a decrease in the transport of antivirals by hOAT1 and hOAT3, suggesting that it can inhibit the renal uptake of antiviral drugs, potentially affecting their pharmacokinetics and efficacy. Finally, understanding the molecular interactions between caffeic acid and these transporters is crucial for predicting and managing drug efficacy in the presence of dietary components (Uwai et al. 2011).

Phenolic acids are bioactive compounds found in various dietary sources and have been linked to numerous health benefits (Croft 1998; Karakaya 2004; Russell and Duthie 2011). Understanding the interactions between phenolic acids and hOATs is crucial for elucidating potential food-drug interactions. Wang and Sweet (2012) investigated the inhibitory effects of nine dietary phenolic acids on hOAT1, hOAT3, and hOAT4, providing insights into their impact on drug transport processes (Wang and Sweet 2012). They aimed to assess the inhibitory effects of dietary phenolic acids on hOAT1, hOAT3, and hOAT4, determine the inhibitory constants and drug-drug interaction indices for potent inhibitors, and evaluate the potential for food-drug interactions involving phenolic acids and OAT substrates. In vitro assays were conducted to measure the inhibitory effects of nine phenolic acids on hOAT1, hOAT3, and hOAT4 using stably expressing cell lines. Increasing concentrations of phenolic acids were applied to determine $IC_{50}$ values for potent inhibitors. Thus, all tested phenolic acids significantly inhibited hOAT3 transport, with specific inhibition of hOAT1 by ferulic, gallic, protocatechuic, vanillic (Fig. 12.4), and sinapinic acids (see Fig. 12.2). Among them, gallic acid showed stronger affinity with hOAT1 ($IC_{50}$ = 1.24 ± 0.36 μM), whereas the other compounds exhibited

$IC_{50}$ values from 7.6 to 18 μM. Regarding hOAT3 inhibition, gallic and ferulic acids demonstrated the best results, with $IC_{50}$ values of 9.02 ± 3.24 and 7.35 ± 3.73 μM, respectively. In contrast, only sinapinic acid exhibited notable inhibition (~33%) of hOAT4 at 100 μM concentration. Phenolic acids, particularly gallic and gentisic acids (Fig. 12.4), demonstrated significant inhibitory effects on hOAT1 and hOAT3, suggesting a strong potential for food-drug interactions (Wang and Sweet 2012). Their findings highlighted the importance of considering dietary phenolic acids in drug therapy to avoid potential interactions that may alter drug pharmacokinetics and efficacy. However, in vivo studies are warranted to validate their in vitro findings and establish safety guidelines for relevant pharmaceutical products.

## 4.4 Human Serum Albumin-Food Interaction

Aiming to understand the impact of food components on drug binding to plasma proteins, such as human serum albumin (HSA, Fig. 12.5), is essential for predicting drug efficacy and potential adverse effects. Wang et al. (2015) focused on the investigation of the influence of flavonoids on the affinity of nifedipine (Fig. 12.5) to HSA (Wang et al. 2015). Then, the interaction between nifedipine, a calcium channel blocker used in the treatment of hypertension, and HSA, a major plasma protein, was studied using spectroscopic techniques. They determined how flavonoids, including rutin and baicalin (Fig. 12.5), affect the binding properties of nifedipine to HSA, elucidated the quenching mechanism involved in the interaction between nifedipine and HSA, and investigated the thermodynamic parameters governing the

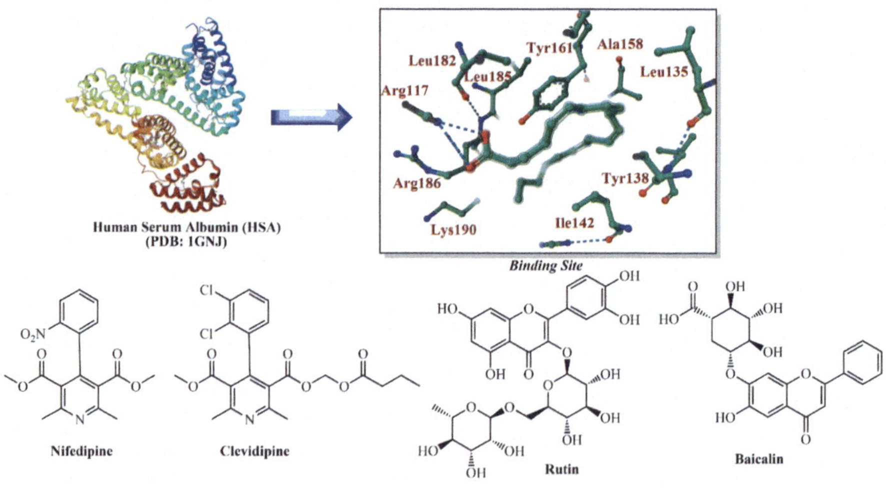


Fig. 12.5 Antihypertensive drugs that have their binding to the human serum albumin (HSA) affected by the presence of baicalin and rutin from food intake. The binding site of HSA (PDB: 1GNJ) is complexed with arachidonic acid by Petitpas et al. (2001)

binding process between nifedipine and HSA. Initially, nifedipine and HSA were prepared in appropriate concentrations for spectroscopic analysis. Then, fluorescence quenching studies were conducted to analyze the interaction between nifedipine and HSA in the presence of flavonoids. Moreover, absorption spectra were recorded to further investigate the binding behavior of nifedipine to HSA. The authors identified that rutin and baicalin were found to modulate the binding affinity of nifedipine to HSA, potentially altering its pharmacokinetics. The quenching mechanism in the nifedipine-HSA interaction was determined to involve both hydrogen bonds and hydrophobic interactions, indicating a complex binding process. The thermodynamic analysis revealed the spontaneity of the binding process between nifedipine and HSA, providing insights into the energetics of the interaction. The binding distance between the amino acid residue of HSA (tyrosine/tryptophan) and nifedipine is 2.6 nm, which indicates that the energy transfer from HSA to nifedipine can occur with high probability. Concluding, flavonoids can alter the binding behavior of nifedipine to HSA, potentially impacting the distribution and efficacy of the drug in the body. Further research on specific drug-food interactions is necessary to enhance the safety and efficacy of pharmacological treatments (Wang et al. 2015).

The antihypertensive drug, clevidipine (Fig. 12.5), is a short-acting dihydropyridine calcium channel antagonist frequently administered in acute hypertension patients (Ericsson et al. 1999; Ndefo et al. 2010). However, it has been reported that grapefruit juice could interfere with its area under the curve (AUC) ratio (Uesawa and Mohri 2012), suggesting a possible drug-food interaction. Regarding this context, Wang et al. (2017) investigated the binding properties of clevidipine to HAS by using several spectroscopic approaches (e.g., molecular fluorescence, UV-vis, and circular dichroism spectroscopies) in the presence of rutin and baicalin, as well as docking studies (PDB: 1GNJ) (Wang et al. 2017). Thus, the authors demonstrated the fluorescence quenching spectra of HSA in different molar ratios of clevidipine, in which it was confirmed an HSA-clevidipine complex. Moreover, it was displayed that hydrophobic interactions ($\Delta S^0 > 0$) and hydrogen bonding ($\Delta H^0 < 0$) represent the major intermolecular forces responsible for the binding of clevidipine to HSA. To identify the binding site of clevidipine to HSA, the authors used warfarin and Ibuprofen drugs as specific binding site markers for site I and site II of HSA, respectively. When these drugs are used after the HSA-clevidipine complex formation, it was observed that Warfarin shifts clevidipine from the binding site, confirming that clevidipine is mainly binding at the HSA site I. Posteriorly, docking studies were performed to display its binding modes at a molecular level. It was demonstrated that clevidipine can be embedded within the hydrophobic pocket of subdomain IIA of HSA (−24.56 kJ/mol), in which two hydrogen bonding interactions are performed. Additionally, it was seen that its chlorine and oxygen atoms could perform H-bonds with $His^{242}$ and $Gln^{196}$ residues, respectively. Finally, the authors demonstrated that baicalin and rutin can decrease the association constant ($K_a$) values of clevidipine binding to HSA, inducing maximum effectiveness and increasing the side effect risks of it (Wang et al. 2017).

# 5 Other Specific Cases of Drug-Food Interactions

## *5.1 Lorcaserin-Food Interaction*

The escalating rates of drug overdose-related fatalities, particularly attributed to potent opioids like fentanyl, underscore the critical need for innovative therapeutic strategies to address opioid use disorders. Current drugs targeting *mu-opioid* receptors have inherent limitations, necessitating the exploration of alternative treatments acting via non-opioid mechanisms (Humphreys et al. 2022; Madras et al. 2020; Volkow and Blanco 2023). Lorcaserin (Fig. 12.6), a selective serotonin 2C (5-$HT_{2C}$) receptor agonist initially approved for obesity management, has emerged as a potential candidate for substance use disorders involving opioids (Thomsen et al. 2008). Maguire (2024) investigated the specific pharmacological interactions between lorcaserin and opioids, focusing on their effects on ventilatory function and food-drug choice behavior in nonhuman primates (Maguire 2024). Then, male and female rhesus monkeys were utilized in the study, engaging in a food versus fentanyl choice procedure. Various doses of lorcaserin, morphine, and fentanyl were administered via different routes to evaluate their impact on ventilation and food-drug choice behavior. The experimental design encompassed forced trials, choice trials, and reinforcer delivery mechanisms. Thus, the author verified that when lorcaserin is administered as monotherapy, it demonstrated a dose-dependent reduction in

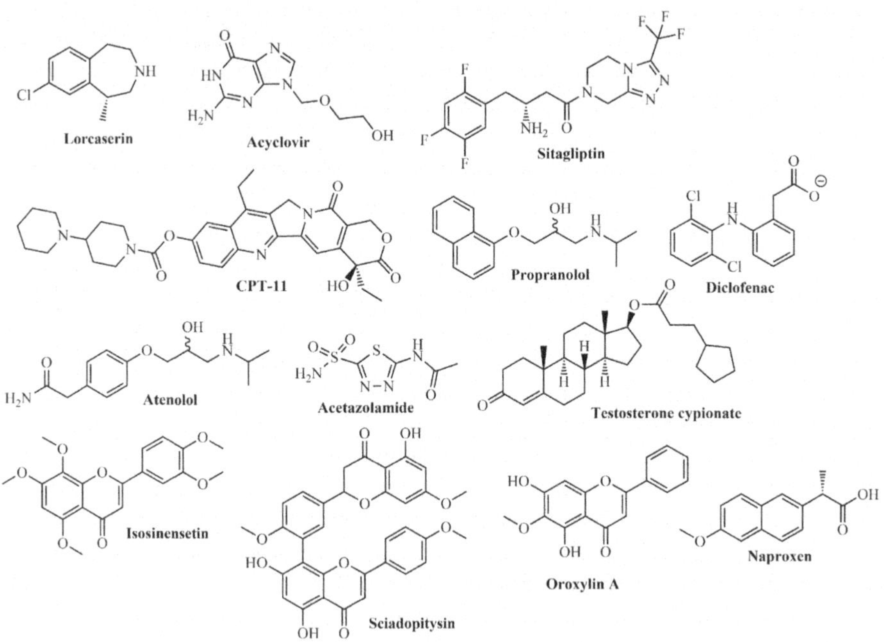


**Fig. 12.6** Drugs of different classes that are affected by dietary intake

minute volume to below 70% of baseline levels. Furthermore, lorcaserin potentiated the ventilatory-depressant effects of morphine and fentanyl, indicating a synergistic impact on respiratory function. The specific target receptors or pathways modulated by the combination of lorcaserin and opioids were not explicitly mentioned in the provided excerpts from the study. Further details regarding the exact receptors or pathways affected by the interaction between lorcaserin and opioids would require a more in-depth analysis of the full study or additional information from the research findings. The concentrations of lorcaserin and opioids significantly influenced ventilatory parameters. The study highlights the potential risks associated with co-administration of lorcaserin and opioids, particularly in exacerbating ventilatory depression. These findings underscore the importance of considering pharmacological interactions in the context of opioid use disorders and respiratory function. The results indicated that lorcaserin did not significantly alter the preference for fentanyl over food in the experimental setup. Future research endeavors should focus on elucidating the intricate molecular mechanisms underlying these interactions to facilitate the development of safer and more efficacious treatments for opioid abuse (Maguire 2024).

## 5.2 *Acyclovir-Food Interaction*

Concerning the drug-food interaction topic, an interesting case is associated with the interplay between acyclovir (Fig. 12.6), a commonly used antiviral medication, and chitosan, a nutraceutical with permeability enhancer properties (Je and Kim 2012). Then, García et al. (2023) performed a study to elucidate the mechanism underlying the unexpected reduction in acyclovir bioavailability observed when co-administered with chitosan. By investigating the interaction between acyclovir, chitosan, and endogenous mucus, this research aimed to provide insights into the specific factors influencing drug absorption in the presence of food supplements (García et al. 2023). Then, a two-compartment distribution/elimination model was constructed using Berkeley Madonna software to simulate the effects of acyclovir and chitosan in healthy humans. Besides, a two-compartmental model of the gastrointestinal tract was also developed, considering factors such as gastric emptying kinetics, mass transfer between intestinal compartments, and absorption coefficients. Subsequently, in vitro and preclinical models were utilized to assess the impact of chitosan on acyclovir bioavailability, including investigations into diffusion, permeation, and bioavailability. Posteriorly, pharmacokinetic modeling was employed to evaluate the clinical relevance of the findings, focusing on changes in acyclovir plasma concentrations and systemic availability. As the results of their studies focused on mucus-chitosan interactions, it was revealed that in situ coacervation between endogenous mucus and chitosan led to the rapid entrapment of a significant portion of acyclovir in the intestinal lumen. Additionally, the entrapment of acyclovir by mucus-chitosan complexes resulted in a local reduction of acyclovir concentration, particularly in the small intestine, leading to decreased drug absorption. The pharmacokinetic modeling demonstrated a decrease in acyclovir $C_{max}$ and AUC when co-administered with chitosan, highlighting the impact of mucus-chitosan interactions on drug bioavailability. This study provided a mechanistic explanation for the reduction in acyclovir bioavailability when co-administered with chitosan, attributing it to the entrapment of acyclovir by mucus-chitosan complexes in the intestinal lumen. These findings underscore the importance of considering mucus interactions in absorption models when assessing the effects of excipients or nutraceuticals on drug absorption. Finally, the authors stated that understanding the specific mechanisms of drug-food interactions, such as the mucus-chitosan-acyclovir interaction, is crucial for optimizing drug formulations and predicting potential alterations in drug pharmacokinetics (García et al. 2023).

### 5.3 Sitagliptin-Food Interaction

Type 2 diabetes (or *mellitus* diabetes) is a relevant public health problem worldwide, in which several different strategies have been used to control it, including antidiabetic drugs along with changes in diet and lifestyle habits (American Diabetes Association et al. 2008; Hawley and Gibala 2012; Mensink et al. 2003). Additionally, it is known that in the presence of glucose, incretins, including glucose-dependent insulinotropic polypeptide (GIP) and glucagon-like peptide-1 (GLP1), are responsible for inducing insulin secretion from pancreatic beta cells (Drucker 2006; Tulipano et al. 2011). Then, dipeptidyl peptidase IV (DPP-IV) inhibitors are a class of drugs used in the management of *mellitus* diabetes by modulating incretin hormone levels, in which the enzymatic degradation of these hormones leads to an in vivo reduction of them (Tulipano et al. 2011). In this context, Nongonierma and Fitzgerald (2013) developed a study focused on sitagliptin (Fig. 12.6), a DPP-IV inhibitor, and whey protein hydrolysate to investigate if it could have some synergic effects on the DPP-IV inhibitors (Nongonierma and Fitzgerald 2013). In this context, the authors decided to elucidate the inhibitory activity of whey protein hydrolysate, regarding its interaction with sitagliptin and gastrointestinal stability during digestion. Thus, the whey protein hydrolysate was produced using a food-grade pancreatic enzyme preparation, and the $IC_{50}$ value was determined. Ultrafiltration was employed to enrich DPP-IV inhibitory peptides in whey protein hydrolysate fractions, whereas the simulated gastrointestinal digestion was conducted to evaluate the stability of whey protein hydrolysate and its fractions to gastrointestinal enzymes. Posteriorly, the interaction between sitagliptin and whey protein hydrolysate, as well as whey-derived dipeptide inhibitors, was investigated to determine the extent of DPP-IV inhibition. Then, it was observed that whey protein hydrolysate displayed an $IC_{50}$ value of $1.34 \pm 0.11$ mg/mL for DPP-IV inhibition. Additionally, the fractionation obtained through ultrafiltration resulted in significantly more potent DPP-IV inhibitors in the permeate and hydrophilic fraction. Still, the DPP-IV inhibitory potency of whey protein hydrolysate increased post-simulated digestion, with an $IC_{50}$ value of $1.02 \pm 0.05$ mg/mL. When investigating the interaction of whey protein hydrolysate with sitagliptin, it was verified that the combined effect of sitagliptin and whey-derived peptides exhibited an additive effect on DPP-IV inhibition, suggesting a potential synergistic interaction. The authors hypothesized that sitagliptin and whey-derived peptides may target different sites or pathways involved in DPP-IV inhibition, leading to a combined effect that surpasses the individual contributions of each component; or the presence of whey-derived peptides may enhance the bioavailability or stability of sitagliptin, thereby increasing its efficacy in inhibiting DPP-IV activity. Furthermore, the competitive inhibition observed indicates a direct interaction of bioactive peptides with the DPP-IV active site. These findings underscore the need for further exploration in human studies to elucidate the implications of optimizing sitagliptin dosage in *mellitus* diabetes management. The study contributed valuable insights into food-drug interactions and their impact on therapeutic outcomes in diabetes management (Nongonierma and Fitzgerald 2013).

## 5.4 Irinotecan-Food Interaction

Lin et al. (2008) investigated the impact of (−)-epigallocatechin-3-gallate (EGCG), a bioactive compound found in green tea, on the pharmacokinetics of irinotecan {7-ethyl-10-[4-(1-piperidino)-1-piperidino]carbonyloxycamptothecin (CPT-11, Fig. 12.6), an alkaloid-derived of camptothecin, which is a compound used as an anticancer agent, and its metabolite, 7-ethyl-10-hydroxycamptothecin (SN-38, see Fig. 12.2) (Lin et al. 2008). Still, the authors aimed to characterize the effects of EGCG on the disposition of these anticancer agents and to elucidate the molecular mechanisms underlying any observed interactions. The experimental design of their research involved the administration of EGCG (20 mg/kg) followed by CPT-11 (10 mg/kg, i.v.) in two groups of Sprague-Dawley rats. Blood and bile samples were collected at specified time points for pharmacokinetic analysis. The concentrations of CPT-11 and SN-38 in plasma and bile were determined using HPLC, and pharmacokinetic parameters were calculated using appropriate software. As a result, it was observed that the pharmacokinetic data exhibited significant alterations in the disposition of CPT-11 and SN-38 in the presence of EGCG. Co-administration of EGCG led to a decrease in the AUC of SN-38 in bile, indicating a potential impact on biliary elimination. Moreover, the AUC ratio of blood to bile decreased after EGCG treatment, suggesting a modulation of the bile efflux transport system. However, the author did not perform in silico studies to obtain information focused on possible targets and their pathways. Thus, this study demonstrated that EGCG can influence the pharmacokinetics of CPT-11 and SN-38, particularly in terms of biliary excretion. The observed changes in drug concentrations and distribution highlight the complex nature of food-drug interactions and underscore the need for further investigation into the molecular mechanisms underlying these effects. Lastly, future research should focus on elucidating the specific molecular targets and pathways involved in these interactions to optimize therapeutic outcomes and ensure patient safety (Lin et al. 2008).

Even with numerous in vitro techniques at our disposal, animal testing persists as the primary benchmark for evaluating preclinical pharmacokinetic factors for drug prototypes. In this context, cassette pharmacokinetics stands out as a reliable non-clinical approach for identifying promising compounds among various new chemical entities in a single trial, despite alternative methods existing for managing drug disposition (Manitpisitkul and White 2004). Up to 2021, according to Kumar et al. (2021), no studies were carried out to identify the occurrence of drug-drug interactions using compounds from all four classes of the biopharmaceutics classification system (BCS). Then, they investigated the effect of sex and food on the pharmacokinetics of propranolol (BCS-I), diclofenac (BCS-II), atenolol (BCS-III), and acetazolamide (BCS-IV) (Fig. 12.6), focusing on the impacting drug bioavailability and efficacy in rats after cassette administration. Still, they aimed to understand the implications for drug development and oral drug delivery (Kumar et al. 2021). The study adhered to the 3Rs (replacement, reduction, and refinement) technique to ensure the welfare of laboratory animals during experimentation, in which the cassette dosing method was employed to simultaneously administer all four BCS drugs to male and female Sprague-Dawley rats in fed and fasting states. Posteriorly, intravenous and oral dosing of the drugs were conducted using the cassette dosing technique to evaluate key pharmacokinetic parameters. Still, pharmacokinetic parameters such as $C_{max}$, $T_{max}$, AUC, and mean residence time (MRT) were determined and compared between different BCS drug classes, sexes, and feeding conditions. Then, the study revealed significant differences in the pharmacokinetic profiles of BCS drugs between male and female rats. These differences were attributed to variations in drug metabolism and distribution influenced by sex-specific factors. Food intake was found to affect the bioavailability of BCS drugs, with differences observed in $C_{max}$ and MRT values between fed and fasting conditions. Atenolol and diclofenac demonstrated significant alterations in pharmacokinetics in the presence of food. Compounds with varying solubility exhibited distinct responses to food intake. In this context, diclofenac and possibly acetazolamide were particularly susceptible to alterations in $C_{max}$ and MRT values in fed conditions. The delayed absorption and extended residence time observed in these compounds highlight the impact of food intake on their pharmacokinetic profiles. Moreover, the cassette dosing technique proved to be an efficient method for assessing the pharmacokinetics of multiple drugs in a single experiment, reducing the number of animals required and minimizing stress and discomfort through venous cannulation. The authors demonstrated that sex and food play significant roles in modulating the pharmacokinetic parameters of BCS drugs, emphasizing the need for tailored dosing regimens based on sex and feeding status. Ultimately, understanding the interplay between sex, food, and drug absorption is essential for optimizing drug formulations and dosing strategies to enhance therapeutic outcomes (Kumar et al. 2021).

## 5.5 *Testosterone-Food Interaction*

It is assumed that nonalcoholic fatty liver disease and dyslipidemias could have multifactorial and complex etiologies, although there is an important genetic factor strongly associated with environmental components and lifestyle, mainly related to the dietary intake profile (Neuschwander-Tetri 2017; Schwenk et al. 2013). Food-drug interactions can have significant implications for health outcomes, particularly in the context of hepatic lipotoxicity and nonalcoholic fatty liver disease. Anabolic steroids are known to affect various metabolic processes, and their interaction with dietary lipids, specifically trans fatty acids, remains poorly understood (Kicman 2008). Then, Santos et al. (2018) developed a study aiming to elucidate the specific effects of anabolic steroids (e.g., testosterone cypionate, Fig. 12.6) on hepatic function when combined with a diet high in trans fatty acids (Santos et al. 2018). In their study, they investigated the impact of anabolic steroids on gene expression related to lipid metabolism, assessed changes in lipid profile and redox status in mice treated with anabolic steroids and a trans fatty acid-rich diet, and determined the development of nonalcoholic fatty liver disease in response to the interaction between anabolic steroids and dietary lipids. With this purpose, seventy-two C57BL/6 mice were divided into six groups and treated with different diets and anabolic steroid dosages. Then, parameters such as dietary intake, energy intake, food efficiency coefficient (FEC), and body mass were monitored. Carcasses were analyzed for fat, protein, and ash content using classical chemical methods, whereas plasma levels of total cholesterol, HDL cholesterol, LDL cholesterol, triacylglycerol, and liver enzymes were quantified. Thus, they demonstrated that the anabolic steroid administration with a standard diet increased lean body mass without hepatic changes. However, in combination with a high-fat diet, anabolic steroids reduced HDL cholesterol levels and induced hepatic damage. Moreover, animals treated with anabolic steroids and a high-fat diet showed reduced HDL cholesterol levels and increased triacylglycerol levels, indicating lipid metabolism alterations. Still, it was seen that the interaction between anabolic steroids and a high-fat diet triggered oxidative stress, potentially exacerbating nonalcoholic fatty liver disease. Furthermore, the combination of anabolic steroids and a high-fat diet worsened hepatic morphofunctional damage, suggesting a synergistic effect on nonalcoholic fatty liver disease progression. This study highlighted the complex interplay between dietary lipids, anabolic steroids, and hepatic health, emphasizing the need for further research on food-drug interactions and their impact on metabolic disorders (Santos et al. 2018).

## 5.6 Flavonoids-Food Interaction

Flavonoids, widely distributed in vegetables, fruits, and herbs, have diverse pharmacological activities, including anticancer, anti-inflammatory, antimicrobial, antioxidant, and antiviral (Xiao and Kai 2012). Then, due to their antioxidant properties, flavonoids are commonly consumed in diets (Brantley et al. 2014). However, their interactions with drug transporters, such as P-glycoprotein, remained understudied. Food-drug interactions pose challenges in clinical practice, emphasizing the need to investigate transporter-mediated interactions. Aiming to fill this gap by exploring the inhibitory effects of flavonoids on P-glycoprotein and their implications for food-drug interactions, Bai et al. (2019) performed an in vitro and in vivo study (Bai et al. 2019). In total, 75 flavonoids were commercially obtained and screened on MDR1-MDCKII cells to evaluate their cytotoxicity and inhibitory effects on P-glycoprotein. Additionally, male Sprague-Dawley rats were used in their in vivo evaluations. Moreover, docking studies were conducted to assess the underlying inhibitory mechanisms of flavonoids toward P-glycoprotein (PDB: 3G60). Furthermore, a pharmacophore model was established to elucidate the structure-activity relationships (SAR) of flavonoids with P-glycoprotein. Their in vitro screening revealed significant inhibition of P-glycoprotein by five flavonoids, being tangeretin ($IC_{50}$:12.66 μM, see Fig. 12.2), sinensetin ($IC_{50}$:18.9 μM, see Fig. 12.2), isosinensetin ($IC_{50}$: 4.2 μM, Fig. 12.6), sciadopitysin ($IC_{50}$: 53.42 μM, Fig. 12.6), and oroxylin A ($IC_{50}$: 78.33 μM, Fig. 12.6). Additionally, wild cancer cells MX-1 and taxol-resistant MX-1/T cells were used to verify the inhibitory effects on P-glycoprotein activity, in which all tested flavonoids were capable of increasing taxol cytotoxicity on these cell lineages. Molecular docking analysis suggested that the inhibitory effects of flavonoids on P-glycoprotein may be related to Pi interactions, in which all of them interacted with $Phe^{974}$ and $Val^{978}$ residues, whereas van der Waals interactions were not observed. Both sciadopitysin and oroxylin A did not perform any hydrogen bonding interactions with P-glycoprotein. Posteriorly, 10 optimal pharmacophoric hypotheses were generated, in which the best of them suggested a great influence of the presence of an aromatic ring, hydrogen bond acceptors, and hydrophobic groups, having the best score for sinensetin. It was elucidated that the hydrophobic groups in the B benzene ring play an essential role in the inhibitory activity, as verified when comparing oroxylin A (moderate inhibitor) with isosinensetin (potent inhibitor). Still, methoxyl or hydrophobic groups at 6- and 7-positions are quite important for the P-glycoprotein inhibition, observed by analyzing sciadopitysin and oroxylin A, respectively (Bai et al. 2019). Lastly, the study demonstrated the inhibitory effects of flavonoids on P-glycoprotein, emphasizing the need to consider herb- and food-drug interactions in clinical practice.

## 5.7 *Naproxen-Food Interactions*

The study of drug-amino acid interactions can provide insights into potential effects on drug efficacy when administered alongside essential amino acid supplements. In this context, Meena and Jain (2024) decided to investigate the interaction of the anti-inflammatory drug Naproxen (Fig. 12.6) with essential amino acids by using a quantitative thermodynamic approach (Meena and Jain 2024). Thus, they quantitatively assessed these interactions based on changes in heat and spectral properties and determined the impact of drug-amino acid interactions on the efficacy of both components. With this purpose, the authors utilized isothermal titration calorimetry (ITC) to measure heat changes in the interaction between Naproxen and various essential amino acids, calculated the standard molar limiting enthalpy of interaction ($\Delta H^{\circ}_{m}$) to understand the nature of interactions based on heat effects, used UV-visible spectroscopy to complement the ITC results and gain mechanistic insights into the interactions, and finally employed computational docking studies to explore potential binding sites and amino acid residues involved in the drug-amino acid interactions. The study revealed exothermic interactions between Naproxen and *L*-arginine, indicating significant polar interactions. Interaction of Naproxen with $100 \times 10^{-3}$ mol/kg of the amino acids exhibits an order of enthalpy change, being *L*-arginine (exothermic) > *L*-lysine (endothermic) > *L*-phenylalanine (endothermic) > *L*-valine (endothermic). In contrast, endothermic heat effects with other amino acids suggested hydrophobic effects and desolvation penalties dominated the association process. Posteriorly, UV-visible results supported the ITC findings, providing additional insights into the nature of interactions between Naproxen and essential amino acids. Computational docking studies may reveal specific binding sites and amino acid residues involved in the drug-amino acid interactions, offering a molecular-level understanding of the interactions. Finally, the study emphasizes the significance of understanding drug-amino acid interactions, especially in the context of administering drugs alongside essential amino acid supplements. Their findings suggest that the interactions between Naproxen and essential amino acids involve a combination of polar, hydrophobic, and desolvation effects, influencing the overall efficacy of both components (Meena and Jain 2024).

## 5.8 *Propranolol-Food Interaction*

While the influence of food on the absorption of immediate-release propranolol has been studied extensively (Byrne et al. 1984; Liedholm and Melander 1986; Melander et al. 1977), the effects on controlled-release formulations remained less understood. Then, El-Arini et al. (1989) investigated the potential food-drug interactions of controlled-release propranolol products by examining the pH-dissolution profiles and the impact of lipid pretreatment on drug release (El-Arini et al. 1989). Thus, four controlled-release propranolol products were selected for the study. Three of the products underwent pretreatment with peanut oil to simulate drug administration with a fatty meal. Also, in vitro dissolution testing was conducted to assess the pH-dissolution profiles of the products. Their results were visualized using 3-D graphs based on the topographical plotting technique. The study found that the influence of pH on the in vitro drug release from the controlled-release formulations was less significant compared to the effect of lipid pretreatment. Lipid pretreatment led to a substantial decrease in dissolution rate across the entire pH range, indicating a significant impact on the products subject to a high first-pass effect. The data suggested that the dissolution rate of the products was drastically decreased by oil pretreatment throughout the testing interval, highlighting the potential influence of fatty foods on in vivo absorption. In summary, their results suggest that the in vivo absorption of these products may be significantly affected by concomitant administration with fatty food, emphasizing the importance of considering food-drug interactions in clinical practice. Ultimately, the authors stated that the drastic reduction in dissolution rate by lipid treatment implies a potential influence of fatty meals on the bioavailability of controlled-release propranolol products (El-Arini et al. 1989).

# 6 In Silico Approaches for Predicting Drug-Food Interactions

## *6.1 Machine-Learning Approaches*

Food-drug interactions can significantly impact the efficacy and safety of pharmacological treatments. Understanding the molecular mechanisms underlying these interactions, particularly to nuclear receptor ligands such as constitutive androstane receptor (CAR) and pregnane X receptor (PXR), which are closely related orphan nuclear receptor proteins that share several ligands and target overlapping sets of genes involved in homeostasis and all phases of CYP450-related drug metabolism (Di Masi et al. 2009), being crucial for optimizing drug therapies. Then, Dring et al. (2010) performed a study focused on the rational quantitative structure-activity relationship (RQSAR) targeting PXR and CAR isoform-specific nuclear receptor ligands, aiming to elucidate the structural features of ligands that interact with CAR1, CAR3, and PXR receptors (Dring et al. 2010). By investigating the specific ligand-receptor interactions, this research provides valuable insights into the development of potential drug-food interactions. Firstly, a total of 60 compounds were selected to represent different structural groups and act as internal controls for key structural features. Then, human hepatoma HuH-7 cells were used in transient transfection luciferase assays. The cells were maintained in Dulbecco's Modified Eagle Medium (DMEM) supplemented with various additives. Subsequently, the compounds were screened in mid-throughput assays using human CAR1, CAR3, and PXR expression plasmids to observe differences in ligand behavior for building their RQSAR models. Terfenadine was excluded due to its higher toxicity on HuH-7 cells. Luciferase activity was measured to assess the effects of the compounds on the receptors. The authors identified structural differences in compounds that act as agonists, antagonists, or inverse agonists of CAR1, CAR3, and PXR receptors. Specific ligands were found to exhibit preferential binding to different isoforms of CAR and PXR, indicating distinct ligand-receptor interactions. For the CAR1 receptor, cresolphthalein, 4-ethylphenol, 4-*tert*-butylphenol, 4-octylphenol, 4-(Octyloxy)phenol, bis(4-hydroxyphenyl)methane, diphenhydramine, and *tert*-butylhydroquinone were identified as antagonists. For the CAR3 receptor, nonylbenzene, doxylamine, and pheniramine were identified as agonists. For the PXR receptor, fexofenadine, bisphenol S [4–4′-sulfonyldiphenol], and bromocresol purple were identified as antagonists, whereas *trans*-Stilbene oxide was found to be an agonist. Their results highlighted the importance of subtle structural variations in determining the activity and specificity of nuclear receptor ligands. Still, structural differences in compounds that exhibit opposite effects on CAR isoforms offer valuable information for pharmacophore development and understanding ligand binding preferences (Dring et al. 2010). However, further studies on specific amino acid residues involved in ligand binding could enhance our understanding of the molecular basis of food-drug interactions mediated by nuclear receptors.

Food-drug interactions can significantly impact the pharmacokinetics and pharmacodynamics of medications, leading to altered treatment outcomes and potential adverse effects (Burdock and Carabin 2004). Understanding specific interactions between food components and drugs is crucial for optimizing therapeutic regimens and patient safety. In this context, Reker et al. (2020) utilized machine-learning models to uncover the biological effects of commonly consumed chemicals, focusing on food and drug additives and elucidating potential food-drug interactions that may have clinical implications (Reker et al. 2020). Initially, 799 inactive ingredients (IIG) generally recognized as safe (GRAS) compounds were extracted from FDA databases and converted into SMILES structures. Subsequently, their physicochemical properties were calculated and compared with approved drugs, in which rotatable bonds, cLogP, and molecular weight parameters were considered in their analyses. Moreover, their biological activity data were obtained from ChEMBL22 for machine-learning model development. Then, in vivo warfarin uptake experiment was conducted in female BALB/c mice, whereas in vitro cell experiments were performed using HepG2 cells cultured in specific conditions. Regarding docking studies, human P-glycoprotein crystal structure (PDB: 6c0v) and UGT2B7 homology model were initially created. Posteriorly, docking simulations were performed for vitamin A palmitate and abietic acid to predict their interactions with P-glycoprotein and UGT2B7, respectively. As a result of their study, the machine-learning model predicted vitamin A palmitate as an inhibitor of P-glycoprotein transport and abietic acid as an inhibitor of UGT2B7 metabolism. Thus, it was verified that vitamin A palmitate inhibited P-glycoprotein-mediated efflux of a fluorescent reporter substrate with an $IC_{50}$ of 2.9 ± 3.6 μM, while abietic acid has an $IC_{50}$ value of 2.2 ± 0.3 μM on UGT2B7. Still, in silico, in vitro, ex vivo, and in vivo validations supported the predicted interactions. Also, it was observed that vitamin A palmitate increases the oral absorption of warfarin in mice by 31%. Docking studies revealed specific binding interactions between vitamin A palmitate and P-glycoprotein at the ATPase binding site and are stabilized by interaction with Arg[1047], as well as abietic acid and UGT2B7, which bind into the cofactor-binding domains. However, the authors did not inform the amino acids' residues involved in this interaction. Furthermore, the authors concluded that machine learning coupled with experimental validation can uncover previously unknown biological effects of food and drug additives, shedding light on potential food-drug interactions. Lastly, the identified inhibitors of P-glycoprotein and UGT2B7 have important implications for drug metabolism and transport, emphasizing the need for further investigation into food-drug interactions in clinical practice (Reker et al. 2020).

Sharma et al. (2023) performed a study to investigate the impact of food on the oral absorption of specific drugs, focusing on key drug properties and physiological factors that influence food-drug interactions (Sharma et al. 2023). Then, they selected a panel of 309 representative drugs known to exhibit food-drug interactions, which were discussed in their previous meta-analysis work (Sharma and Prasad 2021). In addition, gastrointestinal pH, bile acid solubilization, efflux transport saturation, and first-pass metabolism were considered as potential rate-limiting physiological mechanisms involved in the food effect. Still, drug properties such as lipophilicity (logP), dose number, and extraction ratio were analyzed to determine their influence on food effect. The authors employed supervised machine-learning techniques to identify the relative importance of individual drugs and physiological properties in predicting the effect of food on drug absorption. Thus, the study identified some drugs likely affected by different physiological mechanisms, including bile acid solubilization, efflux transport saturation, first-pass metabolism, and pH changes. Also, high logP values, elevated dose numbers, and increased extraction ratios were found to correlate with a higher probability of positive food effects in drugs. Besides, their machine-learning approaches allowed for the identification of key drug properties that significantly impact the effect of food on drug absorption. Lastly, the author stated that further research focusing on in silico modeling and experimental validation of specific drug-food interactions can enhance our understanding of these complex phenomena and improve drug development strategies (Sharma et al. 2023).

## 7 New Prospects for Drug-Non-Drug Interactions—Authors' Viewpoint

In general, the topic of "drug-non-drug interactions" has been extensively studied according to the literature, where it is known that some diseases and foods can alter the pharmacokinetics of drugs, resulting in negative (sub-effective) or positive (over-effective) effects. In contrast, there is much to be discovered and explained, as many works provide a superficial approach to this highly relevant issue for modern days. In this chapter, it becomes clear how essential the theme is for medical areas that deal directly with patients, who may have their treatment directly affected by the pathology in question or indirectly by the effect of their dietary intake. Diseases that affect organs relevant to drug metabolism (e.g., liver and kidneys) can be considered a negative aggravating factor to the pharmacological treatment, causing the patient not to respond as expected. Furthermore, the patient's advancing age should also be taken into consideration when evaluating the pharmacokinetic profile of a drug, as well as its intrinsic pharmacodynamics, where drug bioaccumulation is the main observed effect. Metabolic diseases affecting the CYP450 system (including mutations) or organic anion transporters represent a major threat to patient treatment. However, alternatives involving machine and deep-learning approaches have been developed to detect possible drug interactions with diseases or foods, allowing for a more suitable pharmacological therapy for each patient based on their physiological limitations.

Regarding the interaction of drugs and the digestion of certain types of foods, this represents a current issue that must be considered by the medical team in healthcare services. In this chapter, it is evident that foods affecting pathways involving the CYP family of enzymes are frequently studied and considered promising for elucidating drug-food interactions. In this context, CYP3A4 has been widely investigated as one of the main drug metabolism pathways affected by food ingestion. Preparations that are seemingly considered harmless, such as grapefruit and lemon juice, can cause problems associated with the pharmacokinetics of some drugs, leading to reduced treatment efficacy. In general, flavonoids have been extensively investigated regarding their interactions with drugs due to their frequent presence as food components. Perhaps because they are easily found in various foods of natural origin. Furthermore, the interaction of certain food components may interfere with macromolecules associated with substance transport or detoxification, such as human serum albumin (HSA) and UDP-glucuronosyltransferases (UGTs), respectively. This type of interaction can result in poor pharmacokinetics/efficacy or exacerbated adverse effects.

A major limitation of the studies found in the literature is the lack of greater detail on how food components interact with proteins associated with drug metabolism, affecting their efficacy. Several studies have addressed the issue but works focused on elucidating the binding modes of these compounds to biological targets are very scarce. Thus, many works are superficial and do not provide information on interactions, often concluding only by "suggesting" a possible interference pathway. What is more striking is that, despite all the advances in the fields of machine- or deep-learning or big data mining, no studies are using such techniques to address the problem focusing on the interactions of food-derived compounds and biological targets associated with drugs. *Why? Could it perhaps be considered that if a food component interferes with a medication, it means that it "necessarily" binds to the same receptor/macromolecule? With the same amino acid residues?* Other limitations include the lack of study of cross-reactions between combinations of drugs and foods or drugs and diseases; pharmacokinetic comparisons of drug-food in studies; solid conclusions for the application of results in healthcare services; as well as discussion of possible solutions for unfavorable interactions and not just pointing out the problems and demonstrating the affected parameters.

Finally, it is emphasized that the theme is current and deserves prominence in the medical clinic scenario, where corrective actions could be considered for the development of future public health policies, improving clinical best practices in medical care services, and, consequently, the quality of life of patients.

## 8 Self-Assessment Questions and Answers

### *8.1 Questions*

1. What distinguishes drug-disease interactions from drug-drug interactions, and why are they particularly pertinent in the context of geriatric medicine?
2. Identify key examples of drug-disease interactions highlighted in the text and discuss their clinical implications.
3. How do aging-related physiological changes contribute to the complexity of drug-disease interactions, and what strategies can healthcare providers employ to mitigate their impact?
4. Explain the significance of understanding food-drug interactions in pharmacotherapy. How do these interactions impact medication adherence, efficacy, and safety?
5. Discuss the role of healthcare professionals, particularly pharmacists, in managing food-drug interactions. How can they educate patients and collaborate with other healthcare providers to mitigate risks associated with these interactions?
6. Describe the molecular mechanisms underlying food-drug interactions, focusing on the influence of common food components such as milk, grapefruit juice, and alcohol on drug absorption and metabolism. How can computational tools and in vitro experiments aid in understanding and predicting these interactions at a molecular level?

### *8.2 Answers*

1. Answer: Drug-disease interactions refer to the exacerbation by drugs of preexisting diseases, syndromes, or conditions, affecting both pharmacokinetics and pharmacodynamics parameters. Unlike drug-drug interactions, which involve interactions between pharmaceutical agents, drug-disease interactions involve the complex interplay between medications and physiological states altered by underlying diseases. These interactions are particularly pertinent in geriatric medicine due to the prevalence of multiple chronic pathologies and polypharmacy

among older adults, which heighten the risk of adverse outcomes and therapeutic challenges.
2. Answer: Examples of drug-disease interactions include the interference of β-blockers in depression, opioid analgesics in dementia, and atypical antipsychotics in diabetes. These interactions can lead to exacerbated symptoms, compromised treatment efficacy, and increased risk of adverse events, especially among older adults with multiple comorbidities. Understanding these interactions is crucial for optimizing therapeutic regimens, minimizing adverse outcomes, and ensuring patient safety in clinical practice.
3. Answer: Aging-related physiological changes, such as alterations in organ function and pharmacokinetic parameters, increase the susceptibility of older adults to drug-disease interactions. Strategies to mitigate their impact include comprehensive geriatric assessments to identify vulnerabilities, medication reconciliation to minimize polypharmacy, and personalized therapeutic interventions tailored to individual patient profiles. Additionally, healthcare providers should prioritize patient education and interdisciplinary collaboration to optimize treatment outcomes and enhance patient safety in the face of complex drug-disease interactions.
4. Answer: Understanding food-drug interactions is crucial in pharmacotherapy as these interactions can significantly impact medication adherence, efficacy, and safety. Food-drug interactions can alter the absorption, distribution, metabolism, and excretion (ADME) of medications, leading to changes in drug concentrations and therapeutic effects. Failure to recognize and manage these interactions can result in suboptimal treatment outcomes, compromised patient safety, and increased healthcare costs.
5. Answer: Healthcare professionals, including pharmacists, play a pivotal role in managing food-drug interactions. They can educate patients about potential interactions associated with their medications and provide tailored dietary recommendations to optimize therapeutic outcomes. Pharmacists also collaborate with other healthcare providers to screen for potential interactions, conduct medication reviews, and adjust treatment regimens as needed to mitigate risks and enhance patient safety.
6. Answer: Molecular mechanisms underlying food-drug interactions involve the influence of common food components on drug absorption and metabolism, particularly through the modulation of drug-metabolizing enzymes such as CYP3A and CYP2C9. Computational tools and in vitro experiments aid in understanding and predicting these interactions at a molecular level by simulating drug-food interactions, performing docking studies to predict binding affinities, and validating predictions through experimentation. Alterations in gastrointestinal physiology induced by food intake, such as changes in fluid volumes in the small intestine and stomach, can further impact drug bioavailability, affecting parameters such as maximum plasma concentration ($C_{\max}$) and time to reach maximum concentration ($T_{\max}$).

## References

Amalaradjou MAR, Bhunia AK. Modern approaches in probiotics research to control foodborne pathogens. In: Toldrá F, editor. Advances in food and nutrition research. 62nd ed. Amsterdam: [s.n.]; 2012. p. 185–239.

American Diabetes Association, et al. Nutrition recommendations and interventions for diabetes: a position statement of the American Diabetes Association. Diabetes Care. 2008;31(Suppl 1):S61–78.

Anadón A, et al. Interactions between nutraceuticals/nutrients and therapeutic drugs. In: Nutraceuticals. [s.l.] Elsevier; 2016. p. 855–74.

Ashburn TT, Thor KB. Drug repositioning: identifying and developing new uses for existing drugs. Nat Rev Drug Discov. 2004;3(8):673–83.

Bai J, et al. Inhibitory effects of flavonoids on P-glycoprotein in vitro and in vivo: food/herb-drug interactions and structure–activity relationships. Toxicol Appl Pharmacol. 2019;369:49–59.

Baltes M. Application to drug–food interactions of living cells as in vitro model expressing cytochrome P450 activity: enzyme inhibition by lemon juice. Talanta. 2001;54(5):983–7.

Basheer L, et al. In silico and in vitro inhibition of cytochrome P450 3A by synthetic stilbenoids. Food Chem. 2017;237:895–903.

Bastianetto S, et al. Protective action of resveratrol in human skin: possible involvement of specific receptor binding sites. PLoS One. 2010;5(9):e12935.

Bavaresco L, et al. Wine resveratrol: from the ground up. Nutrients. 2016;8(4):222.

Beceiro A, Tomás M, Bou G. Antimicrobial resistance and virulence: a successful or deleterious association in the bacterial world? Clin Microbiol Rev. 2013;26(2):185–230.

Benet LZ, et al. Understanding drug–drug interaction and pharmacogenomic changes in pharmacokinetics for metabolized drugs. J Pharmacokinet Pharmacodyn. 2019;46(2):155–63.

Boullata JI, Hudson LM. Drug–nutrient interactions: a broad view with implications for practice. J Acad Nutr Diet. 2012;112(4):506–17.

Brantley SJ, et al. Herb–drug interactions: challenges and opportunities for improved predictions. Drug Metab Dispos. 2014;42(3):301–17.

Burdock GA, Carabin IG. Generally recognized as safe (GRAS): history and description. Toxicol Lett. 2004;150(1):3–18.

Bushra R, Aslam N, Khan A. Food drug interactions. Oman Med J. 2011;26(2):77–83.

Byrne A, et al. Stable oral availability of sustained release propranolol when co- administered with hydralazine or food: evidence implicating substrate delivery rate as a determinant of presystemic drug interactions. Br J Clin Pharmacol. 1984;17(S1)

Canet MJ, et al. Modeling human nonalcoholic steatohepatitis-associated changes in drug transporter expression using experimental rodent models. Drug Metab Dispos. 2014;42(4):586–95.

Canio WC. Polypharmacy in older adults. Clin Geriatr Med. 2022;38(4):621–5.

Cannegieter SC, et al. Optimal oral anticoagulant therapy in patients with mechanical heart valves. N Engl J Med. 1995;333(1):11–7.

Cavasotto CN, Di Filippo JI. Artificial intelligence in the early stages of drug discovery. Arch Biochem Biophys. 2021;698:108730.

Chellapandi P, Prisilla A. Clostridium botulinum type A-virulome-gut interactions: a systems biology insight. Hum Microbiome J. 2018;7–8:15–22.

Chen J, et al. Phloretin exhibits potential food-drug interactions by inhibiting human UDP-glucuronosyltransferases in vitro. Toxicol in Vitro. 2022;84:105447.

Chin MH, et al. Appropriateness of medication selection for older persons in an urban academic emergency department. Acad Emerg Med. 1999;6(12):1232–41.

Chong CR, Sullivan DJ. New uses for old drugs. Nature. 2007;448(7154):645–6.

Choonara I, et al. Stereoselective interaction between the R enantiomer of warfarin and cimetidine. Br J Clin Pharmacol. 1986;21(3):271–7.

Clements JD, Jamali F. Pravastatin reverses the down-regulating effect of inflammation on β-Adrenergic receptors: a disease–drug interaction between inflammation, pravastatin, and propranolol. Vasc Pharmacol. 2007;46(1):52–9.
Craig Brater D. Drug-drug and drug-disease interactions with nonsteroidal anti-inflammatory drugs. Am J Med. 1986;80(1):62–77.
Croft KD. The chemistry and biological effects of flavonoids and phenolic acids. Ann N Y Acad Sci. 1998;854(1):435–42.
Das B, Kutsal M, Das R. A geometric deep learning model for display and prediction of potential drug-virus interactions against SARS-CoV-2. Chemom Intell Lab Syst. 2022;229:104640.
Davis AP, et al. A CTD-Pfizer collaboration: manual curation of 88 000 scientific articles text mined for drug-disease and drug-phenotype interactions. Database. 2013;2013:bat080–bat080.
Di Masi A, et al. Nuclear receptors CAR and PXR: molecular, functional, and biomedical aspects. Mol Asp Med. 2009;30(5):297–343.
Dolezelova E, et al. Resveratrol modifies biliary secretion of cholephilic compounds in sham-operated and cholestatic rats. World J Gastroenterol. 2017;23(43):7678–92.
Dos Santos Nascimento IJ, De Aquino TM, Da Silva-Júnior EF. Drug repurposing: a strategy for discovering inhibitors against emerging viral infections. Curr Med Chem. 2021;28(15):2887–942.
Dos Santos Nascimento IJ, et al. Strategies in medicinal chemistry to discover new hit compounds against ebola virus: challenges and perspectives in drug discovery. Mini-Rev Med Chem. 2022;22(22):2896–924.
Dring AM, et al. Rational quantitative structure–activity relationship (RQSAR) screen for PXR and CAR isoform-specific nuclear receptor ligands. Chem Biol Interact. 2010;188(3):512–25.
Drucker DJ. Enhancing the action of incretin hormones: a new whey forward? Endocrinology. 2006;147(7):3171–2.
El-Arini SK, Shiu GK, Skelly JP. An in vitro study of possible food-drug interactions of the controlled-release propranolol products. Int J Pharm. 1989;55(1):25–30.
El-Sheikh AAK, Masereeuw R, Russel FGM. Mechanisms of renal anionic drug transport. Eur J Pharmacol. 2008;585(2–3):245–55.
Ericsson H, et al. Pharmacokinetics of new calcium channel antagonist clevidipine in the rat, rabbit, and dog and pharmacokinetic/pharmacodynamic relationship in anesthetized dogs. Drug Metab Dispos. 1999;27(5):558–64.
Fasinu PS, Bouic PJ, Rosenkranz B. An overview of the evidence and mechanisms of herb–drug interactions. Front Pharmacol. 2012;3(69):1–19.
Fisher CD, et al. Experimental non-alcoholic fatty liver disease results in decreased hepatic uptake transporter expression and function in rats. Eur J Pharmacol. 2009a;613(1–3):119–27.
Fisher CD, et al. Hepatic cytochrome P450 enzyme alterations in humans with progressive stages of nonalcoholic fatty liver disease. Drug Metab Dispos. 2009b;37(10):2087–94.
Fried LP, Walston J. Frailty and failure to thrive. In: Hazzard WR, et al., editors. Principles of Geriatric Medicine and Gerontology. 5th ed. New York: McGraw-Hill; 2003. p. 1487–502.
García MA, et al. Novel food drug interaction mechanism involving acyclovir, chitosan and endogenous mucus. Drug Metab Pharmacokinet. 2023;49:100491.
Ghasemi S, Lakizadeh A. Drug-disease association prediction based on end-to-end multi-layer heterogeneous graph convolutional encoders. Informatics Med Unlocked. 2023;37:101177.
Giron MST, et al. The appropriateness of drug use in an older nondemented and demented population. J Am Geriatr Soc. 2001;49(3):277–83.
Goldberg RM, et al. Drug-drug and drug-disease interactions in the ED: analysis of a high-risk population. Am J Emerg Med. 1996;14(5):447–50.
Gosney M, Tallis R. Prescription of contraindicated and interacting drugs in elderly patients admitted to hospital. Lancet. 1984;324(8402):564–7.
Gougis P, et al. Potential cytochrome P450-mediated pharmacokinetic interactions between herbs, food, and dietary supplements and cancer treatments. Crit Rev Oncol Hematol. 2021;166:103342.

Gougis P, et al. Cytochrome P-450-mediated herb and food–drug interactions can be identified in cancer patients through patient self-reporting with a tablet application: results of a prospective observational study. ESMO Open. 2022;7(6):100650.

Guirguis MS, Jamali F. Disease–drug interaction: reduced response to propranolol despite increased concentration in the rat with inflammation. J Pharm Sci. 2003;92(5):1077–84.

Hanlon JT, et al. Use of inappropriate prescription drugs by older people. J Am Geriatr Soc. 2002;50(1):26–34.

Hawley JA, Gibala MJ. What's new since Hippocrates? Preventing type 2 diabetes by physical exercise and diet. Diabetologia. 2012;55(3):535–9.

Hirotani Y, et al. Effects of Capsaicin on the pharmacokinetics of digoxin phenytoin and theophylline in rats. Medicine. 2007;24(23–28)

Hirsh J, et al. Oral anticoagulants. Chest. 1998;114(5):445S–69S.

Hong SS, et al. Interaction characteristics of flavonoids with human organic anion transporter 1 (hOAT1) and 3 (hOAT3). Pharmacol Res. 2007;56(6):468–73.

Hu G, Agarwal P. Human disease-drug network based on genomic expression profiles. PLoS One. 2009;4(8):e6536.

Hudson EA, et al. Characterization of potentially chemopreventive phenols in extracts of brown rice that inhibit the growth of human breast and colon cancer cells. Cancer Epidemiol Biomarkers Prev. 2000;9(11):1163–70.

Humphreys K, et al. Responding to the opioid crisis in North America and beyond: recommendations of the Stanford–Lancet Commission. Lancet. 2022;399(10324):555–604.

Hyrsova L, et al. Trans-resveratrol, but not other natural stilbenes occurring in food, carries the risk of drug-food interaction via inhibition of cytochrome P450 enzymes or interaction with xenosensor receptors. Toxicol Lett. 2019;300:81–91.

Jandhyala SM. Role of the normal gut microbiota. World J Gastroenterol. 2015;21(29):8787.

Jarada TN, Rokne JG, Alhajj R. SNF–CVAE: computational method to predict drug–disease interactions using similarity network fusion and collective variational autoencoder. Knowl-Based Syst. 2021;212:106585.

Je J-Y, Kim S-K. Chitosan as potential marine nutraceutical. Adv Food Nutr Res. 2012;65:121–35.

Jiang L, et al. Pterostilbene supplements carry the risk of drug interaction via inhibition of UDP-glucuronosyltransferases (UGT) 1A9 enzymes. Toxicol Lett. 2020a;320:46–51.

Jiang L, et al. Piceatannol exhibits potential food-drug interactions through the inhibition of human UDP-glucuronosyltransferase (UGT) in Vitro. Toxicol in Vitro. 2020b;67:104890.

Karakaya S. Bioavailability of phenolic compounds. Crit Rev Food Sci Nutr. 2004;44(6):453–64.

Ketter TA, Post RM, Worthington K. Principles of clinically important drug interactions with carbamazepine. Part I. J Clin Psychopharmacol. 1991;11(3):198–203.

Kicman AT. Pharmacology of anabolic steroids. Br J Pharmacol. 2008;154(3):502–21.

Kim E, Choi A, Nam H. Drug repositioning of herbal compounds via a machine-learning approach. BMC Bioinformatics. 2019;20(S10):247.

Kirschbaum KM, et al. Serum levels of aripiprazole and dehydroaripiprazole, clinical response and side effects. World J Biol Psychiatry. 2008;9(3):212–8.

Kluth D, et al. Modulation of pregnane X receptor-and electrophile responsive element-mediated gene expression by dietary polyphenolic compounds. Free Radic Biol Med. 2007;42(3):315–25.

Koscielny G, et al. Open targets: a platform for therapeutic target identification and validation. Nucleic Acids Res. 2017;45(D1):D985–94.

Koziolek M, et al. The mechanisms of pharmacokinetic food-drug interactions – a perspective from the UNGAP group. Eur J Pharm Sci. 2019;134:31–59.

Kulmatycki KM, Jamali F. Therapeutic relevance of altered cytokine expression. Cytokine. 2001;14(1):1–10.

Kumar S, et al. Effect of sex and food on the pharmacokinetics of different classes of BCS drugs in rats after cassette administration. Int J Pharm. 2021;610:121221.

Landefeld CS, Beyth RJ. Anticoagulant-related bleeding: clinical epidemiology, prediction, and prevention. Am J Med. 1993;95(3):315–28.

Lecun Y, Bengio Y, Hinton G. Deep learning. Nature. 2015;521(7553):436–44.

Lee JW, et al. The emerging era of pharmacogenomics: current successes, future potential, and challenges. Clin Genet. 2014;86(1):21–8.

Li Z, et al. Pharmacokinetic herb-disease-drug interactions: effect of ginkgo biloba extract on the pharmacokinetics of pitavastatin, a substrate of Oatp1b2, in rats with non-alcoholic fatty liver disease. J Ethnopharmacol. 2021;280:114469.

Liedholm H, Melander A. Concomitant food intake can increase the bioavailability of propranolol by transient inhibition of its presystemic primary conjugation. Clin Pharmacol Ther. 1986;40(1):29–36.

Lim H, et al. Large-scale off-target identification using fast and accurate dual regularized one-class collaborative filtering and its application to drug repurposing. PLoS Comput Biol. 2016;12(10):e1005135.

Lin L-C, Wang M-N, Tsai T-H. Food–drug interaction of (−)-epigallocatechin-3-gallate on the pharmacokinetics of irinotecan and the metabolite SN-38. Chem Biol Interact. 2008;174(3):177–82.

Lindblad CI, et al. Clinically important drug-disease interactions and their prevalence in older adults. Clin Ther. 2006;28(8):1133–43.

Madras BK, et al. Improving access to evidence-based medical treatment for opioid use disorder: strategies to address key barriers within the treatment system. NAM Perspectives; 2020.

Maguire DR. Interactions between lorcaserin and opioids: Ventilation and food-versus-drug choice. Pharmacol Biochem Behav. 2024;234:173673.

Mallory EK, et al. Large-scale extraction of gene interactions from full-text literature using DeepDive. Bioinformatics. 2016;32(1):106–13.

Mandery K, et al. Influence of the flavonoids apigenin, kaempferol, and quercetin on the function of organic anion transporting polypeptides 1A2 and 2B1. Biochem Pharmacol. 2010;80(11):1746–53.

Manitpisitkul P, White RE. Whatever happened to cassette-dosing pharmacokinetics? Drug Discov Today. 2004;9(15):652–8.

Mariadoss AVA, et al. Pharmacological aspects and potential use of phloretin: a systemic review. Mini-Rev Med Chem. 2019;19(13):1060–7.

Matsui Y, et al. Extract of Passion fruit (Passiflora edulis) seed containing high amounts of piceatannol inhibits melanogenesis and promotes collagen synthesis. J Agric Food Chem. 2010;58(20):11112–8.

Mayo PR, et al. Decreased dromotropic response to verapamil despite pronounced increased drug concentration in rheumatoid arthritis. Br J Clin Pharmacol. 2000;50(6):605–13.

Mcmahan DA, et al. Risk of major hemorrhage for outpatients treated with warfarin. J Gen Intern Med. 1998;13(5):311–6.

Meena P, Jain A. Interaction of anti-inflammatory drug naproxen with essential amino acids: quantitative thermodynamic insights. J Chem Thermodyn. 2024;190:107211.

Melander A, et al. Enhancement of the bioavailability of propranolol and metoprolol by food. Clin Pharmacol Ther. 1977;22(1):108–12.

Mensink M, et al. Lifestyle intervention according to general recommendations improves glucose tolerance. Obes Res. 2003;11(12):1588–96.

Mooney SD. Progress towards the integration of pharmacogenomics in practice. Hum Genet. 2015;134(5):459–65.

Muszkat M, et al. Warfarin metabolism and anticoagulant effect: a prospective, observational study of the impact of CYP2C9 genetic polymorphism in the presence of drug-disease and drug-drug interactions. Clin Ther. 2007;29(3):427–37.

Napolitano F, et al. Drug repositioning: a machine-learning approach through data integration. J Cheminformatics. 2013;5(1):30.

Naveen Sundar G, et al. An intelligent prediction model for target protein identification in hepatic carcinoma using novel graph theory and ANN model. Comput Model Eng Sci. 2022;133(1):31–46.

Ndefo UA, et al. Clevidipine: a new intravenous option for the management of acute hypertension. Am J Health Syst Pharm. 2010;67(5):351–60.
Neuschwander-Tetri BA. Non-alcoholic fatty liver disease. BMC Med. 2017;15(1):45.
Neuvonen PJ. Drug interactions with HMG-CoA reductase inhibitors (statins): the importance of CYP enzymes, transporters and pharmacogenetics. Curr Opin Investig Drugs (London, England: 2000). 2010;11(3):323–32.
Neuvonen P, Niemi M, Backman J. Drug interactions with lipid-lowering drugs: Mechanisms and clinical relevance. Clin Pharmacol Ther. 2006;80(6):565–81.
Nongonierma AB, Fitzgerald RJ. Dipeptidyl peptidase IV inhibitory properties of a whey protein hydrolysate: influence of fractionation, stability to simulated gastrointestinal digestion and food–drug interaction. Int Dairy J. 2013;32(1):33–9.
Nwose EU, Yee KC. Drug-disease interactions: narrative review of aspirin in gastric ulcer. Expert Opin Drug Metab Toxicol. 2016;12(9):1081–7.
Ötles S, Senturk A. Food and drug interactions: a general review. Acta Sci Pol Technol Aliment. 2014;13(1):89–102.
Palareti G, et al. Bleeding complications of oral anticoagulant treatment: an inception-cohort, prospective collaborative study (ISCOAT). Lancet. 1996;348(9025):423–8.
Papanikolaou N, et al. DrugQuest – a text mining workflow for drug association discovery. BMC Bioinformatics. 2016;17(S5):182.
Parker JL, et al. Molecular basis for selective uptake and elimination of organic anions in the kidney by OAT1. Nat Struct Mol Biol. 2023;30(11):1786–93.
Patel A, Gorman SE. Stockpiling antiviral drugs for the next influenza pandemic. Clin Pharmacol Ther. 2009;86(3):241–3.
Pérez-Cobas AE, et al. Structural and functional changes in the gut microbiota associated to Clostridium difficile infection. Front Microbiol. 2014;5:1–15.
Petitpas I, et al. Crystal structures of human serum albumin complexed with monounsaturated and polyunsaturated fatty acids. J Mol Biol. 2001;314(5):955–60.
Piotrowska H, Kucinska M, Murias M. Biological activity of piceatannol: leaving the shadow of resveratrol. Mutat Res Rev Mutat Res. 2012;750(1):60–82.
Raaska K, et al. Effect of caffeine-containing versus decaffeinated coffee on serum clozapine concentrations in hospitalised patients. Basic Clin Pharmacol Toxicol. 2004;94(1):13–8.
Raish M, et al. Effects of Paeonia emodi on hepatic cytochrome P450 (CYP3A2 and CYP2C11) expression and pharmacokinetics of carbamazepine in rats. Biomed Pharmacother. 2017;90:694–8.
Raish M, et al. Effect of sinapic acid on aripiprazole pharmacokinetics in rats: possible food drug interaction. J Food Drug Anal. 2019a;27(1):332–8.
Raish M, et al. Effects of sinapic acid on hepatic cytochrome P450 3A2, 2C11, and intestinal P-glycoprotein on the pharmacokinetics of oral carbamazepine in rats: potential food/herb-drug interaction. Epilepsy Res. 2019b;153:14–8.
Reker D, et al. Machine learning uncovers food- and excipient-drug interactions. Cell Rep. 2020;30(11):3710–3716.e4.
Relling MV, Evans WE. Pharmacogenomics in the clinic. Nature. 2015;526(7573):343–50.
Rodríguez-Fragoso L, et al. Potential risks resulting from fruit/vegetable–drug interactions: effects on drug-metabolizing enzymes and drug transporters. J Food Sci. 2011;76(4)
Roja B, Chellapandi P. Design and characterization of a multi-epitope vaccine against Clostridium botulinum A3 Loch Maree intoxication in humans. Gene. 2024;892:147865.
Roja B, et al. Inferring molecular mechanisms of host-microbe-drug interactions in the human gastrointestinal tract. Med Omics. 2024;10:100027.
Roujeau J-C, et al. Medication use and the risk of Stevens–Johnson syndrome or toxic epidermal necrolysis. N Engl J Med. 1995;333(24):1600–8.
Russell W, Duthie G. Plant secondary metabolites and gut health: the case for phenolic acids. Proc Nutr Soc. 2011;70(3):389–96.

Samwald M, et al. Semantically enabling pharmacogenomic data for the realization of personalized medicine. Pharmacogenomics. 2012;13(2):201–12.
Santos JDB, et al. Food-drug interaction: anabolic steroids aggravate hepatic lipotoxicity and non-alcoholic fatty liver disease induced by trans fatty acids. Food Chem Toxicol. 2018;116:360–8.
Schmidt LE, Dalhoff K. Food-drug interactions. Drugs. 2002;62(10):1481–502.
Schuffenhauer A, et al. Similarity metrics for ligands reflecting the similarity of the target proteins. J Chem Inf Comput Sci. 2003;43(2):391–405.
Schwenk RW, Vogel H, Schürmann A. Genetic and epigenetic control of metabolic health. Mol Metab. 2013;2(4):337–47.
Scordo M. Influence of CYP2C9 and CYP2C19 genetic polymorphisms on warfarin maintenance dose and metabolic clearance. Clin Pharmacol Ther. 2002;72(6):702–10.
Segler MHS, et al. Generating focused molecule libraries for drug discovery with recurrent neural networks. ACS Cent Sci. 2018;4(1):120–31.
Sevrioukova IF, Poulos TL. Dissecting cytochrome P450 3A4–ligand interactions using Ritonavir analogues. Biochemistry. 2013;52(26):4474–81.
Shapiro DA, et al. Aripiprazole, a novel atypical antipsychotic drug with a unique and robust pharmacology. Neuropsychopharmacology. 2003;28(8):1400–11.
Sharma S, Prasad B. Meta-analysis of food effect on oral absorption of efflux transporter substrate drugs: does delayed gastric emptying influence drug transport kinetics? Pharmaceutics. 2021;13(7):1035.
Sharma S, et al. Analysis of the interplay of physiological response to food intake and drug properties in food-drug interactions. Drug Metab Pharmacokinet. 2023;53:100518.
Smutny T, Pavek P. Resveratrol as an inhibitor of Pregnane X Receptor (PXR): another lesson in PXR antagonism. J Pharmacol Sci. 2014;126(2):177–8.
Stubbins MJ, et al. Genetic analysis of the human cytochrome P450 CYP2C9 locus. Pharmacogenetics. 1996;6(5):429–39.
Sullivan FM, Mcelhatton PR. A comparison of the teratogenic activity of the antiepileptic drugs carbamazepine, clonazepam, ethosuximide, phenobarbital, phenytoin, and primidone in mice. Toxicol Appl Pharmacol. 1977;40(2):365–78.
Swainston Harrison T, Perry CM. Aripiprazole. Drugs. 2004;64(15):1715–36.
Takahashi H, et al. Metabolism of warfarin enantiomers in Japanese patients with heart disease having different CYP2C9 and CYP2C19 genotypes*. Clin Pharmacol Ther. 1998;63(5):519–28.
Takanohashi T, et al. Studies of the toxicological potential of capsinoids, XIII: inhibitory effects of capsaicin and capsinoids on cytochrome P450 3A4 in human liver microsomes. Int J Toxicol. 2010;29(2_suppl):22S–6S.
Tari L, et al. Discovering drug–drug interactions: a text-mining and reasoning approach based on properties of drug metabolism. Bioinformatics. 2010;26(18):i547–53.
Theile D, et al. Clementine juice has the potential for drug interactions – in vitro comparison with grapefruit and mandarin juice. Eur J Pharm Sci. 2017;97:247–56.
Thomsen WJ, et al. Lorcaserin, a novel selective human 5-hydroxytryptamine 2C agonist: in vitro and in vivo pharmacological characterization. J Pharmacol Exp Ther. 2008;325(2):577–87.
Thorn CF, Klein TE, Altman RB. Pharmacogenomics and bioinformatics: PharmGKB. Pharmacogenomics. 2010;11(4):501–5.
Thursby E, Juge N. Introduction to the human gut microbiota. Biochem J. 2017;474(11):1823–36.
Townsend EA, et al. A drug-vs-food "choice" self-administration procedure in rats to investigate pharmacological and environmental mechanisms of substance use disorders. J Neurosci Methods. 2021;354:109110.
Townshend R, et al. Geometric deep learning on biomolecular structure. Biophys J. 2021;120(3):290a.
Tulipano G, et al. Whey proteins as source of dipeptidyl dipeptidase IV (dipeptidyl peptidase-4) inhibitors. Peptides. 2011;32(4):835–8.
Ude C, Schubert-Zsilavecz M, Wurglics M. Ginkgo biloba extracts: a review of the pharmacokinetics of the active ingredients. Clin Pharmacokinet. 2013;52(9):727–49.

Uesawa Y, Mohri K. Quantitative structure-interaction relationship analysis of 1,4-dihydropyridine drugs in concomitant administration with grapefruit juice. Die Pharmazie. 2012;67(3):195–201.
Uwai Y, et al. Inhibitory effect of caffeic acid on human organic anion transporters hOAT1 and hOAT3: a novel candidate for food–drug interaction. Drug Metab Pharmacokinet. 2011;26(5):486–93.
Valdes AM, et al. Role of the gut microbiota in nutrition and health. BMJ. 2018:k2179.
Vazquez M, et al. Text mining for drugs and chemical compounds: methods, tools and applications. Mol Informatics. 2011;30(6–7):506–19.
Viñas P, et al. Directly suspended droplet microextraction with in injection-port derivatization coupled to gas chromatography–mass spectrometry for the analysis of polyphenols in herbal infusions, fruits and functional foods. J Chromatogr A. 2011;1218(5):639–46.
Volkow ND, Blanco C. Fentanyl and other opioid use disorders: treatment and research needs. Am J Psychiatry. 2023;180(6):410–7.
Wang L, Sweet DH. Potential for food–drug interactions by dietary phenolic acids on human organic anion transporters 1 (SLC22A6), 3 (SLC22A8), and 4 (SLC22A11). Biochem Pharmacol. 2012;84(8):1088–95.
Wang D, et al. Intronic polymorphism in CYP3A4 affects hepatic expression and response to statin drugs. Pharmacogenom J. 2011;11(4):274–86.
Wang X, et al. Spectroscopic investigation on the food components–drug interaction: the influence of flavonoids on the affinity of nifedipine to human serum albumin. Food Chem Toxicol. 2015;78:42–51.
Wang X, et al. Studies on the competitive binding of cleviprex and flavonoids to plasma protein by multi-spectroscopic methods: a prediction of food-drug interaction. J Photochem Photobiol B Biol. 2017;175:192–9.
Wang N, et al. Food–drug interactions involving multiple mechanisms: a case study with effect of Capsaicin on the pharmacokinetics of Irinotecan and its main metabolites in rat. J Funct Foods. 2018;40:292–8.
Weiss J, et al. Nobiletin, sinensetin, and tangeretin are the main perpetrators in clementines provoking food-drug interactions in vitro. Food Chem. 2020;319:126578.
Wen M, et al. Deep-learning-based drug–target interaction prediction. J Proteome Res. 2017;16(4):1401–9.
Whitley AC, Sweet DH, Walle T. The dietary polyphenol ellagic acid is a potent inhibitor of hOAT1. Drug Metab Dispos. 2005;33(8):1097–100.
Wong M, et al. Search and visualization of gene-drug-disease interactions for pharmacogenomics and precision medicine research using GeneDive. J Biomed Inform. 2021;117:103732.
Wouters OJ, Mckee M, Luyten J. Estimated research and development investment needed to bring a new medicine to market, 2009–2018. JAMA. 2020;323(9):844.
Wu L, et al. Inhibition of the organic anion-transporting polypeptide 1B1 by quercetin: an in vitro and in vivo assessment. Br J Clin Pharmacol. 2012;73(5):750–7.
Xiao J, Kai G. A review of dietary polyphenol-plasma protein interactions: characterization, influence on the bioactivity, and structure-affinity relationship. Crit Rev Food Sci Nutr. 2012;52(1):85–101.
Yamazaki H, Shimada T. Human liver cytochrome P450 enzymes involved in the 7-hydroxylation of R- and S-warfarin enantiomers. Biochem Pharmacol. 1997;54(11):1195–203.
Yu C, et al. Identification of novel pregnane X receptor activators from traditional Chinese medicines. J Ethnopharmacol. 2011;136(1):137–43.
Yu Z, et al. Predicting drug–disease associations through layer attention graph convolutional network. Brief Bioinform. 2021;22(4)
Zhai X, et al. Food–drug interactions: effect of capsaicin on the pharmacokinetics of simvastatin and its active metabolite in rats. Food Chem Toxicol. 2013;53:168–73.

Zhang J, et al. High-throughput salting-out assisted liquid/liquid extraction with acetonitrile for the simultaneous determination of simvastatin and simvastatin acid in human plasma with liquid chromatography. Anal Chim Acta. 2010a;661(2):167–72.

Zhang Z, et al. The characteristics and mechanism of simvastatin loaded lipid nanoparticles to increase oral bioavailability in rats. Int J Pharm. 2010b;394(1–2):147–53.

Zhang N, Liu Y, Jeong H. Drug-drug interaction potentials of tyrosine kinase inhibitors via inhibition of UDP-glucuronosyltransferases. Sci Rep. 2015;5(1):17778.

Zhou S-F, et al. Metabolic activation of herbal and dietary constituents and its clinical and toxicological implications: an update. Curr Drug Metab. 2007;8(6):526–53.

# Chapter 13
# Relevant of Non-drug Interaction in Complementary and Alternative Medicine for Complex Diseases

**Angela Adamski da Silva Reis and James O. Fajemiroye**

**Abstract** Alternative medicine for complex diseases explores multiple strategies and targets in search of unconventional relief or treatment approaches. Understanding complex diseases is challenging due to the interplay of various comorbidities, genetic, lifestyle, and environmental factors. The limitation of single-drug therapies for complex diseases usually demands drug combinations, complementary (use of a non-mainstream approach in conjunction with conventional medicine), or alternate medicine (use of a non-mainstream approach instead of conventional medicine). Alternate therapeutics that target multiple molecular networks underlying complex diseases could enhance treatment efficacy. However, there are concerns about the accompanying adverse interactions or toxicity in this practice. Clinically significant non-drug interactions from complementary and alternative medicine could impact absorption, distribution, metabolism, excretion, and pharmacological effects. This chapter covers non-drug interactions with a focus on the modalities of alternative medicine.

**Keywords** Complex diseases · Multiple targets · Complementary and alternative medicine

A. A. da Silva Reis
Department of Biochemistry and Molecular Biology, Neurogenetics Research Center – NeuroGene, Institute of Biological Sciences, Federal University of Goiás, Goiânia, Goiás, Brazil
e-mail: angela@ufg.br

J. O. Fajemiroye (✉)
Institute of Biological Sciences, Federal University of Goiás, Goiânia, Goiás, Brazil
e-mail: jamesfajemiroye@ufg.br



J. O. Fajemiroye (ed.), *Fundamentals of Drug and Non-Drug Interactions*,
https://doi.org/10.1007/978-3-031-80107-5_13

## 1 Introduction

Complex disease has features that complicate the detection of its causal factors. Characterization of factors that contribute to complex diseases constitutes challenges. Complex diseases do not obey the single-gene Mendelian inheritance as a result of confounding contributing factors (Schriml et al. 2023). Allergies, cardiovascular, psychiatric, and metabolic disorders are complex etiological readouts of genetic, environmental, and lifestyle-related changes (Virolainen et al. 2023). Complementary and alternative medicine (CAM) provides interesting therapeutic choices. The use of at least one form of CAM is commonplace. In herbal medicine alone, close to 80% of the world's population uses it in primary health care (Posadzki et al. 2013). Herbal medicines provide the basis of primary healthcare for more than four billion people on a global level, comprising over half of the world's population (Marshall 2020). CAM encompasses practices and therapies that fall outside conventional medical approaches. These medical practices can be used individually or in combination, and their effectiveness can vary among patients (Tabish 2008). Some of these practices including dietary supplements, herbal medicine, Traditional Chinese Medicine, Ayurveda medicine, chiropractic, osteopathy, naturopathy, acupuncture, mind-body medicine, or aromatherapy have been used for years. There could be many factors leading to these practices, such as the inability to visit a physician or a hospital, lack of restrictions on prescription, easy access, cultural beliefs, and religious misconceptions.

According to the National Center for Complementary and Integrative Health, CAM represents diverse medical and healthcare practices not considered conventional (Mortada 2024). CAM often offers a more affordable and accessible therapeutic option in localities with low to conventional healthcare resources, and increasingly expensive healthcare services. In this sense, self-medication is an old and universal practice defined as the over-the-counter use of medicinal products that have received marketing authorization, with or without advice from the pharmacist. The increasing cases of self-medication constitute a serious public health concern (Behzadifar et al. 2020). Non-steroidal analgesics/anti-inflammatory drugs (94.2%), drugs indicated for gastrointestinal disorders or used in ear nose throat disorders (63.5%), dietary supplements (54%), drugs indicated in skin disorders (31.8%), and herbal medicine (20.2%) are among the most self-administered drugs. However, the risk factors for significant CAM interactions are like those for conventional medicines. These include aging, frailty, gender, cognition, comorbidities, genetic disposition, high level of medication, substitution of medicines, and exposure to low therapeutic index drugs with limited elimination. Moreover, due to their phytocomplexity, herbal medicines are prone to interactions with metabolic enzymes (Moses et al. 2010).

Non-drug interaction with herbal use could be intensified in complex diseases like cancer with tumor microenvironment in favor of CYP modulations and release of IL-1, IL-6, TNF-$\alpha$, and IFN-$\gamma$ from macrophages, lymphocytes, and neutrophils (Stipp and Acco 2021). The nature of interactions that affect chemotherapy, toxicity, or therapeutic outcome could be determined by CYP mutations, polymorphisms, inhibition, or induction of enzymatic activity (Zhao et al. 2021). The following subtopics summarize the relevance of CAM in the non-drug interactions for complex diseases.

## 2 Unconventional Therapeutic Approaches

Over the past two decades, the use of CAM has dramatically expanded in all modern societies and is growing more quickly than ever in the healthcare sector (Mortada 2024). It is estimated that >30% of patients with heart failure use CAM, and 1 of 5 patients have used herbal therapy annually (Chow et al. 2023). There is still a lack of federal guidance and regulation of CAM products. Complementary therapy is a non-mainstream therapeutic approach combined with conventional/allopathic medicine to supplement or augment the therapeutic outcome. In contrast, an alternative therapeutic approach replaces conventional therapy altogether. Although there is some documented use of these therapies as alternatives, most of the use is complementary. Unconventional medicine is more complex in its definition, as there are many unconventional medical systems and practitioners. The evolution of ideas regarding conventional or unconventional care can be seen through significant shifts and advancements in scientific research and discovery. Increased knowledge led to an emphasis on evidence-based medicine and research to inform clinical practice. Thus, the way "conventional" versus "unconventional" medicine is defined has since changed due to advancements in medical treatments and what is considered "mainstream" practice (Ng et al. 2022).

Integrative medicine (IM) replaced CAM in the 1990s. Methods that had not previously been recognized as legitimate treatments in Western medicine are now referred to as IM. According to 2002 research by the National Center for Health Statistics of the Control and Diseases Center, 62% of individuals utilized IM therapy for health benefits (Mortada 2024). Thus, the term IM is considered another common term prevalent in this field; however, it refers to the integration and thus encompasses conventional and unconventional medicine. The incorporation of complementary and conventional treatments is frequently considered as integrative care (Koithan 2009). Although "traditional medicine" is another common term prevalent in literature, we view it as distinct from integrative medicine in that it describes a system of medicine based on origination from a region/culture, as opposed to describing a system of medicine in competition with conventional medicine (Ng et al. 2022).

Interestingly, the use of the term CAM increased sharply around the same time that the terms "complementary medicine" and "alternative medicine" began to decline. Therefore, the variety of these terms is evolving (Ng et al. 2022). On the other hand, NIH has identified five domains or types of complementary therapies, such as mind-body therapies (MBTs), biological-based therapies (BBTs), manipulative and body-based methods, energy therapies, and whole medical systems

(Koithan 2009). Moreover, there is little overlap between the therapeutic modalities in different domains, except for the alternative medical systems, which draw on therapies from the other four domains (Goldrosen and Straus 2004). CAM practitioners treat the biochemical manifestations of illness while providing nutritional, emotional, social, and spiritual support.

Body-based techniques and mind-body interventions comfort and reduce stress. Furthermore, numerous CAM products can be purchased at health-food shops and supermarkets, and as natural substances that have been used for centuries, they claim to be effective yet safer than drugs. CAM places patients in control of their health (Tangkiatkumjai et al. 2020). For example, cancer patients hope that CAM may cure the illness, lessen the side effects of their treatments, and improve their immune system and well-being to resist cancer (Calcagni et al. 2019). Mind-body interventions are practices (e.g., meditation, internal awareness, anatomic alignment, guided imagery, relaxation, hypnosis, biofeedback, prayer, dance, art, music, and breathwork) focusing on the body, brain, mind, and behavioral interactions where the mind alters the physical function and promotes overall health by reducing anxiety, mood disturbance, and pain (Zheng et al. 2023; Hourston and Atchley 2017). Appropriate MBT interventions can enhance the provider-patient relationship and improve symptoms toward a patient's meaningful experiences (Mayden 2012). As the evidence for efficacy grows, many MBTs previously classified as CAM are now viewed as mainstream treatments. Examples include cognitive behavioral therapy and patient support groups (Koithan 2009). They may benefit patients treated with opioids for acute, procedural, and chronic pain conditions. For much of the twentieth century, opioids were prescribed primarily for postoperative and cancer-related pain (Garland et al. 2020). In cancer patients, sleep disturbance is a common problem. MBTs are emerging as potential tools for addressing the opioid crisis. These therapies are beneficial to patients with autism spectrum disorder (ASD) (Hourston and Atchley 2017). Difficulties in understanding metaphorical or non-literal language, differences in social communication and interaction, slower information processing, and intellectual disability which could co-occur with depression, anxiety, and high stress levels are among the characteristics of ASD.

The use of "substances found in nature" defines biological-based therapies (BBTs). These increasingly popular intervention domains include but are not limited to the use of plants and animal products, prebiotics and probiotics, proteins, vitamins, fatty acids, amino acids, minerals, whole diets, and functional foods (Koithan 2009), coaching intervention and nutritional counseling tailored to clinical laboratory findings and genetic predisposition (Watanabe et al. 2023a, b; Teong et al. 2023). Nutritional supplements and probiotics could build resistance and prevent adverse health outcomes, particularly in chronic diseases with aging (Pitcher et al. 2023). Type 2 diabetes mellitus (T2DM), weight management, and cardiometabolic respond well to diet changes (Diab et al. 2023) or moderate calorie restriction. However, an emerging area of research is the role of meal timing and prolonged fasting in extending the health benefits of calorie restriction (Teong et al. 2023). Over time, various diets have been studied for their effectiveness in cardiovascular disease (CVD) prevention. The main cardiometabolic risk factors are diabetes, hypertension, dyslipidemia, and excess abdominal fat, which are all affected by dietary changes. Lifestyle modifications from exercise and dietary interventions have been well studied and proven effective toward cardiovascular disease (CVD) prevention and management (Diab et al. 2023). The risk of mortality among patients with T2DM and CVD increases with postprandial hyperglycemia (Mineshita et al. 2022). The mediterranean dietary approaches elicit cardioprotective effects in varying degrees. Intermittent fasting and Ketogenic diet still demand more studies (Charkviani et al. 2023). Postprandial glycemic approaches respond with intermittent fasting plus time-restricted eating than with calorie restriction or standard of care (Mineshita et al. 2022).

Manipulative and body-based methods (MBMs) form a crucial domain that aids in managing or alleviating some of the conditions or symptoms of complex diseases. MBMs comprise movement of the body techniques, including bones, joints, soft tissues, and circulatory and lymphatic systems as a means of achieving health and well-being (Chu et al. 2022). Non-medicated therapy approaches such as MBMs are employed; however, the scientific evidence of their efficacy is still unclear (Calcagni et al. 2019). A systematic review described that manipulative CAM may be effective in symptom management in cancer. Massage, reflexology, and acupressure showed positive and significant effects on symptoms of pain and fatigue. The designation of a certain therapy as CAM is evolving, as is the approach to classifying types of CAM into broader domains. CAM seems promising in the fight against diseases (He et al. 2016). The complexity of arthritis, diabetes, heart disease, cancer, and asthma among others often limits the efficacy of commonly used single molecular drug therapeutics. Omics (genomics, transcriptomics, proteomics, metabolomics, and other fields that permit comprehensive study of various aspects of biological molecules) revolution and bioinformatics offer a powerful framework for integrating multi-dimensional data from medical practices. Understanding complex diseases at the molecular level will provide insights into complex molecular networks and facilitate the tailoring of CAM.

## 3 Non-drug Interaction in Naturopathy and Herbal Medicine for Complex Diseases

Non-drug interactions in herbal medicine and naturopathy for complex diseases involve a holistic approach beyond direct chemical interactions that are typical of conventional drug therapies. These multifaceted approaches encompass synergistic effects of multiple treatments, nutritional support, lifestyle modifications, mind-body connections, and patient-centered care. For example, a combination of herbs might stimulate immune function, reduce inflammation, and improve overall organ function. Foods and dietary supplements can interact with the body's biochemical processes, influencing disease progression and treatment outcomes. Exercise, sleep, and hygiene can profoundly impact the body's ability to heal and manage complex diseases. Integrating meditation, acupuncture, yoga, and mind-body therapies can influence the nervous system, hormonal balance, and immune responses, thereby contributing to overall health and disease management. If not tailored to patient needs and health goals, the herbal and naturopathic approaches could provoke non-drug interactions capable of hampering the healing process. Herbs and naturopathic treatments can trigger allergic reactions and interactions with conventional medications. Practitioners of naturopathy and herbal medicine are trained to assess these risks and mitigate them through careful selection and monitoring of treatments.

Herbal extracts and drugs in different combinations have been used for chronic back and musculoskeletal pain (Jin et al. 2016). Inflammation and swelling often accompany painful experiences in patients. With the *Harpagophytum procumbens* (devil's claw), several patients with back pain reported an improvement in relief score without changes in opioid consumption (Ernst et al. 2000). *Allium sativum, Zingiber officinale, Oenothera biennis, Uncaria tomentosa,* and *Boswellia serrata* decrease the pro-inflammatory mediators to improve joint pain (Ernst 2004; Lindler et al. 2020). A combination of N-acetyl D-glucosamine, *Boswellia serrata,* and *Zingiber officinale* has improved clinical outcomes in moderate osteoarthritis (Bolognesi et al. 2016). However, combining herbal or naturopathic treatments with conventional medicine or using them independently could lead to alterations in immune function (suppression or exacerbation), low renal perfusion leading to kidney damage or dysfunction, and inhibition of prostaglandin release among others. Prostaglandins help preserve the stomach lining and prevent peptic ulcers or worsening of existing ulcer conditions.

## 4 Non-drug Interaction in Ayurveda for Complex Diseases

Ayurveda is an old health tradition (Sriranjini et al. 2015; Wallace and Wallace 2021) commonly referred to as the science of life (Ayu = life and Veda = science or knowledge). Ayurveda practices holistic principles primarily focused on personalized health emphasizing health promotion, disease prevention, early diagnosis, and treatment. The control of breathing is primarily associated with practices and techniques within the field of Ayurvedic medicine. Dietary and drug approaches, behavioral modifications, detoxification, and rejuvenation therapies could restore the balance of the body-mind matrix (Gyawali et al. 2021). There seem to be substantial similarities between Ayurveda and the innovative approach of predictive, preventive, and personalized medicine (Patwardhan 2014). A preventive lifestyle includes diet, sleep, exercise, and stress management. The governing principles of physiology, known as doshas, include Vata, Pitta, and Kapha. Vata governs motion, flow, and communication (e.g., blood circulation, heart beating, neurotransmission, etc); Pitta regulates metabolic processes while Kapha controls the body structure. Proper health depends upon keeping these doshas in balance. Aggravation or imbalance leads to disorder and disease (Wallace and Wallace 2021). As an example,

Prakriti-specific prescription may also include supportive therapies, diet, and lifestyle advice to regain physiological balance and relief from the disorder (Patwardhan 2014).

In patients with hypercholesterolemia (CVD risk factor), lifestyle intervention is a primary issue in the treatment. Diet and lifestyle modifications with or before starting the cholesterol-lowering drugs are the primary line of treatment for hypercholesterolemia (Mannarino et al. 2014). However, many times, drugs such as statins and ezetimibe do not even reach the desired reduction in LDL levels, with residual CVD risk persisting. Ayurveda treatments have shown significant effects on hypercholesterolemia (Gyawali et al. 2021). The theoretical assumptions framing the diabetes management strategies are reductionist with a focus on reducing hyperglycemia through targeted molecular drugs. Therefore, a holistic and systemic approach based on the principles of Ayurveda can play a pivotal role in its successful management (Thottapillil et al. 2021). Although Ayurveda uses holistic natural products for human well-being, it is more preventive rather than curative. Extracts, formulations, phytoconstituents, and other preparations have shown effective activities against mammary carcinoma, neuroblastoma, non-small-cell lung carcinoma, lymphocytic leukemia, colorectal adenocarcinoma, Ehrlich ascites carcinoma, prostate adenocarcinoma, glioblastoma astrocytoma, and other malignancies (Vikram et al. 2022). These products have received attention from the pharmaceutical industry. Up to 48 of 74 antihypertensive medications are derived from natural products; 7 out of 10 anti-migraine medicines originated from traditional medicine (Sahu et al. 2023). Some of the herbal formulations include Aswagandharishtam, Saraswatarishtam, Kalyanaka churna, Saraswata churna, Brahmi ghrita, Dadhika ghrita, Kalyanaka ghrita, Maha kalyanaka ghrita, Kushmanda ghrita, Panchagavya ghrita, Maha Panchagavya ghrita, Siddarthaka ghrita, Tiktaka ghrita, and Mahayogaraja guggulu.

Arishtas are decoction of herbs for treating metabolic and digestive disorders; Arjunarishta formulation is being used for cardioprotective activity; Ashwagandharishta formulation possesses antioxidant, anti-atherosclerotic, and anti-stress properties; Ridayarishta, a novel empirical formulation, was prepared to support healthy heart functions and reduce stress; and ayurvedic E1 formulations with promising activities in different cancer models (Pandit et al. 2017; Adhikari et al. 2018; Wallace and Wallace 2021). Thus, recent efforts to develop robust clinical protocols for comparing the effectiveness of complex ayurvedic and conventional treatments are laudable (Patwardhan 2014). As multi-drug combination therapy is now widespread for several diseases, drug-drug, herb-drug, and food-drug interactions are becoming realities in clinical practice (Karamada et al. 2016; Pandit et al. 2017). Hence, interaction parameters could be yardsticks for ayurvedic-conventional treatment comparison.

Antiplatelets, stimulation of the thyroid gland, lowering of cholesterol levels by speeding up the body's process of eliminating cholesterol, reducing the production of cholesterol in the liver, and uses an ion exchange resin in the crude medication to capture cholesterol-containing bile acid and keep it out of the intra-hepatic circulation activities of guggulu (*Commiphora mukul*) are signals of the spectrum of potential interactions with drugs (hypolipidemic, anticoagulants) and non-drugs. Coadministration of guggul lipid reduced the bioavailability of antihypertensive propranolol and diltiazem. Several studies show a correlation between Prakriti and gene expression, thus helping to provide a modern scientific understanding of Ayurveda (Sharma and Keith 2020). The mRNA corresponds with Vata (transmission of information), the tRNA corresponds with Pitta (transformation), and the protein corresponds with Kapha. Nutritional ingredients and phytochemicals are among the most potent regulators of epigenetic function and gene expression (Wallace and Wallace 2021).

Recent trends toward the use of in silico biology (molecular modeling, docking, simulation, etc.) to understand the Ayurveda concept are gaining significant momentum in the field of drug discovery. Though in silico work is not an answer to all questions, it may save time, short-circuit the drug development process, and shed some light on the mode of action of ayurvedic practices (Sahu et al. 2023). Molecular-level studies are expected to provide a deeper view of the interaction of various active principles inside and outside the biological system (Vikram et al. 2022). Extensive preclinical and clinical studies establishing the relationships between medications, diets, and herbs are needed to ascertain the non-drug interactions that are associated with ayurvedic approaches.

## 5 Non-drug Interaction in Traditional Chinese Medicine for Complex Diseases

Traditional Chinese Medicine (TCM) is a medical knowledge system created by the Chinese people. It is a practice defined by the holistic concept of "harmony between man and nature"; thus, it encompasses image information, cognitive activity from the perspective of attribute and relationship, use of the natural resources inside and outside the human body, and regulation and balancing of the body health state (Ma et al. 2021). TCM represents thousands of years of treatment protocols for health, healing, and longevity and has evolved into a complex healthcare system (Wang et al. 2018). One of the most important theoretical concepts of TCM is Yin and Yang, which were recorded in the "Book of Change," dating from 700 before the Current Era. The iconic Yin/Yang symbol represents opposite properties and functions but has complementary qualities. Extreme Yin will transform itself into Yang, and extreme Yang will transform itself into Yin (Marshall 2020). The elements theory has various applications in TCM and the systems of correspondences (organs, tissues, tastes, directions, climates, colors, seasons, etc.). Each of these elements has a corresponding musical tone. TCM acknowledges the therapeutic value of musical tones (Gaur 2024).

TCM could restore the effects of the body's "Qi" to live in harmony with the surrounding energies (the sun, geographical location, plants, moon, soil, water, animals, planets, constellations, and natural formations). TCM aims to treat not only the secondary manifestations ("biao") but also the primary causes ("ben") of several medical conditions in internal medicine, gynecology, pediatrics, traumatology, external medicine, dermatology, emergency medicine, eye, ear, nose, and throat (Matos et al. 2021). Chinese herbal medicine and dietetics follow the same diagnostic principles as acupuncture. The effects of herbs and foods depend on the thermal nature, flavor or "sapor," and organ network. The thermal nature could impact bodily microcirculation as the "sapor" regulates a particular organ network (Matos et al. 2021).

The field of Chinese herbal has drawn increasing interest as a source of novel drugs and drug leads, which is unsurprising considering that many modern drugs are derivatives of herbal medicines and natural products. The awarding of part of the 2015 Nobel Prize in Physiology or Medicine for Tu Youyou's discovery and development of artemisinin, a potent antimalarial derived from the herbal *Artemisia annua*, is a clear example and reminder of the potential in herbal medicine (Wang et al. 2018). The discovery of artemisinin revolutionized antimalaria drug development and research (Su and Miller 2015). Many herbal medicines are traditionally used as nutritional and pharmacological ingredients with health-promoting properties (Tan et al. 2017). In addition, cancer research also progressed with the combination of the Western medicine all-trans retinoic acid (ATRA) and the TCM compound arsenic trioxide (ATO) to treat acute promyelocytic leukemia (APL) (Lam et al. 2019). However, in most such cases, a molecular analysis reveals an underlying PML-RARA fusion transcript formed because of cytogenetically cryptic or complex insertion events. ATRA causes a conformational change of the PML-RARA fusion transcripts, leading to the release of the co-repressors, recruitment of histone acetyltransferases, and relief of transcriptional repression, which causes the treated APL cells to undergo terminal myeloid differentiation and finally apoptosis (Yilmaz et al. 2021). Arsenic trioxide (ATO) is a first-line treatment drug for APL, which is also effective for other kinds of leukemia (Jiang et al. 2023). The synergy of arsenic trioxide and ATRA, which binds the RARA moiety of PML-RARA, has been shown at the biological and clinical levels. Combination therapy can effectively alleviate complex leukemia symptoms. Thus, the introduction of ATO into APL therapy has changed the treatment landscape and allowed the development of a chemotherapy-free regimen with high success rates (Yilmaz et al. 2021).

Excessive production of reactive oxygen species (ROS) can result in oxidative load and stress, leading to various diseases including abnormal cell signal transduction, cardiovascular and neurodegenerative cancer diseases, etc. The active constituents from species (*Ginkgo biloba, Carthamus tinctorius, Salvia miltiorrhiza, Rhodiola rósea, Ginkgo biloba, Glycyrrhiza uralensis, Siberian ginseng, Rehmanniae radix, Fructus lycii, Angelica sinensis, Radix asparagi, Radix ginseng*, etc.) being used in TCM (polysaccharides, quinines, flavonoids, saponins, alkaloids, terpenes, phenolic acids compounds, tannins, etc.) could increase antioxidant enzymes, free radicals scavenging, decrease the generation of lipid peroxide and malondialdehyde, etc. (Liu and Jiang 2012), thus maintaining a redox balance between oxidants and antioxidants. Growing evidence suggests that Chinese herbal medicines reduce oxidative stress (Zhang et al. 2021), maintain the balance of redox activity in the body, and promote overall health. The data showed that these species possess a wide spectrum of biological properties, including antitumor (80%), antioxidant (50%), antidiabetic (40%), antilipemic (40%), anti-aging (40%), antimicrobial (40%), and cardioprotective (40%) activities (Tan et al. 2017). On the other hand, some species used in medicinal diets have been reported to cause allergic or adverse reactions. For instance, *Zanthoxyli Pericarpium* is used to alleviate pain and increase appetite. However, it may cause allergic reactions with itching of the oropharyngeal mucosa; swelling of the tongue, lips, and face; cramping in the stomach; vomiting; diarrhea; breathing problems; and even anaphylactic shock (Tan et al. 2017).

TCM treatment of diabetes in China is known as "Xiaoke," or increased thirst. Among those treatments, the isoquinone alkaloid known as berberine stood out for its clinically demonstrated hypoglycemic and hypolipidemic properties. Berberine is a benzyl tetra isoquinoline alkaloid that naturally occurs in the plant families *Ranunculaceae, Rutaceae, Berberidaceae, Annonaceae, Papaveraceae, and Menispermaceae.* It is being used as an antihyperlipidemic, antidiabetic, antiobesity, anticancer, anti-inflammatory, neuroprotective, cardioprotective, immunomodulatory, antioxidant, gastroprotective, and others (Purwaningsih et al. 2023). The clinical study of berberine as an antidiabetic is still in its infancy (Yilmaz et al. 2021). However, berberine and metformin share many features despite their different chemical structure. Both drugs could treat T2DMs, obesity, cardiac diseases, and tumors. These disorders often share common pathogenic factors and drug targets. Hence, understanding their actions can aid rational use (Wang et al. 2018) and elucidate potential interaction in clinical settings. Interestingly, berberine also modulates liver metabolic function through regulation of gene expression, reduction of intestinal glucose uptake through the inhibition of α-glucosidase, modulation of gut microbiota composition (through antimicrobial activity), and antioxidant properties, among others (Wang et al. 2024).

The antioxidant capacity of berberine is an important mechanism in its neuroprotection properties. This antioxidant activity is via direct scavenging activity, ROS formation, improvement in mitochondrial function by decreasing oxygen consumption and increasing the mitochondrial membrane potential, and upregulation of P13K/AKT/Bcl-2 and Nrf2/HO-1 pathways. The additional mechanisms involved in the neuroprotective role of berberine are antiapoptotic, anti-inflammation, the modulation of neurotransmitter levels, and the regulation of main neurological enzymes. Despite some promising aspects of TCM, its acceptance by the scientific community has been limited. The field of Chinese herbals has drawn increasing interest as a source of novel drugs and drug leads, which is unsurprising considering that many modern drugs are derivatives of herbal medicines and natural products.

## 6 Other Complementary and Alternative Medicine

Non-drug interactions in homeopathy for complex diseases involve a unique approach that focuses on stimulating the body's self-healing mechanisms in complex diseases. Here, a substance (from plants, minerals, or animals) that causes symptoms in a healthy person is used in diluted form to stimulate healing in someone experiencing similar symptoms. The choice of homeopathic is based on an assessment of the patient's physical, emotional, and mental symptoms as well as their overall constitution. This remedy could restore balance to the body's vital force or life energy and support the body's innate ability to heal itself. The dilution process is believed to minimize potential side effects, making it a gentle and generally safe treatment option for patients. However, the appropriate concentration level of the homeopathic substance to be used alongside conventional medical treatments or other complementary therapies remains unclear, as does the duration of homeopathic treatment needed to achieve long-term improvement in health. While homeopathy is debated in scientific communities regarding its mechanisms of action, clinical studies and anecdotal evidence suggest that it can be effective in managing symptoms of complex diseases.

In traditional and complementary medicine, acupuncture remains the most widely used. Acupuncture is traceable to the use of pointed stones and pressure to alleviate illness, rarely included recommendations on acupuncture (Zhang et al. 2022). Acupuncture can complement conventional medical therapies and contribute to improved outcomes by addressing the symptoms and underlying causes of complex diseases. Biochemical changes can modulate pain perception, mood, immune function, and other physiological processes relevant to complex diseases. This treatment modality is performed by inserting fine sterile needles, not much thicker than human hair, in specific locations on the skin, known as acupuncture points—with a higher concentration of Ca, P, K, Fe, Zn, and Mn. Both $Ca^{2+}$ and $K^{+}$ are key mediators of nerve signals. The depth of needle insertion depends on the disease being treated and the location of the acupoints (Marshall 2020). The main stimulation methods include manual acupuncture, electroacupuncture, and transcutaneous acupoint electrical stimulation of nerves and central nervous system.

In this procedural management of pain (neck pain, backache, osteoarthritis, fibromyalgia), specific sites of the body are identified for needle insertion (Wei et al. 2020; Fu et al. 2009) to alter body energy, release endorphins, regulate serotonin and norepinephrine, decrease levels of substance P (released in response to tissue damage), promote vasodilation (improved local and systemic circulation), relax muscles, reduce inflammation, enhance the delivery of oxygen and nutrients to tissues, and modulate autonomic functions among other mechanisms. Although all these can result in pain relief and management of complex disease symptoms (endocrine, metabolic, mental, neurological disorders, circulatory, cutaneous, musculoskeletal, and connective tissue readouts, etc.) (Wang et al. 2023), the risk of non-drug interactions also increases. Non-drug interactions could be minimized by a trained and licensed practitioner using sterile needles.

In hypothesis, acupuncture's impact on the autonomous nervous system enhances parasympathetic or reduced sympathetic activity. This may decrease stress responses and promote immunological homeostasis through the altered brainstem and hypothalamic–neuroendocrine function. Increased vagal stimulation may modulate cholinergic and inflammatory pathways. The brainstem and hypothalamic-driven anti-inflammatory responses may downregulate macrophage activation, tumor necrosis factor, and other pro-inflammatory cytokines. The neurophysiological mechanisms by which acupuncture exerts these effects are complex and still unclear (Matos et al. 2021).

Pain management can also be achieved with massage therapy that offers non-drug interactions that promote. The rhythmic pressure and stroking through massage release tension within the muscles, tendons, and ligaments to relieve sprains, backache, shoulder pain, knee osteoarthritis, fibromyalgia axial spondyloarthropathies, etc. (Null 1999; Danve and Deodhar 2018). In hypothesis, massage could increase lymph flow and blood circulation to painful areas, decreasing nerve compression, reducing stress hormones (like cortisol), promoting the release of endorphins and serotonin, and facilitating the clearance of lactate. These effects may help alleviate symptoms of pain, stiffness, arthritis, fibromyalgia, fibrosis, anxiety, depression, and limited movement associated with complex diseases (Moyer et al. 2004; Field 1998; Bervoets et al. 2015). An improvement in lymphatic circulation could increase the release of natural killer cells and enhance immune function. However, the consequence of medium- and long-term use of massage therapy with acupuncture, chiropractic care, etc. is still lacking. Chiropractic care offers spinal alignment and enhanced circulation to restore mobility while alleviating pain. Potential non-drug interactions with chiropractic care may have implications for conditions influenced by nervous system dysfunction, ergonomics, posture, exercise routines, and lifestyle modifications.

Yoga, meditation, Tai Chi, biofeedback, progressive muscle relaxation, and hypnosis are other forms of unconventional therapies that attenuate pain severity in rheumatoid arthritis and improve mood. The mind-body interventions alter mental and emotional state, balance the mind and body to relieve various symptoms and stress responses, and improve overall well-being. Perception of chronic pain could reflect the state of mind and mood (Hassed 2013). According to Lee et al. (2014), patients with low backache and fibromyalgia could benefit from meditation-induced relief. Yoga in the forms of yogic physical postures, breathing control, meditation, and pranayama improves body flexibility and stamina with reasonable control of mind over body. It improved a variety of musculoskeletal painful conditions and functional and psychosocial outcomes (Ward et al. 2013; McCaffrey 2012) and decreased levels of inflammatory mediators (IL-1α, IL-6, TNF-α) in patients with rheumatoid arthritis (Ganesan et al. 2020), stiffness due to osteoarthritis of the knee, osteoarticular pain, restricted joint function through a graded exercise along with controlled breathing (Lauche et al. 2019). Tai Chi also relieves pain and improves musculoskeletal strength, joint mobility, and joint mechanics (Wang et al. 2009). Relaxation is a training targeting a state of calmness by progressively relaxing the physical tension, stress, or pain through distraction. This process reduces muscle tension, blood pressure, heart rate, oxygen consumption, and cortisol levels (Lee et al. 2014). A trance-like state in hypnosis relieves pain conditions and anxiety. It is achieved by a series of instructions to achieve a high concentration of mind.

## 7 Concluding Remarks

Investigational, modeling, clinical, imaging, and analytical frameworks on non-drug interaction in alternative medicine for complex diseases could help develop novel precision medicines. The combination of artificial intelligence or high computing power that gathers and stores large data is desirable for analysis to provide etiological insights on complex diseases, heterogeneity, and layers of molecular biomarkers and predict a spectrum of interactions with alternate medicine while enhancing treatment precision. There is a need for investment in new technology, research, education, and training of healthcare professionals. Awareness campaigns on non-drug interactions with alternative medicine will be helpful.

## 8 Self-Assessment Questions and Answers

### *8.1 Questions*

1. Identify the correct statement in these options
   - A. Alternative medicine is synonymous with complementary medicine.
   - B. The principle that the body heals itself is irrelevant to complementary and alternative medicine.
   - C. Homeopathy emphasizes large amounts of certain substances to promote healing.
   - D. Prescription drugs may interact with dietary supplements.
2. Which herb from Traditional Chinese Medicine is indicated for malaria treatment?
   - A. *Artemisia annua*
   - B. *Gingko biloba*
   - C. *Valerian officinalis*
   - D. *Pimenta pseudocaryophyllus*
3. The control of breathing is part of which CAM?
   - A. Ayurvedic
   - B. Acupuncture
   - C. Chiropractic
   - D. Osteopathy

4. Which of the following best describes the role of non-drug interactions in CAM for complex diseases?
   A. They primarily replace conventional medical treatments.
   B. They often complement conventional treatments to improve outcomes.
   C. They are rarely effective and should be avoided.
   D. They only work for minor ailments.
5. Identify therapeutic associations with a potential for non-drug interaction in CAM.
   A. Chemotherapy and Antiemetics
   B. Yoga and meditation
   C. Antibiotics and Antineoplastics
   D. Insulin injections and Aspirin
6. How can non-drug interactions in CAM potentially benefit patients with complex diseases?
   A. By causing fewer side effects than conventional treatments.
   B. By completely curing the disease.
   C. By requiring less frequent monitoring by healthcare providers.
   D. By being mainly natural products.
7. What should patients consider when using non-drug interactions in CAM alongside conventional treatments for complex diseases?
   A. They should always stop their conventional treatments immediately.
   B. They should inform all healthcare providers about all treatments they are using.
   C. They should only use CAM treatments recommended by celebrities.
   D. They should only rely on CAM treatments without conventional medical supervision.

## 8.2 Answers

1. D is correct. In option A, alternative medicine is not synonymous with complementary medicine. Alternative medicine refers to treatments used in place of conventional medicine, while complementary medicine refers to treatments used alongside conventional medicine; in option B, the principle that the body heals itself is central to complementary and alternative medicine (CAM). CAM often focuses on supporting the body's natural healing processes; in option C, homeopathy involves the use of highly diluted substances, not large amounts, to stimulate the body's self-healing abilities; in option D, prescription drugs and dietary supplements can interact with each other, which is a well-recognized concern in

healthcare due to potential side effects or reduced efficacy of either substance when taken together.

2. A is correct. *Artemisia annua*, also known as sweet wormwood or Qinghao (in Chinese), has been traditionally used in Traditional Chinese Medicine for treating fevers, including those associated with malaria. Its active compound, artemisinin, is effective against malaria parasites and is widely used in modern medicine as well. Therefore, the correct answer is A. *Artemisia annua.*
3. A is correct. The control of breathing is primarily associated with practices and techniques within the field of Ayurvedic medicine. Ayurveda, an ancient system of medicine originating in India, includes various methods of breath control (pranayama) as a fundamental aspect of its therapeutic approach.
4. B is correct. This option reflects the typical role of non-drug interactions in CAM when used alongside conventional medical treatments for complex diseases.
5. B is correct. Yoga and meditation are examples of non-drug interactions in CAM that can be used as complementary approaches alongside conventional medical treatments for managing complex diseases.
6. A is correct. Non-drug interactions in CAM are often valued for their potential to offer therapeutic benefits with fewer side effects compared to conventional medical treatments.
7. B. Patients must keep all healthcare providers informed about the CAM treatments they are using alongside conventional treatments. This ensures coordinated care and helps prevent any potential interactions or adverse effects between treatments.

## References

Adhikari A, Dey A, Kanjilal S, et al. Gastroprotective and digestive potential of an Ayurvedic asava–arishta preparation. Orient Pharm Exp Med. 2018;18:391–401. https://doi.org/10.1007/s13596-018-0325-y.

Behzadifar M, Behzadifar M, Aryankhesal A, Ravaghi H, Baradaran HR, Sajadi HS, Khaksarian M, Bragazzi NL. Prevalence of self-medication in university students: systematic review and meta-analysis. East Mediterr Health J. 2020;26(7):846–57. https://doi.org/10.26719/emhj.20.052.

Bervoets DC, Luijsterburg PAJ, Alessie JJN, Buijs MJ, Verhagen AP. Massage therapy has short-term benefits for people with common musculoskeletal disorders compared to no treatment: a systematic review. J Physiother. 2015;61:106–116.

Bolognesi G, Belcaro G, Feragalli B. Movardol®(N-acetylglucosamine, Boswellia serrata, ginger) supplementation in the management of knee osteoarthritis: preliminary results from a 6-month registry study. Eur Rev Med Pharmacol Sci. 2016;20:5198–04.

Calcagni N, Gana K, Quintard B. A systematic review of complementary and alternative medicine in oncology: psychological and physical effects of manipulative and body-based practices. PLoS One. 2019;14(10):e0223564. https://doi.org/10.1371/journal.pone.0223564.

Charkviani M, Thongprayoon C, Tangpanithandee S, Krisanapan P, Miao J, Mao MA, Cheungpasitporn W. Effects of Mediterranean Diet, DASH Diet, and Plant-Based Diet on Outcomes among End Stage Kidney Disease Patients: A Systematic Review and Meta-Analysis. Clin. Pract. 2023;3:41–51.

Chow SL, Bozkurt B, Baker WL, Bleske BE, Breathett K, Fonarow GC, Greenberg B, Khazanie P, Leclerc J, Morris AA, Reza N, Yancy CW, American Heart Association Clinical Pharmacology Committee, Failure H, Transplantation Committee of the Council on Clinical Cardiology; Council on Epidemiology and Prevention; and Council on Cardiovascular and Stroke Nursing. Complementary and alternative medicines in the management of heart failure: a scientific statement from the American Heart Association. Circulation. 2023;147(2):e4–e30. https://doi.org/10.1161/CIR.0000000000001110.

Chu SWF, Ng WJ, Yeam CT, Khan RQ, Low LL, Quah JHM, Foo WYM, Seng JJB. Manipulative and body-based methods in chronic kidney disease patients: a systematic review of randomized controlled trials. Complement Ther Clin Pract. 2022;48:101593. https://doi.org/10.1016/j.ctcp.2022.101593.

Danve A, Deodhar AA. Complementary medicine for axial spondyloarthritis: is there any scientific evidence? Curr Opin Rheumatol. 2018;30:310–8.

Diab A, Dastmalchi LN, Gulati M, Michos ED. A heart-healthy diet for cardiovascular disease prevention: where are we now? Vasc Health Risk Manag. 2023;19:237–53. https://doi.org/10.2147/VHRM.S379874.

Ernst E. Musculoskeletal conditions and complementary/alternative medicine. Best Pract Res Clin Rheumatol. 2004;18:539–56.

Ernst E, Chrubasik S, Phyto-anti-inflammatories. A systematic review of randomized, placebo-controlled, double-blind trials. Rheum Dis Clin N Am. 2000;26:13–27.

Field TM. Massage therapy effects. Am Psychol. 1998;53:1270–81.

Fu L-M, Li J-T, Wu W-S. Randomized controlled trials of acupuncture for neck pain: systematic review and meta-analysis. J Altern Complement Med. 2009;15:133–45.

Ganesan S, Gaur GS, Negi VS, Sharma VK, Pal GK. Effect of yoga therapy on disease activity, inflammatory markers, and heart rate variability in patients with rheumatoid arthritis. J Altern Complement Med. 2020;26:501–7.

Garland EL, Brintz CE, Hanley AW, Roseen EJ, Atchley RM, Gaylord SA, Faurot KR, Yaffe J, Fiander M, Keefe FJ. Mind-body therapies for opioid-treated pain: a systematic review and meta-analysis. JAMA Intern Med. 2020;180(1):91–105. https://doi.org/10.1001/jamainternmed.2019.4917.

Gaur R. A brief history: traditional Chinese medicinal system. A brief history. Pharmacol Res Mod Chin Med. 2024;8:100387. https://doi.org/10.1016/j.prmcm.2024.100387.

Goldrosen MH, Straus SE. Complementary and alternative medicine: assessing the evidence for immunological benefits. Nat Rev Immunol. 2004;11:912–21. https://doi.org/10.1038/nri1486.

Gyawali D, Vohra R, Orme-Johnson D, Ramaratnam S, Schneider RH. A systematic review and meta-analysis of Ayurvedic herbal preparations for hypercholesterolemia. Medicina (Kaunas). 2021;57(6):546. https://doi.org/10.3390/medicina57060546.

Hassed C. Mind-body therapies--use in chronic pain management. Aust Fam Physician. 2013;42:112–117.

He B, Lu C, Zheng G, He X, Wang M, Chen G, Zhang G, Lu A. Combination therapeutics in complex diseases. J Cell Mol Med. 2016;12:2231–40. https://doi.org/10.1111/jcmm.12930.

Hourston S, Atchley R. Autism and mind-body therapies: a systematic review. J Altern Complement Med. 2017;5:331–9. https://doi.org/10.1089/acm.2016.0336.

Jiang Y, Shen X, Zhi F, Wen Z, Gao Y, Xu J, Yang B, Bai Y. An overview of arsenic trioxide-involved combined treatment algorithms for leukemia: basic concepts and clinical implications. Cell Death Discov. 2023;9(1):266. https://doi.org/10.1038/s41420-023-01558-z.

Jin SE, Seo C-S, Shin H-K, Ha H. Traditional herbal formulas to as treatments for musculoskeletal disorders: their inhibitory effects on the activities of human microsomal cytochrome P450s and UDP-glucuronosyltransferases. Pharmacogn Mag. 2016;12:241–52.

Karadima V, Kraniotou C, Bellos G, Tsangaris GT. Drug-micronutrient interactions: food for thought and thought for action. EPMA J. 2016;7(1):10.

Koithan M. Introducing complementary and alternative therapies. J Nurse Pract. 2009;5(1):18–20. https://doi.org/10.1016/j.nurpra.2008.10.012.

Lam WC, Lyu A, Bian Z. ICD-11: impact on traditional Chinese medicine and world healthcare systems. Pharmaceut Med. 2019;33(5):373–7. https://doi.org/10.1007/s40290-019-00295-y.

Lauche R, Hunter DJ, Adams J, Cramer H. Yoga for osteoarthritis: a systematic review and meta-analysis. Curr Rheumatol Rep. 2019;21:47.

Lee C, Crawford C, Hickey A. Active self-care therapies for pain (PACT) working group. Mind-body therapies for the self-management of chronic pain symptoms. Pain Med. 2014;15(Suppl 1):S21–39.

Lindler BN, Long KE, Taylor NA, Lei W. Use of herbal medications for treatment of osteoarthritis and rheumatoid arthritis. Medicine. 2020;7:67.

Liu QM, Jiang JG. Antioxidative activities of medicinal plants from TCM. Mini Rev Med Chem. 2012;12(11):1154–72. https://doi.org/10.2174/138955712802762239.

Ma D, Wang S, Shi Y, Ni S, Tang M, Xu A. The development of traditional Chinese medicine. J Tradit Chin Med Sci. 2021;8(Sup 1):S1–9.

Mannarino MR, Ministrini S, Pirro M. Nutraceuticals for the treatment of hypercholesterolemia. Eur J Intern Med. 2014;7:592–9. https://doi.org/10.1016/j.ejim.2014.06.008.

Marshall AC. Traditional Chinese medicine and clinical pharmacology. In: Drug discovery and evaluation: methods in clinical pharmacology. Springer International Publishing; 2020. https://doi.org/10.1007/978-3-319-68864-0_60.

Matos LC, Machado JP, Monteiro FJ, Greten HJ. Understanding traditional Chinese medicine therapeutics: an overview of the basics and clinical applications. Healthcare (Basel). 2021;9(3):257. https://doi.org/10.3390/healthcare9030257.

Mayden KD. Mind-body therapies: evidence and implications in advanced oncology practice. J Adv Pract Oncol. 2012;3(6):357–73.

McCaffrey R. The benefits of yoga for musculoskeletal disorders: a systematic review of the literature. J Yoga Phys Ther. 2012;2:100–22.

Mineshita Y, Sasaki H, Kim HK, Shibata S. Relationship between fasting and postprandial glucose levels and the gut microbiota. Meta. 2022;12(7):669. https://doi.org/10.3390/metabo12070669.

Mortada EM. Evidence-based complementary and alternative medicine in current medical practice. Cureus. 2024;16(1):e52041. https://doi.org/10.7759/cureus.52041.

Moses GM, McGuire TM. Drug interactions with complementary medicines. Aust Prescr 2010;33:181.

Moyer CA, Rounds J, Hannum JW. A meta-analysis of massage therapy research. Psychol Bull. 2004;130:3–18.

Ng JY, Dhawan T, Dogadova E, Taghi-Zada Z, Vacca A, Wieland LS, Moher D. Operational definition of complementary, alternative, and integrative medicine derived from a systematic search. BMC Complement Med Ther. 2022;22(1):104. https://doi.org/10.1186/s12906-022-03556-7.

Null B. Connective tissue massage in the treatment of fibromyalgia. Eur J Pain Lond Engl. 1999;3:235–244.

Pandit S, Kanjilal S, Awasthi A, Chaudhary A, Banerjee D, Bhatt BN, Narwaria A, Singh R, Dutta K, Jaggi M, Singh AT, Sharma N, Katiyar CK. Evaluation of herb-drug interaction of a polyherbal Ayurvedic formulation through high throughput cytochrome P450 enzyme inhibition assay. J Ethnopharmacol. 2017;197:165–72. https://doi.org/10.1016/j.jep.2016.07.061.

Patwardhan B. Bridging Ayurveda with evidence-based scientific approaches in medicine. EPMA J. 2014;5(1):19. https://doi.org/10.1186/1878-5085-5-19.

Pitcher MH, Edwards E, Langevin HM, Rusch HL, Shurtleff D. Complementary and integrative health therapies in whole person resilience research. Stress Health. 2023;39(S1):55–61. https://doi.org/10.1002/smi.3276.

Posadzki P, Watson LK, Alotaibi A, Ernst E. Prevalence of use of complementary and alternative medicine (CAM) by patients/consumers in the UK: systematic review of surveys. Clin Med (Lond). 2013;13(2):126–31. https://doi.org/10.7861/clinmedicine.13-2-126.

Purwaningsih I, Maksum IP, Sumiarsa D, Sriwidodo S. A review of *Fibraurea tinctoria* and its component, berberine, as an antidiabetic and antioxidant. Molecules. 2023;28(3):1294. https://doi.org/10.3390/molecules28031294.

Sahu R, Gupta PK, Mishra A, Kumar A. Ayurveda and in silico approach: a challenging proficient confluence for better development of effective traditional medicine spotlighting network pharmacology. Chin J Integr Med. 2023;29(5):470–80. https://doi.org/10.1007/s11655-022-3584-x.

Schriml LM, Lichenstein R, Bisordi K, Bearer C, Baron JA, Greene C. Modeling the enigma of complex disease etiology. J Transl Med. 2023;21(1):148. https://doi.org/10.1186/s12967-023-03987-x. PMID: 36829165; PMCID: PMC9957692

Sharma H, Keith Wallace R. Ayurveda and epigenetics. Medicina (Kaunas). 2020;56(12):687. https://doi.org/10.3390/medicina56120687.

Sriranjini SJ, Sandhya K, Mamta VS. Ayurveda and botanical drugs for epilepsy: current evidence and future prospects. Epilepsy Behav. 2015;52(Pt B):290–6. https://doi.org/10.1016/j.yebeh.2015.05.039.

Stipp MC, Acco A. Involvement of cytochrome P450 enzymes in inflammation and cancer: a review. Cancer Chemother Pharmacol. 2021;87:295–309. https://doi.org/10.1007/s00280-020-04181-2.

Su XZ, Miller LH. The discovery of artemisinin and the Nobel Prize in physiology or medicine. Sci China Life Sci. 2015;58(11):1175–9. https://doi.org/10.1007/s11427-015-4948-7.

Tabish SA. Complementary and alternative healthcare: is it evidence-based? Int J Health Sci (Qassim). 2008;2(1):V–IX. PMID: 21475465; PMCID: PMC3068720

Tan F, Chen Y, Tan X, Ma Y, Peng Y. Chinese materia medica used in medicinal diets. J Ethnopharmacol. 2017;206:40. https://doi.org/10.1016/j.jep.2017.05.021.

Tangkiatkumjai M, Boardman H, Walker DM. Potential factors that influence usage of complementary and alternative medicine worldwide: a systematic review. BMC Complement Med Ther. 2020;20(1):363. https://doi.org/10.1186/s12906-020-03157-2.

Teong XT, Liu K, Vincent AD, Bensalem J, Liu B, Hattersley KJ, Zhao L, Feinle-Bisset C, Sargeant TJ, Wittert GA, Hutchison AT, Heilbronn LK. Intermittent fasting plus early time-restricted eating versus calorie restriction and standard care in adults at risk of type 2 diabetes: a randomized controlled trial. Nat Med. 2023;29(4):963–72. https://doi.org/10.1038/s41591-023-02287-7.

Thottapillil A, Kouser S, Kukkupuni S, Vishnuprasad CN. An 'Ayurveda-Biology' platform for integrative diabetes management. J Ethnopharmacol. 2021;268:113575). ISSN 0378-8741. https://doi.org/10.1016/j.jep.2020.113575.

Vikram ENT, Ilavarasan R, Kamaraj R. Anti-cancer activities of schedule E1 drugs used in ayurvedic formulations. J Ayurveda Integr Med. 2022;13(2):100545. https://doi.org/10.1016/j.jaim.2022.100545.

Virolainen SJ, VonHandorf A, Viel KCMF, Weirauch MT, Kottyan LC. Gene-environment interactions and their impact on human health. Genes Immun. 2023;24(1):1–11. https://doi.org/10.1038/s41435-022-00192-6.

Wallace RK, Wallace T. Neuroadaptability and habit: modern medicine and Ayurveda. Medicina (Kaunas). 2021;57(2):90. https://doi.org/10.3390/medicina57020090.

Wang C, Schmid CH, Hibberd PL. Tai chi is effective in treating knee osteoarthritis: a randomized controlled trial. Arthritis Rheum. 2009;61:1545–53.

Wang J, Wong YK, Liao F. What has traditional Chinese medicine delivered for modern medicine? Expert Rev Mol Med. 2018;20:e4.

Wang M, Liu W, Ge J, Liu S. The immunomodulatory mechanisms for acupuncture practice. Front Immunol. 2023;14:1147718.

Wang Y, Yang Z. Yao Q. Accurate and interpretable drug-drug interaction prediction enabled by knowledge subgraph learning. Commun Med. 2024;4:59.

Wang J, Wong YK, Liao F. What has traditional Chinese medicine delivered for modern medicine? Expert Rev Mol Med. 2018;20:e4. https://doi.org/10.1017/erm.2018.3.

Ward L, Stebbings S, Cherkin D, Baxter GD. Yoga for functional ability, pain and psychosocial outcomes in musculoskeletal conditions: a systematic review and meta-analysis. Musculoskelet Care. 2013;11:203–17.

Watanabe K, et al. Food as medicine: translating the evidence. Nat Med. 2023a;29:753–4. https://doi.org/10.1038/s41591-023-02330-7.

Watanabe K, Wilmanski T, Diener C, Earls JC, Zimmer A, Lincoln B, Hadlock JJ, Lovejoy JC, Gibbons SM, Magis AT, Hood L, Price ND, Rappaport N. Multiomic signatures of body mass index identify heterogeneous health phenotypes and responses to a lifestyle intervention. Nat Med. 2023b;29(4):996–1008. https://doi.org/10.1038/s41591-023-02248-0.

Wei Y, Yuan N, Ding J. Efficacy and safety of warm needle acupuncture in knee osteoarthritis. Medicine (Baltim). 2020;99(50):e23596.

Yilmaz M, Kantarjian H, Ravandi F. Acute promyelocytic leukemia current treatment algorithms. Blood Cancer J. 2021;11(6):123. https://doi.org/10.1038/s41408-021-00514-3.

Zhang C, Wang N, Feng Y. Chapter 22 – Oxidative stress, Chinese herbals and toxicity: a focused review with examples. In: Patel VB, Preedy VR, editors. Toxicology. Academic Press; 2021. p. 215–24., ISBN 9780128190920. https://doi.org/10.1016/B978-0-12-819092-0.00022-4.

Zhang YQ, Lu L, Xu N, Tang X, Shi X, Carrasco-Labra A, Schünemann H, Chen Y, Xia J, Chen G, Liu J, Liu B, Wang J, Qaseem A, Jing X, Guyatt G, Zhao H. Increasing the usefulness of acupuncture guideline recommendations. BMJ. 2022;376:e070533. https://doi.org/10.1136/bmj-2022-070533.

Zhao M, Ma J, Li M, Zhang Y, Jiang B, Zhao X, Huai C, Shen L, Zhang N, He L, Qin S. Cytochrome P450 enzymes and drug metabolism in humans. Int J Mol Sci. 2021;22(23):12808. https://doi.org/10.3390/ijms222312808.

Zheng RX, Xu JW, Jiang BY, Tang W, Lu CL, Hu XY, Liu JP. Mind-body therapies in traditional Chinese medicine for neuropathic pain: a systematic review of randomized controlled trials. Pain Manag Nurs. 2023;24(2):157–70. https://doi.org/10.1016/j.pmn.2022.10.003.

# Chapter 14
# Remedies Against Plausible Adverse Interactions

**Caroline Vitória de Lima Moreira, Roberto Saavedra-rodríguez, and James O. Fajemiroye**

**Abstract** Adverse interactions that are often underreported remain the bane of morbidity and mortality in health care. Adverse interactions deserve attention as they are frequently responsible for significant mortality and morbidity burdens in older people, pediatric populations, and other vulnerable groups. Ideally, prescribers and patients should know the onset and duration of the spectrum of biological activities and the site of actions to decide on drug choice, dose, and scheduling while predicting adverse drug interactions and possible antidotes. The mechanism of the drug's action and the patient's medical profile with a wide range of lifestyles (consumption, association of medications, nutrients, herbs, vitamins) can provide a helpful remediation (reverse, counteract, treat, or manage) strategy. A careful review of medication and food attributes, patient's medication, nutritional history, education about possible interactions, evidence-based recommendations, interdisciplinary collaboration, and effective monitoring for signs of adverse effects or therapeutic failure are pertinent to mitigating adverse interactions. This chapter explores the mechanism of adverse interactions, associated risks, and relevant regulations to advance plausible remediation strategies.

**Keywords** Remedies · Structure · Mechanisms · Vulnerability · Regulations

---

C. V. de Lima Moreira · J. O. Fajemiroye (✉)
Institute of Biological Sciences, Federal University of Goias, Goiania, Goias, Brazil
e-mail: jamesfajemiroye@ufg.br

R. Saavedra-rodríguez
Department of Basic Science, Universidad Santo Tomas, Campus Puerto Montt, Puerto Montt, Chile

 
J. O. Fajemiroye (ed.), *Fundamentals of Drug and Non-Drug Interactions*,
https://doi.org/10.1007/978-3-031-80107-5_14

# 1 Introduction and a Brief History

Adverse drug and food interactions have been of significant concern throughout history. For example, Egyptians, Chinese, and Greeks have documented using various herbs and substances for medicinal purposes. Widespread cases of poisoning and adverse reactions are attributable to drugs containing unknown and often harmful ingredients. Adverse interactions and effects resulting from these ancient practices are largely neglected due to a lack of pharmacological understanding.

Around the nineteenth century (Fig. 14.1), alcohol was paraphrased as a substance that can save and destroy based on its therapeutic effects, metabolism, and toxicity. The incessant reoccurrence of known and unknown drug fatality provided support for remediation measures, which include increased regulation and legislation, for example, the Pure Food and Drug Act (USA 1906) monitors adulterated and misbranded food and drugs. Coca-Cola was originally formulated by John Pemberton in 1886. Initially, this product contained cocaine derived from the coca leaf. Cocaine was removed from the formulation in 1903 due to public and regulatory concerns about its addictive nature and potential health risks. In the 1960s, some cases related to hypertensive crises were reported in several patients who consumed certain cheeses before antidepressant monoamine oxidase inhibitors (MAOI) treatment. The tyramine, an amine with pressor effects, in these cheeses whose metabolism was inhibited by MAOI inhibitors provoked intense pressor effect in patients (Blackwell 1963; Natoff 1964; Sjöqvist 1965; Horwitz et al. 1964). A patient treated with nialamide following the consumption of Swedish blue cheese with the highest concentration of tyramine died of cerebral hemorrhage (Hammer et al. 1969). In 1963, hypoglycemia was reported after sulfaphenazole and tolbutamide co-administration led to the conclusion that tolbutamide metabolism was inhibited by sulfaphenazole (Christensen et al. 1963). These effects and others reinforced the evidence of drug interactions through the alterations in metabolic enzyme activities. According to Sjöqvist and Böttiger (2010), many interactions are

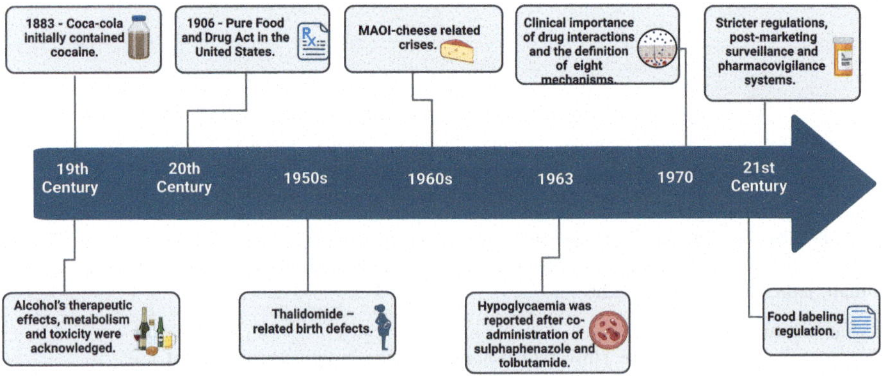


**Fig. 14.1** Timeline for a few historic adverse interactions and regulatory efforts. (Created with BioRender.com)

avoidable if only patients are exposed to one substrate of a particular CYP enzyme, thereby preventing enzyme induction or inhibition. From 1970, the clinical importance of drug interactions led to their classification based on eight mechanisms (absorption, plasma protein binding, tissue distribution, transport, effects on receptors, metabolism, renal elimination, and miscellaneous) (Alexanderson and Sjöqvist 1970).

In the twentieth century, thalidomide, a drug prescribed to pregnant women for morning sickness which led to thousands of birth defects worldwide, provided a basis for the strengthening of drug regulatory agencies and implementation of more rigorous testing protocols, such as the Kefauver Harris Amendment in the United States, which requires proof of efficacy as well as safety for drug approval. In the 1980s, the HIV/AIDS epidemic emerged, leading to the development of antiretroviral drugs. In the twenty-first century, the Vioxx scandal emerged, where the painkiller Vioxx is withdrawn in respect of the high risk of strokes and heart attacks, thereby leading to the implementation of stricter post-marketing surveillance and pharmacovigilance systems to detect and respond to adverse drug reactions more rapidly. The emergence of antibiotic-resistant bacteria due to inappropriate applications of antibiotics now requires prudent use measures (antibiotic stewardship programs and public education campaigns). Allergic reactions to food, such as peanuts, shellfish, and gluten, which become more widely recognized and studied, have resulted in improved food labeling regulation, increased availability of allergen-free alternatives, and advancements in allergy testing and treatment. Throughout history, adverse drug and food interactions have spurred regulatory reforms, advancements in medical research, and public health initiatives aimed at improving safety and reducing harm to consumers. Poor reports on various drug-food interactions, drug-supplement interactions, drug-vitamin interactions, drug-nutrient interactions, and herb-drug interactions are clear evidence of decades of neglect or undocumented remediation efforts regarding drug-nondrug entities interactions (DNIs).

Adverse drug reactions are read out of DNIs (between drugs, drugs with foods, beverages, supplements, or diseases). The risk of adverse drug reactions, unwanted, undesirable, or side effects increases with cocktails or associations of drugs and non-pharmaceutical entities. The reality of unpleasant, harmful, hospitalization, disability, permanent damage, or life-threatening outcomes of DNIs demands a remediation approach before, during, and after patient exposure. The DNIs-related death, years of life lost, adjusted, or lived with a disability that is traceable to medical treatment error provides a basis for the development of error-reporting systems (Havens and Boroughs 2000; Khan et al. 2020). As adverse drug reactions are common and constitute a significant healthcare burden, healthcare professionals must be ready to identify, prevent their occurrence, and treat them (Kommu and Whitfield 2024). Pharmaceutical products could elicit noxious and unintended adverse drug reactions under prophylaxis, diagnosis, or therapeutic doses (Nebeker et al. 2004). According to Bellanca et al. (2023), 30–60% of adverse drug reactions are related to inappropriate prescription, drug chemistry inherent toxicity, long-term treatments of chronic diseases, sex, age, cell-related drug toxicity, or anomalies in drug absorption, distribution, metabolism, and excretion, or DNIs might be predictable and preventable. A risk-benefit assessment of drugs or non-pharmaceuticals in associations is often neglected (Fig. 14.2).

Nowadays, patients (especially elderly ones) frequently take many medications simultaneously. The complexity of therapeutics and an increasingly aging population constitute challenges in managing evitable adverse drug reactions in modern healthcare systems (Coleman and Pontefract 2016). Unpredictable and potential drug interactions may occur by using multiple medications as well as co-treatments with herbal drug products that do not have defined ingredients. The detection of these types of interactions requires patient disclosure (Butterweck and Derendorf 2012). The drug's mechanism of action, individual or patient medical profile with a wide range of lifestyle, age, gender, underlying health conditions, and medication history can determine the severity of interactions. Tailoring recommendations to each patient's unique circumstances could optimize safety, efficacy, and remediation strategy against adverse interactions (Eland et al. 1999). These strategies could involve patient education on the potential risks associated with combining medications with non-pharmaceutical substances, facilitating open communication and proactive management with professionals who in turn should keep receiving continuous training, attending conferences, workshops, and reviewing relevant literature to stay abreast of emerging trends and developments in the field of drug and non-pharmaceutical interactions. The dearth of latest research on DNIs remains a

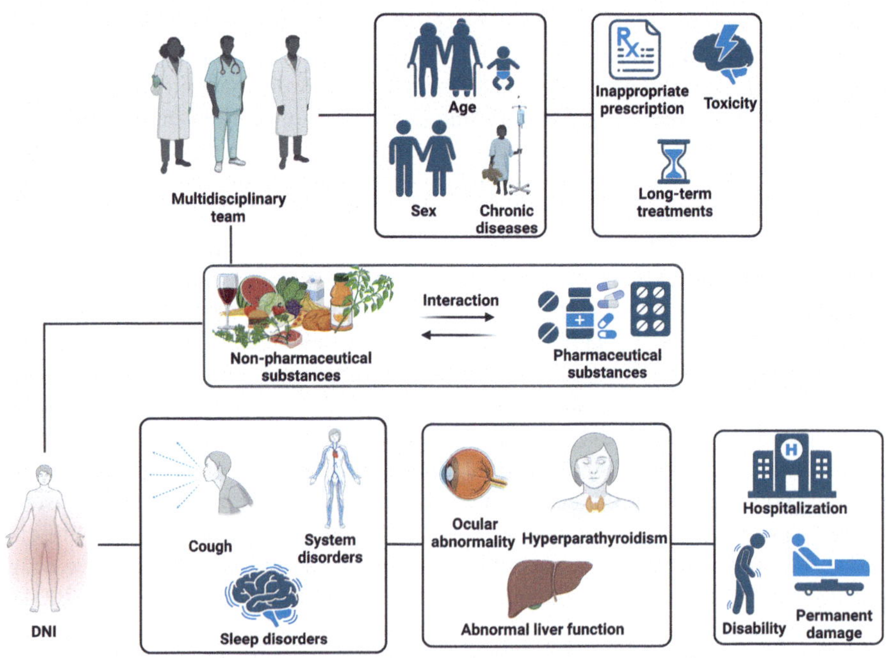


**Fig. 14.2** Drug–nondrug interactions (DNIs) and other risk factors of hospitalization, disability, permanent damage, or life-threatening outcomes. (Created with BioRender.com)

challenge in the search for an authoritative guide toward clinical decision-making and patient counseling. There are increasing need for regular medication reviews, accessibility to information about commonly used non-pharmaceuticals and their potential interactions with prescription drugs, and discussion about the use of herbal supplements, dietary preparations, vitamins, minerals, and other complementary therapies with healthcare professionals to assess the potential for interactions between medications and non-pharmaceutical substances. These remediation approaches may reflect unique/localized practices and adverse interaction reports in different countries or communities.

In India, the rate of adverse effects reporting is less than the global report of 5% (Shukla et al. 2024). The most frequent interactions involving St John's wort, which are widespread and well communicated in the product information leaflet (Jacobsson et al. 2009), reduced the number of interaction cases. Telecommunication, telemedicine, social media and electronic medical records, and artificial intelligence are potential approaches for preventing, monitoring, and reporting adverse impacts in rural areas. Healthcare professionals collaborate to address the complex nature of drugs and non-pharmaceutical interactions. Pharmacists, doctors, nutritionists, and others can contribute their expertise to develop comprehensive management strategies and support patient-centered care. Patients should be encouraged to purchase reputable products from trusted sources, follow recommended dosages and administration instructions, eschew the use of unregulated or potentially harmful substances that lack scientific evidence of efficacy and safety, and report any unusual symptoms or changes in their health status, especially after starting or discontinuing non-pharmaceutical products.

The DNIs can result in a spectrum of central and peripheral nervous system disorders or adverse reactions including myasthenia, myalgia, muscle bleeding, arthralgia, lower limb spasm, osteoporosis associated with musculo-skeletal system. Additionally, they may cause skin and appendage reactions such as itching, urticaria, rash, maculopapular rash, erythema, etc. Occular abnormalities, peripheral neuropathy, coma, grand mal seizure, manic-depressive psychosis and sleep disorders may also occur. The gastrointestinal system may not be spared from nausea, vomiting, abdominal pain, gastrointestinal bleeding, abdominal distention, flatulence, black faeces, diarrhea, hematemesis, etc. Hypotension, hypertension, palpitations, tachycardia, bradycardia, cardiac arrest, arrhythmias, atrioventricular block among other cardiovascular events are possible. Organ and metabolic changes include abnormal liver function, jaundice, elevated liver enzymes, cholestatic hepatitis, biliary cirrhosis, electrolyte abnormality, hyperuricemia, increased blood lactic acid, hypokalemia; hyponatremia, hyperkalemia, hypoglycemia, hyperglycemia, etc. Other adverse outcomes include thyroiditis, hyperparathyroidism, dyspnea, asthma, cough, anemia, leukopenia, granulocytopenia, bone marrow suppression, thrombocytopenia, coagulopathy, hematemesis, hematuria, urinary retention, genital itching, breast enlargement, menstrual disorders, fatigue, allergic reactions, chills, phlebitis, skin necrosis among other hormonal, respiratory, urinary, reproductive and blood cell events (Jiang et al. 2022). A chain of adverse reactions leading to clinical fatality could be provoked by one adverse outcome of DNIs. A common drug adverse reaction like diarrhea (Chassany et al. 2000) can trigger dehydration, and sodium depletion, potentially leading to death (Richardson et al. 2010). Therefore, if the clinical outcomes of adverse reactions to drugs can be predicted and assessed before it is marketed, the risk to patients and the failure of drug development can be reduced before the drugs are released to the market (Setola and Roth 2005).

The existence of complex categories or subcategories of each one of these adverse reactions demands extensive study and research efforts that favor adequate documentation and reporting. As an example here, hypersensitivity reactions are categorized into type I reactions with immediate onset—commonly triggered by beta-lactams, quinolones, and platinum-containing chemotherapeutic agents triggered immunoglobulin E (IgE), mast cells, and basophils activities with the release of vasoactive and proinflammatory mediators that provoke urticarial rash, pruritus, flushing, angioedema, hypotension, and anaphylaxis in severe cases (Abbas et al. 2023); type II reactions including drug-induced hemolytic anemia (cephalosporins, penicillin), thrombocytopenia (heparin, apixaban, quinidine, sulfonamides), and neutropenia (propylthiouracil, flecainide), which late onset after drug exposure involves IgG or IgM cellular or tissue damages (Bajwa and Mohammed 2023); type III reactions with delayed in onset are mediated by immune complexes and complement. In these, a drug forms an immune complex by binding to drug-specific IgG in different tissues (including joints, kidneys, and blood vessels), activating complement, for example, equine-antitoxins and monoclonal antibodies-induced serum sickness, penicillins, cephalosporins, and phenytoin-induced vasculitis, and diphtheria and tetanus vaccines-induced Arthus reactions (Usman and Annamaraju 2023); type IV reactions also with delayed onset include contact dermatitis, acute generalized exanthematous pustulosis, Stevens-Johnson syndrome, hypersensitivity syndrome, and eosinophilia and systemic syndrome) are mediated by T cells, macrophages, eosinophils, and neutrophils (Marwa and Kondamudi 2023). Extensive coverage of each adverse reaction to drugs and non-pharmaceuticals could help overcome under-reporting of adverse drug reactions, which in itself is associated with factors like belief that drugs in the market are safe, guilt of possible or unintended harm, ambitious drive, elaborate reporting requirement, ignorance, clinician indifference to contributing to medical knowledge, procrastination, lack of interest or time, financial incentives to report, and uncertainty in determining drug role in adverse effect (Lopez-Gonzalez et al. 2009). Altogether, the available reports in the literature provide some mechanistic clues and regulations that have proved to be important remedies against adverse interactions involving drugs and nondrug entities.

## 2 Mechanism-Based Remediation

Reversal of adverse interactions, resuscitation of victims, and intensive care are remediation approaches that augment pharmacological antidotes (antagonists, inhibitors, etc). Pharmaceutical and non-pharmaceutical substance-related interactions generally involve a pharmacokinetic mechanism that modifies plasma concentration and subsequently affects the pharmacological action. According to Chacko and Peter (2019), remediants mediate its effect by binding and neutralization to prevent absorption and distribution; by inhibiting conversion or enhancing elimination through urinary alkalinization and hemadsorption.

This type of interaction can be counterbalanced by adjusting the drug's dosage in mild adverse reactions (with close monitoring for potential worsening). At the same time, it is also important to consider that a risky aspect of this interaction is when an herbal supplement or its active ingredients share the same mechanism of action with the co-administered drug. Pharmacokinetic and pharmacodynamic mechanisms could create a competitive scenario between an herbal supplement and a drug before a change in the drug's concentration at the site of action. Pharmacodynamic mechanisms may not necessarily change the concentrations of drugs in circulation. Here, herbal supplements can increase the drug's clinical effects without changing the drug's concentration. As a result, and learning from the previous aspects, clinicians who can differentiate between pharmacokinetic and pharmacodynamic mechanisms will be able to make better clinical decisions by adjusting drug's dosage or suspending herbal supplements. Therefore, to identify pharmacokinetic herb-drug interactions, it is necessary to understand the mechanism underlying ADME processes for a drug and an herbal supplement. The pharmacological antidote for food and drug-induced adverse reactions depends on the specific substance, physicochemical properties, and the nature of the reaction involved.

From a pharmacokinetic mechanism perspective, remedies can be complex and require multiple targets for some drugs. As an example, warfarin (anticoagulant) could trigger complex interactions with other drugs, foods, and herbal medicines as it interferes with the blood clotting process while exhibiting low therapeutic index, discontinued dose-response pattern, high binding rate to plasma proteins, and CYP P450 enzymes. Chelators like dimercaprol, N-acetyl Cysteine (NAC), unithiol, D-penicillamine, zinc or calcium trisodium diethylenetriaminepentaacetate, N-acetyl-D-penicillamine, deferoxamine, calcium disodium ethylenediaminetetraacetate, triethylenetetraamine, and deferiprone with free electrons bind positively charged metal ions and transformed into a new complex. The NAC could remediate the effects of acetaminophen overdose that depletes glutathione reserves. The CYP2E1 and CYP3A4 in the liver metabolize this drug into N-acetyl-4-benzoquinimine-NAPQI, a highly oxidative and hepatotoxic metabolite. Sulfhydryl glutathione or thiol groups from NAC can neutralize or mop up NAPQI through its conversion into cysteine, a substrate for glutathione synthesis and replenishment of its reserves (Aldini et al. 2018). The formation of a stable ring between the sulfhydryl groups of dimercaprol, 2,3-dimercapto-1-propanol; 1,2-dithioglycerol (C3H8OS2); and arsenic, gold, mercury, lead, antimony, or bismuth neutralizes heavy metal-induced toxicity and facilitates their elimination (Vilensky and Redman 2003; Dawn and Whited 2023). In addition, a lipophilic chelator and organic sulfur compound like succimer or meso-2,3-dimercaptosuccinic acid (C4H6O4S2) could reduce adverse interactions by chelating with divalent metals such as mercury, arsenic, cadmium, and gallium arsenide (Flora et al. 2022).

A highly basic and polycationic protein like protamine reduces the risk of adverse effects of strongly anionic anticoagulant unfractionated heparin-induced during dialysis, invasive vascular procedures, acute ischemic strokes, or postoperative bleeding (Ranasinghe et al. 2019). Both form an inactive and stable heparin-protamine complex devoid of anticoagulant properties (Applefield and Krishnan 2023). Fomepizole blocks metabolic acidosis and toxic metabolite formation (e.g., acetic acid and formic acid) by preventing the conversion of ethanol and methanol to acetaldehyde and formaldehyde, respectively. Non-specific binding agents such as activated charcoal have been employed in adverse and life-threatening effects of carbamazepine, dapsone, phenobarbital, quinine, or theophylline (Brent et al. 1999). As a result of its high capacity for adsorption and by interruption of enterohepatic recirculation, activated charcoal inhibits systemic absorption and decreases systemic concentration. An activated carbon attenuates adverse anticholinergic and antihistaminic effects of diphenhydramine (first-generation H1 antihistamine or antagonist of H1 histamine receptors). The muscarinic antagonist, Parkinson's disease medication, antipruritic medicine, local anesthetic, anti dyskinesia, and anti cough agents could reduce gastrointestinal motility and delay absorption (Kobylarz et al. 2023).

Foods, nutrients, beverages, coffee, and milk among other non-pharmaceuticals can significantly affect the pharmacokinetics profile of drugs by altering pH, perfusion, absorbent area, motility along the gastrointestinal tract, gut microflora composition, digestive juice production, gastric emptying, and intestinal transit time. The ingestion of warm, viscous, and fat-rich solid foods could slow down gastric emptying times and delay intestinal absorption (D'Alessandro et al. 2022). The dissolution of the drugs, absorption rate, and bioavailability are determinants of all possible interactions that provide mechanistic clues toward remediation measures. As CYP3A4 isoform detoxifies around 50% of drugs, the inhibitory activity of grapefruit juice on this enzyme could constitute serious adverse interactions and effects. Most pharmacokinetic mechanisms are classified under four large gene superfamilies, which include more than 1000 proteins: the cytochrome P450 (CYP) drug metabolism enzymes; the uridine diphosphate-glucuronosyltransferase (UGT) conjugating enzymes; the adenosine triphosphate–binding cassette (ABC) drug uptake/efflux transporters; and the organic anion-transporting polypeptide (OATP) drug transporters (Zanger and Schwab 2013). The CYP and UGT make drugs more polar and water-soluble, thereby facilitating their elimination from the body, while the ABC (e.g., efflux pump P-glycoprotein [P-gp]) and OATP are transporters of drugs and their metabolites from various body compartments. Altogether, CYP1A2, CYP2C9, CYP2C19, CYP2D6, CYP2E1, CYP3A4, OATP1A1, OATP1A2, OATP2B1, and P-gp are important proteins that account for several clinically relevant interactions.

The herb-drug inhibitory activities on these proteins are partial due to phenolic compounds including naringine and furanocoumarin bergamottine. Naringinine and hesperidine inhibit transmembrane transporters and prevent the passage of the drug molecules from the intestinal lumen. These compounds are found in several fruit juices. Grapefruit juice alone could alter the pharmacokinetic profile of several drugs including atorvastatin, amlodipine, felodipine, manidipine, nicardipine, nifedipine, nimodipine, nisoldipine, nitrendipine, pranidipine, losartan, acebutolol, cyclosporin, tacrolimus, amiodarone, quinidine, disopyramide, propafenone, vinblastine, and erythromycin. As catechins can inhibit the OATP1A2, patients should avoid taking green tea (rich in these compounds) with antibiotics, anticoagulants, oral contraceptives, anxiolytics, and antihypertensive or chemotherapeutic agents (D'Alessandro et al. 2022). The regulatory effects of dietary fibers on intestinal digestion and absorption, transit time, stool formation, and changes in microbiota composition (Gill et al. 2021) provide a mechanistic understanding of the potential benefits of dietary fiber interaction on gastrointestinal health, function, or dysfunction. Although this chapter highlights adverse interactions and effects, there are reports of therapeutic benefits of pharmacokinetic interactions. For example, orlistat for obesity treatment inhibits pancreatic lipases and interferes with lipid absorption as acarbose for the treatment of type 2 diabetes mellitus inhibits competitively and reversibly alpha-glucosidase, increases glucagon-like peptide 1 to interfere with carbohydrate absorption, and elicits a weight-loss effect and cardiovascular protection (Altay 2022). The pancreatic alpha-amylase hydrolyzes complex carbohydrates to oligosaccharides in the small intestine as intestinal membrane-bound alpha-glucosidase metabolizes oligosaccharides, trisaccharides, and disaccharides (sucrose, maltose) to monosaccharides (glucose, fructose). Hence, glucose absorption and postprandial glucose blood concentrations are reduced by acarbose-induced-delayed digestion of carbohydrates (McIver et al. 2024).

From a pharmacodynamic perspective, some medications have specific drugs that can be administered to counteract adverse interactions and effects, especially in cases of overdose. For example, naloxone (N-allylnoroxymorphone—competitive antagonist of μ, δ, and κ receptors) is an antidote for opioid overdose and de-escalation of its adverse effects (Van Dorp et al. 2007); and vitamin K can be used as an antidote for warfarin overdose or excessive anticoagulation. Drug withdrawal could be helpful in the case of (i) macrolide antibiotics that increase theophylline exposure and (ii) respiratory depression that results from coadministered benzodiazepine and opioids. The antidotes for adverse reactions include (i) epinephrine for severe allergic reactions (anaphylaxis); (ii) flumazenil (imidazobenzodiazepine that competes as an antagonist on the benzodiazepine site of GABAA-receptor) blocks benzodiazepine-induced muscle relaxant, sedative, and hypnotic, among other properties (Chacko and Peter 2019; Sharbaf Shoar et al. 2023); (iii) oxygen therapy for carbon monoxide poisoning helps to remove carbon monoxide from the bloodstream and replenish oxygen levels; (iv) peptide, octreotide attenuates adverse effects of KATP channel closure, calcium influx, and insulin release of oral hypoglycemic sulfonylureas (Baumgartner and Devgun 2021). This peptide binds to somatostatin receptors and inhibits cyclic adenosine monophosphate synthesis and voltage-gated calcium channel, thereby eliciting a decrease in calcium influx, pancreatic cells hyperpolarization with a reduction in the secretion of insulin secretion (Glatstein et al. 2012).

Severe antimuscarinic or anticholinergic toxicity often led to the use of physostigmine, an alkaloid from *Physostigma venenosum*, inhibitor of acetylcholinesterase (AChE) and cholinomimetic of indirect action. Strong nucleophile oximes such as pralidoxime, methoxime, trimedoxime, obidoxime, and asoxime are effective

AChE reactivators and antidotes for organophosphate (tabun, sarin, soman, etc) poisoning (Jokanovic 2012; Kuca and Pohanka 2010). An irreversible inhibition of the AChE active site (Ser–His–Glu triad) compromises its critical roles (rapid physiological hydrolysis of acetylcholine (ACh) before a reduction in its synaptic and neuromuscular junction concentration). This leads to the accumulation of ACh, overstimulation of cholinergic receptors, and subsequent multiorgan or system failure (Marrs 1993; Dvir et al. 2010; Pita and Domingo 2014; Hoenig 2007). With the hydroxyl groups in serine and phosphate groups in organophosphate, the occurrence of rapid nucleophilic attacks (Mercey et al. 2012) often results in the release of fluoride ions, formation of a covalent bond between phosphorus atoms and AChE serine, which in turn slows down the hydrolysis of acetylcholine (Korabecny et al. 2014). In this manner, oxime (enzyme reactivator) applications are crucial to restore physiological activities. However, side effects such as liver and kidney damage, nausea, and vomiting are associated with oximes, in addition to their effectiveness as reactivators (Jokanovic 2012; Kuca and Pohanka 2010; Wilson and Ginsburg 1955; Eyer 2003; Worek et al. 2007; Chai et al. 2018).

On the contrary to cholinomimetic drugs, atropine (tropane alkaloid and a competitive antagonist of muscarinic acetylcholine receptors) is a common lifesaving remedy against muscarinic crisis by suppressing secretion and reversing smooth muscle contractions. This antimuscarinic drug promptly relieves bronchial spasms, facilitates optimal air exchange, and sustains cardiovascular function (Kobylarz et al. 2023). The dantrolene, which binds to ryanodine receptors, blocks excessive calcium release from the sarcoplasmic reticulum as a lifesaving postsynaptic muscle relaxant (indication: malignant neuroleptic syndrome or hyperthermia) against depolarizing muscle relaxants or anesthetics. This drug relieves muscle stiffness and increases body temperature as a result of persistent muscle contraction.

An intravenous fatty acid-rich lipid emulsion from soybean oil and egg yolk phospholipids as a supplement for patients with low-fat consumption was effective in treating severe anesthetic drug-induced cardiotoxicity (Weinberg 2003; Neal et al. 2018; Anez-Bustillos et al. 2016), tricyclic antidepressants, calcium channel blockers, beta-blockers, antipsychotics, insecticides, and organophosphates poisonings (Kobylarz et al. 2023). According to Kobylarz et al. (2023), the plasma lipid compartment of lipid emulsion may trap and dissolve lipophilic drugs having octanol to water partitioning coefficient of log $p > 2$ and reduce their free-form plasma concentrations (Ozcan and Weinberg 2012). Lipid emulsion can also increase cardiac contractility and improve vascular resistance, blood pressure, and cardiac output (Sepulveda and Pak 2023). Fat emulsion infusion in persistent cardiovascular collapse is associated with fever, chills, nausea, vomiting, headache, back pain, chest pain, shortness of breath, cyanosis, and other side effects (Geyer 1960). Folic acid antagonists, such as methotrexate, inhibit dihydrofolate reductase that releases tetrahydrofolate molecules (Benedek 2010). As a carbon donor of thymidylate and purine synthesis, the blockade of tetrahydrofolate with methotrexate treatment may result in serious adverse effects. Unlike folic acid, the administration of folinic acid could revert these effects as the conversion of 5-formyltetrahydrofolate to tetrahydrofolate is devoid of dihydrofolate reductase (Howard et al. 2016; Gristan and Moosavi 2023) (Fig. 14.3).

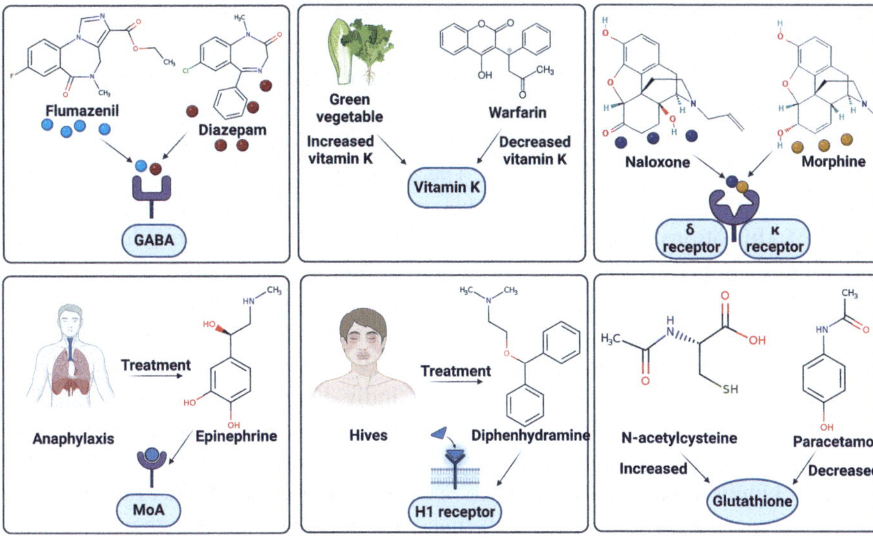


**Fig. 14.3** Representative targets of drugs. (Created with BioRender.com)

Although there are still uncertainties in cancer treatment with supplements (Ambrosone et al. 2020), adverse interactions and effects (fatigue, nausea, vomiting, diarrhea, pain, rash, constipation, shortness of breath, etc) could be managed in cancer patients through concomitant administration of antioxidant dietary supplements (e.g., vitamin C, minerals—selenium, amino acids—n-acetylcysteine, carotenoids—lycopene, botanicals—polyphenols, hormones—melatonin). These dietary supplements can prevent cellular damage by reacting with oxidizing free radicals with chemotherapy, radiotherapy, or biological therapy. Some chemotherapy drugs induce oxidative stress, inflammation, and DNA damage in cancer and healthy cells (Pearce et al. 2017; Wieland et al. 2021). From ethnomedicinal reports and phytochemical analysis, adverse interactions and effects are attenuated by plants with saponins, tannin, alkaloids, sulfur, anthraquinones, terpenoids, oleanolic acid, flavonoids, mannitol, glycosides, ursolic acids, pyrrolidine, formic acids, essential oils, and steroids bioactive components, among other secondary metabolites (Albuquerque et al. 2012; Singha et al. 2012; Teron and Borthakur 2013).

From the literature review, 114 herbal medicines were reported to interact with warfarin (Leite et al. 2021). In total, 84% of the plants increased the warfarin effect and the risk of bleeding. However, ascertaining the mechanism of action of these plants and active ingredients is still challenging. By looking at the possible key targets involved in these interactions (coagulation process, platelet activation, inflammation, etc), effective clinical management of interactions in anticoagulated patients exposed to herbal medicines should be done with caution. A healthcare provider may swiftly opt for discontinuation or adjustment of medication at the first sign of severe adverse effects in cases where drugs or non-pharmaceutical entities are causing or contributing to an adverse interaction. However, adjustment in the drug's dosage scarcely results in a predictable change in clinical outcomes (Asher et al. 2017). Hence, the adjustment of warfarin dose with changes in daily leafy green vegetables could be more beneficial in pharmacodynamic interaction.

Remedies against adverse interactions should also reflect mechanistic changes with aging. The aging process significantly affects the body's homeostasis, thereby inducing pharmacokinetic alterations. The aging impact on drug metabolism and clearance heightens the adverse reaction's risk or altered drug responsiveness (Klotz 2009). Modifications in the distribution volumes of drugs, attributable to a lower body fluid content and differences in lipid distribution, can prolong the half-life of a given drug, thus escalating the probability of toxicity (Bellanca et al. 2023). In patients on polypharmacy, drug metabolism can also be affected by interactions with CYP450 enzymes. A cross-sectional study of 80-year-old institutional and community residents revealed that 72% of individuals exhibited potential CYP drug-to-drug interactions that affected their functional performance and mobility (Bellanca et al. 2023). Change in the levels of steroid hormones with age affects the pharmacokinetic and response profiles of drugs (Tran et al. 1998; Franconi et al. 2007). Increased risk of adverse events that contribute to inappropriate prescriptions is common in older adults taking drugs such as antidiabetic, anticonvulsants, antidepressants, antiparkinson, opioids, anticoagulant therapy, etc. (Saraf et al. 2016; Vetrano et al. 2019; Muhlack et al. 2018; Gage et al. 2005). The risk of adverse events is higher among patients with multiple comorbidities or medications. Older adults taking oral antidiabetic medications are at a heightened risk of hypoglycemia, which can lead to falls and other complications. A significant drop in the circulating glucose (hypoglycemia) leads to dizziness, confusion, and loss of consciousness. Anticonvulsants, antidepressants, and antiparkinson can increase the risk of falls, delirium, cognitive impairment, orthostatic hypotension, and incontinence in older adults. A reduction in blood pressure occurring when standing up from a sitting or lying position can lead to dizziness, fainting, and orthostatic hypotension. Opioid medications can also contribute to an increased risk of falls in older adults, possibly due to dizziness, sedation, and impaired balance. Patients with atrial fibrillation who are at high risk of falls face an elevated risk of intracranial bleeding when taking anticoagulant therapy which increases the risk of bleeding. Hence, healthcare providers need to carefully weigh the benefits of medication therapy against the potential risks in this vulnerable population. Fasting could be a compounding factor as the rates and extent of drug metabolism can vary significantly. In the previous report, the cocktail consisted of caffeine (CYP1A2), metoprolol (CYP2D6), and warfarin (CYP2C9) administered after 36 h of fasting increased systemic clearance of caffeine and metoprolol by 17% ($p = 0.04$) and 13% ($p < 0.01$), respectively, as systemic S-warfarin clearance was decreased by 19% ($p < 0.01$) (Turpault et al. 2009; Lammers et al. 2017, 2019). Hence, close monitoring, regular medication reviews, and patient education are essential to mitigate these risks and ensure the safety and well-being of older adults receiving pharmacotherapy.

## 3 Regulation-Based Remediation

Globally, regulatory frameworks aim to safeguard public health by preventing, identifying, and mitigating adverse interactions between food, drugs, herbs, and vitamins. These involve stringent oversight of quality standards, adverse event monitoring, and public education efforts. Regulatory agencies strive to promote the safe and effective use of pharmaceuticals and non-pharmaceutical products (natural and synthetic) while minimizing potential harmful risks to consumers. In the United States and Europe, reports of hepatotoxicity from herbals or food supplements are increasing, including those advertised for liver diseases. According to Licata et al. (2013), herbal drugs need to be monitored and standardized. In many countries, herbals are sold as food supplements, which drug regulatory agencies cannot control. Herbal drugs are used in children, and pregnant and lactating women for being natural and perceived to be harmless. This misbelief that natural substances are safe seems to be a general perception (Svedlund et al. 2017). The preference for herbal-based slimming agents over conventional dietary and physical activity is evolving. In Italy, the use of non-conventional therapies has been reported for 13.6% of the population, and 3.7% freely use herbal drugs, unaware of the risks associated with a potential interaction with prescription drugs.

In this manner, healthcare professionals should remain vigilant toward these products, especially for signs of interactions during patient check-ups and follow-up assessments. Natural remedies (NRs) are often conceived for self-care use and medicinal products with active ingredients of herbal origin are approved specifically in the European Union (according to the European Directive 2001/83/EC), as herbal medicinal products (HMPs) or as traditional herbal medicinal products (THMPs). In Sweden, complementary and alternative medicinal (CAM) products have a national law that approves natural remedies with active ingredients of natural origin that are of non-herbal origin including a mineral, bacterial culture, and animal (Svedlund et al. 2017). The quality requirements for HMPs, THMPs, and NRs are similar to other medicinal products. For THMPs in comparison with HMPs, the efficacy requirements consist merely of long-standing medicinal use: 15 years for the European Economic Area, and 30 years worldwide. Likewise, THMPs are limited to self-medication, which does not require diagnosis, prescription, or supervision of treatment, as they can be administered orally, externally, or by inhalation (Claeson 2014). The use of THMPs, HMPs, and NRs (Svedlund et al. 2017) and a constant demand for information on adverse interactions indicate that the safety report about the patient and the treatment suspected to have caused the adverse effects is limited. Each report could document the suspected herbal preparation, the adverse reactions, the severity of the adverse reaction (significant incapacity, death, and life-threatening, thereby requiring inpatient hospitalization or prolongation of existing hospitalization), etc. To improve healthcare reporting (Gonzalez-Gonzalez et al. 2013), continuous publications on herbal monographs and the safety of herbal preparations provide important information for approval and the registration of

HMPs, THMPs, and NRs. Healthcare professionals should discuss self-care products and report suspected adverse reactions to herbal medicinal products.

There has been an increasing concern over the potential interaction of prescription medications with herbal dietary supplements (which contain plant extracts that are consumed as powder, tablet, liquid, capsule, or other formulations). The National Center for Complementary and Integrative Health defines dietary supplements as a variety of products, including herbs, vitamins and minerals, and probiotics (Asher et al. 2017). An estimated 40–60% of adults in the United States with chronic disease use dietary supplements while approximately 20–25% of patients taking prescription medications use a dietary supplement concurrently (Asher et al. 2017; Gardiner et al. 2008; Miller et al. 2008). Hence, regulation of food, drugs, dietary supplements, herbal remedies, and vitamins could provide accurate information about allergens and ingredients. Report analysis on adverse drug reactions could help identify potential interactions and ensure the safety of pharmaceutical products. These are important remediation measures being implemented by the U.S. Food and Drug Administration (FDA), the European Medicines Agency (EMA), the European Union (EU), the European Food Safety Authority, the European Food Information Regulation (EU FIC), the Natural and Non-prescription Health Products Directorate, the Therapeutic Goods Administration, the Australian Register of Therapeutic Goods, Brazilian Health Regulatory Agency, among others. Adverse event reporting systems are built to ensure the quality, purity, and safety of dietary supplements and herbal products, thereby allowing healthcare professionals and consumers to report adverse reactions associated with medicines and complementary medicines. Some of these bodies do take enforcement actions against any regulatory violations. As an example, the FDA withdrew from the market in 1997 fenfluramine, an appetite suppressant approved in the United States in 1973 as it provoked potentially fatal cardiovascular diseases (Odi et al. 2021). Implementation of routine monitoring and surveillance could help detect potential adverse interactions between medications and non-pharmaceutical substances. All these attempts can engender safe practices regarding the use of non-pharmaceutical substances with medications (Fig. 14.4).

Several herbal preparations for traditional herbal medicinal products have been in clinical use for many years. The most reported adverse reactions for herbal medicinal products and natural remedies are related to skin and subcutaneous tissue, because of hypersensitivity to herbal treatments, followed by gastrointestinal and nervous system disorders. This finding agrees with the data from the World Health Organization (Farah et al. 2000). Hypersensitivity and allergy activate adverse skin and subcutaneous tissue reactions. Appropriate labeling for medicinal products plays an important regulatory action targeted at reducing risks. Despite an information brochure of products for healthcare professionals and patients, not all patients know about their allergies or the possibility of cross-reactivity. Therefore, adverse reactions related to hypersensitivity and allergies will continue (Svedlund et al. 2017).

In a retrospective study of adverse drug reactions from 2007 to 2015 aimed at approving herbal medicinal products, 116 reports concerning 259 adverse reactions

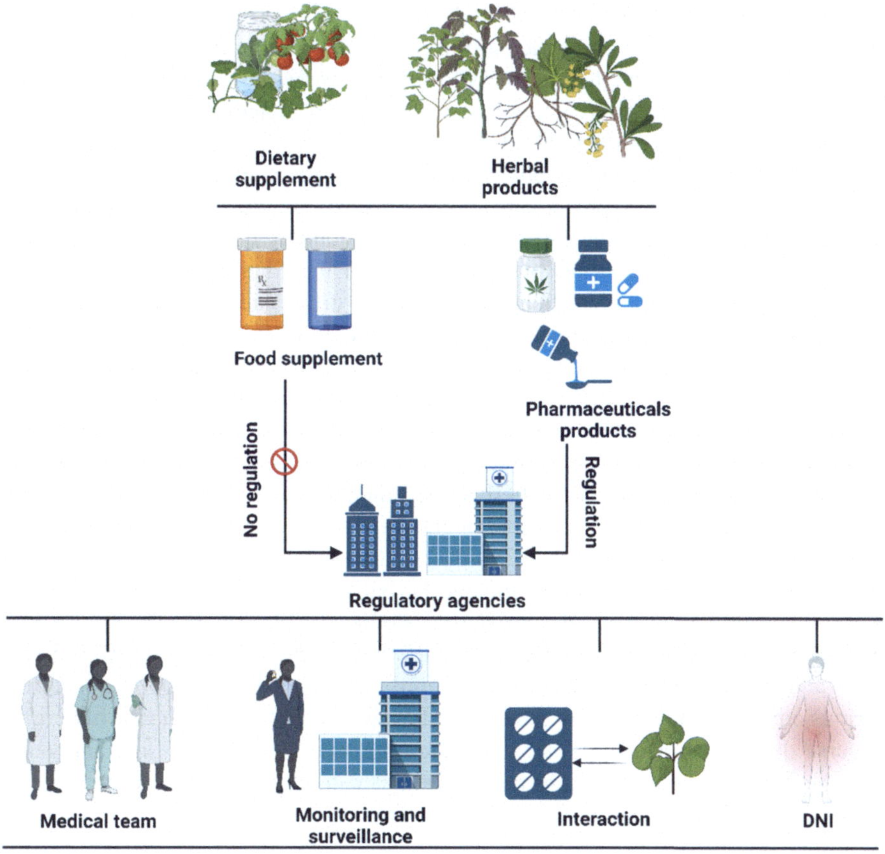


**Fig. 14.4** Regulation-based remediation. (Created with BioRender.com)

were analyzed and found in the Swedish national pharmacovigilance database. Black cohosh rhizome (15 reports), purple coneflower herb (14 reports), and a combination of extracts of pollen (13 reports) are among the most frequently active ingredients suspected during the study period. The use of black cohosh was associated with liver toxicity. However, there is still limited non-clinical and clinical data on black cohosh-induced liver toxicity.

As many herbal preparations in THMPs have been in clinical use for many years before their approval, this system allows ADRs to be acknowledged. Post-approval monitoring of approved medicinal products reduces the risk of self-care products with severe safety problems in the market. Patients, especially the elderly, usually did not disclose their use of herbal self-care products to healthcare professionals (Schnabel et al. 2014; Gardiner et al. 2007).

According to Van Hunsel and Van Grootheest (2013), in the Netherlands during the period 1991–2013, there are reports of suspected adverse reactions to herbal remedies from the Netherlands Pharmacovigilance Centre Lareb. Herbs can cause adverse interactions with medicinal products. Herbal remedies may also be adulterated with regular medicines or illegal or toxic ingredients. In this case, 336 reports of suspected adverse reactions to herbal remedies were shown by the Lareb database. In total, 518 adverse reactions to various herbal remedies were reported and the herbal remedies contained 221 different herbs. Fifty-five reports were classified as serious. Therefore, reporting adverse reactions significantly helps to know more about this subject. Again, physicians and pharmacists should ask patients about the use of alternative medications, especially those containing herbs.

In continuation to the previous analysis from the Pharmacovigilance Centre Lareb, there was also a classification of adverse drug reaction reports on herbal medicinal products and herbal supplements submitted from 1991 to February 2021 based on their regulatory status. The products were categorized based on their registration status and herbal ingredients, and according to the Herbal Anatomical Therapeutic Chemical Classification System. An analysis of the herbal medicinal products and herbal supplements in the Dutch pharmacovigilance database was carried out and showed suspected herbal ingredients. From 789 reports of herbal medicinal products and herbal supplements, a total of 823 herbal products were labeled as suspect for a total of 1727 adverse drug reactions. In total, 229 and 594 products were registered as medicine and herbal supplements, respectively, out of 823 products in the market. Of these herbal products, 522, 256, 27, and 18 reported being concerned with single-herb, combination, vitamin-containing herbal ingredients, and product issues, respectively. Approximately 15% of reports concerned serious adverse drug reactions. Adulterated products present a high risk of serious adverse drug reactions. Thus, a systematic approach to collecting and analyzing adverse drug reactions related to these products can help gain more knowledge to prevent harm (Van Hunsel et al. 2022).

# 4 Final Remarks on Education, Research, and Training

DNIs partly define adverse drug reactions, safety or market withdrawal of drugs, or the need to halt inappropriate associations of drugs and non-pharmaceuticals. The seriousness of DNIs in the clinical context remains largely unclear. Classification of adverse reactions facilitates management measures that prioritize patient protection, care, and quality of life. This could enable prompt medical intervention, guide treatment adjustments, evaluate risk-benefit profiles, and inform regulatory decisions (Zhao et al. 2023). Multidisciplinary approaches that encompass education, research, and training are crucial components in addressing adverse interactions. The public health implications of DNIs are profound and there is a huge knowledge gap among physicians, patients, and policymakers. This situation is hampering the integrated remediation approaches, sustainable clinical outcomes, and health economics.

Medicines and nutritional risk management need integration into effective prediction and mitigation measures. When and how long remedies should be administered requires careful assessments. The benefits of the remediators could be time-dependent, attributed to drug and nondrug entities involved, lag time to presentation, toxicokinetics, dosing, and the mechanism of action, among other uncertain factors. Drug development programs include studies to assess potential interactions between investigational drugs and commonly consumed foods or other medications that could determine safe dosages and identify populations at higher risk of adverse events. A successful remediation approach could include effective pharmacological intervention or remedy with airway, breathing, and circulation management. Remediation of plausible adverse interactions needs a review of medical history, nutrition, medication, and lifestyle in general. Mitigation of adverse interactions involves several measures including the risk assessment of drug-drug, drug-food, and drug-supplement associations; patient education to make informed decisions about their health and dietary choices, on combining medications and foods; substitution of medication or adjustment in terms of régimen, dosage, or discontinuing certain medications altogether as well as regular medical consultation, continuous monitoring, and follow-up appointments for signs and symptoms of adverse interactions; and treatment plan readjustment.

Pharmacogenetics, genomics, in silico, or artificial intelligence tools offer opportunities to understand the best approach against adverse interactions. Digital health interventions could help identify inappropriate prescriptions and promote patient adherence to treatment plans. Multiple approaches that could engender effective remedies to adverse interactions include (i) continuing education programs with updated curricula courses on pharmacology, toxicology, and DNIs; (ii) organizing public awareness campaigns that can help educate the public about the risks of adverse DNIs; (iii) providing information on proper medication use, food safety practices, and allergen awareness; (iv) monitoring adverse drug reactions via surveillance systems to identify potential interactions and safety concerns; (v) investigating the mechanisms behind these interactions and risk factors associated with specific drugs and food combinations; (vi) exploring how food components interact with drugs in the body, affecting absorption, metabolism, and excretion, which could enhance dietary recommendations for individuals taking certain medications; (vii) conducting specialized training programs on medication reconciliation, patient counseling, and decision support tools as well as workshops to equip healthcare professionals with the skills to recognize, prevent, and manage adverse drug and food interactions in clinical practice; and (viii) conducting training programs for food handlers, such as restaurant staff and food service workers, that emphasize safe food handling practices, allergen awareness, and cross-contamination prevention among other measures.

## 5 Self-Assessment Questions and Answers

### 5.1 *Questions*

1. All the options below are relevant to the identification and mitigation of potential adverse interactions except
   A. Assessment of Patient health status by gathering comprehensive medical history, including current medications, any past illnesses, surgeries, allergies, chronic conditions, and any existing health conditions.
   B. Review of all prescribed medications (dosage, frequency, and duration of each medication), over-the-counter drugs, supplements, vitamins, and herbal remedies the patient is currently taking.
   C. Utilize reliable medical databases, and drug interaction checkers, or consult with pharmacists to assess potential interactions between medications.
   D. Adverse interactions normally lead to decreased efficacy, increased side effects, or severe consequences.
2. A 40-year-old man experienced detachment of the epidermis (5% of the body surface area) and mucous membranes for the first time after 5 days of trimethoprim-sulfamethoxazole medication. What kind of adverse drug reaction did she develop?
   A. Toxic epidermal necrolysis
   B. Urticaria
   C. Acute generalized exanthematous pustulosis
   D. Stevens-Johnson syndrome
3. Under-reporting of adverse drug reactions is associated with all these factors except
   A. Belief that drugs in the market are safe
   B. Guilt of possible or unintended harm
   C. Ambitious drive
   D. Mild and common adverse drug reactions are unnecessary

### 5.2 *Answers*

1. D. This option describes the potential outcomes of adverse interactions rather than the steps involved in identifying and mitigating them. The other options (A, B, and C) all describe actions that are directly related to the identification and mitigation of potential adverse interactions, such as gathering comprehensive medical history, reviewing medications, and utilizing resources to assess potential interactions.

2. D. Stevens-Johnson syndrome is characterized by detachment of the epidermis and mucous membranes, typically involving less than 10% of the body surface area. It often occurs within days to weeks after starting a new medication, such as trimethoprim-sulfamethoxazole. Given that the detachment of the epidermis involves only 5% of the body surface area, it fits the criteria for SJS rather than Toxic Epidermal Necrolysis (TEN), which involves detachment of more than 30% of the body surface area.
3. C. Ambitious drive: This seems less directly related to the under-reporting of ADRs. The ambitious drive might push healthcare professionals to excel in their careers, but it is less likely to influence their reporting of adverse reactions.

   A. Belief that drugs in the market are safe: This belief can contribute to under-reporting because individuals may assume that any adverse effects they experience are rare or unrelated to the medication. Therefore, they might not report them. B. Guilt of possible or unintended harm: This can also be a factor. Healthcare professionals may feel guilty or responsible if a patient experiences harm due to a medication they prescribed, which could lead to under-reporting.

   D. Mild and common adverse drug reactions are unnecessary: This attitude can significantly contribute to under-reporting. If healthcare professionals believe that only severe or uncommon adverse reactions are worth reporting, they may not report milder or more common reactions, leading to a skewed understanding of a drug's safety profile.

## References


Abbas M, Moussa M, Akel H. StatPearls [Internet]. Treasure Island: StatPearls Publishing; 2023. Type I hypersensitivity reaction

Albuquerque UP, Melo JG, Medeiros MF, Menezes IR, Moura GJ, Asfora El-Deir AC, Nóbrega Alves RR, de Medeiros PM, de Sousa TAA, Alves MR, et al. Natural products from ethno-directed studies: revisiting the ethnobiology of the zombie poison. Evid Based Complement Alternat Med. 2012;2012:202508.

Aldini G, Altomare A, Baron G, Vistoli G, Carini M, Borsani L, Sergio F. N-Acetylcysteine as an antioxidant and disulphide breaking agent: the reasons why. Free Radic Res. 2018;52:751–62. https://doi.org/10.1080/10715762.2018.1468564.

Alexanderson B, Sjöqvist F. La¨kemedelsinteraktion. FASS; 1970. p. 691–712.

Altay M. Acarbose is again on the stage. World J Diabetes. 2022;13(1):1–4. https://doi.org/10.4239/wjd.v13.i1.1. PMID: 35070055; PMCID: PMC8771266

Ambrosone CB, Zirpoli GR, Hutson AD, et al. Dietary supplement use during chemotherapy and survival outcomes of patients with breast cancer enrolled in a cooperative group clinical trial (SWOG S0221). J Clin Oncol. 2020;38:804–14. https://doi.org/10.1200/JCO.19.01203.

Anez-Bustillos L, Dao DT, Baker MA, Fell GL, Puder M, Gura KM. Intravenous fat emulsion formulations for the adult and pediatric patient: understanding the differences. Nutr Clin Pract. 2016;31:596–609. https://doi.org/10.1177/0884533616662996.

Applefield D, Krishnan S. StatPearls. Treasure Island: StatPearls Publishing; 2023. Protamine

Asher GN, Corbett AH, Hawke RL. Common herbal dietary supplement-drug interactions. Am Fam Physician. 2017;96(2):101–7.

Bajwa SF, Mohammed RH. StatPearls [Internet]. Treasure Island: StatPearls Publishing; 2023. Type II hypersensitivity reaction
Baumgartner K, Devgun J. Toxicology of medications for diabetes mellitus. Crit Care Clin. 2021;37:577–89.
Bellanca CM, Augello E, Cantone AF, Di Mauro R, Attaguile GA, Di Giovanni V, Condorelli GA, Di Benedetto G, Cantarella G, Bernardini R. Insight into risk factors, pharmacogenetics/genomics, and management of adverse drug reactions in elderly: a narrative review. Pharmaceuticals (Basel). 2023;16(11):1542. https://doi.org/10.3390/ph16111542. PMID: 38004408; PMCID: PMC10674329
Benedek TG. Methotrexate: from its introduction to non-oncologic therapeutics to anti-TNF-α. Clin Exp Rheumatol. 2010;28:S3–8.
Blackwell B. Hypertensive crises due to monoamine-oxidaseinhibitors. Lancet. 1963;2:849–50.
Brent J, Jaeger A, McGuigan M, Meulenbelt J, Tenenbein M, Bradberry S, et al. Position statement and practice guidelines on the use of multi-dose activated charcoal in the treatment of acute poisoning. J Toxicol Clin Toxicol. 1999;37(6):731–51. https://doi.org/10.1081/CLT-100102451.
Butterweck V, Derendorf H. Herb-drug interactions. Planta Med. 2012;78(13):1399.
Chacko B, Peter JV. Antidotes in poisoning. Indian J Crit Care Med. 2019;23:S241. https://doi.org/10.5005/jp-journals-10071-23310. PMID: 32020997; PMCID: PMC6996653
Chai PR, Hayes BD, Erickson TB, Boyer EW. Novichok agents: a historical, current, and toxicological perspective. Toxicol Commun. 2018;2:45–8.
Chassany O, Michaux A, Bergmann JF. Drug-induced diarrhoea. Drug Saf. 2000;22:53–72.
Christensen LK, Hansen JM, Kristens M. Sulphaphenazole-induced hypoglycaemic attacks in tolbutamide-treated diabet-ics. Lancet. 1963;2:1298–301.
Claeson P. Requirements on efficacy of herbal medicinal products. J Ethnopharmacol. 2014;158:463–6.
Coleman JJ, Pontefract SK. Adverse drugs reactions. Clin Med. 2016;16(5):481–5.
D'Alessandro C, Benedetti A, Di Paolo A, Giannese D, Cupisti A. Interactions between food and drugs, and nutritional status in renal patients: a narrative review. Nutrients. 2022;14(1):212. https://doi.org/10.3390/nu14010212.
Dawn L, Whited L. Dimercaprol. In: StatPearls. Treasure Island: StatPearls Publishing; 2023.
Dvir H, Silman I, Harel M, Rosenberry TL, Sussman JL. Acetylcholinesterase: from 3D structure to function. Chem Biol Interact. 2010;187:10–22.
Eland IA, Belton KJ, van Grootheest AC, Meiners AP, Rawlins MD, Stricker BH. Attitudinal survey of voluntary reporting of adverse drug reactions. Br J Clin Pharmacol. 1999;48(4):623–7.
Eyer P. The role of oximes in the management of organophosphorus pesticide poisoning. Toxicol Rev. 2003;22:165–90.
Farah MH, Edwards R, Lindquist M, Leon C, Shaw D. International monitoring of adverse health effects associated with herbal medicines. Pharmacoepidemiol Drug Saf. 2000;9(2):105–12.
Flora SJS, Jain K, Panghal A, Patwa J. Chemistry, pharmacology, and toxicology of monoisoamyl dimercaptosuccinic acid: a chelating agent for chronic metal poisoning. Chem Res Toxicol. 2022;35:1701–19. https://doi.org/10.1021/acs.chemrestox.2c00129.
Franconi F, Brunelleschi S, Steardo L, Cuomo V. Gender differences in drug responses. Pharmacol Res. 2007;55:81–95.
Gage BF, Birman-Deych E, Kerzner R, Radford MJ, Nilasena DS, Rich MW. Incidence of intracranial hemorrhage in patients with atrial fibrillation who are prone to fall. Am J Med. 2005;118:612–7.
Gardiner P, Graham R, Legedza AT, Ahn AC, Eisenberg DM, Phillips RS. Factors associated with herbal therapy use by adults in the United States. Altern Ther Health Med. 2007;13(2):22–9.
Gardiner P, Phillips R, Shaughnessy AF. Herbal and dietary supplement—drug interactions in patients with chronic illnesses [published correction appears in Am Fam Physician. 78(7): 808]. Am Fam Physician. 2008;77(1):73–8.
Geyer RP. Parenteral nutrition. Physiol Rev. 1960;40:150–86. https://doi.org/10.1152/physrev.1960.40.1.150.

Gill SK, Rossi M, Bajka B, et al. Dietary fibre in gastrointestinal health and disease. Nat Rev Gastroenterol Hepatol. 2021;18:101–16. https://doi.org/10.1038/s41575-020-00375-4.

Glatstein M, Scolnik D, Bentur Y. Octreotide for the treatment of sulfonylurea poisoning. Clin Toxicol. 2012;50:795–804.

Gonzalez-Gonzalez C, Lopez-Gonzalez E, Herdeiro MT, Figueiras A. Strategies to improve adverse drug reaction reporting: a critical and systematic review. Drug Saf. 2013;36(5):317–28.

Gristan YD, Moosavi L. StatPearls. Treasure Island: StatPearls Publishing; 2023. Folinic acid

Hammer W, Holmberg G, Sjöqvist F, Wicklund PE. Drug interac-tion: lethal interaction between nialamide and blue-veinedcheese. Lakartidningen. 1969;66(Suppl IV):20.5.

Havens DH, Boroughs L. "To err is human": a report from the Institute of Medicine. J Pediatr Health Care. 2000;2000(14):77–80.

Hoenig SL. Compendium of chemical warfare agents. New York: Springer; 2007. ISBN 978-0-387-34626-7

Horwitz D, Lovenberg W, Engelman K, Sjoerdsma A. Monoamineoxidase inhibitors, tyramine, and cheese. JAMA. 1964;188:1108–10.

Howard SC, McCormick J, Pui CH, Buddington RK, Harvey RD. Preventing and managing toxicities of high-dose methotrexate. Oncologist. 2016;21:1471–82. https://doi.org/10.1634/theoncologist.2015-0164.

Jacobsson I, Jonsson AK, Gerden B, Hagg S. Spontaneously reported adverse reactions in association with complementary and alternative medicine substances in Sweden. Pharmacoepidemiol Drug Saf. 2009;18(11):1039–47.

Jiang H, Lin Y, Ren W, Fang Z, Liu Y, Tan X, Lv X, Zhang N. Adverse drug reactions and correlations with drug-drug interactions: a retrospective study of reports from 2011 to 2020. Front Pharmacol. 2022;13:923939. https://doi.org/10.3389/fphar.2022.923939. PMID: 36133826; PMCID: PMC9483724

Jokanovic M. Structure-activity relationship and efficacy of pyridinium oximes in the treatment of poisoning with organophos-phorus compounds: a review of recent data. Curr Top Med Chem. 2012;12:1775–89.

Khan MA, Soteriades ES, King J, Govender R, Hashim MJ, Masood-Husain S, Javaid SF, Debaib Mohammed Saeed Al Darei S, Dahi Al Sheryani S, Nauman J. Global trends and forecast of the burden of adverse effects of medical treatment: epidemiological analysis based on the global burden of disease study. Cureus. 2020;12(3):e7250. https://doi.org/10.7759/cureus.7250. PMID: 32195068; PMCID: PMC7071843

Klotz U. Pharmacokinetics and drug metabolism in the elderly. Drug Metab Rev. 2009;41:67–76.

Kobylarz D, Noga M, Frydrych A, Milan J, Morawiec A, Glaca A, Kucab E, Jastrzębska J, Jabłońska K, Łuc K, Zdeb G, Pasierb J, Toporowska-Kaźmierak J, Półchłopek S, Słoma P, Adamik M, Banasik M, Bartoszek M, Adamczyk A, Rędziniak P, Frączkiewicz P, Orczyk M, Orzechowska M, Tajchman P, Dziuba K, Pelczar R, Zima S, Nyankovska Y, Sowińska M, Pempuś W, Kubacka M, Popielska J, Brzezicki P, Jurowski K. Antidotes in clinical toxicology-critical review. Toxics. 2023;11(9):723. https://doi.org/10.3390/toxics11090723. PMID: 37755734; PMCID: PMC10534475

Kommu S, Whitfield P. Adverse drug reactions. [Updated 2024 Jan 10]. In: StatPearls [Internet]. Treasure Island: StatPearls Publishing; 2024. Available from: https://www.ncbi.nlm.nih.gov/books/NBK599521/.

Korabecny J, Soukup O, Dolezal R, Spilovska K, Nepovimova E, Andrs M, Nguyen T, Jun D, Musilek K, Kucerova-Chlupacova M, et al. From pyridinium-based to centrally active acetylcholinesterase reactivators. Mini Rev Med Chem. 2014;14:215–21.

Kuca K, Pohanka M. Chemical warfare agents. In: Luch A, editor. Molecular, clinical and environmental toxicology, Experientia supplementum, vol. 100. Basel: Birkhäuser Basel; 2010. p. 543–58. ISBN 978-3-7643-8337-4.

Lammers LA, Achterbergh R, Mathôt RAA, Romijn JA. The effects of fasting on drug metabolism. Expert Opin Drug Metab Toxicol. 2019;16(1):79–85. https://doi.org/10.1080/17425255.2020.1706728.

Lammers LA, Achterbergh R, van Schaik RH, et al. Effect of short-term fasting on systemic cytochrome p450-mediated drug metabolism in healthy subjects: a randomized, controlled, crossover study using a cocktail approach. Clin Pharmacokinet. 2017;56:1231–44.

Leite PM, Parreiras MA, Carvalho MG, Oliveira CR. Mechanisms and interactions in concomitant use of herbs and warfarin therapy: an updated review. Biomed Pharmacother. 2021;143:112103.

Licata A, Macaluso FS, Craxì A. Herbal hepatotoxicity: a hidden epidemic. Intern Emerg Med. 2013;8:13–22.

Lopez-Gonzalez E, Herdeiro MT, Figueiras A. Determinants of under-reporting of adverse drug reactions: a systematic review. Drug Saf. 2009;32(1):19–31. https://doi.org/10.2165/00002018-200932010-00002.

Marrs TC. Organophosphate poisoning. Pharmacol Ther. 1993;58:51–66.

Marwa K, Kondamudi NP. StatPearls [Internet]. Treasure Island: StatPearls Publishing; 2023. Type IV hypersensitivity reaction

McIver LA, Preuss CV, Tripp J. Acarbose. In: StatPearls [Internet]. Treasure Island: StatPearls Publishing; 2024. Available from: https://www.ncbi.nlm.nih.gov/books/NBK493214/.

Mercey G, Verdelet T, Renou J, Kliachyna M, Baati R, Nachon F, Jean L, Renard PY. Reactivators of acetylcholinesterase inhibited by organophosphorus nerve agents. Acc Chem Res. 2012;45:756–66.

Miller MF, Bellizzi KM, Sufan M, Ambs AH, Goldstein MS, Ballard-Barbash R. Dietary supplement use in individuals living with cancer and other chronic conditions: a population-based study. J Am Diet Assoc. 2008;108(3):483–94.

Muhlack DC, Hoppe LK, Stock C, Haefeli WE, Brenner H, Schöttker B. The associations of geriatric syndromes and other patient characteristics with the current and future use of potentially inappropriate medications in a large cohort study. Eur J Clin Pharmacol. 2018;74:1633–44. https://doi.org/10.1007/s00228-018-2534-1.

Natoff IL. Cheese and monoamine oxidase inhibitors. Interactionin anaesthetised cats. Lancet. 1964;1:532–3.

Neal JM, Barrington MJ, Fettiplace MR, Gitman M, Memtsoudis SG, Mörwald EE, Rubin DS, Weinberg G. The third American Society of Regional Anesthesia and Pain Medicine practice advisory on local anesthetic systemic toxicity: executive summary 2017. Reg Anesth Pain Med. 2018;2018(43):113–23. https://doi.org/10.1097/AAP.0000000000000720.

Nebeker JR, Barach P, Samore MH. Clarifying adverse drug events: a clinician's guide to terminology, documentation, and reporting. Ann Intern Med. 2004;140(10):795–801. https://doi.org/10.7326/0003-4819-140-10-200405180-00009.

Odi R, Invernizzi RW, Gallily T, Bialer M, Perucca E. Fenfluramine repurposing from weight loss to epilepsy: what we do and do not know. Pharmacol Ther. 2021;226:107866. https://doi.org/10.1016/j.pharmthera.2021.107866. Epub 2021 Apr 22

Ozcan MS, Weinberg G. Intravenous lipid emulsion for the treatment of drug toxicity. J Intensive Care Med. 2012;29(2):59–70. https://doi.org/10.1177/0885066612445978.

Pearce A, Haas M, Viney R, et al. Incidence and severity of self-reported chemotherapy side effects in routine care: a prospective cohort study. PLoS One. 2017;12:e0184360. https://doi.org/10.1371/journal.pone.0184360.

Pita R, Domingo J. The use of chemical weapons in the Syrian conflict. Toxics. 2014;2:391–40.

Ranasinghe T, Mays T, Quedado J, Adcock A. Thrombolysis following heparin reversal with protamine sulfate in acute ischemic stroke: case series and literature review. J Stroke Cerebrovasc Dis. 2019;28:104283. https://doi.org/10.1016/j.jstrokecerebrovasdis.2019.06.041.

Richardson V, et al. Effect of rotavirus vaccination on death from childhood diarrhea in Mexico. N Engl J Med. 2010;362:299–305.

Saraf AA, Petersen AW, Simmons SF, Schnelle JF, Bell SP, Kripalani S, Myers AP, Mixon AS, Long EA, Jacobsen JML, et al. Medications associated with geriatric syndromes and their prevalence in older hospitalized adults discharged to skilled nursing facilities. J Hosp Med. 2016;11:694–700.

Schnabel K, Binting S, Witt CM, Teut M. Use of complementary and alternative medicine by older adults–a cross-sectional survey. BMC Geriatr. 2014;14:38.

Sepulveda EA, Pak A. StatPearls. Treasure Island: StatPearls Publishing; 2023. Lipid emulsion therapy

Setola V, Roth BL. Screening the receptorome reveals molecular targets responsible for drug-induced side effects: focus on 'fen–phen'. Expert Opin Drug Metab Toxicol. 2005;1:377–87.

Sharbaf Shoar N, Bistas KG, Saadabadi A. Flumazenil. In: StatPearls. Treasure Island: StatPearls Publishing; 2023.

Shukla S, Sharma P, Gupta P, Pandey S, Agrawal R, Rathour D, Kumar Kewat D, Singh R, Kumar Thakur S, Paliwal R, Sulakhiya K. Current scenario and future prospects of adverse drug reactions (ADRs) monitoring and reporting mechanisms in the rural areas of India. Curr Drug Saf. 2024;19(2):172–90.

Singha MK, Patel DK, Kanungo VK. Medicinal plants used as antidotes in northern part of Bastar district of Chhattisgarh. J Ecobiotechnol. 2012;4:58–60.

Sjöqvist F, Böttiger Y. Historical perspectives: drug interactions-it all began with cheese. J Intern Med. 2010;268(6):512–5. https://doi.org/10.1111/j.1365-2796.2010.02300.x. Epub 2010 Nov 14. Erratum in: J Intern Med. 2011 Mar;269(3):362

Sjöqvist F. Psychotropic drugs (2). Interaction between mono-amine oxidase (MAO) inhibitors and other substances [Pharma-ceutical Specialitiesin Sweden]. Proc R Soc Med. 1965;58:967–78.

Svedlund E, Larsson M, Hägerkvist R. Spontaneously reported adverse reactions for herbal medicinal products and natural remedies in Sweden 2007–15: report from the medical products agency. Drugs Real World Outcomes. 2017;4(2):119–25.

Teron R, Borthakur SK. Folklore claims of some medicinal plants as antidote against poisons among the Karbis of Assam, India. Pleione. 2013;7:346–56.

Tran C, Knowles SR, Liu BA, Shear NH. Gender differences in adverse drug reactions. J Clin Pharmacol. 1998;38:1003–9.

Turpault S, Brian W, Van Horn R, et al. Pharmacokinetic assessment of a five-probe cocktail for CYPs 1A2, 2C9, 2C19, 2D6 and 3A. Br J Clin Pharmacol. 2009;68(6):928–35.

Usman N, Annamaraju P. StatPearls [Internet]. Treasure Island: StatPearls Publishing; 2023. Type III hypersensitivity reaction

Van Dorp EL, Yassen A, Dahan A. Naloxone treatment in opioid addiction: the risks and benefits. Expert Opin Drug Saf. 2007;6:125–32. https://doi.org/10.1517/14740338.6.2.125.

Van Hunsel FP, van der Kooi D, van de Koppel S, Kroes BH, Woerdenbag HJ. Analysis of reports on adverse drug reactions related to herbal medicinal products and herbal supplements in The Netherlands received by the National Pharmacovigilance Centre Lareb. Drug Saf. 2022;45(6):651–61.

Van Hunsel FP, Van Grootheest AC. Bijwerkingen van kruidenpreparaten: analyse van gemelde bijwerkingen in Nederland [Adverse reactions to herbal remedies: analysis of reported adverse reactions in The Netherlands]. Ned Tijdschr Geneeskd. 2013;157(47):A6615.

Vetrano DL, Palmer K, Marengoni A, Marzetti E, Lattanzio F, Roller-Wirnsberger R, Lopez Samaniego L, Rodríguez-Mañas L, Bernabei R, Onder G, et al. Frailty and multimorbidity: a systematic review and meta-analysis. J Gerontol A Biol Sci Med Sci. 2019;74:659–66.

Vilensky JA, Redman K. British anti-lewisite (Dimercaprol): an amazing history. Ann Emerg Med. 2003;2003(41):378–83.

Weinberg G. Lipid emulsion infusion rescues dogs from bupivacaine-induced cardiac toxicity. Reg Anesth Pain Med. 2003;28:198–202. https://doi.org/10.1097/00115550-200305000-00005.

Wieland LS, Moffet I, Shade S, Emadi A, Knott C, Gorman EF, D'Adamo C. Risks and benefits of antioxidant dietary supplement use during cancer treatment: protocol for a scoping review. BMJ Open. 2021;11(4):e047200. https://doi.org/10.1136/bmjopen-2020-047200. PMID: 33849858; PMCID: PMC8051392

Wilson IB, Ginsburg SA. Powerful reactivator of alkylphosphate-inhibited acetylcholinesterase. Biochim Biophys Acta. 1955;18:168–70.

Worek F, Aurbek N, Thiermann H. Reactivation of organophosphate-inhibited human AChE by combinations of obidoximeand HI 6in vitro. J Appl Toxicol. 2007;27:582–8.
Zanger UM, Schwab M. Cytochrome P 450 enzymes in drug metabolism: regulation of gene expression, enzyme activities, and impact of genetic variation. Pharmacol Ther. 2013;138(1):103–41.
Zhao H, Ni P, Zhao Q, et al. Identifying the serious clinical outcomes of adverse reactions to drugs by a multi-task deep learning framework. Commun Biol. 2023;6:870. https://doi.org/10.1038/s42003-023-05243-w.

# Index




J. O. Fajemiroye (ed.), *Fundamentals of Drug and Non-Drug Interactions*,
https://doi.org/10.1007/978-3-031-80107-5

The manufacturer's authorised representative in the EU is Springer Nature Customer Service Centre GmbH, Europaplatz 3, 69115 Heidelberg, Germany. If you have any concerns regarding our products, please contact ProductSafety@springernature.com

Printed and bound by CPI Group (UK) Ltd, Croydon, CR0 4YY
30/09/2026
02225600-0016